TOBACCO SMOKING AND ATHEROSCLEROSIS
Pathogenesis and Cellular Mechanisms

ADVANCES IN EXPERIMENTAL MEDICINE AND BIOLOGY

TOBACCO SMOKING AND ATHEROSCLEROSIS

Pathogenesis and Cellular Mechanisms

Edited by

John N. Diana

University of Kentucky
Tobacco and Health Research Institute
Lexington, Kentucky

PLENUM PRESS • NEW YORK AND LONDON

Library of Congress Cataloging-in-Publication Data

Tobacco smoking and atherosclerosis : pathogenesis and cellular
  mechanisms / edited by John N. Diana.
       p.   cm. -- (Advances in experimental medicine and biology ; v.
  273)
    Proceedings of an international symposium held Dec. 10-13, 1989.
    Includes bibliographical references.
    Includes index.
    ISBN 0-306-43668-X
    1. Atherosclerosis--Pathogenesis--Congresses. 2. Smoking-
  -Physiological effect--Congresses. 3. Smoking--Health Aspects-
  -Congresses. 4. Vascular endothelium--Pathophysiology--Congresses.
  I. Diana, John N. II. Series.
    [DNLM: 1. Arteriosclerosis--etiology--congresses.
  2. Arteriosclerosis--pathology--congresses. 3. Smoking--adverse
  effects--congresses. W1 AD559 v. 273 / WG 550 T628 1989]
  RC692.T63 1990
  616.1'36071--dc20
  DNLM
  for Library of Congress   90-7922
                                                            CIP

Proceedings of an international symposium on Tobacco Smoking and
Atherosclerosis: Pathogenesis and Cellular Mechanisms,
held December 10-13, 1989, in Lexington, Kentucky

ISBN 0-306-43668-X

© 1990 Plenum Press, New York
A Division of Plenum Publishing Corporation
233 Spring Street, New York, N.Y. 10013

Printed in the United States of America

PREFACE

Atherosclerosis is the principal underlying cause of cardiovascular
and cerebrovascular disease in people of the Western world. Cigarette
smoking has been implicated in both the initiation and exacerbation of
the atherosclerotic process. Data to support this implication derives
primarily from epidemiologic studies where the relationship between the
incidence of atherosclerosis in people who smoke cigarettes has been
shown to have a strong correlation. There are few well established
explanations for this phenomenon, and basic molecular, biochemical, and
cellular mechanisms associated with smoking and the development of
atherosclerosis remain both undefined and virtually unexplored.

Even the epidemiologic correlation between cigarette smoking and the
development of atherosclerosis needs further critical studies. It is
known that individuals who do not smoke cigarettes develop atherosclerosis and it is also known that in people who smoke but have normal
or low blood cholesterol/lipoprotein levels, the incidence of development
of atherosclerosis is no different from that which is found in a non-
smoking population. Answers which explain such observations must address
fundamental biological mechanisms. Toward this end, the purpose of this
volume is to assemble, in a single publication, information which will
address the questions; what basic cellular and/or molecular mechanisms
are associated with the development of atherosclerosis and how does
cigarette smoking influence such mechanisms to initiate or exacerbate the
atherosclerotic process?

Clearly, the development of atherosclerosis is a complex,
multifactorial biological event. The current literature related to
atherosclerosis encompasses studies with a wide range of area emphasis
which includes a multitude of interactive cellular processes; membrane
biology; molecular mechanisms associated with cellular signal transduction and cellular membrane receptor characteristics; biochemical,
biokinetic, and immunologic processes; as well as hemodynamic and
rheologic factors. In large measure the basic mechanisms which are
responsible for the observed responses, and especially the interaction of
such mechanisms between and among each other, are unknown.

Set against this background, the University of Kentucky, Tobacco and
Health Research Institute, organized a symposium on "Tobacco Smoking and
Atherosclerosis: Pathogenesis and Cellular Mechanisms" which was held in
Lexington, Kentucky, on December 11-13, 1989. Scientists from the USA,
Europe, Australia and Asia were assembled to present their research
findings and to interchange information and ideas on this important
health care topic.

Gratitude is expressed to those who participated in the symposium and contributed their time and effort to author the various chapters. The financial support of the Kentucky Tobacco Research Board and the outstanding service support by a dedicated Tobacco and Health Research Institute Staff is also gratefully acknowledged.

Special appreciation is expressed to Ms. Elaine Fisher and Ms. Carol Smith for their unselfish and untiring efforts in preparing this book for publication.

John N. Diana
Lexington, KY
January, 1990

CONTENTS

EFFECT OF CIGARETTE SMOKING
ON PLATELETS, VASCULAR TISSUE AND EICOSANOIDS

EFFECT OF CIGARETTE SMOKING
ON DIET, CHOLESTEROL/LIPOPROTEINS

CIGARETTE SMOKING, HEMODYNAMICS
AND HEMORHEOLOGY

CIGARETTE SMOKING AND
CORONARY ARTERY ATHEROSCLEROSIS

TOBACCO SMOKING AND ATHEROSCLEROSIS: OVERVIEW

John N. Diana

Tobacco and Health Research Institute
University of Kentucky
Lexington, Kentucky 40546

EPIDEMIOLOGY

Epidemiological studies from a variety of countries throughout the
world have clearly established a relationship between smoking cigarettes
and the development and/or progression of the atherosclerotic process
(1). Although the epidemiologic evidence which correlates smoking and
atherosclerosis is extensive, and thereby convincing, from a rigorous
scientific point of view there are confounding variables in such studies
which need further clarification. These confounding variables are
life-style factors which predispose the individual to atherosclerotic
disease in general (and coronary heart disease in particular), and are
typified by sedentary living habits; obesity and/or unrestrained weight
gain; a dietary intake which contains an excessive number of calories,
fat, saturated fat, cholesterol, and salt; personal stress that is not
modified; as well as cigarette smoking. It is not clear just how all of
these risk factors interrelate among or between each other as variables
in the pathogenesis of atherosclerosis. Analytical models which can
unambiguously discriminate between such risk factors for a weighting of
their contribution to the overall atherosclerotic process have yet to be
developed.

GENETIC FACTORS IN STUDIES OF ATHEROSCLEROSIS

Superimposed upon life-style variables may be a genetic
susceptibility for the development of atherosclerosis. For example, it
is an apparent paradox that an individual such as Winston Churchill who
indulged in rich foods, used tobacco, was obese, and had an almost totally
sedentary life-style could live past 90 years, while a tennis star such
as Arthur Ashe who avoided smoking, was thin, and had a very physically
active life-style could have two heart attacks before the age of 40.
Some individuals may have a genetic susceptibility for atherosclerosis
while other individuals may have a genetic protection. It is becoming
increasingly clear that the very large differences which are found
between both individuals and population studies which attempt to correlate life-style risk factors and the development of atherosclerosis may
have a genetic basis. There is a significant contribution of genes to
both risk and "anti-risk" factors in areas such as circulating blood
levels of low density lipoproteins (LDL), high density lipoproteins (HDL),

Tobacco Smoking and Atherosclerosis
Edited by J. N. Diana
Plenum Press, New York, 1990

apolipoproteins B, A-II and A-I (apo B, apo A-II, apo A-I), isoforms of
apolipoprotein B (apo E), very low density lipoproteins (VLDL), and Lp
(a) lipoprotein, all of which could influence the development and
progression of atherosclerosis (2). In this volume recent studies on
smoking and gene interactions which may relate to the development of
atherosclerosis are pointed out (McGill, page 9).

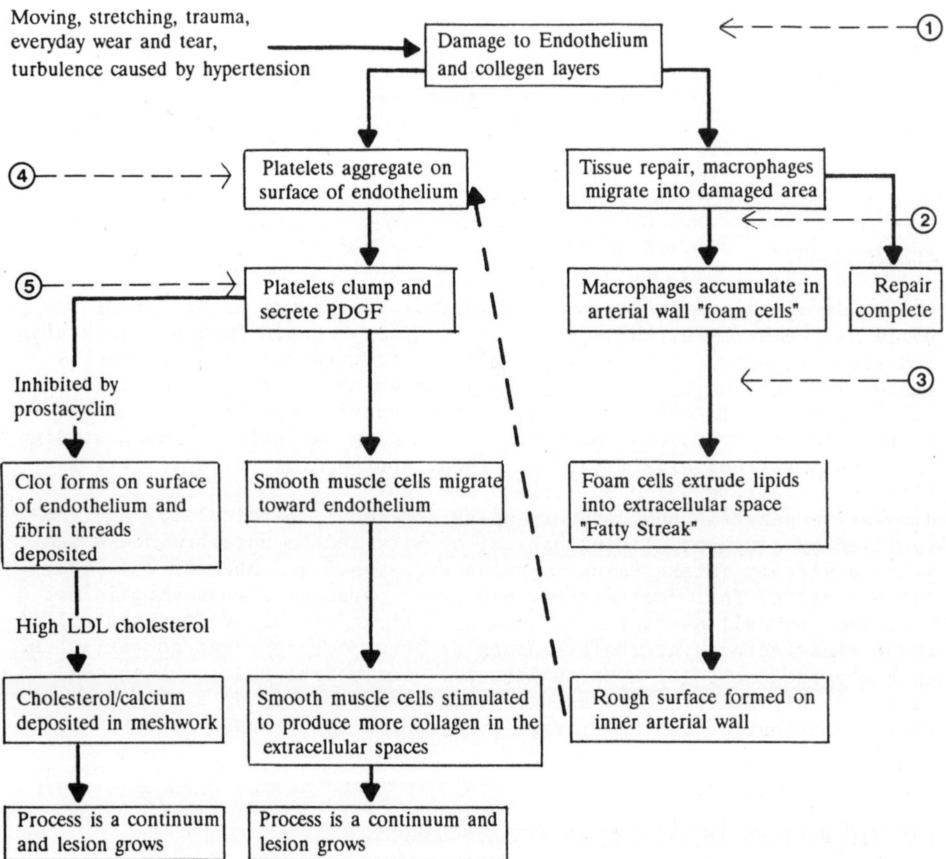

Figure 1. Generalized schema of the pathogenesis of atherosclerosis.
The development of atherosclerosis is a biologically complex
and multi-faceted process. See text for discussion.

    Despite these intrinsic frailties of epidemiologic studies, the
overall association between smoking and the development and/or
progression of atherosclerosis is convincing. A review of the smoking
and atherosclerosis scientific literature provides ample evidence that
this disease process is a complex, multifactorial biological event with
many unconfirmed leads, few well established explanations and a large
amount of incomplete data. A brief review of some factors involved in
the pathogenesis of atherosclerosis and the possible relationship between
smoking and these factors will illustrate this point.

CELLULAR PROCESSES IN THE DEVELOPMENT OF ATHEROSCLEROSIS

Figure 1 presents a schematic diagram of some of the factors involved in the pathogenesis of atherosclerosis. There is a reasonable consensus in the scientific community that the initial event in the atherosclerotic process is injury to endothelial cells [response to injury hypothesis (3)] which line blood vessel walls. The factors and forces which promote such damage are largely undefined and may be physical, metabolic, hormonal, cellular, molecular, or genetic in nature. Once the injury occurs a multitude of biological processes appear to become involved. Although many of the fundamental mechanisms involved with these biologic processes are also undefined and/or unexplored, the pathways which are followed can be generalized. On the right hand side of Figure 1 is shown one pathway which indicates that macrophages migrate into the damaged area of the arterial wall. The macrophages appear to be derived from blood monocytes which initially adhere to the endothelium, principally at the site of endothelial cell junctions. The factors (chemotactic or otherwise) responsible for both adherence to specific sites and migration into the arterial wall by macrophages are unknown. This can lead to healing of the area [injury-repair hypothesis (4)] or it can lead to the accumulation of macrophages in the arterial wall which, in turn, accumulate large amounts of lipids which are in the form of cholesteryl esters. Such "foam cell" macrophages also accumulate abnormal low density lipoproteins (e.g., acetyl LDL) through a scavenger receptor mechanism (5). Continued accumulation of macrophages in the presence of high cholesterol leads to extrusion of lipids into the arterial wall interstitial space and the formation of the fatty streak which has become one of the pathological hallmarks of the atherosclerotic process. Again the biochemical/molecular mechanisms associated with cholesterol uptake and extrusion are not clearly defined.

As formation of the fatty streak continues, the atherosclerotic lesion grows larger. In an as yet unknown manner the antithrombogenic nature of the endothelial cells is modified either chemically or by physical distortion which leads to the accumulation of platelets at the arterial wall site. Either by this pathway (shown by the dashed line) or by some direct modification of the endothelial cell(s) following injury (shown in the middle pathway of Figure 1) the accumulation of platelets and their role in the development of both the atherosclerotic plaque and an atherosclerotic-induced embolitic event has received increasing support from investigative studies. By cellular interactions which presumably involve platelets, endothelial cells, lipoproteins and vascular smooth muscle cells, a variety of chemical mitogens appear to be elaborated at the atherosclerotic site which influence, 1) the synthesis of components of the extracellular matrix (collagen, elastin, and proteoglycans); 2) uptake and then degradation of lipoproteins (e.g., proteases and lipases); and 3) vascular smooth muscle proliferation. For the latter, the secretion from platelets of platelet-derived growth factor (PDGF) appears to be a major factor in promoting the growth of the vascular smooth muscle component of the lesion. The biochemical/ molecular processes related to cellular and/or membrane mechanisms of this aspect of the development of atherosclerosis are poorly understood.

The third pathway shown in Figure 1 (left hand side) is an extension of the endothelial cell-platelet-smooth muscle cell interaction. The endothelial cell, as stated previously, functions to provide a non-thrombogenic surface for circulating blood constituents. It does this by secreting a prostaglandin (PGI_2). Following damage to the endothelial surface and the accumulation of platelets at the injury site (and subsequently accumulation of platelets in the subendothelium) the platelets appear to undergo a physical change. The platelet membrane becomes permeable to calcium which stimulates arachadonic acid synthesis

at the tissue site. Arachadonic acid synthesis, in turn, produces two
products, thromboxane (TXA_2), a prostaglandin, and 12-HETE, a hydroperoxy
acid. Thromboxane promotes platelet aggregation and vascular smooth
muscle constriction. Hence, the normal PGI_2:TXA_2 balance in vascular
tissue is disrupted, which can lead to a loss of endothelial cell
non-thrombogenicity and vascular smooth muscle vasoconstriction. 12-HETE
stimulates smooth muscle migration causing growth of the vascular lesion.

Finally, there is normally a balance between cholesterol influx into
the plasma membrane, which is vital for normal cell function, and efflux
out of the plasma membrane. When influx of cholesterol exceeds efflux
there is an esterification of cholesterol by microsomal
acyl-CoA-cholesterol transferase (ACAT) forming cholesteryl esters which
are stored by the cells. In the presence of hypercholesterolemia low
density lipoproteins, which are the major cholesterol-carrying
lipoprotein, accumulate in arterial walls in proportion to their
concentrations in plasma. With elevated plasma cholesterol/lipoprotein
levels, the deposition of cholesterol in the arterial wall, in
association with the myriad of biological processes described above,
exacerbates the development of the arterial atherosclerotic plaque.

This brief review is presented as an overview of the pathogenesis of
atherosclerosis and makes no attempt to cover an abundant literature
which exists in specific areas of atherosclerosis research such as
membrane phenomena (alterations in transmembrane electrical potential,
ion fluxes, Na^+/K^+ ATPase activity, etc.) or receptor-mediated cellular
signalling and transduction. For additional information the reader is
directed to articles in this volume and other recently published reports
(6).

CIGARETTE SMOKING AND ATHEROSCLEROSIS

Despite the abundant epidemiologic evidence which associates
cigarette smoking with the development of atherosclerosis there is a
paucity of information directed at determining the basic mechanisms
whereby cigarette smoking influences the atherosclerotic process.
Certainly, an impediment to rigorously identifying basic mechanisms
related to smoking and atherogenesis is that an extremely complex smoking
stimulus is superimposed on a very complex biological process. Over
4,000 constituents have been identified in cigarette smoke (7).
Approximately 92% of the mainstream smoke from a burning cigarette is
comprised of 400-500 individual gaseous components. The gas phase
contains oxygen, carbon dioxide, carbon monoxide, nitrogen oxides,
ammonia, hydrogen cyanide, formaldehyde, acrolein, nitroso-compounds and
benzene as major constituents. The particulate phase of cigarette smoke
constitutes 8% of the total in mainstream smoke. The majority of
biologically active teratogenic and mutagenic constituents are in the
particulate phase, which consists of compounds such as nicotine, phenol,
phytosteroids, napthalene, pyrene, benzo(a)pyrene, nitrosamines and trace
metals such as cadmium. Added to the complexity of the stimulus and the
response is that it has been virtually impossible to identify, at least
to date, an appropriate atherosclerotic-smoking animal model.

In spite of these difficulties some experimental information related
to smoking and atherosclerosis is available and can be briefly
summarized. Shown in Figure 1 are numbers 1-5 which are circled. The
physiological and pharmacological response to smoking could influence
atherogenesis at any of these points (and others) in the development of
atherosclerosis. For example, it has been suggested that nicotine may
cause arterial endothelial cell damage thus implicating this compound in

the initial event related to atherogenesis. This topic is covered
extensively in this volume. Cigarette smoking, in humans, has been shown
to produce an elevation in plasma cholesterol and LDL levels and a
reduction in HDL levels (8). This pattern would favor the net flux of
cholesterol into the arterial wall since LDL levels are high and HDL,
which transports used cholesterol out of the artery to the liver, is
low. Nicotine, alone or in conjunction with cigarette smoking, is known
to increase sympathetic nerve activity and release catecholamines from
the adrenal glands. This generalized physiologic response could
potentially influence the atherogenic process by several mechanisms: a)
an increase in circulating blood levels of free fatty acids, b) an
increase in plasma fibrinolytic activity, c) an increase in platelet
aggregation, d) an effect on some steps in the blood coagulation pathway,
e) an increase in blood pressure, flow velocity and blood shear rate, and
f) an increase in blood glucose levels and arterial basic metabolic
rates. All of these processes have been associated with activation of
the sympathoadrenal system and all have been implicated in the
pathogenesis of atherosclerosis. Smoking in general, and nicotine
specifically, has been shown to alter arterial wall prostaglandin
activity. The PGI_2:TXA_2 ratio is disrupted in favor of an increase in
TXA_2 which promotes platelet adhesion to the endothelium of the artery.
It appears now well established that platelets play an essential and
pivotal role in the pathogenesis of the atherosclerotic lesion and have
been strongly implicated in atherosclerotic-induced vascular occlusive
disease. Many of these topics are covered in greater detail in this
volume.

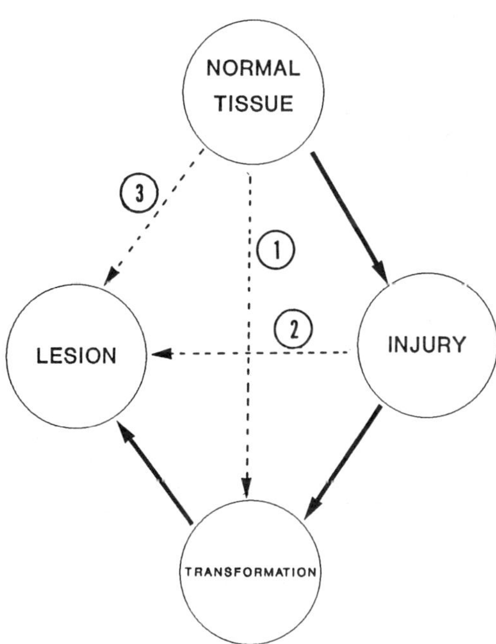

Figure 2. Schematic diagram of events in atherogenesis. Heavy
arrows depict the sequence: normal tissue is injured
which leads to cellular transformation and lesion
development. Some cellular and/or molecular process
which may be associated with these events are depicted
by pathways 1, 2, and 3 shown by dashed lines in the
figure and discussed in the text.

The number of constituents in tobacco smoke which are teratogenic or carcinogenic (or can be converted to carcinogens by metabolizing enzymes in mammalian tissue cells) would suggest a high probability for having a direct effect on basic cellular and molecular events. Figure 2 presents a diagram for the general conceptual scheme of events in the atherosclerotic process (shown by heavy solid lines) which are normal tissue being injured which leads to cellular transformation and development of the lesion. The dashed lines, numbered 1, 2, and 3, indicate interactions between these events. For example, related to pathway 1 (normal tissue to cellular transformation) it has been suggested in the scientific literature that tobacco constituents may a) influence the cellular immune system by depressing suppressor T-cell function, b) promote the Browning Reaction which increases protein glycosylation (especially collagen and lens crystalline), c) decrease the ability of cells to metabolize lipids, d) promote increased cellular lipoprotein levels by hormonal changes, and e) promote cellular growth and differentiation by activation of substances such as PDGF and interferon β. Related to pathway 2 (normal tissue to cellular transformation), it is possible that the constituents in tobacco smoke could promote the aberrant expression of DNA sequences which neoplastically transform cells (oncogenes or transforming genes) which result in abnormal cell growth (9). For example, it is known that catecholamines can activate the cellular c-myc proto-oncogene message to promote abnormal proliferation of cells and that the c-fos proto-oncogene message can be stimulated by angiotensin to promote cellular hypertrophy. Clearly the c-sis proto-oncogene cellular message is stimulated by PDGF to initiate cellular hyperplasia in atherogenesis. In addition, the constituents in tobacco smoke may either directly (e.g., trace metals) or indirectly (hormonal effects) activate the intracellular or membrane related polyinositol phosphate or protein kinase-C transmembrane signalling pathways to influence a variety of intracellular events. Finally, tobacco smoke constituents could influence pathway 3 (normal tissue to lesion) by activating either genetic or non-genetic (stochastic) processes which could lead directly to the development of lesions. It is known that the normal cell population of the arterial wall contains a subpopulation of "stem" cells whose growth may lead to atherogenesis. Very little information is presently available about the potential influence of tobacco smoking on these last two pathways (Figure 2, pathway 2 and 3).

A PHILOSOPHICAL NOTE

In the 500-year history of tobacco usage by people of the world, attempts at abolishing its use have universally failed in every society. Increased excise taxes making the product more expensive have, in most instances, led to an increase in the desirability of the product. Laws enacted to prohibit importation or production of tobacco have only led to the development of a lucrative and extensive black market. The 500-year historical message seems clear: For a great number of possible personal reasons, people choose to use tobacco. Although tobacco use has declined in the U.S. and some European countries, its use appears to be increasing in Asia and third world countries.

All people should be informed of the potential health risks associated with tobacco use. A recent poll taken by the Tobacco and Health Institute in the Commonwealth of Kentucky showed that 95% of those polled (both smokers and non-smokers) believed that smoking cigarettes was harmful to one's health. If people are informed then they are making a conscious decision to balance potential health risks with freedom of choice. This country was founded on the fundamental principle of freedom and the right of self determination.

The association between cigarette smoking and the risk for
development of cardiovascular and pulmonary diseases appears well
established. Clearly, people are placing themselves at risk for
development of these diseases by choosing to smoke. This choice becomes,
then, a life-style factor. Most societies of the world have no
difficulty in allocating resources, medically treating and performing
basic and clinical research on a variety of health problems created by
conscious choice of either the individual or the society. These include,
but are not limited to, such factors as obesity and health, pesticides
and disease, high fat diets and health, alcoholism, drug abuse and AIDS.
In fact, the majority of health problems in our society are at least
partially created by life-style factors. Smoking cigarettes should be
included in this list of life-style factors to be directly addressed with
treatment and research in an attempt to reduce the risk and improve the
health and welfare of the people.

Surely there are answers to the cigarette smoking and health risk
problem. Reducing or correcting the health risks associated with
cigarette smoking would be an important health care contribution to
people in all societies of the world and toward that purpose this volume
is directed.

REFERENCES

1. U.S. Office on Smoking and Health, "The Health Consequences of
    Smoking: Cardiovascular Disease; a Report of the Surgeon
    General." Washington, D.C., U.S. Government Printing Office, pp.
    179-203, 1989.
2. Berg, Kare. Impact of medical genetics on research and practices in
    the area of cardiovascular disease. Clinical Genetics, 36:
    299-312, 1989.
3. Ross, R. and A. Glomset. The pathogenesis of atherosclerosis. New
    Eng. J. Med. 295: 369-432, 1976.
4. Haust, M.D. Injury and repair in the pathogenesis of atherosclerotic
    lesions. In: _Atherosclerosis_, (R.J. Jones, Ed.) Springer-Verlag,
    New York, pp. 12-20, 1970.
5. Brown, M.S. and J.L. Goldstein. Lipoprotein metabolism in the
    macrophage: implications for cholesterol deposition in
    atherosclerosis. Proc. Nat. Acad. Sci. USA 70: 1753-1756, 1983.
6. _Carotid Artery Plaques_: _Pathogenesis, Development, Evaluation,_
    Treatment (M. Hennerici, G. Sitzer, H-D Weger, Eds). Bertelsmann
    Foundation, S. Karger, Publishing Co., New York, N.Y. Basel, 1988.
7. Chemistry and Analysis of Tobacco Smoke. In: IARC Monographs on
    the Evaluation of the Carcinogenic Risk of Chemicals to Humans:
    Tobacco Smoking. International Agency for Research on Cancer,
    Publisher, Vol. 38: 83-126, 1986.
8. Brischetto, C.S., W.B. Connor, S.L. Connor, and J.D. Matarazzo.
    Plasma lipid and lipoprotein profiles of cigarette smokers from
    randomly selected families: enhancement of hyperlipidemia and
    depression of high density lipoprotein. Am. J. Cardiol. 62:
    675-681, 1983.
9. Barrett, T.B. and E.P. Benditt. Platelet-derived growth factor
    gene expression in human atherosclerotic plaques and normal artery
    wall. Proc. Natl. Acad. Sci. 85: 2810-2814, 1988.

SMOKING AND THE PATHOGENESIS OF ATHEROSCLEROSIS

Henry C. McGill, Jr.

Department of Pathology
The University of Texas Health Science Center
San Antonio, TX
         and the
Southwest Foundation for Biomedical Research
P.O. Box 28147
San Antonio, TX

This symposium occurs just a little over 30 years after the landmark study of Hammond and Horn (1958) that, for the first time, conclusively linked smoking with coronary heart disease and other sequelae of atherosclerosis. The effect of that report is difficult to appreciate with our present knowledge, but one measure of its significance is that as late as 1948, coronary heart disease patients were included as controls for studies of smoking and lung cancer (Doll, 1984). Many changes have occurred since 1958: the prevalence of smoking among U.S. adults has declined by more than a half , and the coronary heart disease death rate has declined by one third. The association of smoking with atherosclerotic cardiovascular disease has been strengthened by hundreds of observational studies which have been so unanimous in their results that the question about smoking and atherosclerotic disease is no longer whether, but how. The observation that cessation of smoking reduced risk of coronary heart disease further strengthened the idea of a causal relationship. On the question of mechanisms, we are much less informed. As the presentations in this symposium will probably show, there are many possibilities, but few certainties.

NATURAL HISTORY OF ATHEROSCLEROSIS

Atherosclerosis is a life-long process that begins in infancy and develops through a series of progressive stages to result in clinically manifest disease in middle age or later (Figure). The first grossly detectable lesions are abnormal deposits of lipid in the intima of the aorta, particularly the thoracic aorta, of children about three years of age. These lipid deposits contain cholesterol and its esters, principally cholesteryl oleate, in macrophage foam cells. Fatty streaks increase slowly in extent in the aorta until about age 10, and thereafter increase rapidly until about age 20 (Holman et al., 1958). The increase in extent is paralleled microscopically by the appearance of extracellular lipid and lipid in smooth muscle cells. Fatty streaks first appear in the coronary arteries in the latter part of the second decade of life and increase in extent throughout the third decade (McGill, 1968). Microscopically, clusters of macrophage foam cells appear in the coronary arteries by age 10, and increase rapidly in prevalence with increasing age (Stary, 1989).

Tobacco Smoking and Atherosclerosis
Edited by J. N. Diana
Plenum Press, New York, 1990

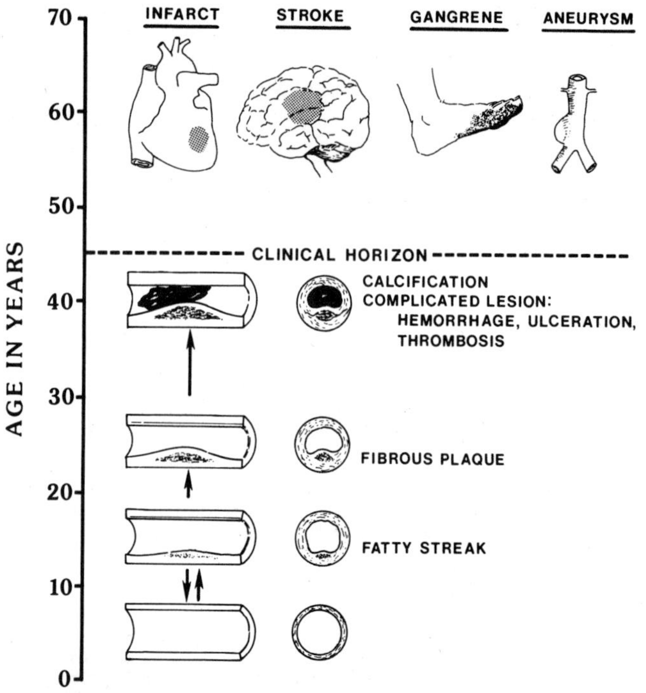

Figure 1. Diagram of the natural history of atherosclerosis. Redrawn from McGill et al., 1963.

At about age 20, a few intimal arterial lesions appear in both the aorta and the coronary arteries that differ from the simple fatty streak. Pale gray nodules project above the surrounding surface of the opened artery. Microscopically, these lesions contain a core of lipid and necrotic debris covered by a cap of smooth muscle and connective tissue, and are called fibrous plaques. In subsequent years, fibrous plaques become more numerous and extensive, particularly in the abdominal aorta and in the coronary arteries, and they undergo vascularization, hemorrhage, calcification, and ulceration. Hemorrhage may cause the plaque to swell, and ulceration may precipitate thrombosis on the intima. Either hemorrhage or thrombosis may cause occlusion and ischemic necrosis of the heart or brain. Other local conditions -- workload, oxygen demand, inflammatory response, collateral circulation, and others -- influence the response of the target tissue to ischemia.

ETIOLOGIC AND PATHOGENETIC MECHANISMS

Hyperlipidemia

The first evidence linking diet and hypercholesterolemia with atherosclerosis came from the discovery in 1908 that rabbits developed lipid-rich arterial lesions when they were fed meat, milk, butter, and eggs (Ignatowski, 1908). A few years later, Anitschov showed that cholesterol was the dietary agent responsible for these effects (Anitschov, 1967). Cholesterol was found to be a prominent component of human atherosclerotic lesions about the same time. The cholesterol-fed rabbit remained a laboratory curiosity until nearly mid-century, when case control studies showed that patients with myocardial infarction had higher serum cholesterol levels than did control subjects and when a number of other animal species turned out also to be sensitive to dietary cholesterol.

Longitudinal epidemiological studies, such as the Framingham Study, soon established the strong predictive power of the serum cholesterol concentration for risk of myocardial infarction (Dawber et al., 1957). The effects of dietary fatty acid saturation on serum cholesterol concentrations were discovered in 1952 and a large number of dietary experiments and epidemiologic surveys established that dietary cholesterol and type of fatty acid were the major environmental determinants of serum cholesterol levels (reviewed by the National Research Council, 1989, Chapter 7).

As new analytic methods became available to analyze the various lipoprotein classes separately, investigators found that low density lipoprotein (LDL) cholesterol was positively associated, and high density lipoprotein cholesterol (HDL) was inversely associated with both clinical atherosclerotic disease and atherosclerosis. The roles of the triglyceride-rich postprandial lipoproteins, chylomicrons, and very low density lipoproteins (VLDL), were uncertain and have remained so.

It had long been suspected that there were genetic determinants of serum cholesterol levels, and knowledge about specific genetic dyslipoproteinemias began to increase in the early 1970's. The most dramatic development was the identification of the LDL receptor and its mutations responsible for familial hypercholesterolemia (reviewed by Brown and Goldstein, 1986). Subsequently, a large number of genetic variants of the apolipoproteins and enzymes involved in lipid and lipoprotein metabolism were identified (reviewed by Lusis, 1988), and many more are likely to be found. These genetic variants appear to be responsible for most of the variation in plasma lipoprotein concentrations among individuals within a population, whereas differences in dietary fats and cholesterol are responsible for most of the variation among populations.

Experimental human evidence can now be added to the animal experimental and human epidemiologic evidence supporting a causal relationship between plasma lipoprotein levels and atherosclerosis. Several large clinical trials, exemplified by the Lipid Research Clinics Coronary Primary Prevention Trial (Lipid Research Clinics Program, 1984), have shown that reduction in serum cholesterol levels, particularly LDL cholesterol, results in lower risk of coronary heart disease events.

The exact cellular and molecular processes by which high plasma LDL cholesterol and low HDL cholesterol levels accelerate the progression of atherosclerosis remain mysterious. Surely, the arterial wall is more than a semipermeable membrane that filters lipid-rich macromolecules from plasma. The most promising current idea is the oxidized LDL hypothesis. Steinberg and his associates have reported a series of experiments beginning in 1981 (Henriksen et al., 1983) showing the LDL can be modified by a number of agents that make it highly susceptible to uptake by macrophages through a scavenger receptor pathway (Goldstein et al., 1979). This pathway, unlike the LDL receptor pathway, is not subject to feedback control. This hypothesis, if true, offers promise of a means of preventing the atherogenic effects of high plasma LDL levels when they cannot be controlled by diet or drugs.

In summary, hyperlipidemia, or more accurately, dyslipoproteinemia, is now understood as a family of variations in lipoprotein metabolism and plasma lipoprotein profiles which influence the rate of progression of atherosclerosis. The various dyslipoproteinemias are determined by an interaction of diet with a number of genetically programmed metabolic characteristics. The major dietary lipids (saturated fatty acids of chain lengths C12 to C16 and cholesterol) have been identified, and many of the mutant genes involved have been cloned and sequenced. Research in this field is focused on the cellular and molecular biology of the genetic-environmental interactions that cause dyslipoproteinemia and thereby accelerate atherogenesis.

Hypertension

Hypertension clearly accelerates atherosclerosis and increases the risk of coronary heart disease and cerebral vascular disease. Hypertension augments atherosclerosis of the cerebral arteries more than it augments coronary atherosclerosis, and this selective effect on cerebral arteries is probably why it is such a strong risk factor for stroke. It is attractive to hypothesize that increased arterial pressure causes plasma to infiltrate the arterial wall more rapidly and thereby augment lipid deposition and atherosclerosis, but this phenomenon has been difficult to demonstrate directly. Despite the proven effectiveness of antihypertensive drugs to reduce blood pressure, lower risk of stroke, and prevent cardiac failure, antihypertensive drug treatment does not substantially reduce risk of coronary heart disease. Some of the more commonly used antihypertensive drugs increase LDL cholesterol concentrations in plasma, and may thereby counteract the beneficial effects of arterial pressure reduction. The availability of new drugs that lower pressure without affecting serum lipid levels may be more effective in protecting against coronary heart disease.

Male Sex

The greater risk of coronary heart disease among men than among women is well known but poorly understood (reviewed by McGill & Stern, 1979). Some, but not all, of the increased risk can be explained by differences in the other established risk factors (hyperlipidemia, hypertension, and smoking). The sex differential is present only in coronary heart disease, and not in cerebral or peripheral vascular disease. Furthermore, the sex differential is not nearly as prominent in nonwhite populations as in white populations. Estrogen produces some effects on plasma lipoproteins that are beneficial, but it is not likely that estrogens explain all the differences because estrogens given to men increased the risk of coronary heart disease. Arterial smooth muscle cells contain receptors for androgens, estrogens, and progestins, and these hormones may directly affect the arterial wall (Shain et al., 1988).

Tobacco Smoking

To examine the mechanisms by which smoking might increase risk of atherosclerotic disease, it is necessary to determine first which stage in the natural history is affected. Does smoking increase the early deposition of cholesterol and its esters in the arterial wall; does it accelerate the formation of the fibrous plaque from fatty streaks; does it influence the incidence of ulceration or hemorrhage in fibrous plaques; does it affect the hemostatic system so as to increase the probability of thrombosis; or does it increase the vulnerability of the target tissue to ischemia? Or does smoking affect all of these stages in the pathogenesis of atherosclerotic disease? Answering these questions has been more difficult than identifying the overall association of smoking with clinically manifest atherosclerotic disease.

Smoking clearly accelerates advanced atherosclerosis, particularly in the coronary arteries and in the abdominal aorta. This topic was reviewed thoroughly in the 1983 Surgeon General's report (U.S. Office on Smoking and Health, 1983). For example, Strong and Richards (1976) examined the arteries of 1,320 autopsied men and determined their lifetime smoking habits by interviewing family members and friends. Heavy smokers, as compared with nonsmokers, had about 50% more raised lesions (fibrous plaques and complicated lesions) in the coronary arteries, and about 100% more lesions in the abdominal aorta. The increase in lesions in the coronary arteries does not seem enough to account for the twofold or greater increase in risk of coronary heart disease among heavy smokers, but the greater proportional increase in lesions in the abdominal aorta is consistent with the well known predisposition of smokers to aneurysms of the abdominal aorta and to peripheral vascular disease. Smoking seems to selectively affect the abdominal aorta. This study found only a slight effect of smoking on lesions in the youngest age group, 25-34 years, and a greater effect between ages 35 and 64 years. No information is available regarding the effect of smoking on lesions before age 25. Such information would be important in assessing the importance of control of smoking among young persons.

A few reports have described a weak association between smoking and atherosclerosis of the cerebral arteries, but this evidence is so limited that a definite conclusion is not possible.

Several investigators have reported attempts to study the effects of tobacco smoke inhalation on experimental atherosclerosis. All results have been negative. Other reports describe tests of the effects of nicotine and carbon monoxide on experimental atherosclerosis with no consistent results. Thus, there is no satisfactory animal model in which to examine the effects of tobacco smoke or its components on atherosclerosis.

A consistent finding in human smokers is a lower HDL cholesterol level (Garrison et al., 1978), now recognized as a strong risk factor for coronary heart disease and associated with more severe atherosclerosis (Solberg and Strong, 1983). This change in plasma lipoproteins might account for some of the effects of smoking on atherosclerosis, but does not seem sufficient to account for the total effect.

Smokers have elevated white blood cell (WBC) counts (Friedman et al., 1973), and elevated WBC counts are strong predictors of myocardial infarction (Friedman et al., 1974). In the Lipid Research Clinics Coronary Primary Prevention Trial, the WBC count was closely correlated with self-reported smoking, and was a stronger predictor of subsequent coronary heart disease events than self-reported smoking (Knoke et. al., 1987). Ernst et al. (1987), in reviewing the role of leucocytes in vascular disease, suggested that WBCs might obstruct blood flow in small arteries or injure endothelial cells. Whatever the mechanism, elevation of the WBC count is a consistent response to smoking that is also associated with myocardial infarction. It has received relatively little attention and should offer another avenue of investigating the smoking-atherosclerosis relationship.

Benditt (1988) has suggested another possible mechanism by which tobacco smoke, through its mutagenic components, might accelerate atherogenesis. This possibly was based on his previous observation that human fibrous plaques are frequently monoclonal, indicating that they may have originated from a single transformed smooth muscle cell. The mutagenic hypothesis would suggest that smoking affects the formation of the fibrous plaque rather than lipid deposition.

The effects of smoking on the hemostatic system -- platelets, fibrinogen, and prostaglandins -- has properly received much attention and will receive more in this symposium. The putative thrombogenic effects of tobacco smoking would provide a mechanism affecting the terminal occlusive episode in coronary heart disease and would explain why clinically manifest coronary heart disease is affected more than would be indicated by the increase in atherosclerotic lesions.

All studies of the effects of smoking on physiologic responses or pathologic end points have observed enormous individual variability. Error in measuring the smoking, the response, or other disease determinants account for some of this variability. However, it is certain that genetic variations will eventually be found to account for much of this variability. As yet, there are only two reports of specific smoking-gene interactions. One concerns a DNA polymorphism in the cholesterol ester transfer protein which is associated with differences in HDL cholesterol concentrations in nonsmokers but not in smokers (Kondo et al., 1989). There was also an interaction of smoking with apolipoprotein H polymorphisms in their effects on HDL cholesterol levels (Kaprio et al., 1989). These are but the first hints of what is likely to be forthcoming as the genetics of lipoprotein metabolism, the hemostatic system, cardiac physiology, and other elements of the cardiovascular system are developed.

The discovery of the structure of apolipoprotein (a) and its homology with plasminogen (reviewed by Scanu, 1988; Utermann., 1989) has suggested the possibility of a direct link between lipoproteins and thrombosis. Apolipoprotein (a) competes with plasminogen for binding to the plasminogen receptor, and may inhibit fibrinolysis in vivo. Since apolipoprotein (a) occurs in a variety of genetically determined isoforms, it is possible that we may find interactions between smoking and these isoforms and thus another genetic-smoking interaction.

SUMMARY

Atherosclerosis begins in childhood as arterial intimal lipid deposits and progresses to occlusive arterial lesions in middle age or later. Dyslipoproteinemia, hypertension, and male sex are major risk factors for atherosclerotic disease and also contribute to atherogenesis. Tobacco smoking is well established as a contributor to atherosclerotic disease, particularly to coronary heart disease and peripheral vascular disease. Smoking augments atherosclerosis of the coronary arteries and probably also increases the risk of thrombosis independently of mural atherosclerosis. Smoking greatly augments atherosclerosis of the abdominal aorta, and is the major cause of abdominal aortic aneurysms. There are many physiologic responses of the body to tobacco smoking that may mediate its effects on atherosclerosis and atherosclerotic disease, but there is little evidence to indicate the importance of these relative to one another. We may anticipate the discovery of many smoking-genetic interactions in the future and these are likely to be helpful in resolving these questions of etiology and pathogenesis.

REFERENCES

Anitschov, N.N., 1967, A history of experimentation on arterial atherosclerosis in animals. in: "Cowdry's Arteriosclerosis: A Survey of the Problem", 2nd ed. Charles C. Thomas, Springfield, Ill. p. 21-44.

Benditt, E.P., 1988, Origins of human atherosclerotic plaques. The role of altered gene expression. Arch. Pathol. Lab. Med. 112:997-1001.

Brown, M.S., and Goldstein, J.L., 1986, A receptor-mediated pathway for cholesterol homeostasis. Science 232:34-47.

Dawber, T.R., Moore, F.E., and Mann, G.V., 1957, Coronary heart disease in the Framingham Study. Am. J. Pub. Health 47(Suppl.):4-24.

Doll, R., 1984, Smoking and death rates. JAMA 251:2854-2857.

Ernst, E., Hammerschmidt, D.E., Bagge, U., Matrai, A., and Dormandy, J.A., 1987, Leukocytes and the risk of ischemic diseases. JAMA 257:2318-2324.

Friedman, G.D., Klatsky, A.L., and Siegelaub, A.B., 1974, The leukocyte count as a predictor of myocardial infarction. N. Engl. J. Med. 290:1275-1278.

Friedman, G.D., Siegelaub, A.B., Seltzer, C.C., Feldman, R., and Collen, M.F., 1973, Smoking habits and the leukocyte count. Arch. Environ. Health 26:137-143.

Garrison, R.J., Kannel, W.B., Feinleib, M., Castelli, W.P., McNamara, P.M., and Padgett, S.J., 1978, Cigarette smoking and HDL cholesterol. The Framingham Offspring Study. Atherosclerosis 30:17-25.

Goldstein, J.L., Ho, Y.K., Basu, S.K., and Brown, M.S., 1979, Binding site on macrophages that mediates uptake and degradation of acetylated low density lipoproteins, producing massive cholesterol deposition. Proc. Natl. Acad. Sci. USA 76:333-337.

Hammond, E.C., and Horn, D., 1958, Smoking and death rates - Report on forty-four months of follow-up of 187,783 men. JAMA 166:1294-1308.

Henriksen, T., Mahoney, E.M., and Steinberg, D., 1983, Enhanced macrophage degradation of biologically modified low density lipoprotein. Arteriosclerosis 3:149-159.

Holman, R.L., McGill, H.C., Jr., Strong, J.P., and Geer, J.C., 1958, The natural history of atherosclerosis. The early aortic lesions as seen in New Orleans in the middle of the 20th century. Am. J Pathol. 34:209-235.

Ignatovski, A.I., 1908, [Influence of animal food on the organisms of rabbits.] Izv. Imp. Voyenno-Med. Akad., S.-Petersburg 16:154-176.

Kaprio, J., Ferrell, R.E., Kottke, B.A., and Sing, C.F., 1989, Smoking and reverse cholesterol transport: evidence for gene-environment interaction. Clin. Genet. 36:266-268.

Knoke, J.D., Hunninghake, D.B., and Heiss, G., 1987, Physiological markers of smoking and their relation to coronary heart disease. The Lipid Research Clinics Coronary Primary Prevention Trial. Arteriosclerosis 7:477-482.

Kondo, I., Berg, K., Drayna, D., and Lawn, R., 1989, DNA polymorphism at the locus for human cholesteryl ester transfer protein (CETP) is associated with high density lipoprotein cholesterol and apolipoprotein levels. Clin. Genet. 35:49 56.

Lipid Research Clinics Program, 1984, The Lipid Research Clinics Coronary Primary Prevention Trial Results. I. Reduction in incidence of coronary heart disease. JAMA 251:351-364.

Lusis, A.J., 1988, Genetic factors affecting blood lipoproteins: the candidate gene approach. J. Lipid Res. 29:397-429.

McGill, H.C., Jr., 1968, Fatty streaks in the coronary arteries and aorta. Lab Invest. 18:560-564.

McGill, H.C., Jr., Geer, J.C., and Strong, J.P., 1963, Natural history of human atherosclerotic lesions, in: "Atherosclerosis and Its Origin," M. Sandler and G.H. Bourne, ed., Academic Press, New York. p. 39-65.

McGill, H.C., Jr., and Stern, M.R., 1979, Sex and atherosclerosis. Atheroscler. Rev. 4:157-242.

National Research Council, Committee on Diet and Health, 1989, "Diet and Health: Implications for Reducing Chronic Disease Risk". Chapter 7: Fats and Other Lipids. Washington, D.C. National Academy Press. p. 159-258.

Shain, S.A., Lin, A.L., and McGill, H.C., Jr., 1988, Steroid receptors in the cardiovascular system. in: "Steroid Receptors and Disease: Cancer, Autoimmune, Bone, and Circulatory Disorders," P.J. Sheridan, K. Blum, and M.C. Trachtenberg, eds. Marcel Dekker, Inc., New York. p. 547-567.

Scanu, A.M., 1988, Lipoprotein(a). A potential bridge between the fields of atherosclerosis and thrombosis. Arch. Pathol. Lab. Med. 112:1045-1047.

Solberg, L.A., and Strong, J.P., 1983, Risk factors and atherosclerotic lesions. A review of autopsy studies. Arteriosclerosis 3:187-198.

Stary, H.C., 1989, Evolution and progression of atherosclerotic lesions in coronary arteries of children and young adults. Arteriosclerosis 9(Suppl):I-19-I-32.

Strong, J.P., and Richards, M.L., 1976, Cigarette smoking and atherosclerosis in autopsied men. Atherosclerosis 23:451-476.

U.S. Office on Smoking and Health, 1983, "The Health Consequences of Smoking: Cardiovascular Disease; a Report of the Surgeon General". Washington, D.C.: U.S. Government Printing Office, 13-62.

Utermann, G., 1989, The mysteries of lipoprotein(a). Science 246:904-910.

SMOKING AS A PREDICTOR OF ATHEROSCLEROSIS IN THE

HONOLULU HEART PROGRAM

Dwayne Reed, Ellen Marcus,
and Takuji Hayashi

National Heart, Lung and Blood Institute
Honolulu Heart Program, Kuakini Medical Center
Honolulu, Hawaii

Cigarette smoking has been implicated in epidemiologic
studies as one of the risk factors for clinical coronary
heart disease. It is not possible to tell from such studies
if smoking is associated with the underlying development of
atherosclerosis or with the precipitation of acute occlusive
events through some totally different mechanisms such as
clot formation.

At the present time, data on the association of risk
factors to atherosclerosis in human subjects has come mostly
from retrospective autopsy studies and coronary
arteriography case series. Both types of studies can suffer
from selection biases[1,2] and the results of arteriography
studies can vary by definition of the control group[3]. In
addition, information concerning risk factors is usually
obtained after the onset of clinical disease, thus adding
another element of bias.

The 20-year follow-up activities of the Honolulu Heart
Program (HHP) provided an opportunity to examine this
question from a different perspective. Whenever possible,
protocol autopsies have been obtained for men in the
original cohort who have died. The advantages of such a
study population are that all risk factors were measured
prior to the onset of clinical disease and that it allows an
estimate of the extent of selection bias through comparison
of men with and without autopsy. The purpose of this study
was to examine the associations of reported baseline
cigarette smoking levels with protocol autopsy determined
levels of atherosclerosis in the aorta, coronary and
cerebral arteries obtained during a 20-year follow-up of the
HHP cohort.

MATERIALS AND METHODS

The HHP is a prospective epidemiologic study of
cardiovascular disease among a cohort of over 8,000 men of

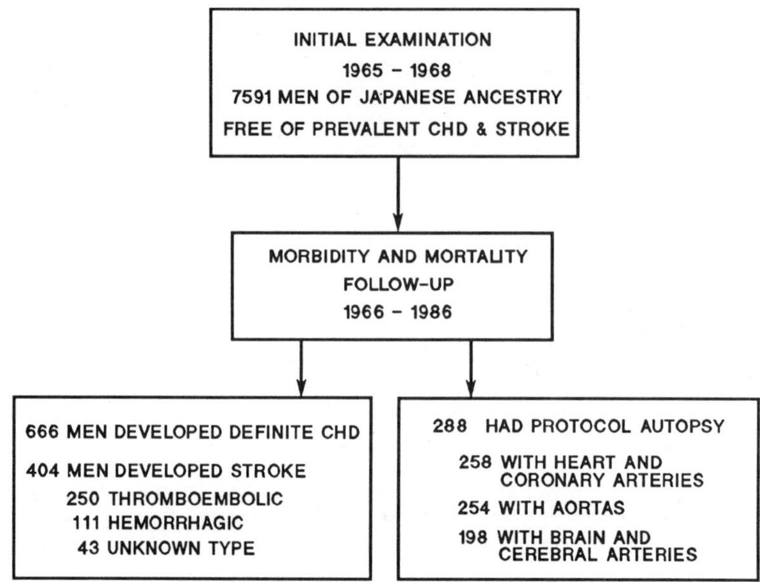

FIGURE 1. STUDY POPULATION OF THE HONOLULU HEART PROGRAM

Japanese ancestry living in Hawaii (Figure 1). During 20 years of follow-up of 7,591 men who were free of prevalent coronary heart disease and stroke at the 1965-1968 baseline examination, 666 men have developed definite coronary heart disease, and 404 men developed definite or probable stroke. Among 288 protocol autopsies of men free of clinical cardiovascular disease at baseline, measures of atherosclerosis were obtained for coronary arteries for 258 men, the aorta for 254 men, and cerebral arteries for 198.

The diagnostic and pathology study methods have been published previously[4-7]. The degree of atherosclerosis in aortas and coronary arteries was estimated by the American Heart Association Panel Method, utilizing a panel of photographs graded from 1 (no atherosclerosis) to 7 (most of the surface of the vessel involved with atherosclerotic lesions or complete stenosis). The worst score of any of the coronary arteries was used for this report as it was related to a greater risk of a clinical event. The cerebral arteries in the circle of Willis and its major branches were scored on a numerical basis with grades from 0 (no atherosclerosis) to 4 (more than 50 per cent lumen narrowing or a plaque involving the entire circumference) in each of 22 sites. For this report, the worst score from a subgroup of large arteries including the internal carotid, the middle and posterior cerebral, the basilar and vertebral arteries were used. In some analyses, a category of men with "severe" atherosclerosis was used. The included men with any coronary artery score of 4 or greater, equivalent to more than 60 per cent of the surface affected with raised lesions, or with any cerebral artery score of 4, equivalent of more than 50 per cent lumen narrowing.

Measures of risk factors were all from the initial
examination described earlier[4]. Cigarette smoking questions
included status as never, past, or current cigarette smoker,
age started, number of years smoked, usual number of
cigarettes a day, type of cigarettes, and use of other types
of tobacco. Cigarette pack years was calculated from the
usual number of cigarettes per day divided by 20 and
multiplied by the number of years smoked. Comparison of
responses given at the first exam with those from exams two
and six years later showed a high degree of consistency.

Statistical analyses included calculation of age-
adjusted atherosclerosis scores or percentages of men with
severe atherosclerosis by measures of cigarette use.
Multivariate analyses used multiple linear regression models
for the atherosclerosis scores and logistic models for the
category of serious atherosclerosis.

RESULTS

In an earlier study we reported that when the men in
the protocol autopsy group were compared to all deaths,
there were no significant differences in the proportion of
major causes of death nor in the mean values of more than 20
risk factors measured at the baseline examination[6]. Thus,
the autopsy group was similar to the group of all deaths.

A more specific measure of selection bias can be
obtained by comparison of the autopsy group with the target
population of 7,591 men free of CHD at baseline. For our
specific interest in cigarette smoking status, Table 1 shows
the percentages of men in the target and autopsy populations
by smoking status, and evidence of incidence CHD during the
follow-up period. There was a general trend for the autopsy
groups to have more smokers and fewer never smokers than the
target population, for both subgroups with and without
incident CHD. These differences, however, were not
statistically significant.

TABLE 1. AGE-ADJUSTED PERCENTAGES OF CHD CASES AND NON-CASES IN THE TARGET
AND AUTPOSIED POPULATION BY BASELINE EXAMINATION SMOKING STATUS

| | Target Population N = 7591 | | Protocol Autopsy N − 288 | |
| | Incident CHD | | Incident CHD | |
Smoking Status	Yes N = 666	No N = 6923	Yes N = 81	No N = 207
Never Smoked (%)	23	29	14	20
Past Smoker (%)	21	25	25	24
Current Smoker (%)	57	46	62	56

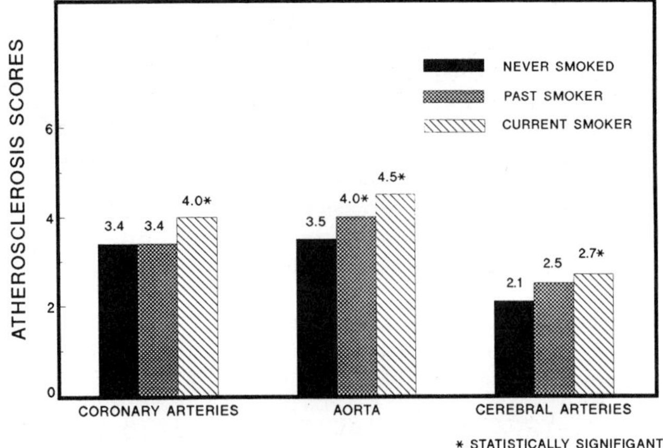

FIGURE 2. AGE-ADJUSTED ATHEROSCLEROSIS SCORES
BY ARTERIES AND SMOKING STATUS

In order to examine the associations between the cigarette smoking and atherosclerosis, the men were first grouped into smoking status categories as never, past and current smokers, and then age-adjusted atherosclerosis scores for the different arteries were calculated (Figure 2). The current smokers had significantly higher atherosclerosis scores in the coronary arteries, the aorta, and the cerebral arteries. There were also dose-response patterns in the aorta and the cerebral arteries but not for the coronary arteries.

Figure 3 shows that when the percentages of men with severe atherosclerosis were grouped by the same categories of smoking status, the percentages were significantly higher for current smokers than never smokers for the coronary arteries and aorta but not in the cerebral arteries. A dose response pattern with this smoking status variable was seen only in the aorta.

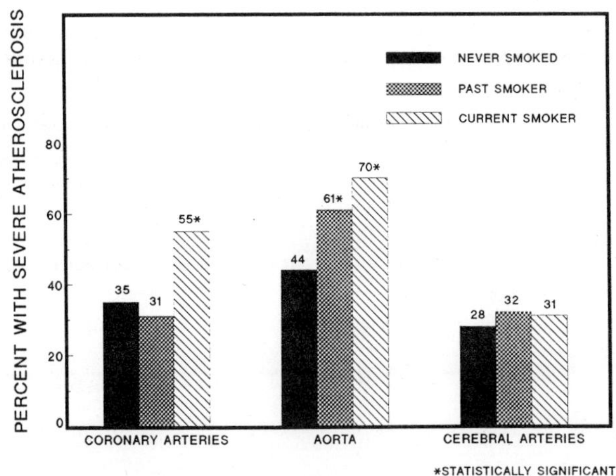

FIGURE 3. AGE-ADJUSTED PERCENTAGE OF MEN WITH SEVERE
ATHEROSCLEROSIS BY ARTERIES AND SMOKING STATUS

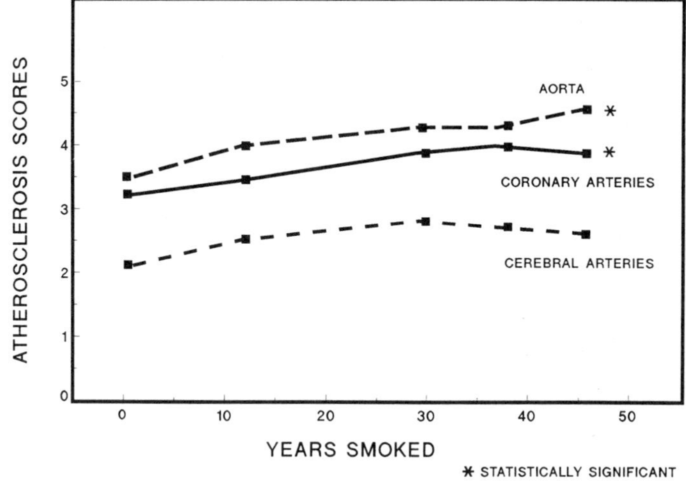

FIGURE 4. AGE-ADJUSTED ATHEROSCLEROSIS SCORES
BY ARTERIES AND YEARS SMOKED

Figure 4 shows age-adjusted atherosclerosis scores by
the number of years of cigarette smoking for all smokers at
baseline examination. There was a general pattern in all
arteries in which the scores tended to increase up to about
30 years of smoking and then to plateau after that. These
associations were statistically significant for the coronary
arteries and the aorta but not for the cerebral arteries.

Figure 5 shows that the age-adjusted percentages of men
with severe atherosclerosis also tended to increase with the
number of years smoking cigarettes for the aorta and for the
coronary arteries but not the cerebral arteries.

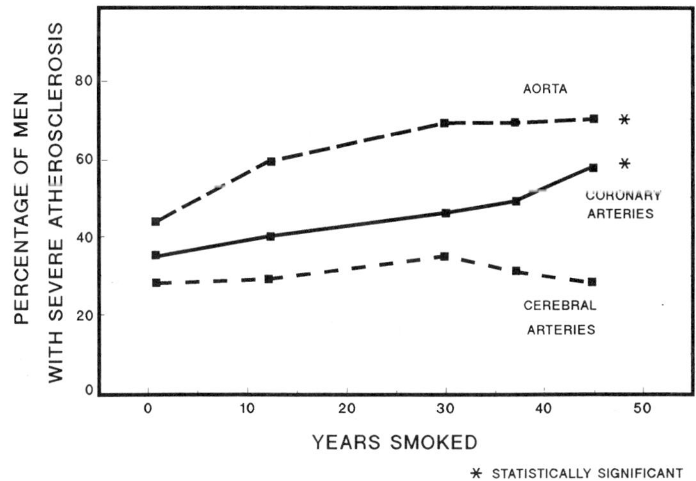

FIGURE 5. AGE-ADJUSTED PERCENTAGES OF MEN WITH SEVERE
ATHEROSCLEROSIS BY ARTERIES AND YEARS SMOKED

Separate analyses (not shown) of number of cigarettes smoked per day for current and past smokers showed weaker levels of associations. They were of borderline significance for atherosclerosis in the aorta and coronary arteries but not significant for the cerebral arteries. In earlier reports[6,7] we also examined the association of the mean atherosclerosis scores with cigarette pack years. The associations of this measure were significant with atherosclerosis in the aorta, of borderline significance in the coronary arteries, but not significant in the cerebral arteries.

The next step of the analysis was to examine the associations of atherosclerosis with cigarette smoking in multivariate models including other variables known to be associated with atherosclerosis in this cohort. Table 2 shows the Z test values from multivariate models relating the measures of atherosclerosis to cigarette smoking. In both types of models, age, systolic blood pressure and serum cholesterol were included. Other models including body mass index, alcohol intake, and serum glucose levels did not differ from the results shown here.

For the aorta, both measures of atherosclerosis were associated with all measures of cigarette smoking. For the coronary arteries, both measures of atherosclerosis were significantly associated with cigarette smoking status and with the years smoked, but not with cigarette pack years. For the circle of Willis there was only one significant association. The atherosclerosis score was associated with smoking status, but not with the other measures of cigarette use.

TABLE 2. Z SCORE FROM MULTIVARIATE MODELS[+] RELATING MEASURE OF ATHEROSCLEROSIS TO CIGARETTE SMOKING

	Coronary Artery	Aorta	Circle of Willis
Atherosclerosis Scores			
Never, Past, Current	2.36[*]	5.44[***]	2.05[*]
Years Smoked	2.10[*]	5.45[***]	1.56
Cigarette Pack Years	1.25	4.28[*]	0.68
Percent of men with Serious Atherosclerosis			
Never, Past, Current	2.96[**]	3.37[***]	0.37
Years Smoked	2.52[*]	3.63[***]	0.25
Cigarette Pack Years	1.49	3.03[***]	0.11

[+] For atherosclerosis scores multiple linear regression models were used.
For percent of men with serious atherosclerosis, multiple logistic models were used.
In both types of models age, systolic blood pressure and serum cholesterol were included.
* p < 0.05   ** p < 0.01   *** p < 0.001

These multivariate analyses were repeated for the subgroup of 182 autopsied men who had no clinical evidence of cardiovascular disease. The Z tests were uniformly higher for measures of atherosclerosis in the aorta and coronary arteries, and all of them were statistically significant, including cigarette pack years for the coronary artery measures. In the same analyses, there was virtually no change in the Z values for the cerebral arteries.

DISCUSSION

The general pattern of these analyses indicate that all measures of cigarette smoking were independently associated with atherosclerosis in the aorta. All measures except cigarette pack years were also consistently associated with atherosclerosis in the coronary arteries. In contrast, only one measure of cigarette smoking was significantly associated with atherosclerosis in the large cerebral arteries. Efforts to avoid some of the effects of autopsy selection bias through analysis of the non-cardiovascular disease subgroup of autopsied men indicated that atherosclerosis measures in both the aorta and coronary arteries were independently associated with all measures of smoking. These analyses, however, did not alter the general lack of association in the large cerebral arteries.

These findings provide stronger and more consistent evidence of the predictive association of cigarette smoking and atherosclerosis in the coronary arteries than we have reported earlier[6] and are similar to the earlier findings of consistent associations in the aorta[6] and weak associations in both large and small cerebral arteries[7]. They also indicate an accumulating dose-response of atherosclerosis in the aorta and coronary arteries for a duration of smoking up to 30 years. This measure appeared to be a better measure of the dose-response than the usual number of cigarettes per day and cigarette pack years. We have no explanation for the relatively weak levels of association of atherosclerosis and cigarette smoking in the cerebral arteries. It is interesting to note, however, that the levels of association in the different arteries follows the temporal development of atherosclerosis; earliest in the aorta, then the coronary arteries, and finally the cerebral arteries.

While both retrospective autopsy studies and arteriography case-series have reported associations of cigarette smoking with measures of atherosclerosis in the coronary arteries and aorta[8,9] the results are less consistent in the few other prospective autopsy studies which have examined this question. Smoking was an independent predictor of atherosclerosis in the aorta in the Puerto Rico cohort[10], but not with the coronary arteries in Puerto Rico[10], Oslo, Norway[11], Malmo, Sweden[12] or Hisayama, Japan[13], nor with cerebral arteries in Oslo[11].

It is difficult to explain this lack of consistency among the different studies. As far as we can tell, none of the prospective studies removed cases with prevalent coronary disease at baseline, nor was there consideration of autopsy selection bias. Furthermore, different methods were used to measure both the level of atherosclerosis and exposure to cigarette smoking. As exemplified in this report, different measures of cigarette smoking or of atherosclerosis can result in different levels of associations.

The original question which stimulated these analyses was whether cigarette smoking is a primary cause of atherosclerosis, or a precipitator of clinical infarction or both. The role of cigarette smoking as a precipitator of clinical events through association with higher fibrinogen levels, hemoglobin concentration and myocardial oxygen supply has been reported[14-16] The consistent associations of cigarette smoking with aortic and coronary atherosclerosis in this study also indicate an atherogenic role. Thus, it appears that cigarette smoking is involved at both stages of cardiovascular disease. It follows that efforts to prevent cardiovascular disease should continue to emphasize curtailment of smoking both at the level of primary prevention of atherosclerosis, and at the level of preventing precipitation of a clinical event in persons with serious atherosclerosis.

ACKNOWLEDGEMENT

This study was supported by National Heart, Lung and Blood Institute, Contract No. NO1-HC-02901. Dr. J.A. Resch provided the measures of atherosclerosis in the cerebral arteries.

REFERENCES

1. T.A. Pearson. Coronary arteriography in the study of the epidemiology of coronary artery disease. Epidemiol Rev 1984;6:140-66
2. D. Mainland. The risk of fallacious conclusions from autopsy data on the incidence of disease with applications to heart disease. Am Heart J 1953;45:644-54
3. L.P. Fried, T.A. Pearson. The association of risk factors with arteriographically defined coronary artery disease: what is the appropriate control group? Am J Epidemiol 1987;125:844-53
4. A. Kagan, B.R. Harris, W. Winkelstein Jr, K.G. Johnson, H. Kato, S.L. Syme, G.G. Rhoads, M.L. Gay, M.Z. Nichaman, H.B. Hamilton, J.L. Tillotson. Epidemiologic studies of coronary heart disease and stroke in Japanese men living in Japan, Hawaii and California: demographic, physical, dietary and biochemical characteristics. J Chronic Dis 1974;27:345-64
5. K. Yano, D. Reed, D. McGee. Ten-year incidence of coronary heart disease in the Honolulu Heart Program: relationship to biological and lifestyle characteristics. Am J Epidemiol 1984;119(5):653-66
6. D.M. Reed, C.J. MacLean, T. Hayashi. Predictors of atherosclerosis in the Honolulu Heart Program.
    1. Biologic, dietary and lifestyle characteristics.
    Am J Epidemiol 1987;126:214-25

7.   D.M. Reed, J.A. Resch, T. Hayashi, C.J. MacLean,
     K. Yano. A prospective study of cerebral artery
     atherosclerosis. Stroke 1988;19;820-5
8.   L.A. Solberg, J.P. Strong. Risk factors and
     atherosclerotic lesions: A review of autopsy studies.
     Arteriosclerosis 1983;3:187-198
9.   T.A. Pearson. Coronary arteriography in the study of
     the epidemiology of coronary artery disease. Epidemiol
     Rev 1984;6:140-66
10. P.D. Sorlie, M.R. Garcia-Palmieri, M.I. Castillo-Staab,
     R. Costa, Jr, M.C. Oalmann, R. Havlik. The relation of
     antemortem factors to atherosclerosis at autopsy. The
     Puerto Rico Heart Health Program. Am J Pathol 1981;
     103:345-52
11. I. Holme, S.C. Enger, A. Helgeland, I. Hjermann, P.
     Leren, P. Lung-Larsen L. Solberg, J.P. Strong.
     Risk factors and raised atherosclerotic lesions in
     coronary and cerebral arteries. Statistical analysis
     from the Oslo Study. Arteriosclerosis 1981; 1;250-6
12. N.H. Sternby. Atherosclerosis, smoking and other risk
     factors, in: "Atherosclerosis V," Gotto AM Jr, Smith
     LC, Allen B, eds., New York: Springer-Verlag, 1980:67-
     70
13. N. Okumiya, K. Tanaka, K. Ueda, T. Omae. Coronary
     atherosclerosis and antecedent risk factors:
     pathologic and epidemiology study in Hisayama, Japan.
     Am J Cardiol 1985;56:62-6
14. W.B. Kannel, R.B. D'Agostino, A.J. Belanger.
     Fibrinogen, cigarette smoking, and risk of
     cardiovascular disease: insights from the Framingham
     Study. Am Heart J 1987;113:1006-10
15. H.C. McGill. Potential mechanisms for the
     augmentation of atherosclerosis and atherosclerotic
     disease by cigarette smoking. Prev Med 1979;8:390-403
16. M.D. Winniford, D.E. Jansen DE, G.A. Reynold,
     P. Apprill, W.H. Black, L.D. Hillis. Cigarette
     smoking-induced coronary vasoconstriction in
     atherosclerotic coronary artery disease and
     prevention by calcium antagonists and nitroglycerin.
     Am J Cardiol 1987;1;59:203-7

CIGARETTE SMOKING AS A RISK FACTOR FOR CORONARY ARTERY DISEASE

William S. Weintraub

Division of Cardiology
Emory University Hospital
1364 Clifton Road
Atlanta, Georgia 30322

Abstract

The epidemiologic evidence linking cigarette smoking to coronary
artery disease has been supported by multiple studies over a period of 30
years. Community based cohort studies such as the Framingham study have
shown that cigarette smokers are at a markedly elevated risk of
developing cardiovascular events. These studies have also shown that
former cigarette smokers have a decline in the excess incidence over a
period of several years to approach that of non-smoking populations.
Uncertainty then appears to exist as to whether cigarette smokers are at
increased risk of cardiovascular events solely because of acute effects
of smoking or whether cigarette smoking is also a risk factor for
atherosclerosis. Evidence that cigarette smoking is a risk factor for
atherosclerosis comes from autopsy and cardiac catheterization
laboratory. While not all of these studies show a relationship, the data
do, for the most part, show a definite relationship between cigarette
smoking and atherosclerosis. Catheterization and autopsy studies suffer
from biases that prevent application of prediction models to the general
population. However, the combined view of the community based,
catheterization and autopsy studies supports a unifying hypothesis of
acute effects that may lead to cardiovascular events in the setting of
established vascular disease as well as vascular damage that may promote
the development of atherosclerosis. Cessation of smoking will rapidly
remove the acute effects, and may lead to inactivation of acute lesions.
The underlying substrate of atherosclerosis and coronary obstructive
lesions appears to be correlated with long term cigarette smoking.

Introduction

Since the first observations of Hammond and Horn (1), multiple
epidemiologic studies have implicated cigarette smoking as a risk factor
for the development of cardiovascular events (2-6). The nature of the
risk posed by cigarettes is uncertain. Do cigarettes lead to increased
atherosclerosis or do cigarettes result in exacerbation of established
coronary artery disease? The importance of this question should not be
minimized. If cigarettes cause coronary disease but to not exacerbate
existing disease, then efforts to stop smoking must be concentrated on
young people before they begin to smoke. In this setting, the impact of
stopping will only be through the prevention of progression. If

Tobacco Smoking and Atherosclerosis
Edited by J. N. Diana
Plenum Press, New York, 1990

cigarettes do not cause coronary disease, but act solely by exacerbating
established disease, then cessation of smoking becomes an urgent issue
for all patients with coronary disease. If, as seems likely, smoking
acts as a risk factor for the development of coronary atherosclerosis and
exacerbates established disease, then it will never be too early nor too
late to begin to influence people not to smoke.

The absolute answer to cigarette smoking and coronary heart disease
would require a data set of the following sort. Randomize many thousands
of people at age 15 to smoking or no smoking, follow the people clinically for life for cardiovascular events, perform frequent arteriograms, and
then autopsy them all as they die decades later. Given that this is
nonsensical, what types of data exist? 1) There are observational data
from large databases followed for prolonged periods of time. 2) There
are cross sectional data from catheterization databases concerning
angiographic evidence of coronary disease. 3) Finally, there are autopsy
data from several large series. The longitudinal observational databases
will largely provide evidence of cardiovascular events such as acute
myocardial infarction, death, or angina pectoris. The catheterization
databases and autopsy databases consider atherosclerosis per se. Each of
these data sets will be reviewed. Taken together, a picture may be
pieced together that resembles our facetious experiment considered above.

<u>Community Based Studies</u>

Compelling epidemiologic evidence for a role for smoking in heart
disease has been provided by studies such as that from Framingham (2),
the Göteborg Primary Prevention Study (3), and the Western Collaborative
Study (6). The Framingham Study has shown that men who smoke have a
relative risk of nonfatal and fatal myocardial infarction that is 2 to 3
times higher than non-smoking men, and 1.5 to 3 time higher in women.
The relative risk of sudden cardiovascular death was 10 times higher in
men and 4.5 times higher in women. The data from the Göteborg Study and
the Western Collaborative Study have been similar. Several studies have
shown evidence that events are dose dependent (4-7). There is not
compelling evidence from the Framingham Study (2) or the Göteborg Study
(8) for the development of angina pectoris independently of acute
myocardial infarction or sudden death. The reason for this is not
clear. It could be that cigarette smoking does not lead to
atherosclerosis, and coronary narrowing which is necessary for angina
pectoris. Cigarettes may then only exacerbate established
atherosclerosis leading to death or infarction. However, angina pectoris
is a relatively soft end point, and the noise may be higher. The
Framingham (9) and Göteborg (10) show that smoking is an independent risk
factor for coronary events, and may even be synergistic with other risk
factors. This was also shown in the MRFIT study (11). For middle aged,
non-smoking, normotensive men, the risk of coronary events per 1000 over
six years rose from 1.6 for a total cholesterol under 182 to 6.4 for a
cholesterol over 245. In contrast, in hypertensive smokers the six year
risk rose from 6.3 if the serum cholesterol was under 182 to 21.4 if the
serum cholesterol was over 245. If cigarette smoking exacerbates
atherosclerosis, but does not cause it, then cessation should remove the
excess risk. Several observational studies (12-18) have found that
cessation of smoking after myocardial infarction results in a
substantial, although not necessarily complete, reduction in risk. One
randomized study (19) of cessation in a healthy population showed only a
trend to decreased risk. The Western Collaborative Study (6) found continuing increased risk in former smokers, although the time since
cessation of smoking was not available. The meaning of the cessation
data is not entirely certain. Risk almost certainly fails with cessation, but whether risk can fall to non-smoking level remains uncertain.

<u>Angiographic Studies</u>

Epidemiologic evidence showing that cigarette smoking leads to cardiovascular events cannot adequately address the question of the role of cigarettes in promoting atherosclerosis. Evidence for an etiologic role for cigarettes in the development of atherosclerosis can best be considered by examining the relationship of cigarette smoking to atherosclerosis either by imaging techniques in vivo or by autopsy.

Angiographic evidence relating cigarette smoking to coronary atherosclerosis is limited. There may be several reasons for this. This may in part relate to most investigators not finding this a meaningful question. This is because risk factors are used to determine the need for catheterization, thus undermining the relationship of the risk factor to angiographically demonstrated disease. While the relationship will change in ways that are hard to predict quantitatively, qualitatively this may be considered. Use of risk factor to determine who will undergo catheterization may be viewed as a form of work-up bias, which will raise sensitivity, lower specificity and remove information content (20). In addition, it may be more difficult to obtain a reliable smoking history in symptomatic people than in a community based observational setting. It is then fair to say that if a relationship is found between angiographic evidence of cigarette smoking and atherosclerosis it is probably real, but it will not be possible to translate that back to the general population as a quantitative relationship.

Despite the problems, cigarette smoking and angiographic evidence of disease has been examined. In 1984 we published a paper that established cigarette smoking as an independent risk factor for coronary disease in a catheterization laboratory (21). In 1,349 patients, the number of pack years was estimated by multiplying the number of years of smoking by the estimated average packs per day. Clearly, this is an imprecise measure. The current packs per day smoked was not a risk factor, while the number of pack years smoked was an independent risk factor for coronary disease. Other risk factors were typical angina pain, male sex, age, and positive family history. A logistic regression model was developed to predict the probability of coronary disease in a catheterization laboratory setting based on these risk factors. These data are summarized in Tables 1 and 2. Table 1 presents the results in women and Table 2 the results in men. In each table, the data are presented for atypical and typical pain, as well as for positive and negative family histories. With increasing pack-years, the relative risk increases, but the relative risk with more pack-years is greater in the younger patients who otherwise would be at relatively low risk. In contrast, older patients with symptoms of angina pectoris have a high risk by virtue of their age and symptoms alone, and are thus unlikely to have a significant increase in risk. An independent association of cigarette smoking over a period of years with the development of coronary artery disease is clearly demonstrated.

These data were compared to similar data stored in the cardiac databank at Emory University. From 1/1/87 through 6/30/89 diagnostic cardiac catheterization was performed on 7874 patients. Data on cigarette smoking are available on 7044. These data are presented in Tables 3 and 4. The percentage of current smokers is slightly higher in the patients with coronary disease. The percentage who have smoked at any time is higher in the patients with coronary disease. The relative risk of having coronary disease is higher in patients who have smoked at any time than in current smokers. The difference between current smokers and people who smoked at some time is quite large, raising the possibility of under-reporting of current smoking. In Table 4 the

Table 1
Probability of Coronary Artery Disease in Women
Pack-Years of Cigarettes

	0	20	40	60	80	100
Negative Family History–Atypical Pain						
Age (yrs)						
30	0.083	0.097	0.113	0.131	0.151	0.174
40	0.132	0.153	0.176	0.202	0.230	0.262
50	0.204	0.233	0.264	0.298	0.335	0.373
60	0.301	0.338	0.376	0.417	0.458	0.500
70	0.420	0.461	0.504	0.546	0.587	0.627
Typical Pain						
Age (yrs)						
30	0.287	0.313	0.351	0.390	0.431	0.473
40	0.393	0.434	0.476	0.518	0.560	0.601
50	0.521	0.563	0.604	0.644	0.681	0.717
60	0.647	0.684	0.719	0.752	0.782	0.810
70	0.755	0.784	0.812	0.836	0.858	0.877
Positive Family History -- Atypical Pain						
Age (yrs)						
30	0.129	0.149	0.172	0.197	0.225	0.256
40	0.199	0.228	0.259	0.292	0.225	0.256
50	0.295	0.331	0.370	0.410	0.451	0.493
60	0.413	0.454	0.496	0.538	0.580	0.620
70	0.542	0.583	0.623	0.662	0.699	0.733
Typical Pain						
Age (yrs)						
30	0.386	0.427	0.496	0.511	0.553	0.594
40	0.514	0.556	0.597	0.637	0.675	0.711
50	0.640	0.678	0.713	0.747	0.777	0.805
60	0.749	0.779	0.807	0.832	0.854	0.874
70	0.834	0.856	0.875	0.893	0.908	0.921

percentages of patients with coronary disease by age ranges for patients
who have and have not smoked at any time are shown. In every age range,
the percentage of patients with coronary disease is higher for smokers.
The relative risk of coronary disease is higher for younger patients who
smoke than for older ones. Thus, these data confirm our previously
published study. As previously noted, these data cannot be translated
into risk of developing coronary disease in smokers. However, since the
prevalence of coronary disease will be markedly elevated in all

catheterization patients relative to the general population, the relative
risk of atherosclerotic coronary disease in the general population may be
considerably higher in smokers than in the data presented here. These
data, however, do reveal that cigarette smoking over a period of time is
associated with angiographic evidence of coronary artery disease.

Their are other studies which support these findings. Vlietstra et
al. (22) and Holmes et al. (23) have shown a risk to cigarette smoking.

Table 2
Probability of Coronary Artery Disease in Men
Pack-Years of Cigarettes

	0	20	40	60	80	100
Negative Family History–Atypical Pain						
Age (yrs)						
30	0.214	0.243	0.276	0.311	0.348	0.387
40	0.313	0.351	0.390	0.431	0.473	0.515
50	0.434	0.476	0.518	0.560	0.601	0.641
60	0.563	0.604	0.644	0.681	0.717	0.750
70	0.684	0.719	0.752	0.782	0.810	0.834
Typical Pain						
Age (yrs)						
30	0.535	0.577	0.618	0.657	0.694	0.728
40	0.660	0.696	0.731	0.763	0.792	0.818
50	0.765	0.794	0.820	0.844	0.865	0.883
60	0.846	0.866	0.885	0.901	0.915	0.927
70	0.902	0.916	0.928	0.938	0.948	0.955
Positive Family History -- Atypical Pain						
Age (yrs)						
30	0.307	0.344	0.383	0.424	0.465	0.508
40	0.427	0.469	0.511	0.553	0.594	0.634
50	0.556	0.597	0.637	0.675	0.711	0.744
60	0.678	0.714	0.747	0.777	0.805	0.830
70	0.780	0.807	0.832	0.854	0.874	0.891
Typical Pain						
Age (yrs)						
30	0.653	0.690	0.725	0.757	0.787	0.814
40	0.760	0.789	0.816	0.840	0.861	0.880
50	0.842	0.863	0.882	0.898	0.913	0.925
60	0.899	0.914	0.926	0.937	0.946	0.954
70	0.938	0.947	0.955	0.961	0.967	0.972

Table 3

Emory Angiographic Data

	Coronary Artery Disease			
	Absent	Present	Relative Risk	P Value
n	2951	4093		
Current Smoker	469(17.3%)	737(19.7%)	1.07	.01
Smoker at Any Time	1647(55.2%)	2866(68.5%)	1.28	p<.0001
Mean Pack Years	14±16	21±18		p<.0001

Vliestra et al. demonstrated results from the CASS registry which are similar to those from Emory. There is an association between ever having smoked and the presence of coronary disease. Present consumption of cigarettes was found to be less important.

The above mentioned studies are supported by a twin study from Finland (24). In this study 49 pair of twins discordant for smoking were studied. Carotid duplex sonography was performed. The total area of carotid plaques was 3.2 times greater in the smoking cotwins. While this study did not concern the coronary arteries, the measure here was atherosclerosis. As atherosclerosis takes years to develop, it is likely that the long term exposure to cigarettes explains the difference between the smoking and non-smoking twin pairs.

Table 4

Emory Angiographic Data

		Coronary Artery Disease		
Age	Non Smoker	Smoker	Relative Risk	P Value
<40	14.9%	38.0%	2.56	<.0001
40-49	29.3%	51.4%	1.75	<.0001
50-59	40.3%	62.1%	1.54	<.0001
60-69	54.3%	69.0%	1.27	<.0001
≥70	67.0%	78.6%	1.17	<.0001

<u>Autopsy Studies</u>

The few studies concerning imaging in patients may be contrasted to autopsy studies of atherosclerosis. Autopsy studies are naturally biased. If smoking leads to death in established disease rather than causing disease, then a relationship of smoking to autopsy severity of atherosclerosis may not be etiologic. This has been addressed in some studies by limiting the study to patients not dying of coronary disease. Another major problem with the autopsy studies is the limited ability to take a smoking history from the family.

Despite the limitations, there have been multiple autopsy studies that have examined the risk of coronary disease in smokers and non-smokers. These studies have generally examined the coronary arteries and aorta for fatty streaks and raised lesions. The results of these studies appear superficially to be inconclusive, with several studies showing evidence that smoking is a risk factor and some failing to show a relationship. Auerback et al. (25) performed post-mortum examinations on 1372 men not dying of coronary artery disease. The population included 126 who never smoked, 893 current smokers, and 353 former smokers. More morphologic evidence of coronary disease was noted in the smokers. A dose response relationship between extent of smoking and severity of disease was noted. In another study Auerback et al. (26) noted that smokers have an unusual hyaline thickening of the arterioles that is rare in non-smokers. This finding was noted in current and former smokers. Only one large study, that by Viel et al. (27) from Chile, failed to show a relationship. In this study smoking was added, as noted by the authors, as something of an afterthought. Ascertainment of smoking status was questionable. Even so, in this study, careful scrutiny of the figures suggests that there are more "raised lesions" in smokers. The study of Holme et al. (28) was part of the Oslo study of coronary artery disease risk. This substudy examined 129 autopsies for coronary atherosclerosis from a total of 16,200 men. There is doubt about ascertainment of smoking status as the database of 16,200 was used. The negative studies of Viel et al. Holme et al. and Sorlier et al. (29) all appeared to assess cigarette usage at one point in time. In fact, none of these negative studies are sufficiently clear as to how smoking status was ascertained. In contrast the study of Strong et al. (30) showed a strong correlation of coronary atherosclerosis with smoking. In this study the methods as applied to understanding the smoking history appear to be more careful. The focus of Strong's study, as with those of Auerback (25,26), concerned smoking as a primary focus, rather than examining it as a peripheral part of an overall larger study. The study of Strong et al. (30) concerns a 10 year history of smoking. Several other autopsy studies have also shown a relationship between smoking and coronary disease (31-33). Thus, overall the results of the autopsy studies support the concept of a positive correlation of coronary atherosclerosis with long term consumption of cigarettes.

<u>Conclusion</u>

The evidence linking smoking to atherosclerosis has been supported by many studies. The studies failing to find a relationship appear to have examined cigarette smoking in a relatively limited manner. However, the epidemiologic relationship between smoking and atherosclerosis is limited by many biases. Autopsy data is limited by difficulty in getting a smoking history from the families of deceased patients. Catheterization laboratories have a high incidence of atherosclerosis. In addition, patients with atherosclerosis may be more inclined to report a smoking history. A reliable history of smoking is hard to take, as any clinician knows. Thus the noise level is very high. Nonetheless, a relationship

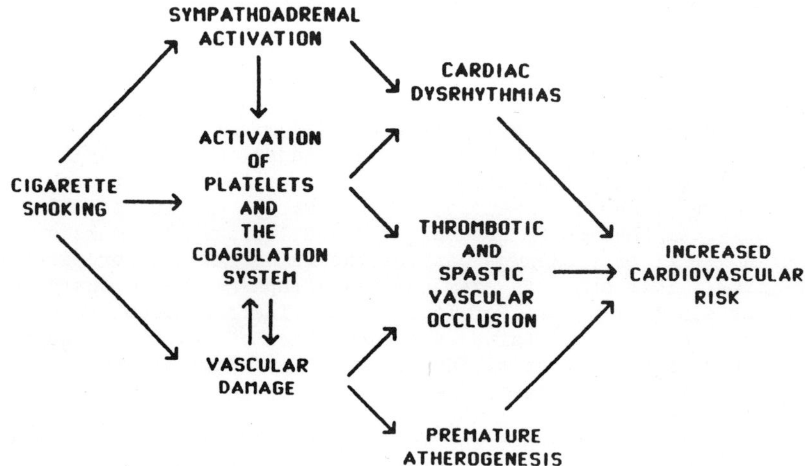

Figure 1. The interelationship of acute effects and chronic vascular damage due to cigarette smoking. With permission, ref. 45.

would appear to be confirmed by an overwhelming mass of data. What is
not clear, is the risk per pack year smoked in the general population.
As opposed to the problem of assessing risk of atherosclerosis, the
incidence of events can be gathered from the Framingham data (10), the
Western Collaborative data (6), and other epidemiologic studies. Neither
the incidence data, the autopsy data, nor the catheterization data can
provide a quantitative estimate of the increased risk of atherosclerosis
due to cigarette smoking in the general population.

The effect of cigarette smoking is multifactorial. Evidence
presented and summarized above establishes that cigarette smoking is, in
fact, related to the development of atherosclerosis. There is also clear
evidence that cigarette smoking has more acute effects as well. These
effects include vasospasm (34), platelet aggregation (35), enhanced
thrombosis (36), elevation of blood pressure and heart rate (37),
lowering of HDL cholesterol (38-40), elevation of the white blood cell
count (41-43), and elevation of plasma catecholamines (43). These
deleterious effects are likely to be synergistic. In another study from
Emory, Pecora et al. (44) studied 43 patients who had non-critical
coronary artery disease. In these patients clustered around a group of
young smokers. The mean age was only 45 and 74% were present smokers.
In contrast, only 16% had elevated cholesterol. Here we note the
interdependence of the premature development of coronary artery disease
coupled with a propensity for thrombosis in a group of patients with a
high percentage of smokers. The implications for the dangers of
cigarette smoking could hardly be more profound.

There is evidence of increased cardiovascular events in smokers.
However, there is also evidence that stopping smoking will reduce risk to
that of non-smokers within several years. A unifying hypothesis is
presented in Figure 1 (45). The effects of cigarette smoking will lead
to vascular damage that will lead to premature or accelerated
atherosclerosis. Effects of the sympathetic nervous system and on
coagulation will lead to increased propensity to events, probably through
enhanced thrombosis. The acute effects will interact synergistically
with the enhanced atherosclerosis. The underlying pathophysiologic
relationships may actually be much more complicated. Thus, cigarette
smoking does appear to have multiple clinically relevant effects. It is
well known that coronary disease may be present but asymptomatic, that is
the lesions are "inactive" and hemodynamically non-obstructive. The
acute effects of cigarettes may then lead to "activation" of coronary
lesions. The lesion activation property of cigarettes may be removed
with cessation of cigarette smoking. Current evidence would then suggest
that over two years risk declines to, or at least towards, normal.
Whether a lifetime of cigarette smoking can be negated by stopping is not
certain. That risk will fall if one stops smoking is certain. The risk
of past smoking may be lost in the confounding effects of other risk
factors. A complete picture thus emerges of an interplay of risk of
premature atherosclerosis and acute effects.

<u>References</u>

1. Hammond EC, Horn D. Smoking and death rates -- report on forty-four
      months of follow-up of 187,783 men. II. Death rates by cause.
      JAMA, 166:1294-1308 (1958).
2. Kannel WB, McGee DL, Castelli WP. Latest perspective on cigarette
      smoking and cardiovascular disease: the Framingham Study. J.
      Cardiac Rehabil, 4:267-277 (1984).
3. Wilhelmsen L. Berglund G, Elmfeldt D. The multifactor primary
      prevention trial in Göteborg, Sweden. Eur Heart J, 7:279-288
      (1986).

4. Pooling Project Research Group. Relationship of blood pressure, serum cholesterol, smoking habit, relative weight and ECG abnormalities to incidence of major coronary events: final report of the Pooling Project. J Chronic Dis, 31:201-306 (1978).

5. Keys A, Taylor HL, Blackburn H, Brozek H, Anderson JT, Simonson E. Coronary heart disease among Minnesota business and professional men followed fifteen years. Circulation, 28:381-395 (1963).

6. Jenkins CD, Rosenman RH, Zyzanski SJ. Cigarette smoking: Its relationship to coronary heart disease and related risk factors in the Western Collaborative Group Study. Circulation, 38:1140-1154 (1968).

7. Doll R. Peto R. Mortality in relation to smoking: 20 years' observations of make British doctors. Br Med J, 2:1525-1536 (1976).

8. Hagman M, Wilhelmsen L, Wedel H, Pennert K. Risk factors for angina pectoris in a population sample of Swedish men. J Chronic Dis, 40:265-275 (1987).

9. Kannel WB, McGee D, Gordon T. A general cardiovascular risk profile: The Framingham Study. Am J Cardiol, 38:46-51 (1976).

10. Wilhelmsen L. Coronary heart disease: Epidemiology of smoking and intervention studies for smoking. Am Heart J, 115:242-249 (1988).

11. Report of the National Cholesterol Education Program Expert Panel on Detection, Evaluation and Treatment of High Blood Cholesterol in Adults. Arch Intern Med, 148:36-39 (1988).

12. Wilhelmsson C, Elmfeldt D, Vedin JA, Tibblin G, Wihelmsen F. Smoking and myocardial infarction. Lancet, 1:415-420 (1975).

13. Aberg A, Bergstrand R. Hohansson S. Cessation of smoking after myocardial infarction -- effects on mortality after 10 years. Br Heart J, 49:416-422 (1983).

14. Mulcahy R, Hickey N, Graham I, McKenzie G. Factors influencing long-term prognosis in male patients surviving a first coronary attack. Br Hear J, 37:158-165 (1975).

15. Mulcahy R. Influence of cigarette smoking on morbidity and mortality after myocardial infarction. Br Heart J, 49:410-415 (1983).

16. Sparrow D, Dawber TR, Colton T. The influence of cigarette smoking on prognosis after a first myocardial infarction. J Chronic Dis, 31:425-432 (1978).

17. Pohjola S, Siltanen P, Romo M. Effect of quitting smoking on the long-term survival after myocardial infarction (abstract) Trans Eur Soc Cardiol, 1:2 (1979).

18. Salonen JT. Stopping smoking and long-term mortality after acute myocardial infarction. Br Heart J, 43:463-469 (1980).

19. Rose G, Hamilton PJS, Colwell L, Shipley MJ. A randomized controlled trial of anti-smoking advice: 10 year results. J Epidemiol Community Health, 36:102-108 (1982).

20. Ransohoff DF, Feinstein AR. Problems of spectrum and bias in evaluating the efficacy of diagnostic tests. N Engl J Med, 299:926-930 (1978).

21. Weintraub WS, Klein LW, Seelaus PA, Agarwal JB, Helfant RH. Importance of total life consumption of cigarettes as a risk factor for coronary artery disease. Am J Cardiol, 55:669-672 (1985).

22. Vlietstra RE, Frye RL, Kronmal RA, Sim DA, Tristani FE, Killip T. Risk factors and angiographic coronary artery disease: a report from the Coronary Artery Surgery Study (CASS). Circulation, 62:254-261 (1980).

23. Holmes DR, Elveback LR, Frye RL, Kottke BA, Ellefson RD. Association of risk factor variables and coronary artery disease documented with angiography. Circulation, 63:293-299 (1981).

24. Haapanen A, Koskenvuo M, Kaprio J, Kesaniemi YA, Heikkila K. Carotid arteriosclerosis in identical twins discordant for cigarette smoking. Circulation, 80:10-16 (1989).

25. Auerbach O, Hammond EC, Garfinkel L. Smoking in relation to athero-
    sclerosis of the coronary arteries. N Engl J Med, 273:775–779
    (1965).
26. Auerbach O, Carter HW, Garfinkel L, Hammond EC. Cigarette smoking
    and coronary artery disease. A macroscopic and microscopic
    study. Chest, 70:697–705 (1976).
27. Viel B, Donoso S, Salcedo D. Coronary atherosclerosis in persons
    dying violently. Arch Intern Med, 122:97–103 (1968).
28. Holme I, Enger SC, Helgeland A. Risk factors and raised athero-
    sclerotic lesions in coronary and cerebral arteries. Statistical
    analysis form the Oslo study. Arteriosclerosis, 1:250–256 (1981).
29. Sorlie PD, Garcia-Palmieri MR, Castillo-Staab MI, Costas JR R,
    Oalmann MC, Havlik R. The relation of antemortem factors to
    atherosclerosis at autopsy: The Puerto Rico Heart Health
    Program. Am J Pathol, 103:345–352 (1981).
30. Strong JP, Richards ML. Cigarette smoking and atherosclerosis in
    autopsied med. Atherosclerosis, 70:697–705 (1976).
31. Lifsic AM. Atherosclerosis in smokers. Bull WHO, 53:631–638 (1976).
32. Rhoads GG, Blackwelder WC, Stemmermann GN, Hayashi T, Kagan A.
    Coronary risk factors and autopsy findings in Japanese-American
    men. Lab Invest, 38:304–311 (1978).
33. Sternby NIH. Atherosclerosis, smoking and other risk factors. In:
    Gotto JR AM, Smith LC, Allen B, eds. International Symposium on
    Atherosclerosis, 5th, Houston, Texas, November 6–9, 1979,
    Proceedings; Atherosclerosis V. New York: Springer-Verlag, 67–70
    (1980).
34. Klein LW, Ambrose J, Pichard A, Holt J, Gorlin R, Teichholz LE.
    Acute coronary hemodynamic response to cigarette smoking in
    patients with coronary artery disease. JACC, 3:879–886 (1984).
35. Glynn MF, Mustard JL, Buchanan JL, Murphy EA. Cigarette smoking and
    platelet aggregation. Can Med Assoc J, 95:549–553 (1966).
36. Ogston D, Bennett NB, Ogston CM. The influence of cigarette smoking
    on the plasma fibrinogen concentration. Atherosclerosis,
    11:349–352 (1970).
37. Trap-Jensen J. Effects of smoking on the heart and peripheral cir-
    culation. Am Heart J, 115:263–267 (1988).
38. Mjos OD. Lipid effects of smoking. Am Heart J, 115:272–275 (1988).
39. Goldbourt U, Medalie JH. Characteristics of smokers, non-smokers and
    ex-smokers among 10,000 adult males in Israel. II. Physiologic
    biochemical and genetic characteristics. Am J Epidemiol,
    105:75–86 (1977).
40. Garrison RJ, Kannel WB, Feinleib M, Castelli WP, McNamara PM,
    Padgett SJ. Cigarette smoking and HDL cholesterol. The Framingham
    Offspring Study. Atherosclerosis, 30:17–25 (1978).
41. Howell RW. Smoking habits and laboratory tests. Lancet, 2:152
    (1970).
42. Corre F, Lellouch J, Schwartz D. Smoking and leukocyte-counts.
    Results of an epidemiological survey. Lancet, 2:632 634 (1971).
43. Friedman GD, Siegelaub AB, Seltzer CC, Feldman R, Collen MF. Smoking
    habits and the leukocyte count. Arch Environ Health, 26:137–143
    (1973).
44. Pecora MJ, Roubin GS, Cobbs BW, Weintraub W, King SB. Myocardial
    infarction in the absence of angiographically significant coronary
    atherosclerotic disease: long-term prognosis. Am J Cardiol,
    62:363–367 (1988).
45. FitzGerald GA, Oates JA, Nowak J. Cigarette smoking and hemostatic
    function. Am Heart J, 115:267–271 (1988).

CIGARETTE SMOKING AND EXTRACRANIAL CAROTID

ATHEROSCLEROSIS

Grethe S. Tell[#], George Howard[#], Gregory W. Evans[#], Michael L. Smith[#], William M. McKinney[@], and James F. Toole[@]

Departments of [#]Public Health Sciences and [@]Neurology
Bowman Gray School of Medicine
300 S. Hawthorne Road
Winston-Salem, NC 27103

INTRODUCTION

Chronic cigarette smoking has been associated with increased risk for atherosclerotic diseases of extracranial carotid arteries in several studies. Angiography was the first method used to image *in vivo* stenosis - narrowing of the arterial lumen - both in the intracranial and extracranial carotid circulation. Stenosis is usually expressed as percent narrowing of the lumen diameter, and is often used as a measure of arterial disease caused by atherosclerotic plaque lesions. Although angiography has been an important tool studying the relation between various risk factors and atherosclerotic disease, the invasive nature of the procedure limits its application to symptomatic subjects and raises ethical concerns for studies in healthy volunteers.

During the last few years, noninvasive procedures employing Doppler ultrasound and high-resolution B-mode scanning have become available to study even mild vascular lesions and wall irregularities of extracranial neck arteries.[1,2] This development has opened up the opportunity to study not only symptomatic patients referred for angiography, but also asymptomatic subjects, thus including a wider range of atherosclerotic lesions than previously possible in studies of risk factors for cerebrovascular disease. Because of their anatomical position, the extracranial portions of the carotid arteries are most easily assessed, either by Doppler ultrasound for the measurement of arterial lumen stenosis or by B-mode ultrasound for the assessment of wall thickness reflecting the extent of atherosclerosis. Thus, whereas angiography and Doppler ultrasound assess the arterial lumen, B-mode ultrasound is capable of measuring wall thickness. Since atherosclerosis may involve a thickening of the wall but because of arterial dilatation no narrowing of the lumen,

Tobacco Smoking and Atherosclerosis
Edited by J. N. Diana
Plenum Press, New York, 1990

lesions apparent on B-mode ultrasound may not show up at angiography or Doppler interrogation.[3]

In this paper, we will first review studies in which the relationship between cigarette smoking and carotid atherosclerotic disease were examined using either angiography, Doppler ultrasound or B-mode ultrasound. Data from our own studies in which this relationship was examined using B-mode ultrasound will be presented, followed by a brief review of the relationship between smoking and clinically overt cerebrovascular disease.

Carotid Stenosis and Atherosclerosis

Angiographic studies. Most studies employing angiography for arterial investigation have found a positive correlation between cigarette smoking and extracranial carotid stenosis. In a case-control study reported by Duncan et al.,[4] patients with angiographically and surgically proven internal carotid artery stenosis had a significantly higher prevalence of a history of smoking compared to a matched control group. An Italian study[5] quantifying the degree of atherosclerosis using an extracranial score based on the number and severity of lesions in 11 arterial segments found that cigarette smoking was the only risk factor to show a strong association with the extracranial score. The smoking effect was independent of the effect of age, hypertension, diabetes and dyslipidemia. A Swiss study[6] of 159 patients with internal carotid artery occlusion or stenosis reported that smoking was more frequent among the patient group than in a matched control group. Another group studying the relationship of cardiovascular risk factors to carotid bifurcation stenosis reported smoking to be significantly more common among patients with abnormal angiograms.[7] The proportion of smokers were 39% in the group with normal angiograms versus 50% among those with abnormal angiograms. On reexamination of the patients, progression of carotid atheroma was not significantly related to smoking; however, the number of patient-years of follow-up was small.

Ultrasound studies. In a recent Swiss study, Müller and Buser assessed the prevalence of smoking, hypertension and diabetes in 221 patients with internal carotid stenosis, and in two sex and age-matched control groups (personal communication). Stenosis was assessed by continuous wave Doppler. Control group one consisted of patients with neurologic symptoms who had no carotid stenosis and control group two of patients with no neurologic symptoms but who may have had clinically silent carotid stenosis (no ultrasound exam was performed on this group). There were 55.7% smokers in the case group, 36.2% in control group one and 44.4% in control group two. When comparing the proportion of people in the three groups who were both smoking and were hypertensive, they found that among cases, 33% were smoking and hypertensive; the corresponding figures for control group one and control group two were 12.2% and 18.6%, respectively. The authors also looked at the relationship of smoking to the degree of stenosis in the carotid arteries. The greatest stenosis was seen among patients who both smoked and were hypertensive, and the least obstruction was seen among patients who were nonsmokers and normotensive. There was also a dose-response effect of smoking, with the greatest degree of stenosis occurring among the heaviest smokers.

Haapanen et al[8] reported on the relationship between smoking and carotid atherosclerosis measured by pulsed Doppler and B-mode ultrasound in a sample of 49 twins, discordant for smoking. Forty-five male and five female pairs were examined. The only significant difference between smoking and nonsmoking cotwins was an almost two times greater alcohol consumption among the smoking cotwins. With stenosis defined as at least 15% narrowing of the arterial lumen, there were significantly more individuals with carotid stenosis among the smoking cotwin group compared with the nonsmoking group. The mean area of all carotid plaques was 3.2 times greater in smoking cotwins than in nonsmoking cotwins, and the thickness of the inner layer of carotid arteries was significantly greater in smoking cotwins. Compared with nonsmokers, the age-adjusted odds ratio among smokers for having carotid stenosis was 5.99 (95% confidence interval (CI): 1.19-30.3). For smokers, the odds ratios of having a total area of carotid plaques $\geq$ 10 mm^2 was 3.97 (95% CI: 1.36-11.6), and for carotid inner layer thickening $\geq$ 1.2 mm 5.91 (95% CI: 1.76-19.8), relative to nonsmokers. These odds ratios remained significantly increased among smokers after adjustment for plasma cholesterol level, diastolic blood pressure and body mass index. They also found a dose-response effect of smoking.

In a study of risk factors for extracranial carotid artery atherosclerosis assessed by B-mode ultrasound, Crouse et al[9] found cigarette smoking to be a strong and significant risk factor for the extent of atherosclerosis. The effect of smoking was independent of the effect of other strong risk factors such as age and smoking.

We have previously reported on studies showing cigarette smoking to be an important risk factor for extracranial carotid atherosclerosis as measured by B-mode ultrasound.[10,11] Here we report on more recent studies from our hospital-based carotid ultrasound registry, again using B-mode ultrasound, in which more complete data on cigarette smoking history are available including information on dose and length of cigarette smoking.

METHODS

Each patient examined in the Clinical Neurosonology Laboratory of the North Carolina Baptist Hospital/Bowman Gray School of Medicine was interviewed to obtain a standard history and risk factor profile. Included in this survey were questions regarding smoking, demographic factors (age, race, and sex), surgical history, and coexisting diseases (myocardial infarction, stroke, diabetes mellitus, hypertension, and hyperlipidemia). Histories of coexisting diseases were based on patient responses to questions asking "Has a physician ever told you that you have ...," and are coded as yes or no. Smoking histories included information on past and current (within 4 months) cigarette use, average number of packs smoked per day, number of years smoked, and, for past smokers, the number of years since quitting smoking. Based on this survey, we classified patients as non-smokers, past smokers, or current smokers. For past and current smokers, we calculated pack years of smoking exposure as the average number of packs smoked per day times the number of years smoked.

We report here on a series of 840 consecutive patients with known smoking histories and risk factor profiles examined in the ultrasound laboratory between

December 1987 and May 1989. Of these 840 patients, 17 (2%) were excluded from subsequent analyses because of previous balloon angioplasty or surgery on the carotid arteries. Another seven patients with missing data on race, sex or age are not included in Table 1, which depicts the age and gender distribution of the patient population.

The ultrasound examination procedure has been described in detail previously.[10] The status of the carotid wall was examined bilaterally at eight sites: the proximal common carotid, mid common carotid, distal common carotid, bifurcation, proximal internal carotid, mid internal carotid, proximal external carotid and mid external carotid. At each of these sites, wall characteristics were categorized as noted in Table 2. For statistical analyses, we assigned a score to each category (also provided in Table 2) which approximated the arterial wall thickness midpoint. We then calculated the average score at each common carotid (across the proximal, mid and distal sites) and at each internal carotid (again across the examined arterial sites). Because our previous studies have shown on average very little plaque thickness in the external carotid,[10] we did not use the values from these areas in analysis presented here. Arterial scores were collapsed into a single number for each patient by computing the average across the two common and internal arteries, and the two bifurcations. If an individual site was not visualized it was not included in the calculation of the average at either step. This process resulted in a single B-mode score for each individual, which served as the primary outcome variable for subsequent analyses.

We evaluated relationships between B-mode scores and smoking status, age, race, gender, coronary heart disease, diabetes, hypertension, stroke history and hyperlipidemia using univariate and multivariate general linear models. In the multivariate linear regression model, a stepwise algorithm was used to limit the predictor variables included to those significantly related to the atherosclerosis score after controlling for other variables in the model. We examined interaction terms for all variables included in the multivariate model.

Table 1. Patient Population

Age Group (Years)	BLACKS Women N (%)	Men N (%)	WHITES Women N (%)	Men N (%)	Total N (%)
≤ 50	14 (1.7)	12 (1.5)	42 (5.2)	42 (5.2)	110 (13.5)
51-60	15 (1.8)	11 (1.4)	78 (9.6)	87 (10.7)	191 (23.4)
61-70	13 (1.6)	14 (1.7)	114 (14.0)	136 (16.7)	277 (34.0)
71-80	11 (1.4)	9 (1.1)	89 (10.9)	87 (10.7)	196 (24.0)
≥ 81	3 (0.4)	1 (0.1)	21 (2.6)	17 (2.1)	42 (5.2)
Total	56 (6.9)	47 (5.8)	344 (42.2)	369 (45.2)	816 (100.0)

Table 2. Ultrasound Categories, Descriptions, and Scores for Analysis

Description	Category	Score used for analysis
No wall abnormalities observed	Normal	0
Wall abnormalities with less than 0.9 mm of plaque	Wall Thickening	1
1.0 to 1.9 mm of plaque	Minimal Disease	1.5
2.0 to 3.9 mm of plaque	Moderate Disease	3
4.0 mm or greater of plaque	Severe Disease	5
No residual lumen	Occluded	6

For past and current smokers, we also examined the impact of pack-years smoked and time since quitting on B-mode scores, again using a general linear models approach. Because both pack-years and time-since-quitting are likely to be large only in older individuals, the relationship between these factors and B-mode scores were evaluated only after adjustment for age. Polynomial forms and interactions between age, pack-years and time-since-quitting were also examined.

RESULTS

Mean age differed between nonsmokers, past smokers and current smokers, as did the proportion of patients who were male or who had a history of coronary heart disease. Mean age was significantly lower for current smokers (59.0 years) compared with either past smokers or nonsmokers (65.5 and 66.0 years, respectively, $p < 0.05$ in both cases). While 69% of past smokers were male, the corresponding figures for current smokers and nonsmokers were 53% and 30%, respectively ($p < 0.05$ for both comparisons). There were also significantly more patients with coronary heart disease among past smokers (44%) compared with current smokers and nonsmokers (35% and 29%, respectively, $p < 0.05$). No differences were found between the three groups in the proportions of patients who were white, hypertensive, diabetic, hyperlipidemic or who had a history of prior stroke.

The mean B-mode score was significantly lower for nonsmokers compared with either past or current smokers (0.87 versus 1.15 and 1.11, respectively, $p < 0.05$ for both comparisons). The univariate impact of smoking on B-mode scores was comparable to that of other significant risk factors including age, hypertension, coronary heart disease, gender, prior stroke, race and diabetes (Table 3). No significant univariate impact of hyperlipidemia on B-mode scores was seen.

Smoking history also had a significant impact on B-mode scores in the multivariate model (Table 3), with mean adjusted B-mode scores increasing from 0.83 for nonsmokers to 1.08 for past smokers and to 1.25 for current smokers ($p < 0.05$). In addition to smoking history, age, hypertension, coronary heart disease, gender and

Table 3. Univariate and Multivariable Regression Results

| | ---UNIVARIATE--- | | ---FINAL MODEL- | |
Factor	Estimated Effect	P-Value	Estimated Effect	P-Value
Cigarette Smoking		<0.0001		<0.0001
Past	0.280		0.263	
Current	0.241		0.408	
Age[*]	0.205	<0.0001	0.236	<0.0001
Hypertension	0.279	<0.0001	0.235	<0.0001
CHD	0.321	<0.0001	0.170	0.0023
Male Gender	0.206	<0.0001	0.195	0.0105
Stroke	0.203	0.0008	0.146	0.0221
Hyperlipidemia	0.054	0.3149		
White Race	0.141	0.0333		
Diabetes Mellitus	0.135	0.0311		

[*]Per 10 year age difference

stroke history continued to be significantly related to B-mode scores in the multivariate model. The effect of current cigarette smoking on adjusted B-mode scores was substantially greater than that for any other risk factor considered and was nearly the equivalent of an additional two decades of life. While the effect of past smoking was less pronounced than that of current smoking, the effect of past smoking was significantly larger than that of other established risk factors, including that associated with an additional decade of life.

Among past and current smokers, both number of pack-years smoked and time-since-quitting (set at 0 for current smokers) were significantly related to the age-adjusted B-mode score ($p \leq 0.001$ in each case). The relationship between the age-adjusted score and time-since-quitting was linear, while the relationship between the age-adjusted score and pack-years was curvilinear, requiring a quadratic term. In a multivariate model containing age, time-since-quitting, pack-years, and pack-years squared, both time-since-quitting and pack-year effects remained significant after adjustment for age ($p \leq 0.03$). The magnitude of the joint effect of time since quitting and pack-years is shown in Figure 1. The impact of pack-years of smoking on B-mode scores decreases with increasing dose, and is fairly level beyond 60 pack-years.

DISCUSSION

Cigarette smoking is widely accepted to have a significant positive association with atherosclerosis in the coronary, aortic, abdominal and peripheral arteries. We feel that the extracranial carotid arteries should now be added to this list of susceptible arteries. The relationship between smoking and carotid atherosclerosis is independent of the effect of other strong cardiovascular disease risk factors such as age and hypertension. The importance of smoking as a risk factor is even more striking when considering our findings that a history of past smoking carries the

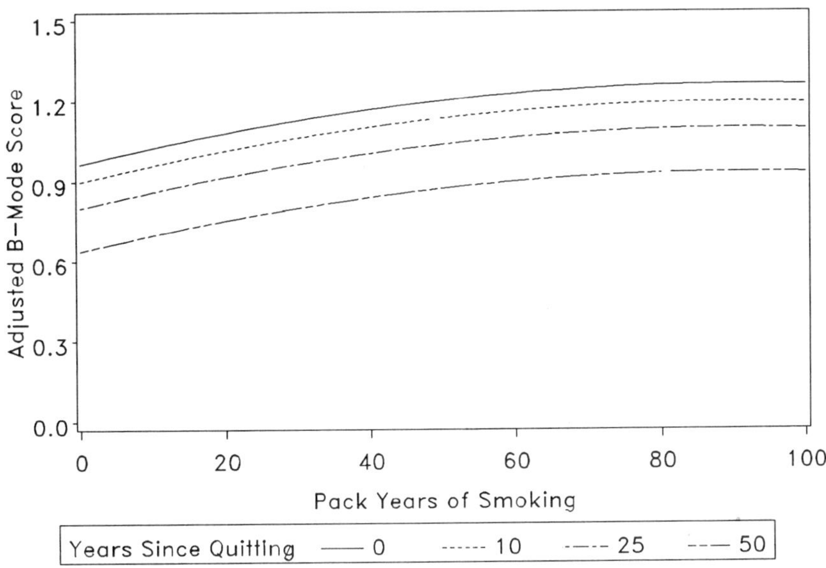

Figure 1. Expected average B-mode score as a function of pack years of smoking, shown for selected levels of years since quitting.

equivalent risk of approximately 10 years of aging, and that the risk of current smoking is equivalent to almost 20 years of aging.

In addition, the statistical significance of smoking in the multivariate model was second only to age, and proved to be a more powerful predictor than other well established risk factors such as hypertension, coronary heart disease and previous stroke. Another important new finding is the observed dose-response effect of smoking which was assessed by pack-years of smoking. The differences in B-mode scores between current and past smokers increases as the time-since-quitting increases for past smokers. Thus, the difference in score between current smokers and those who quit 10 years ago is less than the difference between the two groups who quit 25 years ago and 50 years ago. Because our data are cross-sectional rather than longitudinal, the direct effect of smoking cessation on atherosclerosis progression or regression in individuals cannot be evaluated. However, our data do suggest that quitting smoking may slow the rate of atherosclerosis progression.

Because our studies are hospital-based rather than population-based, the applicability of our findings to the general population is not straightforward. However, recent reports from population-based epidemiologic studies examining the relationship between smoking and carotid atherosclerosis lend support to our findings.

In the population-based Framingham study,[12] a history of smoking correlated significantly with extracranial carotid artery disease measured by Duplex imaging. In a cross-sectional population-based study of Finnish men,[13] cigarette smoking was the strongest single determinant of carotid atherosclerosis measured by B-mode ultrasound, independent of the effects of age, body mass index, serum lipoproteins, hypertension and plasma fibrinogen. In the ARIC Study[14] (an ongoing prospective study of cardiovascular diseases), the proportion of smokers among cases defined as having carotid wall thickening was 20% greater than among controls without wall thickening as measured by B-mode ultrasound at the study's baseline examination.

As part of the German center of the MONICA project, almost 1400 adults were randomly selected to participate in a carotid artery ultrasonographic examination.[15] Among men, a weak association was found between cigarette smoking and presence of plaques, and the frequency of plaques rose slightly with increasing cigarette consumption. Further, the frequency of plaques was marginally correlated to the number of cigarettes smoked per day, although the number of vessels with plaques was not different in current smokers as compared with nonsmokers. No association was found between cigarette smoking and carotid plaques in women.

Thus, three out of the four population-based studies reported so far have shown a strong and independent relation between cigarette smoking and carotid atherosclerosis as measured by ultrasound.

<u>Cigarette Smoking and Stroke</u>

The significance of the extent and severity of extracranial atherosclerosis stems from its potential involvement in clinical cerebrovascular disease, thus, the relation between cigarette smoking and stroke will be discussed briefly.

Cigarette smoking has been related to the development of clinical cerebrovascular events in a number of studies, although not in all. In a meta-analysis review by Shinton and Beevers,[16] the authors pooled the results from 32 studies published before May 1, 1988. Their calculated pooled relative risk (RR) of all strokes in cigarette smokers compared with nonsmokers was 1.51 (95% CI: 1.45-1.58). Their data also showed that there are differences in the relative risks for cigarette smokers between the various subtypes of stroke. Subarachnoid hemorrhage was clearly associated with cigarette smoking (RR=2.93; 95% CI: 2.48-3.46), and cerebral infarction was almost twice as likely to occur in smokers as in nonsmokers (RR=1.92; 95% CI: 1.71-2.16). Cerebral hemorrhage was not associated with smoking. A dose-response effect of cigarette smoking was also found, with the pooled RR estimates ranging from 1.37 in light smokers to 1.82 in heavy smokers. A clear trend was seen in RR as the age at the time of stroke increased, with RR decreasing as age increased, from 2.94 among people under 55 years to 1.11 for those 75 years and older. The authors point out that the relationship between smoking and stroke seems to depend on the ages of the study subjects; no large study has ever shown anything other than a positive association between cigarette smoking and stroke in people under the age of 75 years while several large studies in elderly people have shown little or even a negative association.

The study of the relationship between cigarette smoking and stroke is complicated by confounders and effect modifiers such as age, blood pressure and obesity. Several studies have shown 1) an inverse relationship between smoking and blood pressure and 2) smokers are generally thinner than nonsmokers. Since smoking is related to lower blood pressure and to lower weight, failure to adjust for these factors may lead to an underestimation of the risk associated with cigarette smoking. In fact, after Shinton and Beevers adjusted for age, blood pressure and obesity, the pooled relative risks were substantially higher than those adjusted for age alone.

In an Australian case-control study published after the meta-analysis of Shinton and Beevers, Donnan et al.[17] reported that smoking was associated with elevated risk for ischemic stroke. Unique to this study was the fact that 98% of the strokes included had been verified by CT scans. The estimated relative risks for cerebral ischemia after adjustment for other risk factors such as hypertension and age, were 3.6 for current smokers and 2.0 for past smokers, and both were significant. Hypertension, as expected, was also a significant risk factor, with a relative risk of 4.2. Adjustment for age and hypertension resulted in larger relative risks compared with unadjusted values, in accordance with the findings of Shinton and Beevers. The greatest effect was on the risk for thromboembolic and lacunar stroke (RR=5.7; 95% CI: 2.8-12.0).

In summary, the results of the pooled analysis for the relation between cigarette smoking and stroke provide strong evidence of an excess risk among cigarette smokers, with the risk being strongest for cerebral infarction and subarachnoid hemorrhage. Adding to this is the statement of a recent report by the World Health Organization Task Force on Stroke and Other Cerebrovascular Disorders,[18] where it was concluded that "Evidence of cigarette smoking as a risk factor for all strokes and particularly for ischemic stroke is substantial." This statement was based on a review of "the current world literature".

CONCLUSION

The results of our studies, as well as the findings of most other studies, strongly support a causal role of smoking in the development of extracranial carotid atherosclerosis. We have also presented support for a strong relationship between cigarette smoking and stroke.

In conclusion, numerous studies show that the number one modifiable risk factor for extracranial carotid atherosclerosis and stroke is cigarette smoking. Another important observation is the finding that the difference in mean plaque thickness is smaller between past smokers and nonsmokers than between current smokers and nonsmokers, even after taking into account the effect of other potent risk factors such as age and hypertension. As a corollary to this, the effect of quitting smoking increases with the time since quitting. Because our data are cross-sectional rather than longitudinal, the direct effect of smoking cessation on atherosclerosis progression or regression in individuals cannot be evaluated. However, the data do suggest the possibility of a slower rate of progression of atherosclerosis formation among people who have quit smoking compared with those who continue to smoke. Some of the longitudinal population-based studies now underway may provide answers to this question.

ACKNOWLEDGEMENTS

This study was supported in part by Public Health Service Grant NINCDS-NS06655 and Colonel CC Smith Research Fund. We are indebted to the ultrasonographers at the Clinical Neurosonology Laboratory, North Carolina Baptist Hospital/Bowman Gray School of Medicine, without whose skillful work this study could not have been done: Catherine Nunn, RN, RVT; Lawrence Myers, RDMS,

RVT; Dana Meads, RT-R, RVT. We would also like to thank Loretta W. Sanders and Susan Barefoot for collecting data in patient interviews.

REFERENCES

1. M. Hennerici, G. Reifschneider G, U. Trockel, A. Aulich, Detection of early atherosclerotic lesions by duplex scanning of the carotid artery, J Clin Ultrasound. 12:455 (1984).

2. G. M. von Reutern, Functional and morphological evaluation of the cerebral circulation by ultrasound, in: "Neurology - Proceedings of the XIIIth World Congress of Neurology," K. Poeck, H. J. Freund, H. Ganshirt, eds., Springer-Verlag, New York (1986) pp. 441.

3. S. Glagov, E. Weisenberg, C. K. Zarins, R. Stankunavicius, G. J. Kolettis, Compensatory enlargement of human atherosclerotic coronary arteries, N Engl J Med. 316:1371 (1987).

4. G. W. Duncan, R. S. Lees, R. G. Ojemann, S. S. David, Concomitants of atherosclerotic carotid artery stenosis, Stroke 8:665 (1977).

5. L. Candelise, F. Bianchi, F. Galligoni, V. Albanese, G. Bonelli, L. Bozzao, D. Inzitari, F. Mariani, M. Rasura, F. Rognoni, G. Sangiovanni, C. Fieschi, Italian multicenter study on reversible cerebral ischemic attacks: III. Influence of age and risk factors on cerebrovascular atherosclerosis, Stroke 15:379 (1984).

6. J. Bogousslavsky, F. Regli, G. Van Melle, Risk factors and concomitants of internal carotid artery occlusion or stenosis. A controlled study of 159 cases, Arch Neurol 42:864 (1985).

7. A. Schneidau, M. J. G. Harrison, C. Hurst, H. C. Wilkes, T. W. Meade, Arterial disease risk factors and angiographic evidence of atheroma of the carotid artery, Stroke 20:1466 (1989).

8. A. Haapanen, M. Koskenvuo, J. Kaprio, Y. A. Kesäniemi, K. Heikkilä, Carotid arteriosclerosis in identical twins discordant for cigarette smoking, Circulation 80:10 (1989).

9. J. R. Crouse, J. F. Toole, W. M. McKinney, M. B. Dignan, G. Howard, F. R. Kahl, M. R. McMahan, G. H. Harpold, Risk factors for extracranial carotid artery atherosclerosis, Stroke 18:990 (1987).

10. G. S. Tell, G. Howard, W. M. McKinney, Risk factors for site specific extracranial carotid artery plaque distribution as measured by B-mode ultrasound, J Clin Epidemiol. 42:551 (1989).

11. G. S. Tell, G. Howard, W. M. McKinney, J. F. Toole, Cigarette smoking cessation and extracranial carotid atherosclerosis, JAMA. 261:1178 (1989).

12. D. O'Leary, K. M. Anderson, C. S. Kase, P. A. Wolf, W. B. Kannel, Extracranial carotid atherosclerosis in a general population, the Framingham Study. (Abstract). Stroke 19:143 (1988).

13. R. Salonen, K. Seppänen, R. Rauramaa, J. T. Salonen, Prevalence of carotid atherosclerosis and serum cholesterol levels in Eastern Finland, Arteriosclerosis 8:788 (1988).

14. G. Heiss, ARIC investigators, Association of ultrasonographic measurements of carotid atherosclerosis with cardiovascular risk factors: The ARIC Study. Presented at the 2nd International Conference on Preventive Cardiology, Washington, D.C., June 1989.

15. J. G. Gostomzyk, W. D. Heller, P. Gerhardt P, P. N. Lee, U. Keil, B-scan ultrasound examination of the carotid arteries within a representative population (MONICA Project Augsburg), <u>Klin Wochenschr.</u> 66(Suppl XI):58 (1988).

16. R. Shinton, G. Beevers, Meta-analysis of relation between cigarette smoking and stroke, <u>Br Med J</u>. 298:789 (1989).

17. G. A. Donnan, J. J. McNeil, M. A. Adena, A. E. Doyle, G. C. Neill, Smoking as a risk factor for cerebral ischaemia, <u>Lancet</u> ii:643 (1989).

18. Stroke - 1989. Recommendations on stroke prevention, diagnosis, and therapy. Report of the WHO Task Force on Stroke and Other Cerebrovascular Disorders, <u>Stroke</u> 20:1407 (1989).

SMOKING, CATECHOLAMINES AND THEIR EFFECTS ON ENDOTHELIAL CELL INTEGRITY

Göran Bondjers, Göran Hansson, Gun Olsson, and Knut Pettersson

Wallenberg Laboratory for Cardiovascular Research
University of Göteborg
Göteborg, Sweden

THE RESPONSE-TO-INJURY HYPOTHESIS

Atherosclerosis is not uniformly distributed over the arterial
surface. The preferential localization of atherosclerotic lesions to
areas close to branching points and curvatures suggests that hemodynamic
strain may be involved in the initiation of atherogenesis. The actual
link between hemodynamic strain and atherogenesis has not been defined,
but already Virchow (1856) suggested that injury imposed by the wear and
tear of flow might be of significance. When Virchow's ideas were
revitalized in the later sixties and early seventies (Baumgartner and
Studer 1966, Björkerud 1969, Bondjers and Björnheden 1970, Stemerman and
Ross 1972) the primary emphasis was placed on injury in the form of
endothelial denudation. When the subendothelial tissue was exposed by
removal of the endothelium, platelets adhere and release growth factors.
These growth factors, and in particular PDGF, stimulate underlying smooth
muscle cells to proliferate (review with references: Ross 1981) to form
an intimal thickening. This concept, the original response-to-injury
hypothesis has recently been challenged (reviews with references: Reidy
and Schwartz 1984, Reidy 1985). Thus, it has been difficult to
demonstrate significant endothelial denudation with exposure of
subendothelial tissue in the absence of mechanical injury, and even when
such injury is induced smooth muscle proliferation does not necessarily
follow.

NON DENUDING ENDOTHELIAL INJURY

To explain the preferential localization of atherosclerotic lesions
to areas of hemodynamic strain, other forms of injury have now been
considered. We observed that endothelial cells in the vicinity of
branching points had lost their capacity to exclude anionic dyes, used in
cell culture for evaluation of cell viability (Björkerud and Bondjers
1972). In subsequent studies, we observed that such cells also contained
IgG (Hansson et al. 1979, 1980). Based on these observations Hansson
developed a technique to visualize injured endothelial cells, which could
be applied for quantitative studies (Hansson and Schwartz 1983). Based
on comparisons between the frequencies of IgG-containing cells and
endothelial mitotic activities, Hansson and Schwartz calculated that the
residence time of such cells in the arterial surface might be around 30
hours. Time-lapse cinematography studies on endothelial cells in culture

suggested that IgG-containing cells were detached from the endothelial
mono-layers while being unundermined by surrounding viable cells, leaving
no defect in the endothelium (Hansson and Schwartz 1983). These
observations concurred with earlier predictions concerning the sequence
of events surrounding endothelial cell death (Björkerud and Bondjers
1972). An analysis of the molecular basis for IgG binding to endothelial
cells demonstrated high affinity binding of the Fc fragment of IgG to
endothelial vimentin (Hansson et al. 1984). Such binding might lead to
direct recruitment of monocytes, via their Fc receptors (Hansson et al.
1981), to areas of endothelial cell injury. IgG binding to endothelial
cells might also lead to complement activation (Hansson et al. 1987) with
further leukocyte recruitment. Thus IgG binding to endothelial cells not
only provides a basis for visualization of injured endothelial cells, but
also leads to recruitment of the non-muscle cell population of the
atherosclerotic lesion. Subsequently, these cells may produce cytokines
stimulating smooth muscle proliferation (review with references: Ross
1987).

Apart from the possibility to evaluate endothelial injury through dye
exclusion tests, other possibilities have been considered (discussion
with references: Reidy and Schwartz 1984). Direct morphological studies
on the light microscopical or transmission electron microscopical level
have been difficult to evaluate on a quantitative level. A number of
studies suggesting that endothelial injury is induced by different agents
probably represent methodological variations rather than biological
effects. The possibility that endothelial cells, after injury or death,
may be released into the circulation has also been considered. If such
cells could be quantitated in arterial blood samples they might provide
an accurate measure of endothelial injury. However, the number of
circulating endothelial cells appears to be exceedingly low and it is
unclear how rapidly they are cleared from the circulation. Therefore, it
is not yet possible to evaluate whether this approach may be of value.
At this stage, a more fruitful approach has been to evaluate endothelial
injury indirectly, through an assessment of endothelial mitotic
activity. The rationale for this approach is that injured endothelial
cells must be replaced, and if cell size is constant such replacement can
occur only through proliferation. Schwartz and Benditt (1977)
demonstrated focal regions in the rat aortic endothelium that had much
higher replication rates than the remaining aorta. They also observed
that replication rates increased in acute hypertension, without any
obvious breaks in the endothelium. Therefore, these data support the
concept that injured endothelial cells are replaced without actual
denudation of the subendothelial surface. Various functional properties
of the endothelium or subendothelium may also be used as an indirect
measure of endothelial injury. Of particular interest is the possibility
of using labelled platelets to evaluate adhesion to exposed
subendothelium. Reidy et al. (1984) were able to demonstrate a direct
relationship between the induction of mechanical injury of various sizes
and the presence of Indium[111] -labelled platelets on the arterial
surface. Apart from the denuding form of injury, though, there is little
evidence for increased platelet adhesion as a consequence of endothelial
injury. Also other functional properties of the endothelium, such as the
production of the endothelium derived relaxing factor might change after
injury. Such changes may not only be due to injury, however. Functional
properties might change as an effect of specific stimuli and should
perhaps be regarded as an effect of dysfunction rather than injury.

EFFECTS OF SMOKING ON ARTERIAL ENDOTHELIAL INTEGRITY

Even though smoking is one of the major risk factors for
atherosclerosis, the pathogenic mechanisms linking smoking with

atherosclerosis have been obscure. Already in the early seventies it was
suggested that carbon monoxide could be connected with endothelial injury
(Kjeldsen and Thomsen 1975), but these observations were not confirmed by
other groups. Some years later Astrup's group reported that they
themselves were unable to reproduce the data using more rigorous
methodology (Hugod et al. 1978). Critically evaluated, however, in the
later study the methodology might have been so dull that actual
differences between the animals had been disregarded. No positive
control was included in the study to demonstrate that the methods might
discern a difference. Quite recently, Allen and co-workers (1988)
observed that cigarette smoking and carbon monoxide both increased
arterial permeability to fibrinogen in dogs, whereas nicotine had no such
effects. These observations might be explained through an effect on
arterial endothelial integrity. Therefore, it should still be regarded
as an open question whether increases in COHb levels are connected with
an increase in endothelial injury. Zimmerman and McGeachie (1985, 1987)
reported that nicotine induced ultrastructural changes suggestive of
endothelial injury in mice, and that mitotic activity also increased in
the animals. Tobacco smoke also appeared to increase the numbers of
circulating endothelial cells when compared with non-tobacco smoke,
suggesting a role of nicotine (Davis et al. 1985). Using scanning
electron microscopy, evidence of endothelial injury has been reported in
uterine arteries (Bylock et al. 1978) as well as in umbilical veins
(Asmussen and Kjeldsen 1975). Sieffert et al. (1981) observed similar
effects with scanning electron microscopy in rats after exposure to
tobacco smoke. Davis and co-workers (1987) reported an increase in the
number of circulating endothelial cells in coronary heart patients
immediately after smoking. The increase was not correlated with plasma
nicotine levels and could not be inhibited by aspirin. As discussed
above, the methodology used in these studies is insecure, and the
causative agent is unknown. In conclusion, studies in experimental
animals as well as observations in man have indicated that tobacco
smoking, and carbon monoxide and/or nicotine, might induce structural
changes in the endothelium and that these changes might interfere with
important functional properties.

In order to establish whether smoking at all might induce injury in
the arterial endothelium, we decided to expose guinea pigs to main-stream
cigarette smoke using the methodology developed by Rylander and Hellström
(1973). Endothelial injury was evaluated with four different methods –
scanning electron microscopy, dye exclusion test, the IgG peroxidase
method, and mitotic index (Bondjers et al. 1989). For all analyses we
used coded specimens and the code was not broken until all material was
analyzed. We were unable to discern any difference between the various
groups with scanning electron microscopy, using criteria like cellular
swelling, exposure of subendothelium or the presence of intercellular
clefts. However, with the IgG immunoperoxidase method, a tenfold
increase in the frequency of IgG positive cells was observed in smoking
animals when compared with different control animals (Figure 1).
Significant differences in mitotic activity were also observed,
supporting the suggestion that smoking had induced endothelial injury
(Figure 1). Even the dye exclusion test indicated effects in the same
direction but with far larger variations (Figure 1). All these effects
were obvious already after one day's exposure to tobacco smoke. The
early effects of smoking on mitotic activity were unexpected as increases
in endothelial mitotic activity are barely discerned 24 hours after
endothelial denudation through small experimental injury (Reidy and
Schwartz 1981). In earlier phases after injury, exposed subendothelial
tissue is covered through cell migration. After larger trauma, however,
a dramatic increase in ^{3}H-thymidine incorporation was observed already
after 24 hours. Thus, in this respect the effects of tobacco smoking

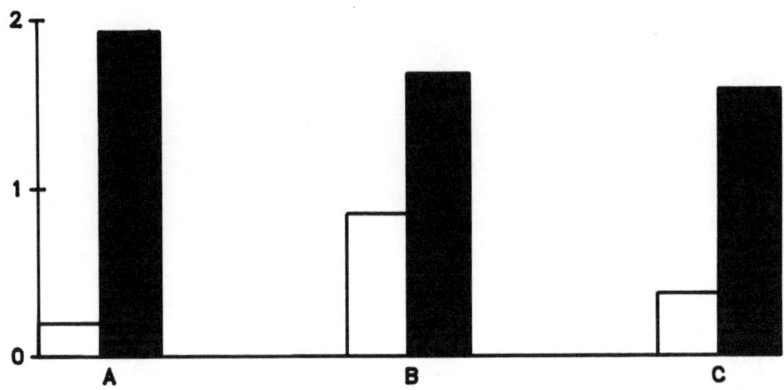

Fig. 1. Effects of smoking on arterial endothelial integrity in guinea
pigs as measured with different techniques. A: IgG immunoperoxidase
technique. B: Mitotic index. C: Dye exclusion test. Empty bars:
Controls. Filled bars: Animals exposed to tobacco smoke.

resemble those of wide-spread trauma. An alternative explanation for the
early increase in mitotic activity is a direct stimulation of endothelial
cell mitosis by some component of tobacco smoke. What component of
tobacco smoke might have this effect is unclear. Likewise it is unclear
whether the injurious effect of tobacco smoke is a direct one induced by
some component of the tobacco or an indirect one through some of the
pharmacological effects. Our inability to demonstrate any differences
related to smoking with scanning electron microscopy probably represents
the qualitative nature of this technique rather than an actual absence of
quantitative changes.

EFFECTS OF INCREASED CATECHOLAMINE LEVELS ON ARTERIAL ENDOTHELIAL
INTEGRITY

Behavioral and psychosocial factors have been implicated in the
pathogenesis of atherosclerosis for decades. The significance of such
factors has been disputed, however (review with references: Manuck et al.
1986). More recently, a special behavioral pattern, called type A
behaviour, has been discussed as a risk factor for coronary heart disease
(Dimsdale 1988). Recent studies in cynomolgus monkeys support the notion
that particular behavioral patterns may be of importance for the
development of atherosclerosis (Kaplan et al. 1983). The effects
appeared to involve an increase in sympathetic nervous activity, as it
could be totally inhibited by beta-adrenoceptor blockade (Kaplan et al.
1987). Several other studies in various experimental animal systems have
also demonstrated the antiatherogenic effect of beta-blockade (Åblad et
al. 1988). More recently it has been suggested that the effect is
mediated via $beta_1$-adrenoceptors, as the $beta_1$-selective antagonist
metoprolol is also an effective antiatherogenic drug in rabbits (Östlund-
Lindqvist et al. 1988). It has been unclear, though, what mechanistic
links connect beta-adrenoceptor activation and atherogenesis. In view of
the possibility that endothelial injury may be an important initiating
factor in atherogenesis we decided to test if sympathetic activation
might lead to endothelial injury, and if this effect could be inhibited
by $beta_1$ adrenoceptor blockade.

In one series of experiments, we used the increase in catecholamine
levels induced by chloralose, an anaesthetic drug, as a model system
(Pettersson et al. 1990). As a control for non-catecholamine mediated
effects of the drug, we included a $beta_1$-adrenoceptor antagonist,
metoprolol, in one group of chloralose-treated rabbits. Chloralose

anaesthesia increased heart rate, blood pressor, and plasma noradrenaline
levels, indicating sympathetic activation. These effects could be
inhibited by metoprolol, further supporting this notion. The frequency
of IgG positive endothelial cells increased tenfold both in non-branching
parts of the aorta and in areas close to branching points (Figure 2).
The increases could be inhibited by metoprolol, indicating that they were
mediated via beta$_1$-adrenoceptors. These observations may explain the
accelerated development of atherosclerosis associated with psychosocial
stress, and they may also contribute to explain the antiatherogenic
effects of beta-adrenoceptor antagonists.

In another series of experiments in Cynomolgus monkeys, we used a
psycho-social stress model to evaluate whether beta$_1$-adrenoceptor
blockade might decrease endothelial injury (Strawn et al. 1990). A
significant decrease in IgG positive endothelial cells, as well as a
significant decrease in mitotic activity was observed when the monkeys
were given metoprolol. These observations suggest that catecholamines
had contributed to induce endothelial injury also in this more
"physiological" stress model.

These experiments indicate that psycho-social stress and increased
catecholamine levels lead to an increased frequency of injured
endothelial cells in the aorta of experimental animals. The mechanisms
involved in these effects are unclear, but it seems reasonable to assume
that an increase in heart rate and blood pressure may be of
significance. Experiments aiming to test this hypothesis appear to be
feasible.

EFFECTS OF SMOKING ON CATECHOLAMINE LEVELS

Tobacco smoking induces a number of rapid changes in the
cardiovascular system. Heart rate, blood pressure, and cardiac output
increase during smoking (Spohr et al. 1979). This effect appears to be
mediated by nicotine (review with references: Epstein and Jennings
1986). A number of pharmacological effects of nicotine appear to be
involved in these changes: withdrawal of vagus stimulation,
alpha-adrenergic stimulation and beta-adrenergic stimulation. As we
observed rather similar effects induced by smoking and by increased

Fig. 2. The effects of increased catecholamine levels, induced by
chloralose on arterial endothelial integrity. Triplets of siblings were
used and frequency of IgG positive cells in chloralose (open bar) or
chloralose + metoprolol-treated (closed bar) animals expressed as per
cent of untreated sibling. Unbranched area of the aorta (left) and
branching points (right).

catecholamine levels we decided to test the hypothesis that the effects
of smoking on endothelial integrity we had observed were mediated by
catecholamines.

In this experiment, guinea pigs were exposed to mainstream tobacco
smoke as described above. In one group of animals, a mini-osmotic pump
with metoprolol had been introduced one week before the experiment. We
observed an increased frequency of IgG positive cells in the animals
which had been exposed to tobacco smoking compared with non-smoking
controls (Figure 3). This increase could be inhibited by metoprolol,
supporting the hypothesis that the increased frequency of injured
endothelial cells had been mediated by $beta_1$-adrenoceptor activation.
The hypothesis that beta adrenergic mechanisms might be of significance
for the development of cardiovascular disease in smokers recently gained
support from results of the MAPHY study (Wikstrand et al. 1989),
indicating that metoprolol decreased the incidence of clinical
complications to atherosclerosis in hypertensive individuals, when
compared with other patients treated with diuretics.

GENERAL CONCLUSIONS

Clinical complications to atherosclerosis are more common in smokers
than in non-smokers. The mechanisms behind this difference have been
unclear, even though an increase in endothelial injury induced by some
component in tobacco smoke has been implicated. Quantitating endothelial
injury through the IgG immunoperoxidase technique we verified the notion
that tobacco smoking might lead to endothelial injury. As an increase in
catecholamine levels induced by chloralose anaesthesia or psycho-social
stress leads to similar effects, we hypothesized that the effect of
smoking might be mediated by increased catecholamine levels. Actually,
the endothelial injury induced by smoking could be inhibited by
metoprolol supporting this hypothesis. It is unclear what component in
tobacco smoke mediates the increase in catecholamines leading to
endothelial injury, but other adrenergic effects of tobacco smoking have
been connected with nicotine. It is also unclear whether the present
results, gained in guinea pigs, monkeys, and rabbits in rather acute
experiments are relevant for the situation in man. Therefore, further
experiments, using nicotine by itself in more chronic models are of
significance.

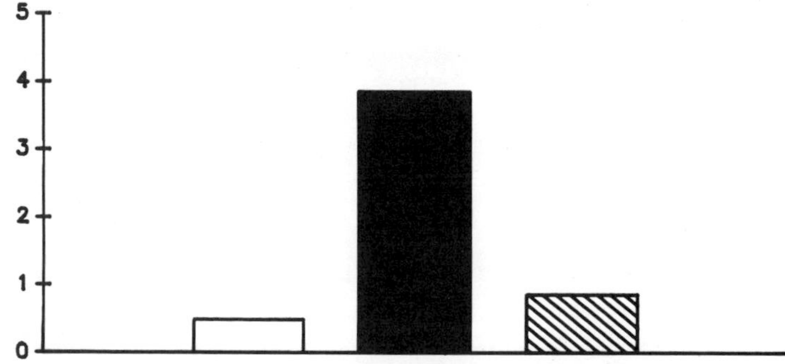

Fig. 3. Effects of catecholamines on the increase in endothelial injury
induced by tobacco smoking. Controls: left bar; animals exposed to
tobacco smoke: middle bar; metoprolol-treated smoking animals: right bar.

REFERENCES

Åblad, B., Björkman, J.-A., Gustafsson, D., Hansson, G., Östlund-
    Lindqvist A.-M. and Pettersson, K. The role of sympathetic
    activity in atherogenesis: Effects of beta-blockade, Am. Heart J.
    116: 322 (1988).
Allen, D. R., Browse, N. L., Rutt, D. L., Butler, L., and Fletcher, C.
    The effect of cigarette smoke, nicotine and carbon monoxide on the
    permeability of the arterial wall. J. Vasc. Surg. 7: 139 (1988).
Asmussen, I., and Kjeldsen, K. Intimal ultrastructure of human umbilical
    arteries. Circ. Res. 36: 579 (1975).
Baumgartner, H. R., and Studer, A. Folgen des Gefässkathererismus am
    normo- under hypercholesterinaemischen Kaninchen. Pathol.
    Microbiol. 29: 393 (1966).
Björkerud, S. Atherosclerosis initiated by mechanical trauma in normo-
    lipidemic rabbits. J. Atheroscler. Res. 9: 209 (1969).
Björkerud, S., and Bondjers, G. Endothelial integrity and viability in
    the aorta of the normal rabbit and rat as evaluated with dye
    exclusion tests and interference contrast microscopy.
    Atherosclerosis 15: 285 (1972).
Bondjers, G., and Björnheden, T. Experimental atherosclerosis induced by
    mechanical trauma in rats. Atherosclerosis 12: 301 (1970).
Bondjers, G., Bylock, A., Hansson, G., Jonasson, L., and Olofsson, S.-O.
    The effects of cigarette smoking on aortic endothelial integrity
    in guinea pigs, as evaluated by scanning electron microscopy, dye
    exclusion tests, immunohistochemical techniques and mitotic
    activity. In preparation (1989).
Bylock, A., Bondjers, G., Jansson, I., and Hansson, H.-A. Surface
    ultrastructure of human arteries with special reference to the
    effects of smoking. Acta Path. Microbiol. Scand. Sect. A: 87: 201
    (1978).
Davis, J. W., Shelton, L., Eigenberg, D. A., and Hignite, C.E. Lack of
    effect of aspirin on cigarette smoke-induced increase in circulat-
    ing endothelial cells. Haemostrasis 7: 66 (1987).
Davis, J. W., Shelton, L., Eigenberg D. A., Hignite, C. E., and
    Watanabe, I. S. Effects of tobacco and non-tobacco cigarette
    smoking on endothelium and platelets. Clin. Pharmacol. Ther. 37:
    529 (1985).
Dimsdale, J. E. A perspective on type A behaviour and coronary disease.
    N. Engl. J. Med. 318: 110 (1988).
Epstein, L. H., and Jennings, J. R. Smoking, stress, cardiovascular
    reactivity and coronary heart disease. In Handbook of Stress,
    Reactivity and Cardiovascular Disease. Edited by Matthews, K. A.,
    Weiss, S. M., Detre, T., Dembroski T. M. Falkner, B., Manuck, S.
    B., Williams, R. B. Wiley and Sons, New York, p. 291 (1986).
Hansson, G. K., and Schwartz, S. M. Evidence for cell death in the
    vascular endothelium in vivo and in vitro. Amer. J. Path. 112:
    278 (1983).
Hansson, G. K., Bondjers, G., and Nilsson, L. Å. Plasma protein accumu-
    lation in injured endothelial cells. Exp. Mol. Pathol. 30: 12
    (1983).
Hansson, G. K., Bondjers, G., G., Bylock, A., and Hjalmarsson, L. Ultra-
    structural studies on the localization of IgG in the aortic
    endothelium and subendothelial intima of atherosclerotic and non-
    atherosclerotic rabbits. Exp. Mol. Path. 33: 301 (1980).
Hansson, G. K., Björnheden, T., Bylock, A., and Bondjers, G. Fc-
    dependent binding of monocytes to areas with endothelial injury in
    the rabbit aorta. Exp. Mol. Path. 34: 264 (1981).

Hansson, G. K., Starkebaum, G. A., Benditt, E. P., and Schwartz, S. M. Fc-mediated binding of IgG to vimentin-type intermediate filaments in vascular endothelial cells. Proc. Natl. Acad. Sci. 81: 3103 (1984).

Hansson, G. K., Lagerstedt, E., Bengtsson, A., and Heideman, M. IgG binding to cytoskeletal intermediate filaments activates the complement cascade. Exp. Cell Res. 170: 338 (1987).

Hugod, C., Hawkins, L. H., Kjeldsen, K., Thomsen, H. K. and Astrup, P. Effect of carbon monoxide exposure on aortic and coronary intimal morphology in the rabbit. Atherosclerosis 30: 333 (1978).

Kaplan, J. R., Manuck, S. B., Clarkson, T. B., Lusso, F.M., Taub, D. B. and Miller, E. W. Social stress and atherosclerosis in normo-cholesterolemic monkeys. Science 220: 733 (1983).

Kaplan, J. R., Manuck, S. B., Adams, M. R., Weingard, K. W., and Clarkson, T. B. Inhibition of coronary atherosclerosis by propranolol in behaviourally predisposed monkeys fed an atherogenic diet. Circulation 76: 1364 (1987).

Kjeldsen, K., and Thomsen, H. K. The effect of hypoxia on the fine structure of the aortic intima in rabbits. Lab. Invest. 33: 533 (1975).

Manuck, S. B., Kaplan, J. R., and Matthews, K. A. Behavioral antecedents of coronary heart disease and atherosclerosis. Arteriosclerosis 6: 2 (1986).

Östlund-Lindqvist, A. M., Lindqvist, P., Bräutigam, J., Olsson, G., Bondjers, G., and Nordborg, C. Effect of metoprolol on diet-induced atherosclerosis in rabbits. Arteriosclerosis 8: 40 (1988).

Pettersson, K., Bejne, B., Björk, H., Strawn, W. B. and Bondjers G. Experimental sympathetic activation causes endothelial injury in the rabbit thoracic aorta via $beta_1$-adrenoceptor activation. Circulation Res. submitted (1990).

Reidy, M. A reassessment of endothelial injury and arterial lesion formation. Lab. Invest. 53: 513 (1985).

Reidy, M. A., and Schwartz, S. M. Endothelial regeneration. III. Time course of intimal changes after small defined injury to the rat aortic endothelium. Lab. Invest. 44: 301 (1981).

Reidy, M. A., and Schwartz, S. M. Arterial endothelium – assessment of in vivo injury. Exp. Mol. Path. 41: 419 (1984).

Reidy, M. A., Harker, L. A., Yoshida, K., and Schwartz, S. M. Vascular injury: detection of focal endothelial denudation using indium$_{111}$- labeled platelets. J. Clin. Invest. 49: 565 (1984).

Ross, R. Atherosclerosis: A problem of the biology of the arterial wall cells and their interactions with blood components. Arterio-sclerosis 1: 293 (1981).

Ross, R. The pathogenesis of atherosclerosis – an update. N. Engl. J. Med. 314: 488 (1986).

Rylander, R., and Hellström, P. A. Versatile cage for environmental protection housing of research animals. Lab. Animal Science, 23: 876 (1973).

Schwartz, S. M., and Benditt, E. P. Aortic endothelial cell replication. Effects of age and hypertension in the rat. Circ. Res. 41: 248 (1977).

Sieffert, G. F., Keown, K., and Moore, W. S. Pathologic effect of tobacco smoke inhalation on arterial intima. Surg. Forum 32: 333 (1981).

Spohr, U. and Hoffman, W. Evaluation of smoking-induced effects on sympathetic, hemodynamic and metabolic variables with respect to plasma nicotine and COHb levels. Atherosclerosis 33: 271 (1979).

Stemerman, M. B., and Ross, R. Experimental arteriosclerosis I. Fibrous plaque formation in primates, an electron microscopic study. J. Exp. Med. 136: 769 (1972).

Strawn, W. B., Bondjers, G., Kaplan, J. R., Manuck, S. B., Hansson, G. K.,
     and Clarkson, T. B. Psychosocial stress and endothelial integrity
     in Cynomolgus monkeys. In preparation (1990).
Wikstrand, J., Warnold, I., Olsson, G., Tuomilehto, J., Elmfeldt, D., and
     Berglund, G. Primary prevention with metoprolol in patients with
     hypertension. Mortality results from the MAPHY. J A M A 259:
     1976 (1988).
Virchow, R. Der ateromatose prozess der arterien. Wien Med. Wochenschr
     6: 825 (1856).
Zimmerman, M. J. and McGeachie, J. The effect of nicotine on aortic
     endothelium. A quantitative ultrastructural study.
     Atherosclerosis 63: 33 (1987).
Zimmerman, M. J. and McGeachie, J. The effect of nicotine on aortic
     endothelial cell turnover – an autoradiographic study.
     Atherosclerosis 58: 39 (1985).

CIGARETTE SMOKING AND ENDOTHELIAL INJURY: A REVIEW

R. Michael Pittilo

School of Life Sciences
Kingston Polytechnic
Penrhyn Road
Kingston upon Thames
Surrey, KT1 2EE, U.K.

INTRODUCTION

Epidemiological considerations

There is overwhelming epidemiological evidence that an association exists between cigarette smoking and cardiovascular disease, in particular atherosclerosis (Auerbach et al., 1965, 1976; Auerbach and Garfinkel 1980; Dawber et al., 1959; Doll and Peto, 1976; Eastcott, 1962; Kannel et al., 1976; Laing et al., 1981; Murphy and Mustard, 1966; Sackett et al., 1968; Spain and Bradess, 1970; Strong and Richards, 1976). Despite these considerable epidemiological data, the component or components of cigarette smoke responsible for this relationship and the mechanisms through which they mediate their effects remain unknown.

Cigarette smoking and atherosclerosis

The pathogenesis and pathology of atherosclerosis have been the subject of many studies and reviews (see Ross, 1986; Woolf, 1987) and are not within the realm of this paper other than to note that there is still controversy as to what constitutes the earliest lesion of the disease, and that the fibro-lipid plaque may develop by a variety of routes (Haust, 1971; Woolf, 1987). Ross (1986) has proposed a response to injury hypothesis for the pathogenesis of atherosclerosis that takes account of recent experimental evidence. Two main pathways form the basis of this hypothesis and one of these or a variation of it may be important in relation to cigarette smoking as well as diabetes and hypertension. It is hypothesised that direct stimulation of endothelium may result in it releasing growth factors promoting the migration and proliferation of smooth muscle cells which in themselves may also release growth-factors on stimulation (Ross, 1986).

Endothelium and Endothelial Injury

Vascular endothelium has been the subject of many reviews

Tobacco Smoking and Atherosclerosis
Edited by J. N. Diana
Plenum Press, New York, 1990

(see for example, Petty and Pearson, 1989; Pittilo, 1988a; Bull
1988) and a consideration of its morphology and known
functional roles is outside the scope of this review.

Reidy (1985) has pointed out that historically endothelial
injury has been considered to be the absence of endothelium and
the presence of platelets. We now know that functional
alteration of intact endothelium can influence the process of
atherogenesis and now have to consider that intact endothelium
may be functionally abnormal and represent an injured state
(see Reidy, 1985). There are data to suggest that cigarette
smoking may mediate its effect on the vessel wall partly
through modifying endothelial cell function (see below).

Although this review is concerned with endothelial cell
injury in relation to cigarette smoking, it is important not to
forget that interactions between endothelial cells and other
cells, such as blood platelets, may to a large extent be due to
factors independent of the endothelial cell.

Morphological Limitations in Investigating Endothelial Injury

It has seemed logical to assume that the primary target for
cigarette smoke components absorbed into the blood would be the
vascular endothelium. However, from a morphological viewpoint
it is not a straightforward matter to distinguish between
normal and injured endothelium. It has been suggested that
loss of endothelial cells might be a common consequence of
injury with replacement of the lost cells by neighbouring cells
(Reidy and Schwartz, 1984). These events could easily escape
morphological detection. There is good evidence to indicate
that endothelial cell loss can occur without denudation and
consequent exposure of the sub-endothelium. Reidy and
Schwartz (1983) used endotoxin to increase the turnover and
replication of arterial endothelial cells and demonstrated
desquamating endothelial cells without denudation.
Furthermore, it would seem reasonable to speculate that
endothelial injury might occur without there being any
morphological manifestations (Ross, 1986). In other words, the
endothelium might be modified to secrete growth factors or
perhaps altered amounts of prostacyclin, without morphological
change.

Examining the morphology of vascular endothelium in order to
detect change resulting from injury is fraught with another
problem. It is well established that vascular endothelium can
show considerable morphological variation depending on the
preparative methodologies adopted (Hollweg and Buss, 1980).
Pittilo (1988a) advocated caution when examining vascular
endothelium in view of the ease by which it can be altered
artefactually and the fact that structures considered normal
now, may be shown in the future to result from poor
preparation.

Despite the dual cautions that endothelial injury might not
have a morphological manifestation, and the ease through which
endothelium can be altered artefactually during preparation,
morphological studies have proven useful in demonstrating that
endothelial damage can result from exposure to smoke or smoke
components. In conjunction with available biochemical
evidence, they have provided some insight into how cigarette

smoke may be influential as a risk factor for cardiovascular
disease.

THE EFFECTS OF CIGARETTE SMOKE ON ENDOTHELIAL MORPHOLOGY

<u>Human Studies</u>

Ideally the human model would represent the most relevant
system for studying the effects of cigarette smoking in
relation to atherosclerosis.   There have been a number of
reports in relation to the effects of smoking on human
endothelium.   Irregular swollen cells possessing blebs of the
luminal membrane, containing dilated granular endoplasmic
reticulum and contracted basal filaments were a feature of the
umbilical artery endothelium from smoking mothers in a study
involving 15 non-smokers and 13 smokers (Asmussen and Kjeldsen,
1975).   Sub-endothelial oedema was observed and there was
opening of the intercellular junctions and thickening of the
basement membrane (Asmussen and Kjeldsen, 1975).

Nuclear chromatin changes consisting of fine granular spots
have been reported in the umbilical arteries of smoking mothers
and have been used to discriminate between tissue obtained from
smoking and non-smoking mothers (Asmussen, 1982a).   These
granular spots were also observed within intimal smooth muscle
cells (Asmussen, 1982a).   Increased numbers of mitochondria
are reported to exist in the endothelium of umbilical arteries
from smoking mothers (Asmussen, 1984).   Using a double blind
methodology, Asmussen (1982b) was able to distinguish
ultrastructurally between the umbilical arteries from 9 smokers
and fourteen non-smokers.   The features used included those
already alluded to such as distension of the endoplasmic
reticulum and increased numbers of mitochondria but also the
accumulation of glycogen and lipid within medial smooth muscle
cells (Asmussen, 1982b).

These studies and others (Asmussen, 1977, 1978a, 1978b,
1980, 1982c; Van der Veen and Fox, 1982) which have focussed on
the foeto-placental cardiovascular system have provided
evidence to suggest that endothelial alteration at the
ultrastructural level can occur in relation to smoking.

Bylock et al. (1979) have examined the surface morphology of
human uterine arteries from women smoking more than fifteen
cigarettes a day and compared them with the same vessels from
non-smokers.   They reported that holes in the endothelium were
more common on the vessels of smokers than non-smokers (Bylock
et al., 1979).

Although the above studies have provided evidence that
morphological abnormalities compatible with injury of the
vascular endothelium can result from smoking, they are open to
criticism.   Asmussen and Kjeldsen (1975) have pointed out that
the competition between carbon monoxide and oxygen will be more
pronounced in the umbilical artery than in the wall of maternal
arteries and this may be responsible for the marked
ultrastructural changes seen.   They also considered that the
morphological dissimilarities between foetal and adult
endothelium might be reflected in the latter being more
sensitive to the effects of hypoxia (Asmussen and Kjeldsen,

1975).   More recently Stauseband et al. (1984) have reported
similar ultrastructural features in umbilical cord vessels from
non-smokers, including adherent thrombi, and have concluded
that the umbilical cord is not a suitable model for studying
the damaging effects of atherosclerotic risk factors.   There
can be little doubt that the endothelium of umbilical vessels
will also be sensitive to the fixation regime and any attempt
to compare vessels from smokers and non-smokers will only be
valid if a precise methodology concerning the preparation is
adhered to.

The study on the uterine arteries also raises the
possibility of fixation artefact as inevitably there would be
some delay following interruption of the blood supply and
transport of the specimens to the laboratory albeit in
Ringer's glucose (Bylock et al., 1979).   In our own laboratory
Pittilo, Clarke and Woolf (unpublished) studied the superficial
circumflex iliac artery obtained from smoking and non-smoking
males but were not able to distinguish between the two groups.
Although the specimens were quickly obtained and perfused with
fixative,we considered that the preparative process almost
certainly induced artefacts that would mask any moderate
endothelial differences that might have been apparent between
the two groups.

In an attempt to move away from animal models and avoid the
difficulties associated with preserving human vessels removed
surgically, Pittilo et al. (1985) investigated the effects of
plasma obtained from volunteers after smoking on cultured rat
peritoneal mesothelial cells.   These cells were used because
they were routinely available within our laboratory, easy to
culture (Nicholson et al., 1984) and had been shown to be
closely related to endothelium and compatable with flowing
blood both, in vivo (Clarke et al., 1984) and when cultured
(Nicholson et al., 1984; Blow, 1988).   Plasma obtained from
volunteers who were normally non-smokers had little effect on
cultured mesothelial cells.   However, following smoking, the
plasma produced marked morphological alteration of the cells
including blebbing of the luminal membrane (Pittilo et al.,
1985).   Busacca et al. (1984) have reported that umbilical
endothelial cells from mild and heavy smokers are more
difficult to grow in comparison with those obtained from non-
smokers.

Pittilo et al. (1984) exposed non-abraded rabbit
endothelium to human blood taken from male volunteers who were
normally non-smokers before and after smoking two cigarettes.
The post-smoking blood sample resulted in the deposition of
large numbers of platelets on the endothelial surface.
Whether this was due to an effect of the plasma on the
endothelial surface or a change in the platelets was not clear
although we were unable to detect any change in ADP or collagen
induced aggregation of platelets in the post-smoking samples
(Pittilo et al., 1984).   Factor VIII related antigen (Factor
VIII:RAg) was increased in six out of eight subjects after
smoking.   As both platelets and endothelial cells contain
Factor VIII:RAg it is possible that it might have been released
from both these sources as a direct toxic effect from the smoke
or in relation to the stress of smoking (Pittilo et al., 1984).
The increased levels of Factor VIII:RAg might have contributed

to the increased platelet adhesion to the endothelial surface
seen after smoking.

Animal Studies

Cigarette smoking has been shown to result in morphological
alteration to the endothelium of the rabbit and the rat (Woolf
and Wilson-Holt, 1981; Pittilo et al., 1982) (see Figures 1-4).
A similar picture emerged from both these studies. The
endothelial cells appeared swollen and irregular and there were
surface morphological abnormalities including the formation of
blebs and microvillous-like projections on the luminal
membranes of the cells. Large numbers of apertures were also
seen on the luminal membranes of the endothelial cells
following smoking and these were thought to represent the
openings of plasmalemmal vesicles suggesting that the
endothelium had become more permeable (Woolf and Wilson-Holt,
1981; Pittilo et al., 1982). In both these studies, platelets
were found adhering to what appeared to be intact endothelium
and Woolf and Wilson-Holt (1981) considered that they might
represent early mural thrombi. It was possible that the
endothelium in these sites was injured but morphologically it
appeared normal and intact. Marked endothelial changes
including cell loss was reported in the rat by Sieffert et al.
(1981). They also observed platelet adhesion both where
endothelial denudation had taken place and where the intima
appeared intact. Cigarette smoking has also been shown to
result in a small decrease in the cell coat of rat endothelial
cells (Roberge et al., 1983).

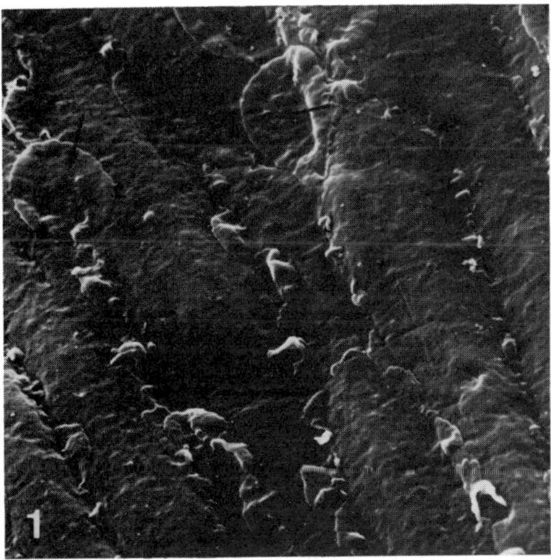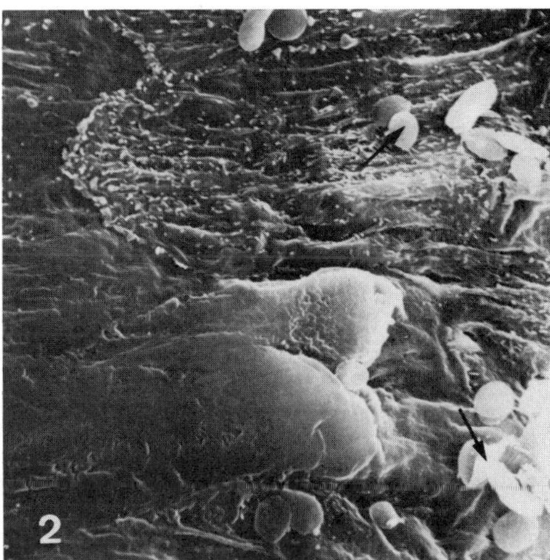

Figure 1. SEM (scanning electron micrograph) of rabbit aortic
endothelium. There are few projections from the luminal
surface and cell flaps (arrowed) are evident. No formed
elements of the blood are in contact with the surface, X6,300.

Figure 2. SEM of rat aortic endothelium following smoke
exposure. The endothelial surface is irregular and platelets
(arrowed) are in contact with the luminal membrane of some
cells. Swollen cells can be seen. X2,730.

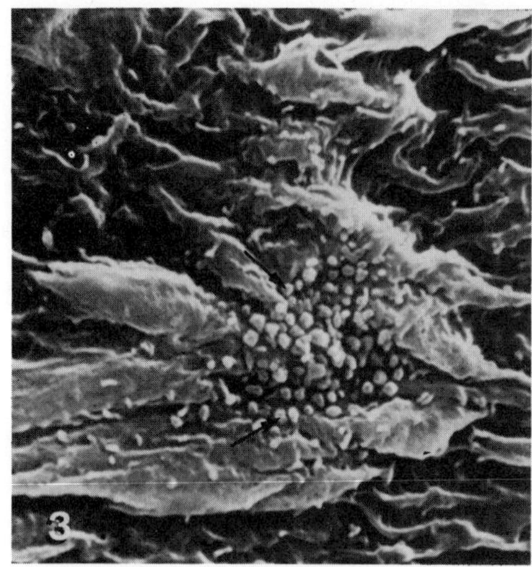

Figure 3.   SEM of rat aortic endothelium following smoke exposure.   Many bleb-like projections (arrowed) can be seen on the surface of an endothelial cell.   The luminal membrane of the cell is also very irregular. X2,730,

Figure 4.   SEM of rat aortic endothelium following smoke exposure.   This micrograph is from a region close to an intercostal artery branch.   Irregular cells with microvillus-like projections can be seen (arrowed).   In association with the endothelium are many platelets (PL).   Although the endothelium is in all probability "injured", it is not possible to discern whether or not sub-endothelial collagen has been exposed. X1,890.

        Cigarette smoking has been shown to increase the permeability of rat aortic myocardial endothelium to peroxidase (Bazin et al., 1981).   In the case of the myocardial endothelium the increased permeability was not accompanied by morphological change (Bazin et al., 1981).   Although only one electron micrograph was published, Allen et al. (1988) did not observe any endothelial changes in the dog following smoke exposure.   They did however find that cigarette smoking resulted in increased permeability of the vessel wall to ^{125}I labelled fibrinogen.   Presumably there may be some morphological correlation to this increased endothelial permeability.   Allen et al. (1988) considered that the morphological changes seen by other workers (Woolf and Wilson-Holt, 1981) following smoke exposure might not have had time to develop as a result of the short smoke exposure in their study.   An alternative explanation might relate to the species difference between the studies.

THE EFFECTS OF NICOTINE ON ENDOTHELIAL MORPHOLOGY

Animal Studies

    Not surprisingly, individual components of smoke have been examined to see their effects on endothelial morphology.

Nicotine administered to rabbits orally resulted in endothelial
abnormalities in the aortic arches which included "ruffling" and
microvillus formation (Booyse at al., 1981a). These abnormal
areas correlated with increased Evans blue uptake (Booyse et
al., 1981a). Zimmerman and McGeachie (1987) reported that oral
administration of nicotine to mice resulted in endothelial
abnormalities including cytoplasmic vacuolation, mitochondrial
swelling and sub-endothelial oedema. However, only two electron
micrographs were published in this study (Zimmerman and
McGeachie, 1987) and as in the previous study (Booyse et al.,
1981a) no plasma nicotine levels were given.

Fraser et al. (1988) reported a reduction in the porosity of the
rat liver sieve in animals given nicotine in their drinking
water. This was brought about by a reduction in the diameter of
the endothelial fenestrae and the authors postulated that this
might explain the higher serum cholesterol and triglyceride
levels seen in the rats that had ingested nicotine (Fraser et
al., 1988).

We were not able to demonstrate aortic endothelial
morphological changes in rats receiving nicotine acutely by
subcutaneous injection even when plasma levels of 2600 ng/ml
were obtained (Pittilo and Bull, 1988) (see Figures 5 and 6).

However sub-acute subcutaneous administration of nicotine
over a 7 day period in which plasma nicotine levels were around

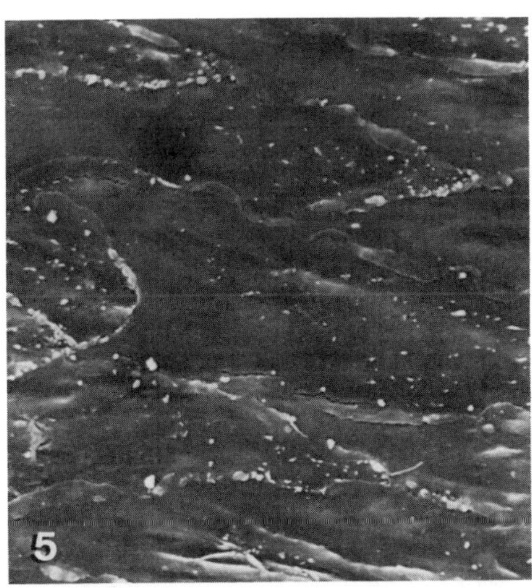 

Figure 5. SEM of rat aortic endothelium from an animal 20
minutes after receiving a subcutaneous saline injection. The
endothelial surface appears normal with few projections from the
cells and no adhering formed elements of the blood. X2,100.

Figure 6. SEM of rat aortic endothelium from an animal 20
minutes after receiving a subcutaneous injection of nicotine at
10mg/kg. The endothelial cells appear smooth with few
perjections and no formed elements of the blood are in contact
with the vessel wall. This morphological appearance is
indistinguishable from normal rat aortic endothelium. X2,100.

400ng/ml resulted in endothelial cell abnormalities which included microvillous-like formations of the luminal membrane, platelet adhesion and cell separation (Pittilo and Bull, 1988). Allen et al. (1988) were not able to detect any endothelial morphological changes or permeability changes in the dog following administration of intravenous nicotine.

Intravenous administration of nicotine to the rat has been reported to result in increased numbers of endothelial cell carcasses within the blood (Hladovec, 1978).

Cell Culture Studies

The effects of nicotine on endothelial cells has been investigated in cell culture. Concentrations of nicotine between 10^{-9}M and 10^{-3}M had no effect on human umbilical endothelial cell morphology whereas exposure to 10^{-2}M resulted in the appearance of large translucent vesicles within the endothelial cells (Bull et al., 1988). Booyse et al. (1981b) reported vacuolation of cultured bovine endothelial cells at 10^{-4}M. These studies suggest that nicotine has no effect on endothelial morphology, or indeed prostacyclin production (see below), at concentrations comparable to the plasma levels found within smokers.

However, Csonka et al. (1985) found that nicotine stimulated the synthesis and polymerisation of endothelial cell cytoskeletal proteins and Blaes et al. (1986) detected morphological changes in human umbilical cells treated with nicotine. They found that nicotine induced an elongated morphology and the cells contained hardly any Weibel-Palade bodies in comparison with controls (Blaes et al., 1986). There was a reduction in the levels of FVIII/vWF and fibronectin within the nicotine treated cells and the secretion of the former was raised although the secretion of fibronectin was reduced (Blaes et al., 1986).

THE EFFECTS OF CARBON MONOXIDE ON ENDOTHELIAL MORPHOLOGY

Animal Studies

The literature in relation to the effects of carbon monoxide on endothelial morphology present a confusing picture (see Pittilo, 1988b). Morphological changes including platelet adhesion to the endothelium have been recorded in some studies as a result of carbon monoxide exposure (Kjeldsen et al., 1972; Astrup, 1972; Astrup and Kjeldsen, 1973; Marshall and Hess, 1981) and it has been associated with the induction and development of atherosclerosis (Thomsen, 1974; Webster et al., 1970). However, other studies have failed to show any relationship between carbon monoxide exposure and endothelial morphology or the development of atherosclerosis. Hugod et al. (1978) were unable to differentiate between control or treated aortas or coronary arteries in relation to carbon monoxide exposure. Armitage et al. (1976) did not observe any enhancing effect of carbon monoxide in relation to development of atherosclerosis in normocholesterolaemic White Carneau pigeons although carbon monoxide had an initial enhancing effect in hypercholesterolaemic birds.

Although no electron micrographs were published in relation
to carbon monoxide exposure, Allen et al. (1988) reported
normal endothelial cell morphology for the dog following carbon
monoxide exposure at both the light and ultrastructural level,
but did find an increased permeability of the vessel wall to
^{125}I-labelled fibrinogen. Presumably this does correlate with
some morphological abnormality but further work is required to
clarify the position with regard to the relationship between
carbon monoxide and endothelial injury.

<u>Cell Culture</u>

Levene et al. (1985) examined the effects of hypoxia and
carbon monoxide on collagen synthesis in cultured porcine and
bovine aortic endothelial cells. This study provided further
cautionary evidence against drawing conclusions between species
in that species differences to the effects of carbon monoxide
were apparent. They found that hypoxia depressed the cell
numbers of cells of both species, but while carbon monoxide
enhanced the fall with porcine cells it counteracted the fall
with bovine cells (Levine et al., 1985). This study clearly
demonstrated that carbon monoxide and hypoxia have differing
actions and different effects depending on the species being
investigated (Levene et al., 1985). Ultrastructurally,
hypoxia resulted in a reduction in the numbers of free
ribosomes, degranulation of the rough endoplasmic reticulum and
accumulation of dense lysosomal bodies. Normoxic cells
however showed essentially normal ultrastructure when treated
with carbon monoxide (Levine et al., 1985).

THE EFFECTS OF CIGARETTE SMOKE AND SMOKE COMPONENTS ON
ENDOTHELIAL CELL PROSTACYCLIN PRODUCTION

The balance between prostacyclin (PGI_2) and thromboxane A_2
(TXA_2) is known to be important in maintenance of the
anticoagulated state of blood vessels and these vascular
eicosanoids are implicated in atherogenesis (see Pomerantz and
Hajjar, 1989). There are considerable data in relation to the
effects of nicotine, cigarette smoke and cigarette smoke
condensates on the ability of endothelium to produce PGI_2 and
on the ability of platelets to produce TXA_2).

Pittilo et al. (1982) produced experimental evidence to
implicate cigarette smoking in reducing PGI_2 synthesis by
aortic endothelial cells. Aortic rings obtained from
anaesthetised rats exposed to fresh cigarette smoke produced
less PGI_2 in vitro in comparison with controls (Pittilo et al.,
1982). Dadak et al. (1981) have shown that the umbilical
arteries in neonates of smoking mothers produce less
prostacyclin in comparison with those from non-smokers.

Support for these data has come from a number of studies.
Nadler et al. (1983) reported reduced urinary PGI_2 levels in
chronic smokers after smoking nicotine containing cigarettes.
Smoking of nicotine free-cigarettes did not have this effect
and volunteers who were normally non-smokers did not show
reduced urinary PGI_2 with either type of cigarette (Nadler et
al., 1983). Mileikowsky et al. (1988) have reported reduced
levels of the stable metabolite of prostacyclin in the urine of
oral contraceptive users who smoked for five years or more in
comparison with non-smokers or those smoking for less than five

years.   Umbilical arteries from smokers have also been
reported to produce less prostacyclin in comparison with non-
smokers by Ahlsten et al. (1986) and endothelial cells cultured
from the umbilical veins of mothers who smoked were less able
to produce prostacyclin in comparison with controls (Busacca et
al., 1984).

     Lubawy et al. (1986) also reported depressed PGI_2 production
by rat aortic tissue following cigarette smoking.   They only
observed this depression following whole smoke exposure as
removal of the particulate matter by a Cambridge filter and
exposure of the animals to the gas phase did not have any
effect (Lubawy et al., 1986).   Whole smoke also resulted in
increased production of TXA_2 by rat platelets (Lubawy et al.,
1986).   This has been supported by Toivanen et al. (1986) who
reported cigarette smoking, but not pipe smoking, to result in
increased serum TXA_2 levels in vivo.

     However, Sherratt et al. (1987) did not observe any effects
of smoking, regardless of nicotine content, on rat aortic PGI_2
production but did observe an increase in platelet TXA_2
production.   Sherratt et al. (1988) found that cigarette smoke
condensate administered by infusion pumps depressed rat aortic
PGI_2 synthesis.   This is in agreement with Reinders et al.
(1986) who reported that cigarette smoke condensate inhibited
basal human endothelial cell PGI_2 release.   However production
of PGI_2 from exogenous arachidonate was not impaired suggesting
that nicotine exerted its effects by inhibiting the
mobilization of arachidonate from cellular phospholipids rather
than interfering with cyclooxygenase or prostaglandin
synthetase (Reinders et al., 1986).   In this study no
morphological abnormalities were seen nor was endothelial cell
von Willebrand factor release affected.

     Jeremy et al. (1985a) found a dose-dependent inhibition of
PGI_2 production in human umbilical artery, rabbit aorta, rat
aorta and rat lung with cigarette smoke extracts but nicotine
at concentrations of up to 1 g/l was without effect.   They
considered that cigarette smoke inhibited PGI_2 synthesis at the
level of cyclooxygenase or beyond it (Jeremy et al., 1985a).

     Decreased PGI_2 production by rat gastric mucosae has been
associated with cigarette smoking (Balint and Varro, 1986) and
cigarette smoke extracts but neither nicotine or cotinine
inhibits rat urinary bladder PGI_2 synthesis (Jeremy et al.,
1985b).   Cigarette smoke extracts induce a dose-related
inhibition of methacholine stimulated PGI_2 production in the
rat penis (Jeremy et al., 1986).

     Using a Langendorff technique, Wennmalm (1978a, b)
demonstrated that nicotine infusion of rabbit hearts resulted
in reduced prostacyclin synthesis.   He demonstrated an
inhibitory effect of nicotine on hypoxia and arachidonate
induced PGI_2 release by rabbit hearts and suggested that
nicotine interferes with the enzymatic conversion of
arachidonate to PGI_2 (Wennmalm, 1980; see Wennmalm, 1982).
Alster and Wennmalm (1983) reported competitive inhibition of
rat aortic PGI_2 and postulated that this effect resulted from
competitive inhibition of cyclo-oxygenase rather than an effect
on PGI_2 synthetase.   Sonnenfeld and Wennmalm (1980) suggested
an inhibitory effect on the formation of PGI_2 in rabbit aorta

and human umbilical vein.   Alster and Wennmalm (1984) found a difference in sensitivity to nicotine between cyclo-oxygenase in rabbit aorta and platelets.   Nicotine impaired dose-dependently the ability of rabbit aortic rings to convert exogenous arachidonic acid to PGI_2 but did not impair the ability of platelets to convert the exogenous substrate to TXA_2 (Alster and Wennmalm, 1984).

Bull et al. (1985) found that single subcutaneous injections of nicotine in rats had no effect on PGI_2 production by aortic rings.   However, sub-acute exposure of rats to nicotine delivered by infusion pumps over one week did result in depressed PGI_2 production.   The data suggested that nicotine was not acting as an inhibitor of cyclo-oxygenase, but in sub-acute exposure is capable of altering the endothelium rendering it unable to produce normal amounts of PGI_2 (Bull et al., 1985).   Support for this comes from another study in which concentrations of nicotine between $10^{-9}M$ and $10^{-4}M$ had no effect on cultured human umbilical cells ability to release PGI_2 (Bull et al., 1988).   Acute but not sub-acute exposure to nicotine at $10^{-3}M$ did depress PGI_2 release and both acute and sub-acute exposure to $10^{-2}M$ resulted in inhibition of PGI_2 release.

Nicotine has been shown to depress human endocardial PGI_2 production (Alster et al., 1986; Effeney, 1977).   Carbon monoxide however, raised PGI_2 production by rabbit hearts and a combination of nicotine and carbon monoxide further raised the levels of production (Effeney, 1987).   However,   Toivanen et al., (1986) observed a stimulatory effect of nicotine on PGI_2 production in vitro but a depressor effect on TXA_2 production and suggested the effect of nicotine was not at the cyclooxygenase level.

Stoel et al. (1982) reported a direct depressive effect of nicotine on PGI_2 production in umbilical arteries.   However, Ylikorkala et al., ((1985) did not observe any effect of nicotine on umbilical artery PGI_2 production but did observe reduced TXA_2 synthesis by foetal platelets both at the baseline level and with exogenous arachidonic acid.   They suggested that nicotine might inhibit cyclooxygenase and/or TXA_2 synthetase.

CONCLUSIONS

Notwithstanding the cautions expressed earlier in this review, there is good experimental evidence to suggest that exposure to cigarette smoke can result in vascular endothelial modification.   Interference with the normal homeostasis of these cells is evidenced by changes in their morphology, permeability and levels of prostanoid synthesis.

At the simplest level these changes could result in the vessel wall being more permeable to lipid or having reduced antiplatelet properties.   More subtle effects could include an effect on the procoagulant and prothrombotic nature of endothelium and the anticoagulant properties of these cells. Although there are no published data, it is possible that cigarette smoking could result in altered production of endothelial derived growth factors which include a smooth muscle mitogen.   Furthermore altered endothelium could present as a more attractive surface for the attachment of other cells such as platelets and monocytes which could contribute to the release of growth factors. Although the experimental evidence to date in relation to the effects of cigarette smoking on vascular endothelium is in places contradictory and requires consolidation, there are good grounds to suppose that endothelial injury can occur. The data supports the pathway involving direct stimulation of the endothelium in the revised response-to-injury hypothesis of atherogenesis proposed by Ross (1986).

Attempts to identify the components of cigarette smoke responsible for the observed endothelial alteration observed have concentrated in the main on carbon monoxide, nicotine and cigarette smoke condensates. The data for carbon monoxide requires further examination and there is experimental evidence to suggest that nicotine may not be as important as other, as yet unknown, components of cigarette smoke as far as endothelial injury is concerned.

Cigarette smoke is an extremely complex mixture and it is not easy to predict which of the many known components, never mind those not yet described, may ultimately be shown to be responsible for the epidemiological link between cigarette smoking and cardiovascular disease. We are focussing our attention on free radicals for three reasons. Cigarette smoke is known to contain free radicals (Nakayama et al., 1984). Morphologically similar effects to some of these described above in relation to cigarette smoke exposure are seen in free radical induced lipid peroxidation in cardiac myocytes (Noronha-Dutra and Steen, 1982). Finally, it has been known for many years that atheromatous plaques contain lipid peroxides and lipid peroxidation products (Glavind et al., 1952).

REFERENCES

Ahlsten, G., Ewald, U., and Tuvemo, T., 1986, Maternal smoking reduces prostacyclin formation in human umbilical arteries. A study on strictly selected pregnancies, Acta. Obstet. Gynecol. Scand., 65:645.

Allen, D.R., Browse, N.L., Rutt, D.L., Butler, L., and Fletcher, C., 1988, The effect of cigarette smoke, nicotine, and carbon monoxide on the permeability of the arterial wall, J. Vasc. Surg., 7:139.

Alster, P., and Wennmalm, A., 1983, Effect of nicotine on prostacyclin formation in rat aorta, Eur. J. Pharmacol., 86:441.

Alster, P., and Wennmalm, A., 1984, Effect of nicotine on the formation of prostacyclin-like activity and thromboxane in rabbit aorta and platelets, Br. J. Pharmacol, 81:55.

Alster, P., Brandt, R., Koul, B.L., Nowak, J., and Sonnenfeld, D.T., 1986, Effect of nicotine on prostacyclin formation in human endocardium in vitro, Gen. Pharmacol, 17, 441.

Armitage, A.K., Davies, R.F., and Turner, D.M., 1976, The

effects of carbon monoxide on the development of atherosclerosis in the white carneau pigeon, Atherosclerosis, 23:333.

Asmussen, I., 1977, Ultrastructure of the human placenta at term. Observations on placentas from newborn children of smoking and nonsmoking mothers, Acta. Obstet. Gynecol. Scand., 56:119.

Asmussen, I., 1978a, Ultrastructure of human umbilical veins. Observations on veins from newborn children of smoking and nonsmoking mothers, Acta. Obstet. Gynecol. Scand., 57:253.

Asmussen, I., 1978b, Arterial changes in infants of smoking mothers, Postgrad. Med. J., 54:200.

Asmussen, I., 1980, Ultrastructure of the villi and fetal capillaries in the placentas delivered by smoking and nonsmoking mothers, Br. J. Obstet., 87:239.

Asmussen, I., 1982a, Chromatin changes of endothelial cells in umbilical arteries in smokers, Clin. Cardiol., 5:653.

Asmussen, I., 1982b, Ultrastructure of the umbilical arteries from new-born smoking and nonsmoking mothers, Acta. Path. Scand., 90:375.

Asmussen, I., 1982c, Ultrastructure of the umbilical artery from a newborn delivered at term by a mother who smoked 80 cigarettes per day, Acta Path. Scand., 90:397.

Asmussen, I., 1984, Mitochondrial proliferation in endothelium. Observations on umbilical arteries from newborn children of smoking mothers, Atherosclerosis, 50:203.

Asmussen, I., and Kjeldsen., K., 1975, Intimal ultrastructure from newborn children of smoking and non-smoking mothers, Circ. Res., 36:577.

Astrup, P., 1972, Some physiological and pathological effects of moderate carbon monoxide exposure, Br. Med. J., 4:447.

Astrup, P., and Kjeldsen, K., 1973, Carbon Monoxide, smoking and atherosclerosis, Med. Clins. N. Amer., 58:323.

Auerbach, O., Carter, H.W., Garfinkel, L., and Hammond, E.C.,1976, Cigarette smoking and coronary artery disease: A macroscopic and microscopic study, Chest, 70:697.

Auerbach, O., and Garfinkel, L., 1980, Atherosclerosis and aneurysm of the aorta in relation to smoking habit and age, Chest, 78:805.

Auerbach, O., Hammond, E.C., and Garfinkel, L., 1965, Smoking in relation to atherosclerosis of the coronary arteries, New Engl. J. Med., 273:775.

Balint, G.A., and Varro, V., 1986, The effect of cigarette smoke on gastrointestinal mucosal endogenous prostacyclin level (experimental and clinical observations), Agents Actions, 19:224.

Bazin, M., Turcotte, H., Lagace, R., and Boutet, M., 1981, Effects cardiovasculaires de la fumee de cigarette chez le rat. Etude de la permeabilite endotheliale aortique et capillaire myocardique chez le rat, Rev. Can. Biol., 40:263.

Blaes, N. Piovella, F., Samaden, A., Boutherin-Falson, O., and Ricetti, M., 1986, Nicotine alters fibronectin and factor VIII/vWF in human vascular endothelial cells, Br. J. Haematol., 64:675.

Blow, C.M., 1988, Mesothelium as a non-thrombogenic surface, in, "Platelet-Vessel Wall Interactions," R.M. Pittilo and S.J. Machin, eds., Springer-Verlag, Berlin.

Booyse, F.M., Osikowicz, G., and Quarfoot, A.J., 1981a, Effects of chronic oral consumption of nicotine on the rabbit aortic endothelium, Am. J. Pathol., 102:229.

Booyse, F.M., Osikowicz, G., and Radek, J., 1981b, Effect of nicotine on cultured bovine aortic endothelial cells, Thromb. Res., 23:169.

Bull, H.A., 1988, The use of cultured endothelial cells in the study of platelet-vessel wall interactions, in, "Platelet-Vessel Wall Interactions," R.M. Pittilo and S.J. Machin, eds., Springer-Verlag, Berlin.

Bull, H.A., Pittilo, R.M., Blow, D.J., Blow, C.M., Rowles, P.M., Woolf, N., and Machin, S.J., 1985, The effects of nicotine on PGI_2 production by rat aortic endothelium, Thromb. Haemostas., 54:472.

Bull, H.A., Pittilo, R.M., Woolf, N., and Machin, S.J., 1988, The effect of nicotine on human endothelial cell release of prostaglandins and ultrastructure, Br. J. exp. Path., 69:413.

Busacca, M., Balconi, G., Pietra, A., et al., 1984, Maternal smoking and prostacyclin production by cultured endothelial cells, Am. J. Obstet. Gynecol., 148:1127.

Bylock, A., Bondjers, G., Jansson, I., and Hansson, H.A., 1979, Surface ultrastructure of human arteries with special reference to the effects of smoking, Acta. Path. Microbiol. Scand. Sect. A, 87:201.

Clarke, J.M.F., Pittilo, R.M., Machin, S.J. and Woolf, N., 1984, A study of the possible role of mesothelium as a surface for flowing blood, Thromb. Haemostas., 51:57.

Csonka, E., Somogyi, A., Augustin, J., Haberbosch, W., Schettler, G., and Jellinek, H., 1985, The effect of nicotine on cultured cells of vascular origin, Virchows Arch (pathol anat), 407:441.

Dadak, C., Leithner, C., Sinzinger, H., and Silberbauer, K., 1981, Diminished prostacyclin formation in umbilical arteries of babies born to women who smoke, Lancet, 1, 94.

Dawber, T.R., Kannel, W.B., Revotskie, N., Stokes, J., Kagan, A., and Gordon, T., 1959, Some factors associated with the development of coronary heart disease: Six years follow-up experience in the Framingham study, Am. J. Pub. Health., 4:1349.

Doll, R., and Peto, R., 1976, Mortality in relation to smoking: 20 years' observations on male British doctors, Br. Med. J., 4:1525.

Eastcott, H.H.G., 1962, Rarity of lower limb ischaemia in non-smokers, Lancet, 2:1117.

Effeney, D.J., 1987, Prostacyclin production by the heart: effect of nicotine and carbon monoxide, J. Vasc. Surg., 5:237.

Fraser, R., Clark, S.A., Day, W.A. and Murray, F.E.M., 1988, Nicotine decreases the porosity of the rat liver sieve: a possible mechanism for hypercholesterolaemia, Br. J. exp. Path., 69:345.

Glavind, J., Hartmann, F., Clemmensen, J., Jessen, K.E., and Dam, H., 1952, Studies on the role of lipoperoxides in human pathology. II. The presence of peroxidized lipids in the atherosclerotic aorta, Acta Path. Scand., 30:1.

Haust, M.D., 1971, The morphogenesis and fate of potential and early atherosclerotic lesions in man, Hum. Pathol., 2:1.

Hladovek, J., 1978, Endothelial injury by nicotine and its prevention, Experientia, 34:1585.

Hollweg, H.G., and Buss, H., 1980, Problems with the preparation of blood vessels for scanning electron microscopy. A critical review., Scanning, 3:3.

Hugod, C., Hawkins, L.H., Kjeldsen, K., Thomsen, H.K., and Astrup, P., 1978, Effect of carbon monoxide exposure on aortic and coronary intimal morphology in the rabbit, Atherosclerosis, 30:333.

Jeremy, J.Y., Mikhailidis, A.M., and Dandona, P., 1985a, Cigarette smoke extracts, but not nicotine, inhibit prostacyclin (PGI_2) synthesis in human, rabbit and rat vascular tissue, Prostaglandins Leukot. Med., 19:261.

Jeremy, J.Y., Mikhailidis, A.M., and Dandona, P., 1985b, Cigarette smoke extracts inhibit prostacyclin synthesis by the rat urinary bladder, BR. J. Cancer, 51:837.

Jeremy, J.Y., Mikhailidis, A.M., Thompson, C.S. and Dandona, P., 1986, The effect of cigarette smoke and diabetes mellitus on muscarinic stimulation of prostacyclin by the rat penis, Diabetes Res., 3:467.

Kannel, W.B., McGee, D. and Gordon, T., 1976, A general cardiovascular risk profile - The Framingham Study, Am. J. Cardiol., 38:4.

Kjeldsen, K., Astrup, P., and Wanstrup, J., 1972, Ultrastructural intimal changes in the rabbit aorta after a moderate carbon monoxide exposure, Atherosclerosis, 16:67.

Laing, S.P., Greenhalgh, R.M. and Taylor, G.W., 1981, The prevalence of cigarette smoking in patients with arterial disease, in: "Smoking and Arterial Disease," R.M. Greenhalgh, ed., Pitman Medical, London.

Levene, C.I., Bartlet, C.P., Fornieri, C., and Heale, G., 1985, Effect of hypoxia and carbon monoxide on collagen synthesis in cultured porcine and bovine aortic endothelium, Br. J. exp. Pathol., 66:399.

Lubawy, W.C., Culpepper, B.T., and Valentovic, M.A., 1986, Alterations in prostacyclin and thromboxane formation by chronic cigarette smoke exposure: temporal relationships and whole smoke vs. gas phase, J. Appl. Toxicol., 6:77.

Marshall, M., and Hess, H., 1981, Acute effects of low carbon monoxide concentrations on blood rheology, platelet function, and the arterial wall in the minipig, Res. Exp. Med., 178:201.

Mileikowsky, G.N., Nadler, J.L., Huey, F., Francis, R., and Roy, S., 1988, Evidence that smoking alters prostacyclin formation and platelet aggregation in women who use oral contraceptives, Am. J. Obstet. Gynecol., 159:1547.

Murphy, E.A,, and Mustard, J.F., 1966, Tobacco and Thrombosis, Am. J. Pub. Health, 56:1061.

Nadler, J.L., Velasco, J.S., and Horton, R., 1983, Cigarette smoking inhibits prostacyclin production, Lancet, i:1248.

Nakayama, T., Kodama, M., and Nagata, C., 1984, Generation of hydrogen peroxide and superoxoide anion radical from cigarette smoke, Gann, 75:95.

Nicholson, L.J., Clarke, J.M.F., Pittilo, R.M., Machin, S.J., and Woolf, N., 1984, The mesothelial cell as a non-thrombogenic surface, Thromb. Haemostas., 52:102.

Noronha-Dutra, A.,A., and Steen, E.M., 1982, Lipid peroxidation as a mechanism of injury in cardiac myocytes, Lab. Invest., 47:346.

Petty, R.G., and Pearson, J.D., 1989, Endothelium - the axis of vascular health and disease, J. Royal. Coll. Phys., 23:92.

Pittilo, R.M., Clarke, J.M.F., Harris, D., Mackie, I.J., Rowles, P.M., Machin, S.J., and Woolf, N., 1984, Cigarette smoking and platelet adhesion, Brit. J. Haematol., 58:627.

Pittilo, R.M., Mackie, I.J., Rowles, P.M., Machin, S.J., and Woolf, N., 1982. Effects of cigarette smoking on the ultrastructure of rat thoracic aorta and its ability to produce prostacyclin, Thromb. Haemostas., 48:173.

Pittilo, R.M., Nicholson, L.J., Clarke, J.M.F., Blow. C.M., and Woolf, N., 1985, Cigarette smoke-induced injury of peritoneal mesothelial cells, Br. J. Exp. Path., 65:365.

Pittilo, R.M., 1988a, Endothelium and the vessel wall, in, "Platelet-Vessel Wall Interactions,", R.M. Pittilo and S.J. Machin, eds., Springer-Verlag, Berlin.

Pittilo, R.M., 1988b, Smoking and platelet-vessel wall interactions, in, "Platelet-Vessel Wall Interactions,", R.M. Pittilo and S.J. Machin, eds., Springer-Verlag, Berlin.

Pittilo, R.M., and Bull, H.A., 1988, The effects of nicotine on endothelial morphology and prostacyclin production, in, "The Pharmacology of Nicotine,"M. Rand, K. Tharau, IRL Press.

Pomerantz, K.B., Hajjar, D.P., 1989, Eicosanoids in regulation of arterial smooth muscle cell phenotype, proliferative capacity, and cholesterol metabolism, Arteriosclerosis, 9:413.

Reidy, M.A., 1985, Biology of disease. A reassessment of endothelial injury and arterial lesion formation, Lab. Invest., 53:513.

Reidy, M.A. and Schwartz, S.M., 1983, Endothelial injury and regeneration. IV. Endotoxin: a non-denuding injury to aortic endothelium, Lab. Invest., 48:25.

Reidy, M.A. and Schwartz, S.M., 1984, Recent advances in molecular pathology: arterial endothelium - assessment of in vivo injury, Exp. Mol. Pathol., 41:419.

Reinders, J.H., Brinkman, H.J.M., van Mourik, J.A., and de Groot, P,G., 1986, Cigarette smoke impairs endothelial cell prostacyclin production, Arteriosclerosis,, 6:15.

Roberge, S., Bazin, M., and Boutet, M., 1983, Endothelial cell modifications in rat thoracic aorta. Effects of ovariectomy and cigarette smoke, Experientia, 39:72.

Ross, R. 1986, The pathogenesis of atherosclerosis - an update, New Engl. J. Med., 314:488.

Sackett, D.L., Gibson, R.W., Bross, I.D.J., and Ickren, J.W., 1968, Relation between aortic atherosclerosis and the use of cigarettes and alcohol, New Engl. J. Med., 279:1413.

Sherratt, A.J., Culpepper, B.T., and Lubawy, W.C., 1987, Alterations in prostacyclin and thromboxane formation following chronic exposure to high and low nicotine cigarettes in rats, Artery, 14:216.

Sherratt, A.J., Culpepper, B.T., and Lubawy, W.C., 1988, Prostacyclin and thromboxane formation following chronic exposure to cigarette smoke condensate administered via osmotic pumps in rats. A method for chronic administration of particulates of whole smoke, J. Pharmacol. Methods, 20:47.

Sieffert, G.F., Keown, K., and Moore, W.S., 1981, Pathologic effect of tobacco smoke inhalation on arterial intima, Surgical Forum, 32:333.

Sonnfield, T., and Wennmalm, A., 1980, Inhibition by nicotine of the formation of prostacyclin-like activity in rabbit and human vascular tissue, Br. J. Pharmacol., 71:609.

Spain, D.M., Bradess, V.A., 1970, Sudden death from coronary heart disease: Survival time, frequency of thrombi, and cigarette smoking, Chest, 58:107.

Stoel, I., Giessen, W.J. vd, Wolsman, E., Verheugt, F.W.A., Ten Hoor, F., Quadt, F.J.A., and Hugenholtz, P.G., 1982, Effect of nicotine on production of prostacyclin in human umbilical artery, Br. Heart J., 48: 493.

Staubesand, J., Hugod, C., Mau, G., and Seydewitz, V., 1984, Intimal ultrastructural morphology of umbilical cord vasculature in babies born to cigarette smoking mothers. Vasa, 13:138.

Strong, J.P., and Richards, M.L., 1976, Cigarette smoking and atherosclerosis in autopsied men, Atherosclerosis, 23:451.

Toivanen, J., Ylikorkala, O., and Viinikka, L., 1986, Effects of smoking and nicotine on human prostacyclin and thromboxane production in vivo and in vitro, Toxicol. Appl. Pharmacol., 82:301.

Thomsen, H.K., 1974, Carbon monoxide induced atherosclerosis in primates - an electron microscopic study on the coronary arteries of Macaca irus monkeys, Atherosclerosis, 20:233.

Van der Veen, F., and Fox, H., 1982, The effects of cigarette smoking on the human placenta: a light and electron microscopic study, Placenta, 3:243.

Webster, W.S., Clarkson, T.B., and Lofland, H.B., 1970, Carbon monoxide-aggravated atherosclerosis in the squirrel monkey, Exp Molec Pathol, 13:36.

Wennmalm, A., 1978a, Nicotine inhibits release of 6-keto prostaglandin F_1 from isolated perfused rabbit heart, Acta Physiol. Scand., 103:107.

Wennmalm, A., 1978b, Effects of nicotine on cardiac prostaglandin and platelet thromboxane synthesis, Br. J. Pharmac., 64:559.

Wennmalm, A., 1980, Nicotine inhibits hypoxia and arachidonic acid induced release of prostacyclin-like activity in rabbit hearts, Br. J. Pharmacol., 69:545.

Wennmalm, A., 1982, Interaction of nicotine and prostaglandins in the cardiovascular system, Prostaglandins, 23:139.

Woolf, N., and Wilson-Holt, N., 1981, Cigarette smoking and atherosclerosis, in: "Smoking and Arterial Disease", R.M. Greenhalgh, ed., Pitman Medical, Bath, England.

Woolf, N., 1987, The pathology of atherosclerosis with particular reference to the effects of hyperlipidaemia, Eur. Heart J., 8:3.

Ylikorkala, O, Viinikka, L., and Lehtovirta, P., 1985, Effect of nicotine on fetal prostacyclin and thromboxane in humans, Obstet. Gynecol., 66:102.

Zimmerman, M., and McGeachie, J., 1987, The effect of nicotine on aortic endothelium. A quantitative ultrastructural study, Atherosclerosis, 63:33.

THE EFFECTS OF NICOTINE ON AORTIC ENDOTHELIAL CELL TURNOVER

AND ULTRASTRUCTURE

Matthew Zimmerman* and John K. McGeachie**

*Resident Medical Officer
Queen Elizabeth II Medical Centre
Perth, Western Australia

**Associate Professor
Department of Anatomy and Human Biology
The University of Western Australia

INTRODUCTION

Endothelial cell injury is one of the earliest detectable changes in
the initiation of arterial intimal hyperplasia and atherosclerosis (Ross
and Glomset, 1976; Moore, 1981; Ross, 1986). These early pathological
changes are associated with the development of lesions which result in
clinical conditions such as myocardial infarction, stroke and peripheral
vascular disease. Cigarette smoking has been linked to arterial disease
for over 30 years (Surgeon General USA, 1964; 1979) and is a
well-recognized risk factor. However the direct relationship between
smoking and the pathogenesis of arterial disease has not been precisely
defined.

These experimental studies test the hypothesis that exposure to
nicotine (one of the toxins of cigarette smoke) induces structural
changes in vascular endothelial cells: changes which may be indicative
of early pathology. Moreover, there is evidence that endothelium
responds to such insults by increasing mitotic activity and cell turnover
(Florentin et al., 1969; Payling-Wright et al., 1975). The present study
quantitates both the mitotic and ultrastructural changes that occur in
endothelial cells in mouse aortae, after they had been exposed to
nicotine (via the drinking water) for a period of 5 weeks.

MATERIALS AND METHODS

Animals and Treatments

Forty-three adult male Swiss mice (35-40) were used in these
experiments, in accordance with guidelines of the National Health and
Medical Research Council of Australia; and with the formal approval of
the Animal Experimentation Ethics Committee of the University of Western
Australia. Twenty-three mice were given a daily dose of nicotine in the
drinking water over a 5-week period. The dose was equivalent to a human
smoking 50-100 cigarettes per day. The dose of nicotine was 5mg/kg body

Tobacco Smoking and Atherosclerosis
Edited by J. N. Diana
Plenum Press, New York, 1990

weight per day and the model was based on a previous study which showed
that the physiological changes induced by such a dose, administered via
the drinking water, were the same as if the nicotine has been inhaled
with smoke (Booyse et al., 1981).

The other 20 control mice were housed under the same standard
laboratory conditions but not exposed to nicotine.

Autoradiography

Thirty-five days after the first administration of nicotine, 9
nicotine-treated and 8 control mice were injected intraperitoneally with
tritiated thymidine (3H-TdR) with a specific activity of 5Ci/m mol at the
dosage of 1 µCi/g body weight. Twenty-four hours later these 17 mice
were anaesthetized with ether and perfused with heparinized saline,
followed by Karnovsky's fixative (Karnovsky, 1965) via the heart.

Following perfusion, the aortae were excised and enface preparations
made as follows (Zimmerman and McGeachie, 1986a). Ringlets of aorta
5-8mm long were cut then divided in half; each segment was rinsed in 0.1M
cacodylate buffer and dried with filter paper. The lumenal side of the
segment was then pressed against a gelatin film (9% concentration) on a
standard microscope slide. The outer side of the aortic segment was
covered with a coverslip and gentle pressure applied to the preparation
for 2-3 hours. Then, 24 hours later the media and adventitia were
removed from the intima, using a scalpel blade: this left the endothelium
adherent to the gelatin film on the slide.

These en-face preparations were coated with Kodak AR 10
autoradiographic stripping film and exposed in light-tight boxes at -20°C
for 12 weeks. Standard 1 µm Araldite plastic sections of intestinal
mucosa were also prepared for autoradiography and placed on the same
slides as above, to act as autoradiographic controls. Autoradiographs
were developed in Kodak D 19 for 5 minutes at 16 C, washed in 0.5% acetic
acid, fixed in 25% sodium thiosulphate at 16 C for 5 minutes, washed in 5
changes of distilled water at 16 C for 20 minutes total and air dried
overnight. Prior to staining they were soaked in distilled water for 5
minutes and stained in Harris' Haematoxylin for 4 minutes, washed in
distilled water and dried in air.

Analysis of Autoradiographs

This was done using a 100 x oil immersion lens with a 10 x ocular
lens. Random fields on the preparations were selected, the numbers of
labelled and unlabelled endothelial cells recorded and the percentages of
labelled cells (labelling indices) calculated. At the time of analysis
the slides were coded so that the observer did not know which were
control and which were nicotine-treated specimens.

Electronmicroscopy

The remaining 26 mice (14 nicotine-treated and 12 control) were also
left for 35 days after the commencement of the study before being
sacrificed and the tissues fixed by perfusion, as above. Segments from
the aortic arch, thoracic aorta, mid-abdominal aorta and the iliac
bifurction were taken, rinsed in 0.1M cacodylate buffer, dehydrated
through graded ethanols and embedded in Epon/Aralidite resin. Blocks
from each aortic segment were pooled and equal numbers of blocks were
selected randomly. Ultra-thin sections from two blocks per mouse were
cut and stained with uranyl acetate and lead citrate (Zimmerman and
McGeachie, 1987). Five electron micrographs (x 20,300) from each sample
block were taken systematically to insure independence of the sampled

regions (Weibel, 1979). An additional set of micrographs were taken at a magnification of x 32,400. As with the autoradiographic analysis all electron micrographs were coded to eliminate observer bias.

Analysis of Electron Micrographs

Stereological analysis was used with custom-made grids (Weibel, 1979). Point counts were performed with a double square lattice L (q^2 = 4, d = 0.483 µm). Intersection counts were made with a double square lattice (q^2 - 6.25 d = 0.493 µm). Stereological indices were calculated using methods described by (Weibel, 1979). Aortic endothelial junctional morphology was assessed by a protocol developed by Zimmerman and McGeachie (1986b). All endothelial cell indices were measured from the x 20,300 micrographs except plasmalemmal volume fractions which were measured from x 32,400 micrographs.

The indices for each mouse were pooled and data from the nicotine-treated and control aortae were compared statistically (Snedcore and Cochran, 1980). Because volume fractions are ratios and are not distributed normally, arcsine transformations were used prior to applying statistical analyses (Snedcore and Cochran, 1980).

RESULTS

Detailed data from these experimental protocols have been published previously (Zimmerman and McGeachie, 1985, 1986a, 1986b, 1987).

Autoradiography

The intestinal mucosal preparations from all mice (nicotine-treated and control) contained cells labelled with ^{3}H-TdR. Likewise endothelial cells were found in both nicotine-treated and control aortae (Fig. 1), however there was a significantly hither level of (p<0.01, x^2 test) labelling in nicotine-treated endothelium (0.46 $\pm$ 0.11%) compared with controls (0.14 $\pm$ 0.06%). See Fig. 2A.

The endothelial cell densities did not differ significantly between the nicotine-treated (2130 $\pm$ 140 cells/mm^2 and control endothelium (1990 $\pm$ 120 cells/mm^2). See Fig. 2B.

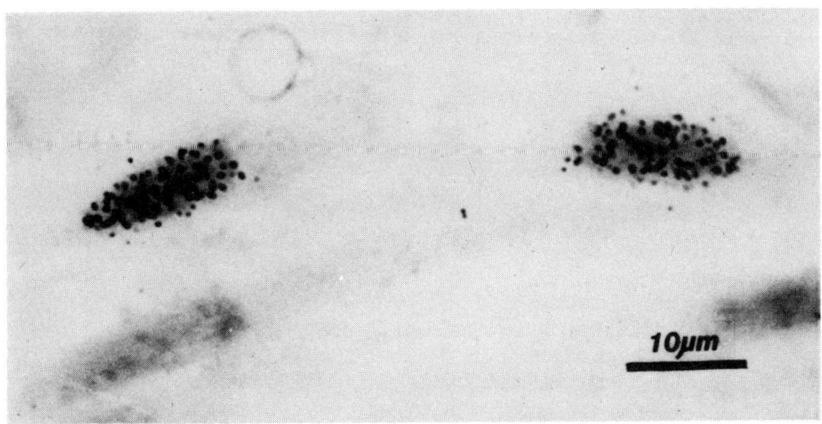

Fig. 1. Aortic endothelial cell nuclei labelled autoradiographically with ^{3}H-TdR. This specimen is from a nicotine-treated mouse that was injected with ^{3}H-TdR 24 hours prior to sampling the aorta.

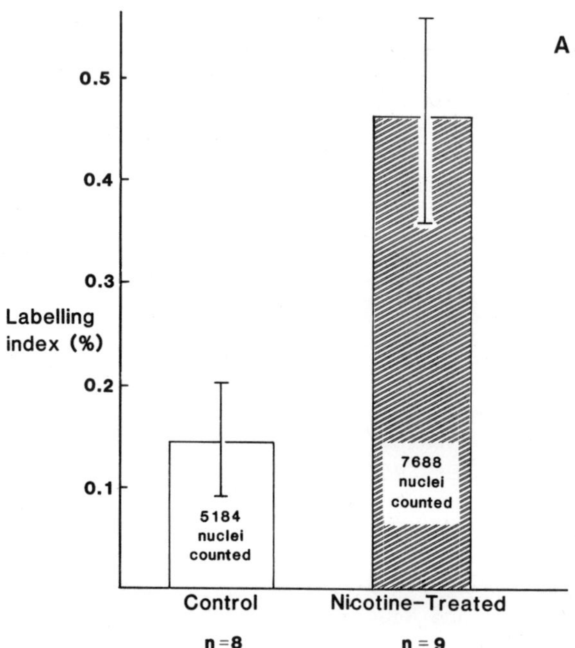

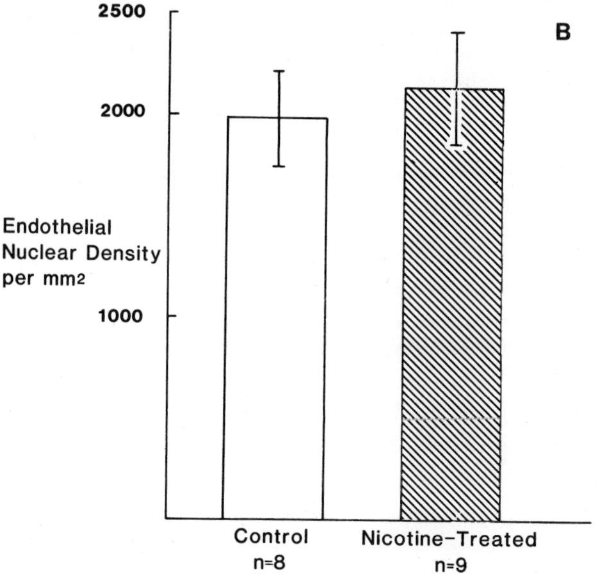

Fig. 2. Data from autoradiographic studies of control and nicotine-treated aortic endothelium.

    A. Aortic endothelial cell labelling levels (mean ± SEM). These are significantly different ($p < 0.01$).

    B. Aortic endothelial cell densities (cells/mm^2). These are not significantly different.

<u>Electronmicroscopy</u>

Endothelial cells in the nicotine-treated aortae were slightly
thicker (but not significantly different) than in controls. However
there was a substantial increase in the sub-endothelial space in
nicotine-treated aortae (almost twice the thickness of controls [p<
0.25]). There were variations in the sub-endothelial space in both
groups. Areas where there was extensive widening of the space (where
endothelial cells were attached to the internal elastic lamina by thin
basal projections) were more evident in the nicotine-treated group. Such
focal attachments have been described as "arcades" by Tsao and Glagov
(1970). Small concentrations of smooth muscle were evident in the intima
in about 3% of samples from both nicotine-treated and control aortae.
This is similar to the Diffuse Intimal Thickening described in human
vessels.

<u>Endothelial Cytoplasm</u>

There was a significantly greater ratio (p<0.025) of the outer
mitochondrial membrane surface area:to the inner membrane area, in
nicotine-treated endothelium, compared with controls. This indicated
that nicotine had altered mitochondrial structure and that they were less
condensed (more swollen) than normal. Using other parameters (Zimmerman
and McGeachie, 1987) it was shown that mitochondria in nicotine-treated
endothelial cells were very significantly (p<0.005) different from
controls. It was also noted that the mitochondrial matrix was less dense
and that the cristae were less obvious in nicotine-treated endothelium.

Cytoplasmic vacuoles occupied a significantly greater volume
(p<0.025) or endothelial cells influenced by nicotine, although the
distribution and actual size of the vacuoles were similar to those in
controls. Other features of the cytoplasm: plasmalemmal vesicles,
Weibel-Palade bodies and cytoplasmic inclusions were similar in both
nicotine-treated and control endothelium.

<u>Endothelial Intercellular Junctions</u>

Four different configurations of these interface areas between
adjacent endothelial cells have been described (Schwartz and Benditt,
1972), and it has been shown that the more simple junctional interfaces
are permeable to blood-borne macromolecules (Zimmerman and McGeachie,
1986b). These four different junctional profiles are shown as insets in
Fig. 3.

All four junctional types were seen in both nicotine-treated and
control endothelium, however there was a highly significant increase
(p<0.005) in the occurrence of simple (permeable) junctions in
nicotine-treated aortae (Fig. 3).

DISCUSSION

Cigarette smoking is a well established risk factor in the
development of arterial disease. The early pathological changes in
atherogenesis include structural damage to endothelial cells (Ross and
Glomsett, 1976; Ross, 1986) and invasion of the sub-endothelial space by
macrophages (Rhoden, 1984). Whilst the "response to injury" hypothesis
is currently controversial, there is no doubt that endothelial cell
injury is an important event in the early pathogenesis of arterial
disease. Exposure to cigarette smoke produces demonstrable damage to
endothelial cells (Asmussen and Kjeldsen, 1975; Asmussen, 1979; Bylock et
al., 1979; Woolf and Wilson-Holt, 1981; Pittilo et al., 1982) and also

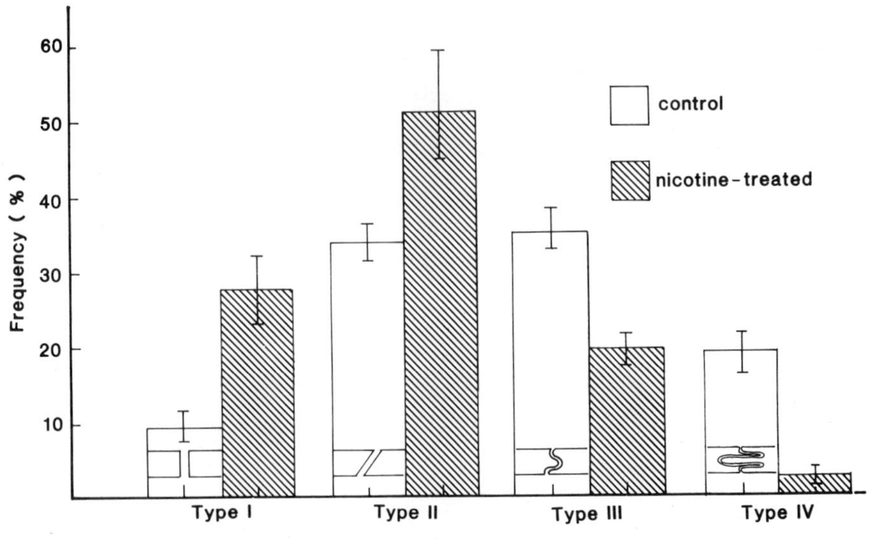

Fig. 3.   The distribution (mean ± SEM) of the four recognized edendo-
thelial cell junctional types (shown in the control columns)
between nicotine-treated and control aortic endothelium. The
frequency distribution in nicotine-treated endothelium was
significantly different (p < 0.005); junctions were simpler
and thus more permeable.

increases endothelial permeability to blood-borne macromolecules (Boutet
et al., 1980). Whilst there is ample evidence that cigarette smoke is
injurious to endothelium, and promotes the development of arterial
disease, there is a paucity of detailed studies to identify which
components of cigarette smoke are responsible for these pathogenic
processes.

Nicotine induces a range of physiological effects including:
increased blood pressure, raised blood glucose and lactate levels, and
stimulates the release of catecholamines (Surgeon General USA, 1964). In
combination with an atherogenic diet and stress, nicotine induces
atherosclerosis in experimental animals (Stefanovich et al., 1969; Scott
et al., 1970). Endothelial cells grown in vitro are damaged by exposure
to nicotine (Pollack, 1969; Tulloss and Booyse, 1978). Thus there is a
substantial body of evidence to show that nicotine is cytotoxic to
endothelial cells and helps to promote arterial disease. The objectives
of the studies presented in this paper were to measure the specific
effects of nicotine on aortic endothelium, with respect to the level of
mitotic activity and cell turnover (Autoradiography) and the
ultrastructural changes in endothelial cells (Electronmicroscopy).

<u>Autoradiography</u>

Increased mitotic activity and cell loss are characteristic features
of endothelial injury. This phenomenon has been reported after
endothelial cell damage from endotoxinaemia (Gaynor, 1971; Reidy and
Schwartz, 1983), chemical insult (Spaet and Lejnieks, 1968) anaphylactic
shock (Payling-Wright and Giacometti, 1972) and hypercholestorolaemia

84

(Florentin et al., 1969; Payling-Wright et al., 1975; Weber et al.,
1979). Thus, it was hypothesized in the present study that nicotine
would have a mitogenic effect on aortic endothelium and thus increase
cell turnover. This would be indicative of "endothelial injury."

Mitotic activity is difficult to quantitate precisely because visible
mitosis (prophase, metaphase, etc.) occupies about one hour of the cell
replication cycle. Even with the most rapidly proliferating cells in the
body, with a turnover time of 12 to 18 hours, the observation of a cell
in mitosis is an inaccurate indication of the actual level of cell
replication that is actually occurring in the population. Measuring DNA
synthesis in cells (which takes place over about 12 hours of the cell
replication cycle) by labelling DNA with radioactive thymidine (^{3}H-TdR)
gives a much more accurate measure of the level of cell division.
Moreover, cell turnover levels can be determined by labelling cells and
leaving them to divide subsequently, allowing the ^{3}H-TdR, to be carried
through to daughter cell populations. This was the basis of the present
autoradiography study.

We calculate that 1.76% of endothelial cells enter mitosis per day in
"normal" aortae. This is based upon the labelling levels 24 hours after
injection of ^{3}H-TdR, where 0.14 $\pm$ 0.06% of endothelial nuclei were
labelled; this represents the progeny of cells pulse-labelled at the time
of injection, 24 hours previously. ^{3}H-TdR is available for about 30
minutes after injection therefore it labels only the cells synthesizing
DNA at that time. Thus if one assumes that cells are equally likely to
enter DNA synthesis at any time during the 24 hours, then the figure of
1.76% of cells turning over is calculated. Our control data are in direct
agreement with similar studies in rats (Schwartz and Benditt, 1977).

By contrast to the relatively low level of endothelial cell turnover
in control aortae (1.76% over 24 hours) nicotine-treated endothelial
cells had a significantly higher level of turnover (5.53%), based on the
same calculations as above.

Nicotine may be mitogenic to endothelial cells, but this is unlikely
in view of the evidence of its cytotoxic effects in vitro. The increased
turnover of endothelial cells in nicotine-treated aortae was probably a
response to the toxic injury. However, there was no increase in endothelial cell density in nicotine-treated endothelial cells did not add to
the cell population; rather, cells must have been lost from the population. Thus it was concluded that cell turnover in nicotine-treated endothelium was at least doubled. This is presented as evidence for injurious effect of nicotine and its possible role in the pathogenesis of arterial disease. Was this toxic effect on endothelial cells also associated
with cell structural changes which would also promote arterial disease?

<u>Electronmicroscopy</u>

Several experimental studies have shown ultrastructural evidence of
endothelial cell injury; characterized by swelling, irregularity of the
lumenal surface, mitrochondrial swelling, cytoplasmic vacuolation, and
subendothelial oedema (Tsáo and Glagov, 1970; Bjorkerud and Bondjers,
1972; Goode et al., 1977, Svedsen, 1978; 1979). These features of
endothelial damage, or early pathological changes, have been observed in
endothelial cells of umbilical vessels from 1975; Bylock et al., 1979;
Asmussen, 1982). Likewise, other models of endothelial cell research
show evidence of pathological changes when the cells are exposed to
cigarette smoke (Woolf and Wilson-Holt, 1981; Pittilo et al., 1982). In
view of this evidence, and the evidence cited above (in the
Autoradiography section) that nicotine caused pathological changes in

aortic endothelium, we proposed that nicotine would induce the specific
ultrastructural changes, characteristic endothelial cell injury.

The results of our study show clear quantitative changes in
endothelial cells exposed to nicotine (Zimmerman and McGeachie, 1987).
There was a significant increase in cytoplasmic vacuolation,
sub-endothelial oedema, and mitochondrial swelling in nicotine-treated
endothelium, compared with control endothelium.

Another highly significant result of our study is the finding that
nicotine induced changes in endothelial cell junctional morphology
(Zimmerman and McGeachie, 1987). The junctions were more simple than in
controls, showing clear evidence of increased permeability to blood-borne
macromolecules (Zimmerman and McGeachie 1986b). It is well established
that such increased permeability may facilitate the accumulation of
undesirable lipids in the vessel wall. Increased endothelial
permeability also occurs after acute haemodynamic injury (Stephens,
1978), mechanical injury (Moore, 1981), and activated complement injury
(Forbes and Guttman, 1982).

This evidence, together with the observation that nicotine also
induces increased endothelial cell turnover (evidence of endothelial
"stress"), clearly demonstrates that nicotine does damage endothelium and
predisposes to, or promotes, the development of arterial disease,
specifically atherosclerosis.

REFERENCES

1. Assmussen I. Ultrastructure of human umbilical arteries from newborn
    children of smoking and non-smoking mothers. Acta Path Microbiol
    Immunol Scand Sect A, 90: 375-383 (1982).
2. Asmussen I, and Kjeldsen. Intimal ultrastructure of human umbilical
    arteries. Observations on arteries from newborn children of
    smoking and non-smoking mothers, Circ Res, 35: 579-589 (1975).
3. Bjorkerud S, and Bondjers G. Endothelial integrity and viability
    in the aorta of the normal rat and rabbit as evaluated with dye
    exclusion tests and interference contrast microscopy. Athero-
    sclerosis, 15: 285-300 (1972).
4. Booyse FM, Osikowicz G, and Quarfoot A. Effects of chronic oral con-
    sumption of nicotine on rabbit aortic endothelium. Amer J Path,
    102: 229-238 (1981).
5. Bouter M, Bazin M, Turcotte T, and Legrace R. Effects of cigarette
    smoke on rat thoracic aorta. Artery, 7: 56-72 (1980).
6. Bylock A, Bondjers G, Jansson I, and Hansson H. Surface ultra-
    structure of human arteries with special reference to the effects
    of smoking. Acta Path Microbiol Scand Sect A, 87: 201-209 (1979).
7. Florentin RA, Nam S, Lee K, and Thomas D. Increased mitotic activity
    in aortas of swine. Arch Path, 88:a 463-476 (1969).
8. Forbes R, and Guttman R. Evidence for complement-induced endothelial
    injury in vivo. Amer J Pathol, 106: 378-387 (1982).
9. Gaynor E. Increased mitotic acitivity in rabbit endothelium after
    endotoxin. Lab Invest, 24: 318-321 (1971).
10. Goode TB, Davies P, Reidy MA, and Bowyer DE. Aortic endothelial cell
    morphology observed in situ by scanning electron microscopy during
    atherogenesis in the rabbit. Atherosclerosis, 27: 235-251 (1977).
11. Hoff HE. Vascular Injury: a review in: Vascular Factors and
    Thrombosis. Ed Brinkhous K, Thromb, Diath, Haemorrh, Suppl 40
    (1970).

12. Karnovsky MJ. A formaldehyde-glutaraldehyde fixative of high
     osmolarity for use in electron microscopy. J Cell Biol, 27: 137A
     (1965).
13. Moore S. Injury mechanisms in atherogenesis. In: S Moore (Ed).
     Vascular Injury and Atherosclerosis, Marcel Dekker, New York, pp
     25-52 (1981).
14. Payling-Wright H, Giacometti NJ. Circulating endothelial cells and
     arterial endothelial mitosis in anaphylactic shock. Brit J Exp
     Path 52: 1-4 (1972).
15. Pittilo R, Mackie I, Rowles P, Machin S, and Woolf N. Effects of
     cigarette smoking on the ultrastructure of rat thoracic aorta and
     its ability to produce prostacyclin. Thromb Haemostas (Stuttgart)
     48: 173-176 (1982).
16. Pollack OJ. Tissue Culture (Monographs on Atherosclerosis Vol 1,
     Karger, Basel pp 101-106 (1969).
17. Reidy MA, and Schwartz SM. Endothelial injury and regeneration.
     IV. Lab Invest 48: 25-34 (1969).
18. Rhodin JAG. Handbook of Physiology, Section 2. The Cardiovascular
     System, Volume IV. Architecture of the vessel wall. Eds Geiger
     SR, Renkin EM, Michel CC. Waverly Press, Baltimore, Maryland pp
     7-10 (1984).
19. Ross R. The pathogenesis of atherosclerosis - an update. N Eng J
     Med 314: 488-499 (1986).
20. Ross R, and Glomset JA. The pathogenesis of atherosclerosis. N Eng
     J Med 295: 420-425 (1976).
21. Schwartz SM and Benditt EP. Studies on aortic intima: I: Structure
     and permeability of rat thoracic intima. Amer J Path 66: 241-264
     (1972).
22. Schwartz SM, Benditt EP. Aortic endothelium cell replication -
     effects of age and hypertension. Circ Res 41: 248-255 (1977).
23. Scott R, Henson D, and Hemmens A. Relationship between nicotine
     induction of arteriosclerotic thromboarteritis and nicotine
     induced rise in serum free fatty acids. Am J Pathol, 59: 72a
     (1970).
24. Snedcore G and Cochran W. Statistical methods, Iowa State University
     Press, Ames, Iowa (1980).
25. Spaet TH, and Lejnieks I. Mitotic activity of rabbit blood vessels.
     Proc Soc Exp Biol Med 125: 1197-1201 (1968).
26. Stebbens WE. Endothelial permeability in experimental aneurysms and
     arterio-venous fistulas demonstrated by the uptake of Evan's Blue
     dye. Atherosclerosis, 30: 343-349 (1978).
27. Stefanovich V, Gore I, Kajiyama G, and Iwanaga. The effect of
     nicotine on dietary atherogenesis in rabbits. Exp Mol Pathol 11:
     71-81 (1969).
28. Surgeon General USA. Report on Smoking and Health, US Department
     of Health Education and Welfare, pp 315-333 (1964).
29. Surgeon General USA. Report on Smoking and Health, US Department of
     Health Education and Welfare, Cardiovascular Disease pp 4-77
     (1979).
30. Svendsen S, and Jorgensen. Focal "spontaneous" alterations and loss
     of endothelial cells in rabbit aorta. Acta Path Microbiol Scand
     Sect A, 86: 1-13 (1978).
31. Svendsen S. Focal endothelial cell injury in rabbit aorta.
     Aggravation of injury by 2 days of cholesterol feeding. Acta Path
     Microbiol Scand Sect A, 87: 123-130 (1979).
32. Ts'ao CH, and Glagov S. Basal endothelial attachment. Lab Invest 23:
     510-516 (1970).
33. Tulloss J, and Booyse FM. Effects of various agents and physical
     damage in bovine endothelial cultures. Microvasc Res 16: 51-58
     (1978).

34. Weber G, Losi M, Toti P, and Vatti R. Circulating endothelial cells
     in artery peripheral blood of hypercholesterolemic rabbits, Artery
     5: 29 (1979).
35. Weibel ER. Stereological methods – Practical Methods for Biological
     Morphometry, Academic Press, London (1979).
36. Woolf N, and Wilson-Holt NJ. Cigarette smoking and arterial disease,
     Pitamn Press, Bath pp 46-59 (1981).
37. Zimmerman MJ, McGeachie JK. The effect of nicotine on aortic endo-
     thelial cell turnover: an autoradiographic study.
     Atherosclerosis 58: 39-47 (1985).
38. Zimmerman M, and McGeachie JK. An autoradiographic technique for en-
     face preparations of aortic endothelium. J Pathol, 150: 65-68
     (1986a).
39. Zimmerman M, and McGeachie JK. The effect of nicotine on aortic
     endothelium. A quantitative ultrastructural study.
     Atherosclerosis 63: 33-41 (1987).

ULTRASTRUCTURAL EVENTS ASSOCIATED WITH ENDOTHELIAL CELL CHANGES DURING THE INITIATION AND EARLY PROGRESSION OF ATHEROSCLEROSIS

Richard G. Taylor, W. Gray Jerome, and Jon C. Lewis

Department of Pathology
Bowman Gray School of Medicine
of Wake Forest University
Winston-Salem, North Carolina

INTRODUCTION

Vascular endothelial cells maintain homeostasis by anti-adhesive and anti-fibrinolytic properties. Endothelial cell dysfunction is associated with the pathogenesis of disease processes such as inflammation and atherosclerosis. The study of endothelial cell function in the initiation and progression of atherosclerosis has focussed upon endothelial cell injury. It is now recognized that endothelium is altered by many means, including flow perturbations, hypercholesterolemia, endotoxin or immune complexes (Engelberg, 1989); and that a nondenuding injury, or dysfunction (Reidy, 1985) is likely to be associated with atherogenesis.

This review will discuss the changes in endothelium during the initiation and progression of atherosclerosis in White Carneau pigeons. Ultrastructural techniques have been used to examine endothelium at normal, lesion and prelesion sites of coronary arteries and the aorta using both qualitative and quantitative analysis. The two questions which have directed these morphologic evaluations are: 1) do alterations in endothelial cell structure precede lesion formation? and 2) what changes in endothelium accompany the progression of atherosclerotic lesions?

EXPERIMENTAL MODEL

The White Carneau pigeon was developed as a model for atherosclerosis by Clarkson and Prichard (Prichard et al., 1964). The advantages of this model are that pigeons develop naturally occurring atherosclerotic lesions which progress to complex plaques, the morphology of the lesions is very similar to those found in man, and that the incidence of atherosclerosis is 100% by 4 years of age (Prichard et al., 1964). In addition, pigeons develop mural thrombi and die of myocardial infarction even when fed a cholesterol free diet (Prichard et al., 1963). A great advantage for morphologic studies is that pigeon lesions develop at specific anatomic sites in both the aorta (Jerome and Lewis, 1984) and coronary (Lewis et al., 1982) arteries. The lesion site studied in the coronary artery is near the ostia at the aortic arch, and the highest predilection site in the aorta is 1 mm cephalad to the celiac bifurcation. We have studied these sites in prelesion and lesion states to detect endothelial cells alterations.

Tobacco Smoking and Atherosclerosis
Edited by J. N. Diana
Plenum Press, New York, 1990

When White Carneau pigeons are fed a diet containing cholesterol and lard, the rate of lesion formation is markedly accelerated such that at a total plasma cholesterol of 700 mg/dl (normal TPC is 240-280 mg/dl) 100% of pigeons have lesions in the aorta after 12 weeks (Jerome and Lewis, 1984). Most of the studies discussed herein used pigeons fed cholesterol, however naturally occurring lesions were used for comparison. The initial lesions which develop either naturally or with cholesterol feeding are raised foam cell lesions characterized by extensive monocyte adhesion, endothelial cell proliferation, and change in endothelial cell shape.

INITIATION

Structural studies of endothelium in pigeons were undertaken to elucidate changes which precede lesion formation. These studies examined the effect of hypercholesterolemia on the morphology of endothelial cells at lesion prone and lesion resistant sites both before and after lesions formed. The integrity of the endothelial cells glycocalyx was probed using ruthenium red (Lewis et al., 1982) and it was found that cholesterol feeding decreased glycocalyx thickness. Similar observations have been made for other animal models (Weber, 1977) suggesting that a thinner glycocalyx is a consistent feature of endothelial cells altered by hypercholesterolemia. The charge of endothelial cells, as indicated by cationized ferritin binding, also decreased with cholesterol feeding in pigeons (Lewis et al., 1982); and in rabbits (Sarphie, 1986) this was concomitant with an increase in LDL binding (Sarphie, 1986). Although these changes are systemic rather than site specific they do demonstrate that hypercholesterolemia alters endothelial cell structure.

Hypercholesterolemia also induced an alteration in endothelial cell shape in pigeon coronary arteries (Lewis et al., 1982) and aorta (Jerome and Lewis, 1984). The change was characterized by a deviation from the normal elongated morphology to cells having a more cuboidal shape. In the both the aorta and coronary this was most dramatic at sites with a high predilection for atherosclerosis. The altered shape resembled that found over lesions where endothelial cells are cuboidal and have fewer microvilli. Prelesion sites in the pig also have cuboidal shaped endothelial cells (Gerrity et al., 1977). Shape change at these sites may indicate an alteration in the endothelial cell basement membrane which prevents the cells from maintaining the normal alignment with blood flow. Examination of endothelial cell basement membrane composition with immunocytochemical techniques will be required to explore this potential relationship.

The hallmark characteristic of initial lesions in pigeons is the adhesion of monocytes to endothelium, and under all experimental conditions studied approximately 80% of adherent cells have been identified as monocytes and 60% of adherent cells are extensively spread (Jerome and Lewis, 1984). To determine if monocyte adhesion at lesion sites preceded lesion formation, two types of experiments were performed. In the first set of experiments lesion prone sites were studied in control pigeons and pigeons fed cholesterol. Cholesterol feeding did not increase monocyte adhesion to nonlesion regions in either the coronary artery or aorta suggesting that adhesion is a lesion related event not an initiating event.

Another way to examine the association of events with lesion formation is to examine the forming edge of more advanced lesions. Pigeon lesions grow rapidly from their lateral edges and there is sharp demarcation at the edge between lesion and nonlesion regions. The edge of lesions typically observed as a region approximately five endothelial cells wide where the endothelial cells have more microvilli and are cuboidal in shape. In addition to enumerating cells over the lesions and in general nonlesion areas, we have quantitated nonlesion regions adjacent to the edge i.e., regions which will become lesion in a matter of weeks. Monocyte adhesion in nonlesion regions is 10-100/mm^2 surface area as compared to 1000/mm^2 for lesions and 2000/mm^2 for the lesion edge (Lewis et al., 1985; Jerome and Lewis, 1984; Taylor and Lewis, 1986). In addition, it was found that monocyte adhesion was the same for naturally occurring and diet induced lesions. These studies demonstrated that monocyte adhesion does not increase due to hypercholesterolemia nor in prelesion sites.

The rate of endothelial cell proliferation as indicated by DNA labelling with ^{3}H-thymidine was done to determine if endothelial cell proliferation was induced by hypercholesterolemia and if proliferation was greater at prelesion sites. Quantitation was done using scanning electron microscopic autoradiography. It was found that nonlesion regions of normocholesterolemic pigeons had the same rate of proliferation as nonlesion sites in pigeons fed cholesterol-supplemented diet (Taylor and Lewis, 1986). However, the proliferation rate was increased fivefold or more at the lesion edge and over lesions. Examination of nonlesion regions adjacent to the lesion edge demonstrated that proliferation did not precede lesion initiation. Endothelial cell proliferation has been quantitated in other animal models, and in general high endothelial cell proliferation is a unique feature of lesions (Stary, 1974). This is not a new observation in that more than three decades ago Duff observed increased mitotic figures in endothelial cells associated with rabbit lesions (Duff et al., 1957). Contrary to these observations in pigeons, increased endothelial cell proliferation does occur at prelesion stages in pigs (Caplan and Schwartz, 1973), and endothelial cell proliferation in pigs is increased with cholesterol feeding (Florentin et al., 1969).

PROGRESSION

The progression of atherosclerosis in pigeons follows stages similar to that for man: The initial macrophage foam cell lesions eventually gives rise to lesions comprised of smooth muscle cells, a highly fibrous matrix and extracellular lipid. Atherosclerotic lesions in man and pigeons also develop a fibrous cap, mineralization, neovascularization, necrotic core and thrombi (Prichard et al., 1964). Our interest in the role of endothelial cells in lesion progression was investigated at the stage of transition from the macrophage foam cell stage to an early SMC proliferative lesion with increased matrix. The approach was identical to that described for leukocyte studies of initiation. We compared the endothelial cell structure, leukocyte adhesiveness and proliferation of endothelial cells associated with progressing lesions in control and cholesterol fed birds.

Naturally occurring lesions in pigeons 7 years of age cover 11% of the aortic surface compared to 32% coverage after 6 months of cholesterol feeding (St. Clair, 1983). We anticipated that if adhesion and proliferation were associated with lesion progression the two events should be much greater for cholesterol exacerbated lesions. Surprisingly, naturally occurring lesions in the aorta had the same magnitude of adhesion (Taylor and Lewis, 1986; Jerome and Lewis, 1984) and proliferation. Thus, the rapid progression of lesions following cholesterol feeding is associated with monocyte adhesion and endothelial cell proliferation, but may be initiated by factors undetectable by ultrastructural techniques. It warrants mention that the extent of adhesion and proliferation on lesions in the coronary artery is several fold less than values for the aorta (Lewis et al., 1985), and progression of coronary lesions is also much slower. In both the coronary and aorta, adhesion and proliferation could only be detected at the same time, i.e. one did not precede or cause the other.

Careful ultrastructural evaluation of pigeon endothelial cells to which monocytes were adherent failed to reveal any signs of endothelial cell necrosis or alterations (Lewis et al., 1982; Jerome and Lewis, 1985). Even when monocytes were passing through endothelial cell junctions, the adjacent cells were unaltered. This corroborates the landmark studies of Duff in rabbits and is consistent with observations made of progressing lesions in other animal models including the rat (Joris et al., 1983) and baboon (Schwartz et al., 1985) as well as man (Haust, 1971).

Pigeon lesions, both naturally occurring and cholesterol exacerbated, have a hundred fold or greater increase in cholesteryl ester content as compared to normal aorta (St. Clair, 1983). We used quantitative autoradiography to examine the short term uptake of ^{125}I-LDL in pigeons that had been fed cholesterol for one year. The purpose of these studies was to determine if the relative accumulation of LDL was greater in lesion and lesion edge sites as compared to nonlesion sites. In vitro studies have shown that proliferating endothelial cells have more LDL receptors (Vlodavsky and Gospodarowicz, 1979) and are more permeable (Constantinides, 1977). A bolus injection of ^{125}I-LDL was administered intravenously. The pigeons were sacrificed after 30 minutes and the vessels prepared for EM autoradiography. When the results were expressed as the percentage of grains beneath an equivalent number of endothelial cells it was found that 32.5% of grains were beneath lesion endothelial cells, 65% beneath lesion edge endothelial cells and 2.5% beneath nonlesion endothelial cells. These results documented that lesions are more permeable than nonlesion regions to LDL, and that LDL accumulation does not precede other lesion related events such as endothelial cell turnover or monocyte infiltration. Recent studies in the rabbit suggest that the important event is not LDL permeability per se but LDL retention (Schwenke and Carew, 1989). This suggests that an altered intimal matrix may trap more LDL.

CONCLUSIONS

Morphologic alterations in endothelium do precede lesion formation at lesion prone sites. These changes were detected as a thinning of the glycocalyx, a reduction in surface charge and a change in shape. Perhaps more significant is that initiation is not preceded by endothelial cell necrosis, LDL accumulation, monocyte adhesion, or endothelial proliferation. Many of these changes are an important feature of lesion progression especially monocyte adhesion, endothelial cell proliferation and LDL accumulation. It is also significant that naturally occurring lesions do not differ from cholesterol exacerbated lesions with regards to monocyte adhesion, endothelial cell proliferation and general ultrastructural features.

REFERENCES

Caplan, B.A, and Schwartz, C.J., 1973, Increased endothelial cell turnover in areas of in vivo Evans Blue uptake in the pig aorta, Atherosclerosis 17:401.

Constantinides, P., 1977, The morphological basis for altered endothelial permeability in atherosclerosis, Adv. Exp. Biol. Med. 82:969.

Duff, G.L., McMillan, G.C., and Ritchie, A.C., 1957, The morphology of early atherosclerotic lesions of the aorta demonstrated by the surface technique in rabbits fed cholesterol, Am. J. Pathol. 33:845.

Engelberg, H., 1989, Endothelium in health and disease, Sem. Thrombos. Hemostas. 15:178.

Florentin, R.A., Nam, S.C., Lee, K.T., Lee, K.J., and Thomas, W.A., 1969, Increased mitotic activity in aortas of swine after three days of cholesterol feeding, Arch. Path. 88:463.

Gerrity, R.G., Richardson, M., Somer, J.B., Bell, F.P., and Schwartz, C.J., 1977, Endothelial cell morphology in areas of in vivo Evans Blue uptake in the aorta of young pigs, Am. J. Pathol. 89:313.

Haust, M.D., 1971, The morphogenesis and fate of potential and early atherosclerotic lesions in man, Human Path. 2:1.

Jerome, W.G., and Lewis, J.C., 1984, Early atherogenesis in White Carneau pigeons. I. Leukocyte margination and endothelial alterations at the celiac bifurcation. Am. J. Pathol. 116:56.

Jerome, W.G., and Lewis, J.C., 1985, Early atherogenesis in White Carneau pigeons. II. Ultrastructural and cytochemical observations, Am. J. Pathol. 119:210.

Joris, E., Zand, T., Nunnari, J.J., Krolikowski, E.J., and Majno, G., 1983, Studies on the pathogenesis of atherosclerosis. I. Adhesion and emigration of mononuclear cells in the aorta of hypercholesterolemic rats, Am. J. Pathol. 113:341.

Lewis, J.C., Taylor, R.G., Jones, N.D., St. Clair, R.W., Cornhill, J.F., 1982, Endothelial surface characteristics in pigeon coronary artery atherosclerosis. I. Cellular alterations during the initial stages of dietary cholesterol challenge, Lab. Invest. 46:123.

Lewis, J.C., Taylor, R.G., and Jerome, W.G., 1985, Foam cell characteristics in coronary arteries and aortas of White Carneau pigeons with moderate hypercholesterolemia, Ann. N.Y. Acad. Sci. 454:91.

Prichard, R.W., Clarkson, T.B., Lofland, H.B., and Goodman, H.O., 1963, Myocardial infarcts in pigeons, Am. J. Pathol. 43:651.

Prichard, R.W., Clarkson, T.B., Goodman, H.O., and Lofland, H.B., 1964, Aortic atherosclerosis in pigeons and its complications, Arch. Pathol. 77:244.

Reidy, M.A., 1985, Biology of Disease. A reassessment of endothelial injury and arterial lesion formation, Lab. Invest. 53:513.

Sarphie, T.G., 1986, A cytochemical study of the surface properties of aortic and mitral valve endothelium from hypercholesterolemic rabbits, Exp. Molec. Pathol. 44:281.

Schwartz, C.J., Sprague, E.A., Kelley, J.L., Valente, A.J., and Suenram, C.A., 1985, Aortic intimal monocyte recruitment in the normo and hypercholesterolemic baboon (Papio cynocephalus). An ultrastructural study: Implications in atherogenesis, Virchows Arch [Pathol Anat] 405:175.

Schwenke, D.C., and Carew, T.E., 1989, Initiation of atherosclerotic lesions in cholesterol-fed rabbits. II. Selective retention of LDL vs. selective increases in LDL permeability in susceptible sites of arteries, Arteriosclerosis 9:908.

St. Clair, R.W., 1983, Metabolic changes in the arterial wall associated with atherosclerosis in the pigeon, Fed. Proc. 42:2480.

Stary, II.C., 1974, Proliferation of arterial cells in atherosclerosis, Adv. Exp. Med. Biol. 43:59.

Taylor, R.G., and Lewis, J.C., 1986, Endothelial cell proliferation and monocyte adhesion to atherosclerotic lesions of White Carneau pigeons, Am. J. Pathol. 125:152.

Vlodavsky, I., and Gospodarowicz, D., 1979, Structural and functional alterations in the surface of vascular endothelial cells associated with the formation of a confluent cell monolayer and with the withdrawal of fibroblast growth factor, J. Supramolec. Struct. 12:73.

Weber, G., 1977, The influence of hypercholesterolemia upon endothelial glycocalyx. Adv. Exp. Med. Biol. 82:975.

THE EFFECT OF CIGARETTE SMOKE, NICOTINE AND CARBON MONOXIDE ON ARTERIAL

WALL PERMEABILITY AND ARTERIAL WALL UPTAKE OF ^{125}I-FIBRINOGEN

D.R. Allen and N.L. Browse

Department of Surgery
St. Thomas' Hospital
London, England

Atherosclerosis is essentially a disease of humans in which lipid and fibrinogen accumulate in the subintimal layers of large elastic arteries and medium sized muscular arteries. Although not a new disease, the prevalence and importance of atherosclerosis in the Western World have increased dramatically this century; a change, which coincides with an increase in the usage of cigarettes.

To the vascular surgeon the risks of smoking are clear. In 1962, Eastcott noted the rarity of lower limb ischaemia in non-smokers (1) and more recently Laing and Greenhalgh have found that over 97% of patients presenting with peripheral vascular disease are smokers (2). The association between cigarette smoking and the development of atherosclerosis, in particular peripheral vascular disease, is strong; however, the mechanisms which result in cigarettes being such a potent risk factor are by no means clear.

Ross and Glomsett (3,4) have put forward the "response to injury" hypothesis, which links endothelial damage and smooth muscle proliferation to the incorporation of lipids and fibrin into the arterial wall. If this hypothesis is correct, a likely target on which cigarette smoke could exert its effect is on the endothelium.

Woolf has shown that after a single brief exposure to cigarette smoke, endothelial cells become swollen and that longer exposure causes the appearance of sharply demarcated holes which may be the stomata of active pinocytotic vesicles; however, the endothelium maintained its integrity (5). There is strong evidence from both human and animal studies that, elsewhere in the body, cigarette smoke increases epithelial permeability; Mason et al. and Jones et al. have used technetium labelled diethylene triamine penta acetate to measure clearance from the lungs of smokers and non-smokers. They found that the clearance rate was up to five times greater in smokers than in non-smokers and that this change was reversible after stopping smoking for one week (6,7). Interestingly this increase in permeability could not be reproduced by administering nicotine alone (8).

It is therefore likely that cigarette smoke alters endothelial cells and that these changes may affect the permeability of the endothelium. If this is the case, then it is possible that such a change in the

Tobacco Smoking and Atherosclerosis
Edited by J. N. Diana
Plenum Press, New York, 1990

permeability of the arterial wall would result in an influx of lipids and
other plasma macromolecules into the arterial wall.

Cigarette smoke contains over 4,000 constituents but of these,
nicotine and carbon monoxide have attracted most interest and have been
most extensively investigated.

Plasma nicotine levels in cigarette smokers may increase to a level
between 4 and 70 ng/ml (9). Nicotine is known to have a desquamating
effect on endothelium (10,11). It is also known to increase heart rate
and blood pressure and to have a vasoconstrictive effect on small blood
vessels producing a 42% reduction in blood velocity in fingers after
smoking one cigarette (12).

Nicotine could also be implicated in the development of atherosclerosis through several of its known actions on the cardiovascular
system in that hypertension and raised plasma cholesterol are both
associated with atheroma. Although the pharmacological effects of
nicotine are both small and transient, in smokers, they are repeated many
times each day. It would seem therefore that nicotine has many potential
mechanisms through which it could exert an atherogenic effect.

When diluted by air during inhalation, the concentration of carbon
monoxide (CO) in cigarette smoke is approximately 400 ppm. Atmospheric
pollution, largely from petrol combustion, results in COHb levels of
between 1.4% and 3% in non-smoking taxi drivers in London (13). COHb
levels above 3% are indicative that the individual smokes tobacco, and
indeed there is a correlation between the number of cigarettes smoked and
COHb levels, which can rise to over 12% in heavy smokers. In a study of
950 Danish smokers, subjects with a COHb level of 5% or more had a 21
times greater incidence of ischaemic heart disease than those with a COHb
level of less than 5% (14). Despite these clinical correlations, animal
experiments have failed to demonstrate an atherogenic effect of CO when
given alone.

Ross has suggested that at least two pathways may lead to intimal
smooth muscle proliferation, which he still considers to be the key event
in the development of atheroma (4). One mechanism involves monocyte and
platelet interactions with an intact endothelium, resulting in the
release of growth factors to promote smooth muscle proliferation. A
second pathway involves direct stimulation of the endothelial cells,
which may themselves release a growth factor responsible for smooth
muscle proliferation or a chemotactic factor causing the migration of
smooth muscle cells into the intima. The key to both hypotheses is the
evidence that the endothelium can be damaged without desquamation or cell
loss and that this damage might alter the permeability of the endothelial
barrier.

To investigate arterial wall permeability, the Department of Surgery
at St. Thomas' Hospital, London, England, has developed a polystyrene
capsule that can be placed around the femoral artery of an anaesthetised
dog. The capsule has been used to investigate the effects of cigarette
smoke, CO, and nicotine on arterial wall permeability to 125I-fibrinogen.

AN INVESTIGATION OF ARTERIAL WALL PERMEABILITY

The polystyrene capsule is 2 cm long and is injection-moulded in two
parts. At each end of each half is a semicircular channel. After
applying polystyrene cement to the edge of each half, the bottom half of
the capsule is placed beneath the artery and the upper half is aligned
over the bottom half of the capsule using locating pins and is held in

place with modified artery forceps until the cement has set. When the
two halves of the capsule are together, the semicircular grooves in each
half form a circular tunnel which is 4 mm in diameter at the proximal end
of the capsule and 3 mm at the distal end. The tunnels are 3 mm long and
thus "grip" the artery sufficiently at each end of the capsule to form a
watertight seal without causing undue compression of the vessel. If
necessary, larger vessels (6 mm diameter or larger) can be accommodated
by 'reaming' out the tunnels prior to placement of the capsule. In the
top half of the capsule are two ports through which silicone tubing
(external diameter 2 mm) is pulled. The tubing in the proximal port is
connected to a calibrated 2-ml syringe filled with warmed (37 C) Ringer's
solution. The tubing in the distal port is left open to the atmosphere
and thus acts as an air vent. The capsule can therefore be filled with
Ringer's solution, which bathes the entire outer surface of the artery
that lies within the capsule.

Once in place, the capsule provides a means of collecting molecules
arriving at the adventitia after they have crossed the arterial wall. To
measure permeability, the 125 I-human fibrinogen was injected
intravenously and the rate at which it crossed the arterial wall was
recorded by counting the radioactivity in the Ringer's solution (Figure
1).

One vial of 125I-fibrinogen (1.2mg 4.1MBq) was injected intravenously
via a peripheral vein after filling each arterial capsule with Ringer's
solution. Measurement of the quantity of labelled fibrinogen that had
crossed the arterial wall was made over a period of 3 hours by
withdrawing the Ringer's solution at 30 minute intervals with a 2 ml
syringe attached to the proximal silastic tubing. The Ringer's was
transferred to a plastic test tube and placed in a well scintillation
counter where the counts for 300 seconds were recorded using a
counter/ratemeter. Two recordings were made for each sample. The
capsule was refilled immediately with a fresh volume of Ringer's using a
fresh syringe. The exact volume used to refill the capsule and the
volume subsequently emptied 30 minutes later were recorded.

The plasma radioactivity of arterial blood was also measured every 30
minutes with 10 ml of arterial blood withdrawn from the arterial line and
placed in a lithium heparin bottle. After centrifugation for 10 minutes,
2 mls of supernatant plasma was placed in a plastic test tube and the
radioactivity counted for 100 seconds in the well counter. Two
recordings are made for each sample. Background readings from the well
counter were made before and after each experiment.

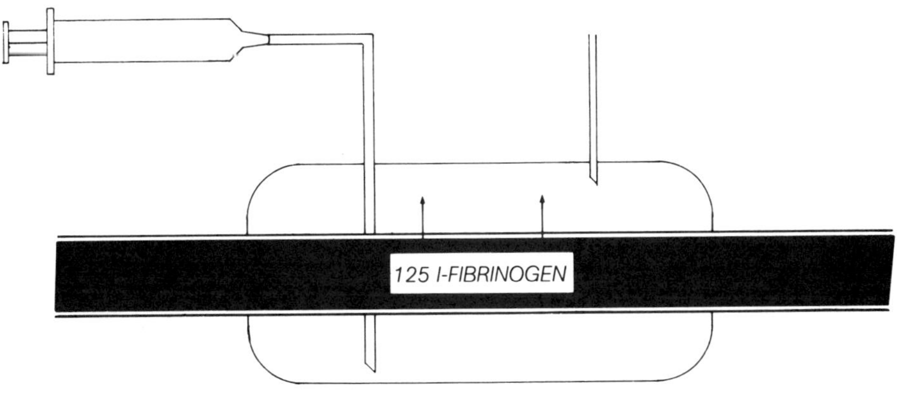

Fig. 1

Calculation of the permeability of the arterial wall:

A measurement of permeability has been called the permeability coefficient (PC) and it is calculated as follows:

Ringer's solution: after subtraction of the background count, for each 30 minute period, the count in the Ringer's solution was converted to counts per ml per minute per square centimetre of artery contained within the capsule, using the following equation:

$$C \text{ Ringer's} = \frac{C \times 60 \times 1 \times 1}{300 \times V \times SA}$$

 C = actual counts of Ringer's solution in 300 seconds minus background.
 V = volume of Ringer's solution counted.
SA = surface area of the artery (π x radius cm x intracapsular length cm).

Plasma: The plasma count is corrected to counts per ml per minute after subtraction of the background count, using the following equation:

$$C \text{ plasma} = \frac{C \times 60 \times 1}{100 \times 2}$$

The permeability coefficient (PC): is then derived using the equation:

$$PC \text{ (units. cm-2)} = \frac{C \text{ Ringer's}}{C \text{ plasma}}$$

<u>Methods</u>

All dogs were anaesthetized using intravenous sodium pentobarbitone (30 mg/kg) and anaesthesia was maintained with 120 mg "top ups" as required.

N/Saline was given at a rate of 200 ml/hour through an intravenous line in a cephalic vein. The central venous pressure and arterial pressure were recorded every 30 minutes throughout the experiment using a pressure transducer. A cuffed endotracheal tube (size 7.5) was inserted and all dogs were allowed to breathe spontaneously.

Incisions were made in both groins extending from the inguinal ligament to a point below the proximal caudal femoral artery. All branches of the artery were tied with 6/0 silk. The diameter of the artery was measured with calipers at a point which was estimated would lie in the middle of the arterial capsule, following which the capsule was placed around the artery. The capsule was filled with Ringer's solution and, after intravenous injection of the fibrinogen, permeability measurements were commenced.

At the end of the experiments all dogs were killed using intravenous, saturated potassium chloride.

<u>Cigarette Smoke Inhalation Studies</u>

The cuffed portex endotracheal was connected to an "Ambu valve." A polystyrene and metal adaptor which held a cigarette was inserted into the inlet side of the valve. The dogs breathed spontaneously inhaling cigarette smoke through the inlet of the valve and exhaling it through the outlet of the valve (Figure 2).

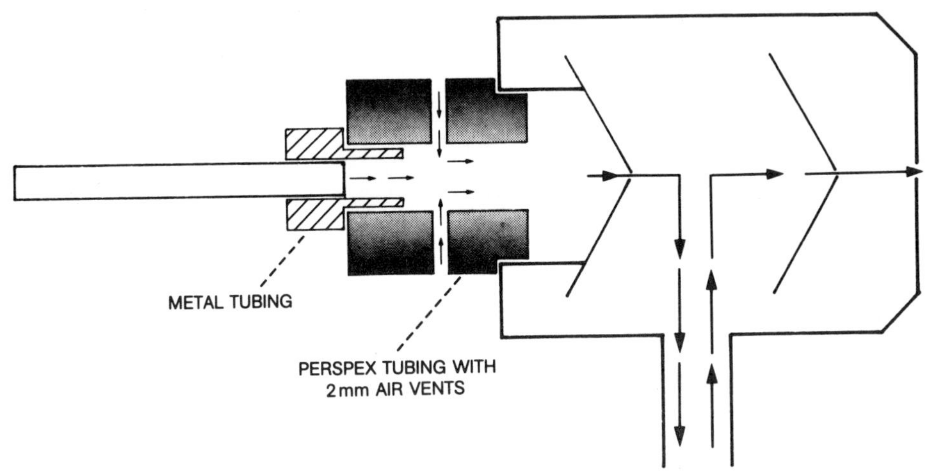

Fig. 2

The dogs inhaled the smoke from 4 cigarettes an hour for two hours,
at which point, labelled fibrinogen was injected and permeability
measurements were started. The dogs then continued to inhale smoke
throughout the three-hour permeability experiment and had therefore
consumed 20 cigarettes at the end of the five-hour period. The
cigarettes were high tar and high nicotine content.

Arterial blood samples were taken in a heparinized 5 ml syringe every
30 minutes throughout the experiment. The samples were then analyzed for
% COHb using a CO Oximeter analyzer adapted to measure dog blood.

Nicotine Studies

An intravenous infusion of nicotine was given at a rate 2.5
ug/kg/min. This was infused for 90 minutes before commencing
permcability measurements and then continued throughout the three-hour
experiment. Plasma nicotine levels were measured 90 minutes and 270
minutes after starting the infusion. Measurement of nicotine levels was
performed using a flame ionizing gas chromatography technique with
nitrogen sensitive detectors, which produce a sensitivity of 0.1 ng/ml of
nicotine in plasma (15).

Carbon Monoxide Studies

Carbon monoxide in a concentration of 400 ppm was inhaled through the
Ambu valve. Inhalation was commenced 2 hours prior to injection of the
fibrinogen and then continued throughout the experiment. Blood COHb
levels were recorded before inhalation and then at 30 minute intervals
during the 3 hour experiment.

Results

There was no significant difference between the animals used in each
of the four experiments in terms of weight, size of arteries and
anaesthetic requirements.

The permeability of 10 arteries was measured in the control group; 11
arteries were investigated in the smoke inhalation group and 8 arteries
in both the carbon monoxide group in the nicotine infusion group.

There was no significant difference between the arterial pressure
recordings in the control, smoke inhalation or CO inhalation groups.
There was a significant increase in both arterial pressure and pulse rate
in the nicotine infusion group.

In the smoke inhalation group the mean carboxyhaemoglobin rose from a
resting level of 1.5% ± 0.1% to a mean level of 7.4% ± 0.9% after
inhaling the smoke from 8 cigarettes over a period of 2 hours. At the
end of the five-hour experiment the mean carboxyhaemoglobin level had
risen to 12.8% ± 0.8%.

Levels of 47.6 ng/ml and 54 ng/ml were recorded at the end of the
experiment in two of the dogs in the smoke inhalation group. These
levels are within the range found in human smokers (16).

In the carbon monoxide group, the COHb rose from a resting level of
2.0% ± 0.2% to 10.0% ± 0.8% after inhaling CO for 2 hours. At the end of
the experiment the mean level had risen to 19.8% ± 1.4%.

In the nicotine infusion group the mean plasma nicotine level at 90
minutes was 59.2 ± 12.5 ng/ml and at 270 minutes was 48.8 ± 18 ng/ml.

The permeability coefficients for the four groups are shown in Figure
3. Statistical analysis has been made using the Mann Whitney U test.

<u>Discussion of the Method</u>

The animal model:

The animals used were ex-racing greyhounds. Although dogs are not
normally prone to developing atheroma, the anatomy of their large elastic
and muscular arteries has been shown to be similar to man's with a
similar distribution of the vasa vasorum (17) and it was therefore felt
that this was a valid model for measurement of arterial permeability.

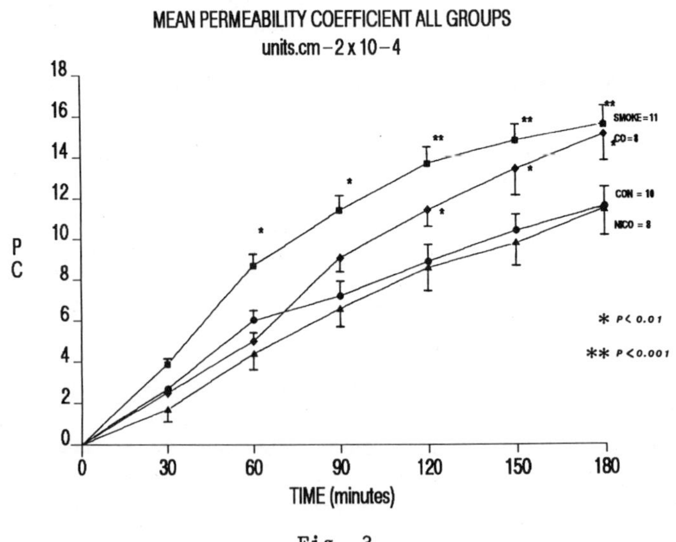

Fig. 3

The isotope:

125 I-fibrinogen was chosen because it is a large molecule (MW
340,000), which is found as a component of fibrous and complicated

atheromatous plaques. The molecule is available with the isotope 125 I
attached and this isotope has a long half life (60 days).

The capsule:

The aim of using this technique was to study the permeability of a
length of arterial wall under as near to physiological conditions as
possible. The capsule itself is likely to disturb the artery for a
number of reasons:

  (a) the artery has to be dissected from surrounding tissues to
       allow placement of the capsule.
  (b) in order to produce a watertight seal around the artery at
       each end of the capsule, the tunnels must "grip" the artery.
       This will lead to a slight stenosis of the artery at each
       tunnel and also the arterial wall itself may be damaged from
       pressure exerted by the tunnels.

Histology of the intra-capsular artery revealed no evidence of damage
to the arterial wall. The 3-mm length of artery lying within the tunnel
did have some changes indicating damage of the arterial wall. However,
this piece of artery was not bathed by Ringer's solution and therefore
did not contribute to the transmural transport of labelled fibrinogen.
It must be accepted, however, that artery is disturbed and this must be a
limitation of the technique.

To be certain that the 125 I-fibrinogen is actually crossing the full
thickness of the arterial wall rather than leaking out from the vasa
vasorum, autoradiography of the intracapsular arterial wall was
performed. After completion of the permeability experiment, two arteries
were washed in saline to remove adherent blood. The adventitia, media
and intima were carefully separated with sharp dissection and the three
components of the arterial wall were stored on an X ray plate at - 70 C
for six weeks. The plates showed radioactivity in all three components,
suggesting that the labelled fibrinogen has crossed the full thickness of
the arterial wall.

To ensure that the radioactivity measured in the Ringer's solution
was attached to fibrinogen molecules, electrophoresis of the Ringer's
solution was performed after the Ringer's had been concentrated times
twenty using an Amicon filter. The concentrated Ringer's was then placed
on a cellulose acetate strip with a diluted solution of 125 I-fibrinogen
from the original vial as the control. Following electrophoresis, the
acetate strip was stained for fibrinogen and both the control solution
and the test solutions had moved the same distance from the origin.

Discussion of the Results

The aim of these experiments was to investigate the effects of
cigarette smoke, nicotine and carbon monoxide, on the permeability of the
arterial wall to 125 I-fibrinogen. There is normally a permeation of
plasma macromolecules through the arterial wall (18) and the permeability
of the arterial wall is increased if the endothelium is damaged with or
without desquamation of the endothelial cells (19). We have used an in
vivo animal model to test the hypothesis that cigarette smoke increases
the permeability of the arterial wall thinking that this might, in the

long-term, result in the deposition of fibrinogen and lipids in the
intima. The model has also been used to investigate the effects of
nicotine and carbon monoxide on arterial wall permeability. Both the
smoking and carbon monoxide groups have a significant increase in
permeability over the control group and there is no significant
difference between the control group and the nicotine group. It would
appear that cigarette smoke does increase the flux of fibrinogen across
the arterial wall and this may be one of the mechanisms by which
cigarettes exert their atherogenic effect.

Histological examination of arteries within the capsule was performed
in all groups. Light microscopy and transmission electron microscopy did
not reveal any signs of damage to the endothelial cells. The short
duration of exposure of the arteries to cigarette smoke, carbon monoxide
or nicotine may mean that morphological changes have not had time to
develop. However, the permeability of the arterial wall is increased
after only a few hours exposure to cigarette smoke and carbon monoxide.
The mechanism producing this change is not clear but it seems likely that
it is occurring at a sub cellular level to interfere with transmural
transport.

Another mechanism through which atherogenic risk factors might act on
the arterial wall is to impair the clearance of insudated molecules from
the artery. To test this hypothesis, we have used the same animal model
to investigate the effects of cigarette smoke, nicotine and carbon
monoxide on the amount of fibrinogen that is retained within the arterial
wall of carotid arteries.

AN INVESTIGATION OF ARTERIAL WALL UPTAKE OF 125 I-FIBRINOGEN

<u>Methods</u>

Measurements of arterial wall uptake of fibrinogen:

One vial of 125 I-fibrinogen (1.2 mg 4.1 MBq), was injected
intravenously and the time recorded as t = 0. Plasma radioactivity was
measured 15 minutes after the injection of the fibrinogen and thereafter
every 30 minutes until time t = 195 minutes.

The results from the seven plasma samples collected during the
three-hour experiment were plotted against time and the linear regression
of the "biological decay" of the injected fibrinogen calculated. The
estimated plasma radioactivity for the time t = 0 was calculated from the
regression of analysis using the equation y = mx + c:

        where: y = plasma value
                m = slope of line
                x = time
                c = y axis intercept

Three hours after injection of the fibrinogen, a midline incision was
made in the neck and both carotid arteries were exposed. A 3 cm length
of artery was marked at each end and the diameter of the midpoint was
measured using calipers. Ligatures were placed at the proximal and
distal ends of the artery and the marked length was excised. The artery
was opened with a longitudinal incision and pinned to a polythene board,
intima facing down. The adventitia was dissected off the media using
scissors and was discarded. The remaining artery was then washed in
eight ice cold saline washes to remove any adherent blood. The artery
wall was snap frozen to -70°C in solid carbon dioxide. The then brittle
sheet of artery was placed in a teflon cup containing a stainless steel

ball bearing. The cup was clamped onto the vibrating cage of a
Mikro-Dismembrator, which shook the teflon cup at high frequency for two
minutes. The steel ball within the cup shattered the arterial sheet into
tiny fragments which were suspended in 2 ml of saline. The radioactivity
of the suspension was counted in a well-scintillation counter for 300
seconds: two counts were made for each sample.

The count was converted to counts. cm-2. min-1 using the equation:

$$\text{C Wall (counts. cm-2 min-1} = \frac{C \times 60 \times 1}{300 \times SA}$$

where SA = surface area of the artery derived from the 'in vivo'
measured length and diameter.

Wall uptake of 125 I-fibrinogen was calculated as:

$$\text{Wall uptake} = \frac{\text{C Wall (counts. cm-2. min-1)}}{\text{C Plasma @ t = 0 (counts. ml-1. min-1)}}$$

There were four groups:

  i) A control group of 6 arteries.
 ii) A smoke inhalation group of 7 arteries.
iii) A carbon monoxide inhalation group of 8 arteries.
 iv) A nicotine infusion group of 8 arteries.

The methods for smoke inhalation, carbon monoxide inhalation and the
nicotine infusion were identical to those used in the permeability
experiments.

RESULTS

There was no significant difference between the four groups with
respect to weight or anaesthetic requirements. The carbon monoxide and
nicotine groups had a higher mean arterial pressure throughout the
experiment when compared to the control and smoking groups and these
differences were significant.

Carboxyhaemoglobin levels

The mean resting levels of blood COHb were 1.9% ± 0.2% in the control
group, 1.8% ± 0.3% in the nicotine group, 2.0% ± 0.6% in the smoke
inhalation group and 2.0% ± 0.4% in the CO group. There was no
significant difference between these resting levels.

In the smoke inhalation group, the plasma COHb had risen to 6.4% ±
0.3 when the fibrinogen was injected and was 13.5% ± 0.8 at the end of
the experiment. In the carbon monoxide group the COHb was 9.3% ± 0.5
after inhaling CO for two hours and 18.2% ± 0.9 at the end of the
experiment.

Nicotine levels

The mean plasma nicotine levels in the smoking group at the end of
the experiment was 50.8 ng/ml ± 4.5 ng/ml. In the nicotine group, the
mean plasma nicotine level 90 minutes after commencing the nicotine
infusion was 59.2 ng/ml ± 12.5 ng/ml and at the end of the experiment the
level was 48.0 ng/ml ± 18.0 ng/ml. There is no significant difference
between the mean nicotine levels at the end of the experiment in the
smoking and the nicotine groups.

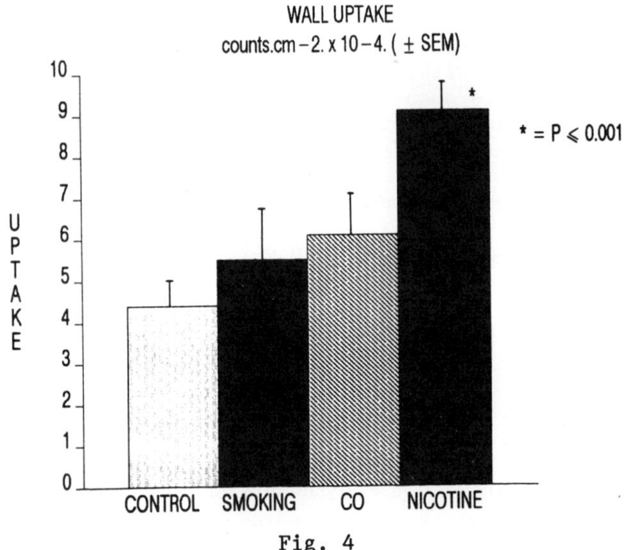

Fig. 4

Arterial wall uptake of 125 I-fibrinogen:

The mean arterial wall uptake of 125 I-fibrinogen in units cm-2 x 10-4 was 4.4 ± 1.4 in the control group, 5.5 ± 2.8 in the smoking group, 6.2 ± 2.7 in the CO group and 9.1 ± 2.1 in the nicotine group. These results are shown in the form of a histogram in Figure 4. There is a slight, but insignificant difference between the smoking and CO groups when compared to the control group but a highly significant increase in the uptake of fibrinogen in the nicotine group (p < 0.001) using the Mann-Whitney U test.

Discussion

Any conclusions made from the investigation of arterial permeability to and of arterial wall uptake of fibrinogen should be interpreted carefully with regards to the effects of long-term exposure to cigarettes in humans. However, because of the time span over which arterial disease develops and the difficulties in separating the various aetiological mechanisms, long-term studies of the effects of the individual 'risk factors' are not practicable and it is therefore necessary to accept the limitations of short-term animal experiments.

If there is a constant insudation of plasma macromolecules into the arterial wall, it is likely that there are clearance mechanisms to remove these molecules from the arterial wall. The walls of muscular arteries are well supplied with vasa vasorum and lymphatics (20). Elsewhere in the body, the role of the lymphatics is to clear large molecules from the interstitial fluid (21) and therefore it is not unreasonable to suggest that they might play a similar role in the arterial wall, although evidence to support this hypothesis is scanty. The vasa vasorum are thought to be responsible for providing the nutrition of the outer layers of the artery wall, but the vena vasorum may serve a clearance role. It is possible that risk factors associated with the development of atheroma might upset the balance between the ingress of plasma macromolecules and their clearance from the arterial wall thus resulting in the deposition of lipids and fibrinogen in the intima.

This experiment was designed to investigate whether or not short-term exposure to cigarette smoke would increase the amount of fibrinogen that is retained within the arterial wall when compared to a control group.

In these experiments, cigarette smoke and carbon monoxide produced a marginal but insignificant increase in the retention of fibrinogen by the arterial wall, but nicotine alone produced a significant increase in wall uptake. The plasma nicotine levels in the nicotine group, and the blood COHb of the CO group were similar to the levels found in the smoking group, which, in turn, were both within the range found in human smokers.

The different effect of nicotine alone is difficult to explain. One explanation is that plasma nicotine levels in the smoking group are likely to have fluctuated considerably between cigarettes because of the short half life of nicotine (5-10 minutes) when tissue saturation has not occurred, whereas in the nicotine infusion group, the plasma nicotine level would have been continuously elevated.

The mechanism through which nicotine might increase the uptake of fibrinogen by the artery is not clear, particularly as we have shown that nicotine does not increase the permeability of the femoral artery. However, the two experiments were performed under different conditions in that the permeability experiment required dissection of the artery and placement of a rigid capsule around the vessel before injecting the fibrinogen whereas in the wall uptake study, the carotid arteries were not disturbed until the end of the experiment. The capsule undoubtedly had an effect on the femoral arteries and this is reflected in the fact that retention of fibrinogen in these arteries was increased by up to 50% when compared to the present study of the carotid arteries. If nicotine does not increase the rate at which 125 I-fibrinogen enters the arterial wall, then an alternative explanation for the increase in wall uptake of fibrinogen by arteries in the nicotine group might lie in the clearance mechanisms of the arterial wall. Nicotine is known to be a potent constrictor of small vessels; if the small vessels of the arterial wall, the vasa vasorum, were similarly constricted by nicotine, and if the vasa do serve as a clearance mechanism for the arterial wall, such vasospasm is likely to reduce the rate of clearance of molecules such as fibrinogen, thus resulting in their accumulation within the arterial wall. If this is the case, it is possible that both the carbon monoxide and the nicotine in cigarette smoke might each produce an atherogenic effect through different pathways - CO affecting permeability and nicotine affecting clearance.

REFERENCES

1. Eastcott HHG. Rarity of lower limb ischaemia in nonsmokers. Lancet, 2:1117 (1962).
2. Laing SP, Greenhalgh RM, Taylor GW. The prevalence of cigarette smoking in patients with arterial disease. In: _Smoking and Arterial Disease_, (Greenhalgh RM, ed.) London, Pitman Medical, 1-3 (1981).
3. Ross R Glomset JA. The pathogenesis of atherosclerosis. N Engl J Med, 295:369-76, 420-5 (1976).
4. Ross R. The pathogenesis of atherosclerosis - an update. N Engl J Med, 314:488-500 (1986).
5. Woolf N. and Wilson Holt NJ, Cigarette smoking and atherosclerosis. In: _Smoking and Arterial Disease_, (Greenhalgh RM, ed), London, Pitman Medical, 46-59 (1981).
6. Mason GR, Uszler JM, Effros RM, Reid E. Rapidly reversible alterations of pulmonary epithelial permeability induced by smoking. Chest, 1:6-11 (1983).

7. Jones JG, Lawler P, Crawley JCW, Minty BD, Hulards G, Veal N. Increased alveolar epithelial permeability in cigarette smokers. Lancet 1:66–68 (1980).

8. Minty BD, Royston D, Jones JG, Hulands GH. The effect of nicotine on nicotine on pulmonary epithelial permeability in man. Chest 72–86 (1984).

9. Russel MAH, Jarvis M, Iyer R, Feyerabend C. Relation of nicotine yield of cigarettes to blood nicotine concentrations in smokers. Br Med J, 3:972–6 (1980).

10. Hladovec J. Endothelial injury by nicotine and its prevention. Experimentia, s34:1585–6 (1978).

11. Booyse FM, Osikowiez G, Quarfoot AJ. Effects of chronic oral consumption of nicotine on the rabbit aortic endothelium. Am J Pathol, 102:229–238 (1981).

12. Sarin CL, Austin JC, and Nickel WO. Effects of smoking on digital blood flow velocity, J Am Med Assoc, 229:1327 (1974).

13. Jones RD, Commins BT, Cernick AA. Blood lead and carboxyhaemoglobin levels in London taxi drivers. Lancet, 2:302–3 (1972).

14. Wald N, Howard S, Smithe PG, Kjeldsen K. Association between atherosclerotic diseases and carboxyhaemglobin levels in tobacco smokers, Br Med J, 1:761–5 (1973).

15. Feyerabend C, and Russell MAH. Improved gas–chromatographic method and microextraction technique for the measurement of nicotine in biological fluids. J Pharm Pharmacol, 31:73 (1979).

16. Issac PF, and Rand MJ. Cigarette smoking and plasma levels of nicotine. Nature, 236:308 (1972).

17. Wolinsky H. and Glagov S. Nature of species differences in the media distribution of aortic vasa vasorum in mammals. Circ Res, 20:409 (1967).

18. Smith EB and Slater RS. The relationship between low density lipoprotein in aortic intima and serum lipid levels, Lancet, 1:463 (1972).

19. Bondjers G, Bjorkerud S. Arterial repair and atherosclerosis after mechanical injury (part 3). Cholesterol accumulation and removal in morphologically defined regions of aortic atherosclerotic lesions in the rabbit. Atherosclerosis, 17:85–94 (1973).

20. Johnson RA. Lymphatics of the blood vessels. Lymphology, 2:44 (1969).

21. Yoffey JM and Courtice FC. Lymphatics, Lymph and Lymphomyeloid Complex. Academic Press Inc., London (1970).

SOME ACUTE EFFECTS OF SMOKING ON

ENDOTHELIAL CELLS AND PLATELETS

James W. Davis

VA Medical Center
Kansas City, Missouri and
Department of Medicine
University of Kansas
Kansas City, Kansas

INTRODUCTION

Cigarette smoking is well established as a risk factor for atherosclerosis,[1] occlusive peripheral vascular disease[2] and coronary artery
disease.[2,3] Endothelial damage and platelet activation appear to be among
the more important mechanisms involved in the pathogenesis of atherosclerosis and arterial thrombosis.[4,5] Studies done more than a decade ago using
three different methods in five different laboratories were reported to
show acute enhancement of platelet aggregation by cigarette smoking.[6-10]
Subsequently, two reports indicated no acute effect of cigarette smoking on
platelet aggregation in the platelet-rich plasma of small groups of subjects
whose duration of abstinence from tobacco before experimental smoking was
not stated.[11,12] Hladovec and Rossmann[13] described a method for the
isolation of anuclear carcasses of endothelial cells from blood. The cells
presumably lost their nuclei when they detached from vessel walls. We[14]
showed that the carcasses fluoresced after incubation with fluorescein-
labeled anti-factor VIII related antigen antibody, while simultaneously
incubated sections of skin showed no fluorescence of the epithelium, tending
to confirm an endothelial origin of the anuclear carcasses isolated from
blood. Prerovský and Hladovec[15] reported that counts of the anuclear
carcasses of endothelial cells increased after volunteers smoked two
cigarettes. Hladovec[16] described an increase in circulating anuclear
carcasses of endothelial cells in rats after administration of nicotine
suggesting a potential role of this compound as a mediator of the smoking-
induced increase in endothelial cell count (ECC). In this chapter I will
summarize some of our recent work which shows acute effects of both active
and passive smoking on the endothelium and platelets of various groups of
subjects.

METHODS

Study Designs

Each study utilized 20-minute experimental periods preceded by 12 hours
of abstinence from tobacco. Nonsteroidal anti-inflammatory agents and
dipyridamole were forbidden from 10 days before the first experimental
period of each study until its completion except when prescribed by the

Tobacco Smoking and Atherosclerosis
Edited by J. N. Diana
Plenum Press, New York, 1990

study protocol. Two cigarettes were smoked during each experimental smoking period. Antecubital venous blood was obtained before and after each experimental period.

Endothelial Cell Counts (ECC)

Counts of circulating anuclear carcasses of endothelial cells were determined by the method of Hladovec and Rossmann.[13] Nine ml of venous blood were collected in a siliconized glass centrifuge tube containing 1 ml of 3.8% trisodium citrate solution and mixed. Centrifugation at 395 x g (middle of tube) for 20 minutes removed the erythrocytes and leukocytes. One ml of the supernatant was mixed with 0.2 ml of adenosine-5'-diphosphate, disodium salt (1 mg/ml) and mechanically shaken for 10 minutes. Another centrifugation at 395 x g for 20 minutes removed the platelet aggregates. The supernatant was then centrifuged at 2,100 x g for 20 minutes. The resulting sediment was suspended in 0.1 ml of physiologic saline by stirring with a siliconized glass rod. Four Neubauer chambers were filled with the suspension and the endothelial cells counted using phase-contrast microscopy. Results were expressed as the mean cell count of the four 0.9-μl chambers.

Plasma Nicotine

A portion of the platelet-rich plasma prepared in the first centrifugation in the procedure for endothelial cell counting was prepared for gas chromatography by the method of Feyerabend and Russell.[17] Chromatography was done with an Aerograph 1400 (Varian Instruments Division) equipped with a nitrogen-phosphorus detector. The column temperature was 150°C. Its length was 3 feet. Means of duplicate assays of each sample of plasma were used for statistical analysis.

Carboxyhemoglobin

A heparinized syringe was used to obtain blood for determination of carboxyhemoglobin by a spectrophotometric method using an I1-282 CO-oximeter (Instrumentation Laboratories).

Platelet Aggregate Ratio

A modification[10] of the method of Wu and Hoak[18] was used. The method is based on the ratio of the platelet count of platelet-rich plasma prepared from blood mixed immediately after venipuncture with a solution containing ethylenediaminetetraacetic acid (EDTA) and formaldehyde to that of platelet-rich plasma prepared in the same manner except for the absence of formaldehyde. Wu and Hoak[18] theorized that platelet aggregates are fixed when blood is drawn into a solution containing formaldehyde and EDTA and break apart when blood is drawn into a solution containing EDTA without formaldehyde. A decrease in the platelet aggregate ratio should reflect increased platelet aggregate formation.

α–Granule Proteins

Blood was taken into a precooled plastic syringe containing 1/5 volume of 88mM EDTA and 5mM aminophylline and mixed by inversion five times. It was transferred to a precooled plastic tube and centrifuged at 4°C and 4,300 x g for 1 hour. Plasma was removed with a plastic Pasteur pipette, the tip of which was placed 1 cm below the top of the plasma. Commercial kits were used for radioimmunoassays of β-thromboglobulin (Amersham Corporation) and platelet factor 4 (Abbott Laboratories). Means of duplicate assays of each sample of plasma were used for statistical analysis. Assays of all samples from a given subject were done in the same batch.

RESULTS

Some Effects of Smoking Tobacco and Non-Tobacco Cigarettes[14]

Twenty healthy subjects (10 men and 10 women), who said that they had
never smoked on a daily basis, participated in each of two experimental
periods. Tobacco cigarettes were smoked on one occasion. On the other,
the subjects smoked cigarettes which were made from wheat, cocoa and citrus
plants and contained no nicotine. The subjects were asked to smoke each
cigarette down to the filter and were not asked to inhale. Ten subjects
(six men and four women) returned for a third period during which they sham
smoked by sucking air through straws. The mean ECC$\pm$SD per chamber was
2.6$\pm$0.6 before and 2.8$\pm$0.7 after sham smoking. The means did not differ
significantly. The mean platelet aggregate ratio was 0.78$\pm$0.08 before and
0.78$\pm$0.07 after sham smoking.

The mean ECC of the four chambers rose after each subject smoked two
tobacco cigarettes and increased in 18 of the 20 subjects after smoking
non-tobacco cigarettes. The mean ECC of the 20 subjects more than doubled
when tobacco cigarettes were smoked and increased by 20% ($P<0.0002$) when
non-tobacco cigarettes were smoked. For each subject the increase after
smoking tobacco cigarettes was greater than that after smoking non-tobacco
cigarettes. Mean increases were approximately the same in men and women.

The platelet aggregate ratio of each subject decreased when tobacco
cigarettes were smoked. The mean platelet aggregate ratio was 0.80$\pm$0.06
before and 0.65$\pm$0.07 after smoking tobacco cigarettes ($P<0.0002$). The
change in platelet aggregate ratio was statistically significant for each
gender. When the 20 subjects smoked non-tobacco cigarettes, the mean
platelet aggregate ratio decreased from 0.81$\pm$0.10 to 0.78$\pm$0.10($P=0.004$).
The ratio decreased when nine of the men and five of the women smoked non-
tobacco cigarettes. The difference between the means before and after
smoking non-tobacco cigarettes was statistically significant for the men
and was not significant when the women smoked. For each subject there was
a greater decrease in the platelet aggregate ratio when tobacco cigarettes
were smoked than when non-tobacco cigarettes were smoked.

The median plasma nicotine concentration increased from zero before
smoking to 6.2 ng/ml after smoking tobacco cigarettes. During tobacco
smoking the changes in concentrations of nicotine in the plasma of the 20
subjects were not significantly correlated with either the changes in the
ECC or the changes in the platelet aggregate ratios. There was no significant correlation between the changes in ECC and the changes in platelet
aggregate ratios.

Aspirin, Smoking and Endothelial Cells in Men with Coronary Artery Disease[19]

The effects of two different doses of aspirin were compared with those
of placebo in a random-order, double-blind crossover study. Each of 17
male habitual smokers with coronary artery disease underwent three experimental smoking periods separated by 2 weeks. Each man took a tablet containing 150 mg of aspirin, 300 mg of aspirin or a placebo 12 hours before
each smoking period. After taking the placebo, the mean ECC$\pm$SD was 2.7$\pm$0.8
before smoking and 4.5$\pm$0.9 after smoking ($P<0.001$). At the same time, the
mean plasma nicotine concentration increased approximately three-fold. For
the placebo period, the ECC did not correlate significantly with the plasma
nicotine concentrations before or after smoking, and the changes in ECC did
not correlate significantly with the changes in plasma nicotine concentrations. Neither dose of aspirin affected the mean presmoking ECC or the mean
smoking-induced change in ECC.

Table 1. Endothelial Cell Counts Before and After Cigarette Smoking

Group (n)	Placebo Period			Rutoside Period		
	Before	After (mean±SD)	P	Before	After (mean±SD)	P
Habitual smokers (24)	4.0±1.7	5.4±1.9	<0.001	3.7±1.2	5.1±1.7	<0.001
Nonsmokers (22)	3.3±1.2	4.7±1.2	<0.001	3.4±1.1	4.3±1.5	<0.005

Table 2. Platelet Aggregate Ratios Before and After Cigarette Smoking

Group (n)	Placebo Period			Rutoside Period		
	Before	After (mean±SD)	P	Before	After (mean±SD)	P
Habitual smokers (24)	0.81±0.08	0.68±0.09	<0.001	0.81±0.09	0.70±0.10	<0.001
Nonsmokers (22)	0.82±0.06	0.73±0.06	<0.001	0.81±0.06	0.73±0.08	<0.001

Hydroxyethylrutosides Do Not Prevent Effects of Smoking on Endothelium or Platelets in Habitual Smokers or Nonsmokers[20]

A random-order, double-blind crossover design was used to compare effects of hydroxyethylrutosides with those of placebo on the ECC and platelet aggregate ratio before and after 24 healthy male habitual smokers and 22 healthy male nonsmokers smoked during each of two experimental periods separated by 1 week. Each subject was asked to take four doses of 600 mg of O-(β-hydroxyethyl)-rutosides or its placebo with the last dose 2 hours before smoking. The mean ECC (Table 1.) increased and the mean platelet aggregate ratio (Table 2.) decreased with experimental smoking after each group of men took rutosides or placebo. Rutosides significantly affected neither the mean presmoking values of nor the mean smoking-induced changes in the ECC, platelet aggregate ratio and plasma nicotine concentration. After smoking, plasma nicotine concentrations were not correlated significantly with the other variables of either group of men for either placebo or rutoside periods. The mean plasma nicotine concentration of the habitual smokers was 19.6 ng/ml and that of the nonsmokers was 4.1 ng/ml after smoking in the placebo period.

The differences between the mean ECC of habitual smokers and nonsmokers were not statistically significant before or after smoking in the placebo period. Table 2. shows that the mean platelet aggregate ratios of the two groups were nearly the same before smoking. After smoking, the mean platelet aggregate ratio of the habitual smokers was less than that of the nonsmokers (P<0.01) in the placebo period.

Some Effects of Smoking on Endothelium and Platelets: Lack of Prevention by Dipyridamole and Aspirin in Men with Coronary Artery Disease[21]

The effects of dipyridamole and dipyridamole plus aspirin were compared with those of placebo in a random-order, double-blind crossover study. Twelve male habitual smokers with coronary artery disease took dipyridamole 75 mg and aspirin 324 mg, dipyridamole 75 mg and placebo for aspirin, or a placebo for each drug three times daily for 1 week before each of three experimental smoking periods separated by 2 weeks. During each period there were increases in the mean values of the endothelial cell count, the plasma concentrations of β-thromboglobulin, platelet factor 4 and nicotine and the blood level of carboxyhemoglobin. The mean platelet aggregate ratio decreased during each period. After ingestion of placebos for both dipyridamole and aspirin, the respective mean values before and after smoking were 4.2 and 5.4 per counting chamber (ECC), 27.5 and 30.1 ng/ml (β-thromboglobulin), 7.4 and 8.2 ng/ml (platelet factor 4), 3.7 and 15.7 ng/ml (nicotine), 5.0 and 6.6% (carboxyhemoglobin) and 0.80 and 0.68 (platelet aggregate ratio). Each of the differences between the means before and after smoking was statistically significant ($P \leq 0.02$). Neither dipyridamole alone nor in combination with aspirin significantly affected the mean smoking-induced change in any of these variables. After smoking, the 36 plasma nicotine concentrations from the three experimental periods correlated significantly with those of β-thromboglobulin (Fig. 1.) and platelet factor 4 ($r=0.50$; $P=0.002$). After smoking, plasma nicotine concentrations did not correlate significantly with ECC or platelet aggregate ratios for any study period and carboxyhemoglobin levels did not correlate significantly with any of the other variables for any study period.

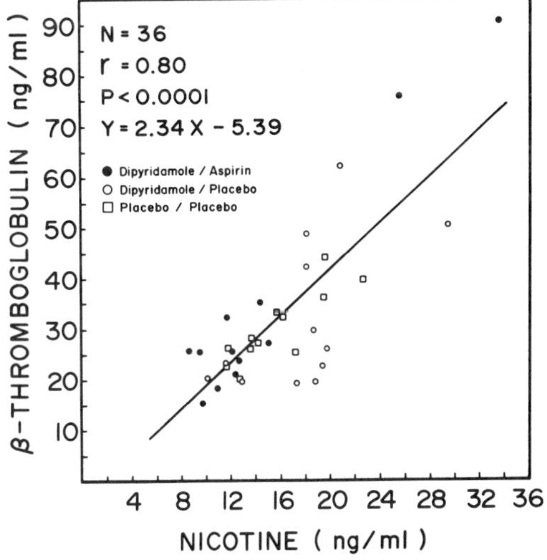

Fig. 1. Relation between plasma concentrations of β-thromboglobulin and nicotine after the three experimental smoking periods.

Table 3. Mean Values ($\pm$SD) of Variables Before and After Control Period

Variable	Before	After
Endothelial cell count	2.2 ($\pm$0.8)	2.3 ($\pm$1.0)
Platelet aggregate ratio	0.88 ($\pm$0.05)	0.88 ($\pm$0.04)
Plasma nicotine concentration, ng/ml	0	0
Blood carboxyhemoglobin level, %	1.1 ($\pm$0.6)	1.2 ($\pm$0.7)

Passive Smoking, Endothelium and Platelets[22]

Ten healthy male nonsmokers participated in two experimental periods separated by 1 week, with five men having a passive smoking period first and five men having a control period first. The periods took place in the morning before work to avoid exposure to environmental smoke. Control periods consisted of sitting in the laboratory where smoking was prohibited. Passive smoking periods consisted of sitting where several patients were smoking of their own accord in chairs located against the length of one wall opposite elevators where a 37 by 16-foot atrium with an 8-foot ceiling connected with corridors at each end by a 7 1/2-foot opening. Usually there was an unoccupied seat between two people who were smoking, and the experimental subject sat there. Sometimes, there was no unoccupied seat between two smokers, and the subject then sat beside only one person who was smoking.

Table 3. shows that the mean values of the ECC, platelet aggregate ratio, plasma nicotine concentration and blood carboxyhemoglobin level were either the same or nearly the same at the beginning and the end of the control period.

Fig. 2. shows that the ECC of each subject was higher after than before passive smoking. The mean ECC$\pm$SD per chamber was 2.8$\pm$0.9 before and 3.7$\pm$1.1 after passive smoking.

Fig. 3. shows that the platelet aggregate ratio of each subject was lower after than before passive smoking. The mean platelet aggregate ratio $\pm$SD was 0.87$\pm$0.06 before and 0.78$\pm$0.07 after passive smoking.

Nicotine was not detected in the plasma of any subject before passive smoking and was present in the plasma of all but one after passive smoking when the mean concentration was 2.8$\pm$1.2 ng/ml (P=0.004). After passive smoking, the carboxyhemoglobin level was higher in all but one subject whose value was unchanged. Mean values were 0.9$\pm$0.3% before and 1.3$\pm$0.6% after passive smoking (P=0.004). After passive smoking, neither the plasma nicotine concentrations nor the carboxyhemoglobin levels correlated significantly with the ECC or the platelet aggregate ratios.

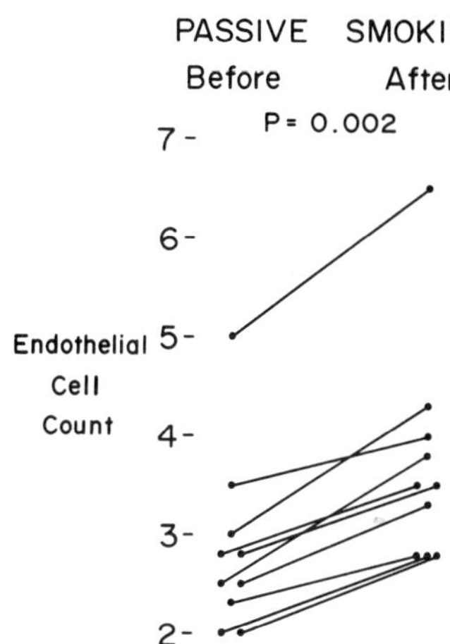

Fig. 2.   ECC before and after passive
smoking.

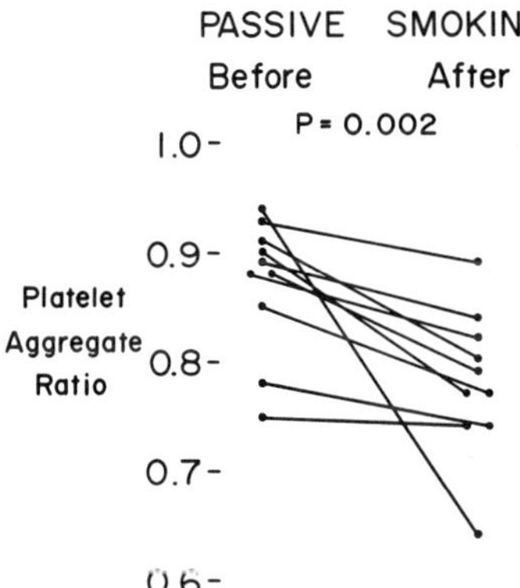

Fig. 3.   Platelet aggregate ratios before
and after passive smoking.

The subjects of this study were eight male habitual smokers ranging in age from 38 to 54 years. They were apparently healthy with the exception of one, who had an old myocardial infarct. Antecubital venipuncture was done immediately before, immediately after, 55 minutes after and 2 hours after completion of an experimental period during which two Winston King Size cigarettes were smoked down to the filter. In this study the method for ECC was modified in that formalin was added to the preparation before the final centrifugation. Nicotine was not determined by gas chromatography as in our previous studies. Both plasma nicotine and cotinine concentrations were determined by radioimmunoassay at the American Health Foundation of Valhalla, New York.

In Fig. 4. the mean ECC is plotted for the four occasions on which each subject was tested. Immediately after smoking, the mean ECC was approximately twice the value before smoking. Two hours after smoking the mean ECC was nearly the same as the presmoking value. Using the 24 sets of data obtained from the eight subjects on the three postsmoking occasions, Pearson correlation coefficients were calculated. The coefficient between ECC and nicotine was 0.56(P<0.005) and between ECC and cotinine was 0.47(P<0.025).

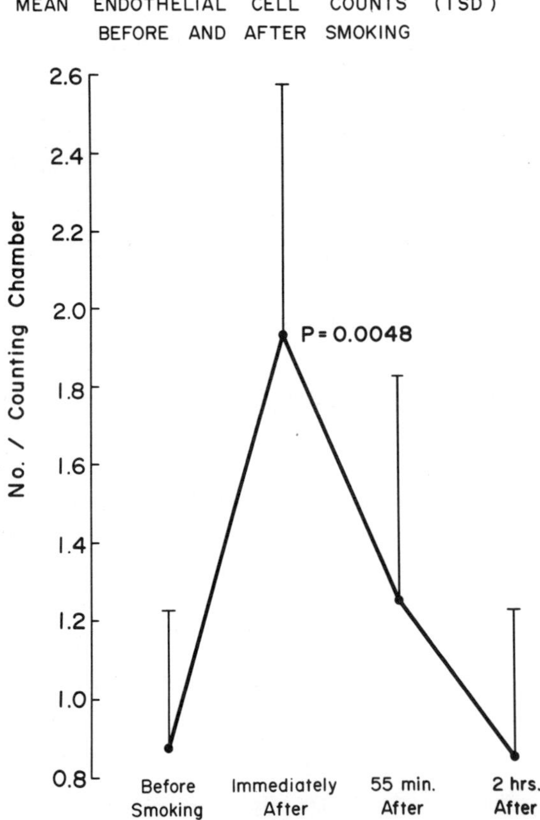

Fig. 4. Paired t-tests were used to determine the significance of the differences between the mean presmoking ECC and that on each of the three occasions after smoking. The only statistically significant difference is indicated by the P value.

DISCUSSION

In 1979 Prerovský and Hladovec[15] reported a desquamating effect of cigarette smoking on human endothelium manifested by an increase in the concentration of circulating anuclear carcasses of endothelial cells. Their observations have been confirmed by several studies in our laboratory[14,19-21] using three different brands of tobacco cigarettes. Although there was a statistically significant increase in the mean ECC of male and female naive smokers after smoking non-tobacco cigarettes, it was much less than that after the same subjects smoked tobacco cigarettes.[14] After smoking tobacco cigarettes, male habitual smokers had the same mean increase in ECC as naive smokers although their mean nicotine concentration after smoking was much greater than that of the naive smokers.[20] Hladovec[23] reported that several vasoactive compounds caused an increase in ECC. If the increase in ECC after tobacco smoking is due to the release of catecholamines which occurs with tobacco smoking,[11] it may be speculated that small amounts of catecholamines also are released when naive smokers smoke non-tobacco cigarettes and that lower plasma concentrations of nicotine may be required to release catecholamines in naive smokers than in habitual smokers after smoking tobacco. Perhaps, even the small increases in plasma nicotine concentrations from undetectable levels to a mean of 2.8 ng/ml which occurred during passive smoking by nonsmokers,[22] led to release of catecholamines. Unfortunately, catecholamine concentrations were not determined in these studies.

Prerovský and Hladovec[15] reported that hydroxyethylrutosides prevented endothelial desquamation when people smoked, and Hladovec[24] reported that aspirin protected rats against endothelemia induced by the injection of citrate. However, we found no effect of hydroxyethylrutosides,[20] aspirin alone,[19] dipyridamole alone[21] or aspirin plus dipyridamole[21] on the smoking-induced increase in the ECC.

The platelet aggregate ratio consistently decreases when cigarettes are smoked.[10,14,20,21,25,26] We have also observed an increase in the mean plasma concentration of β-thromboglobulin when either healthy habitual smokers[20] or habitual smokers with coronary artery disease[21] smoked and an increase in the mean plasma concentration of platelet factor 4 when habitual smokers with coronary artery disease[21,26] smoked. Although aspirin prevented a smoking-induced decrease in the mean platelet aggregate ratio in healthy male and female habitual smokers,[25] it did not prevent a smoking-induced decrease in the mean platelet aggregate ratio or a smoking-induced increase in the mean plasma concentration of platelet factor 4 in male habitual smokers with coronary artery disease.[26] Neither dipyridamole alone nor in combination with aspirin prevented smoking-induced changes in the platelet aggregate ratio or plasma concentrations of β-thromboglobulin and platelet factor 4 in male habitual smokers with coronary artery disease.[21] Thus it appears that cyclooxygenase-independent pathways are involved in activation of platelets by cigarette smoking in men with coronary artery disease. Such pathways might involve platelet aggregation induced by catecholamines in combination with other agents.[27] Folts and Bonebrake,[28] who used a dog model of coronary artery stenosis, showed that inhalation of cigarette smoke or infusion of nicotine was followed by increased concentrations of epinephrine in plasma and increased cyclic reductions in blood flow due to platelet aggregate formation, which could be inhibited by phentolamine. Their work suggests that release of catecholamines by nicotine may be a mediator of smoking-induced platelet activation. Renaud et al.[29] observed that the addition of nicotine to human platelet-rich plasma to attain a concentration of 20 ng/ml enhanced thrombin-induced platelet aggregation by 65% and adenosine diphosphate-induced platelet aggregation by 55%. Since nicotine concentrations of approximately 20 ng/ml often occur after smoking cigarettes,[19-21,26,29] it seems that nicotine absorbed from cigarette smoke should be capable of activating platelets independently of catecholamine release, which might potentiate the effects of nicotine and other agents on platelets.

Work from two laboratories shows that collagen-induced platelet aggregation is enhanced after normal people smoke cigarettes.[29,30] Patients receiving hemodialysis for chronic renal failure most often die of cardiovascular disease. Recent data from our laboratory shows that collagen-induced platelet aggregation is enhanced no less when men receiving chronic hemodialysis smoke than when healthy men smoke.[30] This suggests that men undergoing chronic hemodialysis may not be protected against thrombosis by impaired platelet function.

The experimental use of snuff by male habitual smokers, who were naive to the use of smokeless tobacco, increased the ECC and lowered the platelet aggregate ratio.[31] In these men the acute effects of snuff were less than those of experimental cigarette smoking on the plasma nicotine concentrations as well as on the ECC and platelet aggregate ratios.[31]

The demonstration of an effect of passive smoke inhalation on endothelium (Fig.2.) and platelets (Fig.3.) raises the question of whether repeated episodes of passive exposure to tobacco smoke over a period of years might increase morbidity and mortality from cardiovascular disease. A review[32] of the epidemiological literature on passive smoking by Wells suggests that passive smoking does increase mortality from heart disease. A subsequent report[33] indicated that the adjusted relative risk for mortality from ischemic heart disease associated with passive smoking in Scotland (2.01) was nearly as great as that associated with active smoking (2.27).

ACKNOWLEDGEMENTS

Tables 1. and 2. are used with permission of the British Journal of Experimental Pathology. Fig. 1. is used with permission of the American Journal of Cardiology. Figs. 2. and 3. and Table 3. are used with permission of the Archives of Internal Medicine.

Work from the author's laboratory described in this article was supported by the Veterans Administration and by grants from Zyma, SA, the American Heart Association, Kansas Affiliate, and the Missouri Kidney Program.

REFERENCES

1. Strong JP, Richards ML. Cigarette smoking and atherosclerosis in autopsied men. Atherosclerosis 1976;23:451-76.

2. Kannel WB. Update on the role of cigarette smoking in coronary artery disease. Am Heart J 1981;101:319-28.

3. Doll R, Peto R. Mortality in relation to smoking: 20 years' observations on male British doctors. Br Med J 1976;2:1525-36.

4. Ross R. The pathogenesis of atherosclerosis—an update. N Engl J Med 1986;314:488-500.

5. Fuster V, Chesebro JH. Antithrombotic therapy: role of platelet-inhibitor drugs. I. Current concepts of thrombogenesis: role of platelets. Mayo Clin Proc 1981;56:102-12.

6. Hawkins RI. Smoking, platelets and thrombosis. Nature 1972;236:450-2.

7. Levine PH. An acute effect of cigarette smoking on platelet function: a possible link between smoking and arterial thrombosis. Circulation 1973;48:619-23.

8. Grignani G, Gamba G, Ascari E. Cigarette-smoking effect on platelet function. Thromb Haemostas 1977;37:423-8.

9. Bierenbaum ML, Fleischman AI, Stier A, Somol H, Watson PB. Effect of cigarette smoking upon in vivo platelet function in man. Thromb Res 1978;12:1051-7.

10. Davis JW, Davis RF. Acute effect of tobacco cigarette smoking on the platelet aggregate ratio. Am J Med Sci 1979;278:139-43.

11. Siess W, Lorenz R, Roth P, Weber PC. Plasma catecholamines, platelet aggregation and associated thromboxane formation after physical exercise, smoking or norepinephrine infusion. Circulation 1982;66:44-8.

12. László E, Káldi N, Kovács L. Alterations in plasma proteins and platelet functions with aging and cigarette smoking in healthy men. Thromb Haemostas 1983;49:150.

13. Hladovec J, Rossmann P. Circulating endothelial cells isolated together with platelets and the experimental modification of their counts in rats. Thromb Res 1973;3:665-74.

14. Davis JW, Shelton L, Eigenberg DA, Hignite CE, Watanabe IS. Effects of tobacco and non-tobacco cigarette smoking on endothelium and platelets. Clin Pharmacol Ther 1985;37:529-33.

15. Prerovský I, Hladovec J. Suppression of the desquamating effect of smoking on the human endothelium by hydroxyethylrutosides. Blood Vessels 1979;16:239-40.

16. Hladovec J. Endothelial injury by nicotine and its prevention. Experientia 1978;34:1585-6.

17. Feyerabend C, Russell MAH. Improved gas-chromatographic method and micro-extraction technique for the measurement of nicotine in biological fluids. J Pharm Pharmacol 1979;31:73-6.

18. Wu KK, Hoak JC. A new method for the quantitative detection of platelet aggregates in patients with arterial insufficiency. Lancet 1974;2:924-6.

19. Davis JW, Shelton L, Eigenberg DA, Hignite CE. Lack of effect of aspirin on cigarette smoke-induced increase in circulating endothelial cells. Haemostasis 1987;17:66-9.

20. Davis JW, Shelton L, Hartman CR, Eigenberg DA, Ruttinger HA. Smoking-induced changes in endothelium and platelets are not affected by hydroxyethylrutosides. Br J Exp Pathol 1986;67:765-71.

21. Davis JW, Hartman CR, Shelton L, Ruttinger HA. A trial of dipyridamole and aspirin in the prevention of smoking-induced changes in platelets and endothelium in men with coronary artery disease. Am J Cardiol 1989;63:1450-4.

22. Davis JW, Shelton L, Watanabe IS, Arnold J. Passive smoking affects endothelium and platelets. Arch Intern Med 1989;149:386-9.

23. Hladovec J. Circulating endothelial cells as a sign of vessel wall lesions. Physiol Bohemoslov 1978;27:140-4.

24. Hladovec J. Vasotropic drugs—a survey based on a unifying concept of their mechanism of action. Arzneimittelforschung 1977;27:1073-6.

25. Davis JW, Davis RF, Hassanein KM. In healthy habitual smokers acetylsalicylic acid abolishes the effects of tobacco smoke on the platelet aggregate ratio. Can Med Assoc J 1982;126:637-9.

26. Davis JW, Hartman CR, Lewis HD Jr, Shelton L, Eigenberg DA, Hassanein KM, Hignite CE, Ruttinger HA. Cigarette smoking-induced enhancement of platelet function: lack of prevention by aspirin in men with coronary artery disease. J Lab Clin Med 1985;105:479-83.

27. Lauri D, Cerletti C, de Gaetano G. Amplification of primary response of human platelets to platelet-activating factor: aspirin-sensitive and aspirin-insensitive pathways. J Lab Clin Med 1985;105:653-8.

28. Folts JD, Bonebrake FC. The effects of cigarette smoke and nicotine on platelet thrombus formation in stenosed dog coronary arteries: inhibition with phentolamine. Circulation 1982;65:465-70.

29. Renaud S, Blache D, Dumont E, Thevenon C, Wissendanger T. Platelet function after cigarette smoking in relation to nicotine and carbon monoxide. Clin Pharmacol Ther 1984;36:389-95.

30. Davis JW, Arnold J, Wiegmann TB. Cigarette smoking increases collagen-induced platelet aggregation of men with chronic renal failure. Clin Res 1988;36:889A.

31. Davis JW, Shelton L. Comparative effects of snuff and cigarettes on endothelium and platelets. Clin Pharmacol Ther 1985;37:190.

32. Wells AJ. An estimate of adult mortality in the United States from passive smoking. Environ Int 1988;14:249-65.

33. Hole DJ, Gillis CR, Chopra C, Hawthorne VM. Passive smoking and cardiorespiratory health in a general population in the west of Scotland. Brit Med J 1989;299:423-7.

PLATELETS IN THE PATHOGENESIS OF ATHEROSCLEROSIS

Wolfgang Siess

Institut für Prophylaxe und Epidemiologie der
Kreislaufkrankheiten der Universität München
Pettenkoferstr. 9, D 8 München 2
Federal Republic of Germany

The major function of platelets is to maintain the hemostatic
integrity of the blood vessel and to stop bleeding after injury.
Platelets show different responses that can be studied separately in
vitro, but that are closely linked during hemostasis in vivo: adhesion
and shape change, aggregation and secretion. Many diverse substances can
activate platelets, and platelets are very sensitive to these agents and
react within seconds of exposure to these stimuli. The most important
physiological stimuli are von Willebrand factor and collagen, which
mediate platelet adhesion; thrombin, which is generated by the
coagulation system; prostaglandin endoperoxides, thromboxane, and ADP,
which are released from activated platelets; platelet-activating factor,
which is liberated from stimulated neutrophils and macrophages, and
catecholamines, which are present in the circulation and show high levels
after stress and cigarette smoking.

Mechanisms of Platelet Activation

Although platelet stimuli and platelet responses are quite diverse,
evidence is emerging that platelet activation is regulated by a distinct
number of mechanisms (1). Common to all the physiological stimuli is
their binding to specific receptors on the platelet surface. The signal
emitted from the activated receptor is transduced through the plasma
membrane by guanine nucleotide-binding proteins. These transducing
mechanisms stimulate and regulate specific effector systems such as phospholipase C-induced inositol phospholipid hydrolysis and ion channels.
The effector systems modulate the levels of intracellular second
messengers such as inositol 1,4,5-trisphosphate, calcium, and
diacylglycerol. These messengers in turn trigger the physiological
response by changes in phosphorylation, enzymatic activities and
structural properties of proteins.

Platelet activation occurs probably in two steps (1). The first
physiological response is platelet shape change which is caused by
inositol 1,4,5-trisphosphate formation, calcium mobilization, and myosin
light chain phosphorylation. Mobilization of calcium also primes protein
kinase C by placing the enzyme into a strategic position in the plasma
membrane without, however, significantly activating the enzyme. The
second step involves the activation of protein kinase C by diacylglycerol, the bulk of which is generated later by hydrolysis of
phosphatidylinositol. Protein kinase C activation is a positive signal

Tobacco Smoking and Atherosclerosis
Edited by J. N. Diana
Plenum Press, New York, 1990

for platelet aggregation and secretion, but inhibits signal transduction
between activated receptor and phospholipase C (2).

In addition to the calcium pathway and protein kinase C-pathway, we
have identified a third pathway that is selectively activated by
epinephrine (1,3). This pathway involves G_i2 protein dissociation and an
effector system which is still unknown. Protein kinase C-activation,
calcium mobilization and the pathway induced by G_i2 protein dissociation
are alone able to trigger a certain degree of platelet aggregation. It
is likely that under physiological conditions these pathways synergize in
the induction of platelet response. Synergism has been demonstrated for
each pair of these pathways: calcium and protein kinase C (1,2),
epinephrine and protein kinase C (3), and epinephrine and calcium (4).
Each pathway can synergize with the other independently of the third
pathway.

Besides these intracellular pathways of platelet activation,
platelets possess two strong positive feedback loops: the formation of
prostaglandin endoperoxides/thromboxane A_2 and the secretion of ADP.
These substances, released from activated platelets, act as local
hormones, bind to specific receptors on the platelet surface and trigger
a chain of molecular events similar to the other stimuli. Both the
cyclooxygenase pathway and the ADP pathway mediate secretion and
irreversible aggregation evoked by many platelet stimuli. They are,
however, not involved in the primary and reversible platelet responses
such as adhesion, shape change, and primary aggregation (1).

Platelets and Experimental Atherosclerosis

What is the evidence that platelets are involved in the development
of atherosclerosis? Several studies in rabbits have shown that experimentally induced thrombocytopenia inhibits the intimal thickening after
baloon-de-endothelialization or catheter injury of arteries (5). Also,
pigs with severe von Willebrand disease that have a strong defect in
platelet adhesion developed less atherosclerosis of the abdominal aorta
than did normal control pigs. However, similar evidence that genetic
platelet defects can prevent the development of atherosclerosis in humans
is lacking. In humans, it is highly likely that platelets are involved
in the early lesion formation in catheter-induced injury, in peri-
anastomotic lesions of vascular grafts and in other traumatic vascular
lesions. The primary role of platelets after nondenuding injury is less
certain.

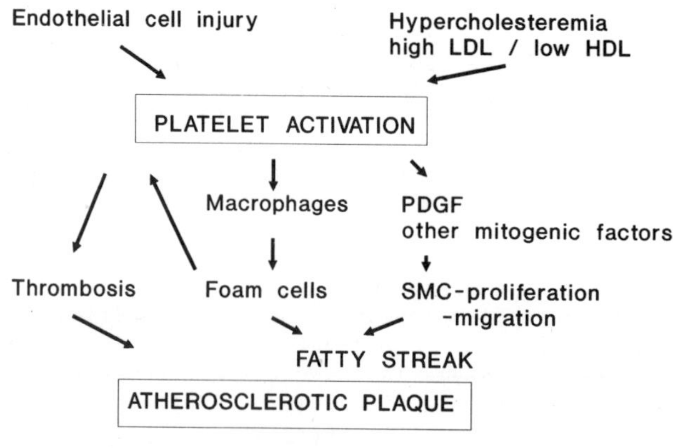

Fig. 1

<u>How are Platelets Involved in the Pathogenesis of Atherosclerosis?</u>

As shown in Fig. 1, platelets are in many ways involved in the pathogenesis of atherosclerosis. Atherosclerotic plaques are often the end result of an organized and subsequently endothelialized thrombus. Ulceration, plaque disruption or subintimal hemorrhage in atherosclerotic vessels stimulate the above described mechanisms of platelet activation and lead ultimately to thrombus formation. The rapidly progressing atherosclerosis occurs most frequently in the carotid, coronary and femoral arterial beds, leads to significant stenosis or occlusion, and is associated with mural thrombosis (6). The development of thrombosis in relation to preexisting atherosclerotic plaque is the most important complicating event in the natural history of atherosclerosis and is responsible for many clinical manifestations such as myocardial infarction, stroke, and peripheral gangrene.

Recent studies have shown that platelets are also involved at early steps of the atherosclerotic process. The "response to injury" hypothesis involves the interaction of platelets with monocytes/ macrophages after injury to the endothelium (7). The loss of endothelial integrity is considered as the primary step in the development of athero- sclerosis and can be caused by many factors. Important are mechanical forces at sites of flow effects, subintimal cholesterol deposition and foam cell infiltration, bacterial toxins, viral infections, oxygen radicals and immunological injury. Furthermore, hypercholesteremia, the main risk factor of coronary heart disease, and LDL render platelets more sensitive to aggregating agents. Mitogens such as PDGF, secreted from platelet α-granules, stimulate the migration of smooth muscle cells into the intima and enhance their proliferation. The interaction of platelets with endothelial cells and monocytes is complex and will be considered in more detail.

<u>Platelets and the Vessel Wall</u>

The integrity of the vascular endothelium is also very important in another aspect. Thrombin and products released from aggregating platelets (ADP, serotonin) evoke in intact vessels an endothelium- dependent relaxation (8). However, the absence or dysfunction of the endothelium favors the constriction of blood vessels resulting in vasospasm. Recent studies have identified the endothelium-derived relaxing factor (EDRF) as nitric oxide (9). In atherosclerotic vessels less EDRF is produced, and atherosclerotic vessels exhibit vaso- constriction in response to thrombin and ADP rather than the usual vasodilation (10). Vasospasm may contribute significantly to the thrombus formation in atherosclerotic vessels. Another important effect of nitric oxide is its inhibition of platelet aggregation and adhesion (11). Nitric oxide is freely membrane-diffusible, enters the platelet cytosol, and activates the soluble guanylate cyclase. The resulting increase of cyclic GMP levels causes platelet inhibition. Nitric oxide is considered to be responsible for the antiadhesive property of intact endothelium.

Prostacyclin (PGI_2), also formed by the endothelium, is the most powerful antiaggregating substance known (12). In addition, it causes vasodilation. PGI_2 increases cyclic AMP levels in platelets, activates cyclic AMP-dependent protein kinase and prevents stimulus-induced shape change, aggregation and secretion. PGI_2 seems to be less effective in inhibiting adhesion than nitric oxide. Interestingly, nitric oxide and PGI_2 synergize in inhibiting platelet activation (13). Both substances are important for preventing the formation and/or counterbalancing the

action of the potent aggregating and vasoconstrictive substance
thromboxane A_2 which is released from activated platelets.

Platelets, Hypercholesterolemia and Lipoproteins

Platelets of patients with primary hypercholesterolemia show increased platelet adhesion, aggregation and secretion (14). The platelet
membranes of these patients contain increased cholesterol. Also cholesterol feeding in both animals and man or cholesterol-loading of platelets
in vitro resulted in an enhanced platelet sensitivity to aggregating
agents (14,15). Therapeutic reductions in cholesterol levels by diet and
medication have shown to depress platelet function. Increased amounts of
cholesterol in the platelet membrane could facilitate signal transduction
mechanisms or the activation of effector systems.

Platelets interact with low and high density lipoproteins (LDL, HDL)
(14). Specific binding sites exist on platelets for LDL and HDL. These
have been recently identified as glycoproteins IIb/IIIa, which also form
the fibrinogen receptor (16). LDL at hormonal concentrations (>2g
protein/1) activates platelets whereas HDL (specifically HDL_2) has the
opposite effect. Platelet activation induced by LDL is associated with
phospholipase C-induced phosphoinositide hydrolysis, calcium mobilization, and protein kinase C-activation (17). However, direct platelet
activation induced by LDL may certainly not occur in vivo. Other plasma
factors such as fibrinogen may bind more effectively to glycoprotein
IIb/IIIa than LDL. It is possible, however, that in vivo increased LDL
concentrations render platelets more sensitive to aggregating agents.

More important than the direct activation of platelets by LDL seems
to be the ability of platelets to degrade LDL and to incorporate the
LDL-cholesterol. Platelets do not engulf lipoproteins by endocytosis:
they accumulate LDL in the open canalicular system and on the plasma
membrane. The LDL-cholesterol is then transferred to the platelet
membrane (18). Binding of LDL to platelets has been found to result in
an increase in platelet cholesterol. Also the fatty acid content of the
platelet membrane quantitatively paralleled that of LDL.

Platelets and Monocytes/Macrophages

One of the earliest events in experimental atherosclerosis, induced
by hypercholesteremia, is increased adhesion of monocytes to an
apparently intact endothelium (19,20). The monocytes squeeze themselves
in between the endothelial cells and reach the intima, where they become
macrophages, take up large amounts of cholesterol and cholesterol esters
and transform into foam cells. Foam cells are the hallmark of early
fatty streak lesions, that are still reversible. Interestingly, foam
cell formation in experimental studies is only observed, when thrombi
containing platelets and not just blood clots are introduced into the
arteries (6). Platelets and platelet derived products are often found
phagocytosed in foam cells and sometimes in macrophages (21).

Platelets interact with monocytes/macrophages in several ways.
Platelets have the ability to modify LDL to MDA-LDL and to other LDL
forms that promote cholesterol ester accumulation in macrophages (14).
Cholesterol ester and cholesterol are taken up by macrophages not via LDL
and classical LDL-receptor, but in the form of modified LDL, such as
acetyl-LDL or malon dialdehyde (MDA)-LDL via a scavenger receptor on the
macrophage surface.

Platelets enhance also in the absence of LDL cholesterol ester
accumulation into macrophages or monocytes. The cholesterol-donating

activity is found in the 1000 g supernatant of thrombin- or collagen-activated platelets (22,23). The activity does not seem to be cholesterol per se which is found in the supernatant, but rather the cholesterol present in lipid-protein particles shed by the activated platelets. In that context, it has been shown that an integral membrane protein of the platelet α-granules ("platelet activation dependent granule external membrane protein," PADGEM, GMP-140) is expressed on the surface of activated platelets and is the receptor that mediates the interaction of activated platelets with monocytes (24). The same protein might also be present and expressed in the microparticles shed by activated platelets, since α-granule proteins are found in these microparticles (25). Platelets do not only enhance cholesterol ester accumulation in macrophages, but also in aortic smooth muscle cells (26).

It has been shown that in turn foam cells can activate platelets. Foam cells that had left the intima and re-entered the circulation were found to be surrounded by platelet aggregates (7). The molecular mechanisms of this interaction are unknown.

Mitogens in Platelet α-Granules

Platelets contain several mitogens in their α-granules that are secreted into the medium upon activation. Interestingly, with several physiological platelet stimuli, α-granule proteins are secreted at lower concentrations of the stimulating agent than are required for release of dense granule constituents (27). This might be of significance for the role of platelets in the development of atherosclerosis: just platelets adherent to the subendothelium without undergoing full aggregation and release reaction might be able to secrete mitogens.

The most prominent and intensively studied mitogen is PDGF that stimulates smooth muscle cells to migrate and proliferate, and has an important function in the early intima proliferation (28). Other mitogens include the platelet-derived endothelial cell growth factor (PD-ECGF) that stimulates growth and chemotaxis of endothelial cells in vitro and has angiogenic activity in vivo (29); an epidermal growth factor-like molecule and TGFB that regulate the proliferation of fibroblasts.

Platelet Inhibition as Strategy for the Prevention of Atherosclerosis

STRATEGIES FOR THE PREVENTION OF
ATHEROSCLEROSIS AND CORONARY
HEART DISEASE

- Physical exercise

- Diet: low cholesterol and saturated fatty
  acids high n-3 polyunsaturated
  fatty acids

- Drugs: acetylsalicylic acid

Fig. 2

<u>Physical Exercise</u>

Even a little physical exercise seems to confer significant protection from cardiovascular disease. Physical exercise has a positive
effect on many cardiovascular risk factors. Its effect on platelets
might be an increased production of antiaggregating PGI_2 and probably
nitric oxide.

<u>Diet</u>

A diet low in saturated fatty acids and cholesterol will lower blood
cholesterol and platelet cholesterol, LDL and, as outlined above, favor
platelet inhibition.

Epidemiological, clinical, and experimental studies suggest that high
intake of n-3 polyunsaturated fatty acids contained in fish and fish oil
and low intake of n-6 polyunsaturated fatty acids have a beneficial
effect in the prevention of atherosclerosis. A decade has passed since
Dyerberg and Bang published their findings about diet, blood lipids,
platelet function, and cardiovascular disease in the Eskimos of Western
Greenland (30,31). We had shown one year later that Caucasians can
become somewhat Eskimo-like when the mackerel diet is followed (32).

Eicosapentaenoic acid gets incorporated into the platelet
phospholipids at the expense of arachidonic acid, and is liberated upon
platelet activation. The high eicosapentaenoic/arachidonic acid ratio
was found to be related to the reduction of platelet aggregation and
associated thromboxane formation induced by collagen and ADP (32). Since
these initial findings, a large number of investigators have examined the
effect of fish and fish oil or its main active components,
eicosapentaenoic acid, and docosahexaenoic acid on the pathogenesis of
atherosclerosis and thrombosis (33). An epidemiological study from
Holland indicated that a diet containing greater than 30g fish per day
resulted in a reduced incidence of death from coronary heart disease
(34). In a recent study, the subjects advised to eat fatty fish two to
three portions a week had a 29% reduction in all-cause mortality (35).
These epidemiological studies are supported by experimental studies in
hyperlipidemic swine and monkeys, in which dietary fish oil retarded the
development of coronary atherosclerosis (36,37). The mechanisms
underlying the beneficial effects of dietary n-3 fatty acids are not
restricted to platelets, but involve also monocyte/macrophage function,
blood pressure regulation, vasorelaxation, fibrinolytic activity and
hyperlipemia (33).

<u>Drugs: Aspirin</u>

Aspirin in low doses, such as 350 mg every other day, irreversibly
acetylates the platelet cyclooxygenase and, hence, prevents the formation
of thromboxane A_2, a potent platelet- aggregating agent and vasoconstrictor. Platelet dense and α-granule secretion and irreversible
aggregation induced by many (but not all) physiological platelet stimuli
are reduced or inhibited (1). Low doses of aspirin affect only a little
the vascular PGI_2 synthesis. Also the effect of aspirin is maintained
for the lifespan of the platelet, whereas the cells of the vessel wall
have the capacity to synthesize more cyclooxygenase and, hence, continue
to produce PGI_2. Aspirin has been proven of significant benefit (a) for
the prevention of myocardial infarction in patients with angina pectoris
(38), (b) for the reduction of mortality after myocardial infarction, if
given within the first six hours after the event (39), (c) for the
improvement of graft patency of aorto-coronary bypass (38). Recently, in
a large study on healthy American physicians carried out to examine

aspirin for primary prevention of coronary heart disease, aspirin has
been shown to reduce the risk of myocardial infarction by almost 50% (40).

<u>Conclusion</u>

Platelets are in multiple ways involved in the pathogenesis of
atherosclerosis. The reactivity of platelets to aggregating agents is
influenced by the blood cholesterol level and lipoproteins. The
integrity of the endothelial cell is important for the production of
platelet-inhibiting PGI_2 and nitric oxide.

Endothelial injury not only induces platelet activation by the
exposure of subendothelial tissue, but also leads to vasospasm instead of vasodilation in response to products released by
aggregating platelets. The interaction of platelets with monocytes/
macrophages and smooth muscle cells, key cells in the development of
atherosclerosis, has been discussed in more detail.

Platelet inhibition as strategy for the prevention of atherosclerosis
and its main complication, coronary heart disease, seems to be successful.
A diet, low in cholesterol and saturated fatty acids, and high in n-3
fatty acids is expected to delay the development of atherosclerosis.
Aspirin has been proven of significant benefit for the prevention of
atherosclerotic complications such as angina pectoris and myocardial
infarction, in which platelet activation and thrombus formation are
prominent features.

<u>Acknowledgement</u>

Supported by the Deutsche Forschungsgemeinschaft (Si 274).

<u>References</u>

1. Siess W. Molecular mechanisms of platelet activation. Physiol.
    Rev. 69:58-178 (1989).
2. Siess, W., Lapetina, E.G. Calcium primes protein kinase C in human
    platelets. Biochem. J. 255:309-318 (1988).
3. Siess W., Lapetina E.G. Platelet aggregation induced by
    α2-adrenoceptor- and protein kinase C activation: A novel
    synergism. Biochem. J. in press (1989).
4. Olbrich C., Siess W. Epinephrine and the Ca^{2+}-ionophore A 23187
    synergistically induce platelet aggregation without protein kinase
    C activation. FEBS Lett. 243:275-279 (1989).
5. Friedman J., Stemerman M.D., Wenz B., Moore S., Gauldie J., Gent M.,
    Tiell M.L., Spaet T.H. The effect of thrombocytopenia on
    experimental arteriosclerotic lesion formation in rabbits. J.
    Clin. Invest. 60:1191-1201 (1977).
6. Schwartz C.J., Valente A.J., Kelley J.L., Sprague E.A., Edwards
    E.H. Thrombosis and the development of atherosclerosis:
    Rokitansky revisited. Semin. Thrombos. Hemostas. 14:189-195
    (1988).
7. Ross R. The pathogenesis of atherosclerosis - an update. N. Engl.
    J. Med. 314:488-500 (1986).
8. Furchgott R.F., Zawadzki J.V. The obligatory role of endothelial
    cells in the relaxation of arterial smooth muscle by
    acetylcholine. Nature 288:373-376 (1980).
9. Palmer R.M., Ferrige A.G., Moncada S. Nitric oxide release accounts
    for the biological activity of endothelium- derived relaxing
    factor. Nature 327:524-526 (1987).

10. Freiman P.C., Mitchell G.C., Heistad D.D., Armstrong M.L., Harrison
     D.G. Atherosclerosis impairs endothelium-dependent vascular
     relaxation to acetylcholine and thrombin in primates. Circ. Res.
     58:783-789 (1986).
11. Radomski M.W., Palmer R.M.J., Moncada S. Endogenous nitric oxide
     inhibits human platelet adhesion to vascular endothelium. Lancet
     2:1057-1058 (1987).
12. Moncada S., Gryglewski R., Bunting S., Vane J.R. An enzyme isolated
     from arteries transforms prostaglandin endoper-oxides to an
     unstable substance that inhibits platelet aggregation. Nature
     263:663-665 (1976).
13. Radomski M.W., Palmer R.M.J., Moncada S. The antiaggregating
     properties of vascular endothelium: interaction between
     prostacyclin and nitric oxide. Br. J. Pharmacol. 92:639-646
     (1987).
14. Brook G.J., Aviram M. Platelet lipoprotein interactions. Semin.
     Thrombos. Hemostas. 14:258-265 (1988).
15. Shattil S.J., Anaya-Galindo R., Bennett J., Colman R.W., Cooper
     R.A. Platelet hypersensitivity induced by cholesterol
     incorporation. J. Clin. Invest. 55:636-643 (1975).
16. Koller E., Koller F., Binder B.R. Purification and identification
     of the lipoprotein-binding proteins from human blood platelet
     membrane. J. Biol. Chem. 264:12412-12418 (1989).
17. Andrews H.E., Aitken J.W., Hassal D.G., Skinner V.O., Bruckdorfer
     K.R. Intracellular mechanisms in the activation of human
     platelets by low-density lipoproteins. Biochem. J. 242:559-564
     (1987).
18. Block L.H., Knorr M., Vogt E., Locher R., Vetter W., Grosgurth P.,
     Qiao B-Y., Pometta D., James R., Regenass M., Pletscher A. Low
     density lipoprotein causes general cellular activation with
     increased phosphatidylinositol turnover and lipoprotein
     catabolism. Proc. Natl. Acad. Sci. 85:885-889 (1988).
19. Faggiotto A., Ross R., Harker L. Studies of hypercholesterolemia in
     the nonhuman primate. I. Changes that lead to fatty streak
     formation. Arteriosclerosis 4:323- (1984).
20. Munro M.J., Cotran R. The pathogenesis of atherosclerosis:
     atherogenesis and inflammation. Lab. Invest. 58:249-261 (1988).
21. Sevitt S. Platelets and foam cells in the evolution of
     atherosclerosis. Atherosclerosis 61:107-115 (1986).
22. Curtiss L.K., Black A.S., Takagi Y., Plow E.F. New mechanisms for
     foam cell generation in atherosclerosis lesions. J. Clin. Invest.
     80:367-373 (1987).
23. Mendelsohn M.E., Loscalzo J. Role of platelets in cholesterol
     ester formation by U-937 cells. J. Clin. Invest. 81:62-68 (1988).
24. Larsen E., Celi A., Gilbert G.E., Furie B.C., Erban J.K., Bonfanti
     R., Wagner D.D., Furie B. PADGEM Protein: A receptor that
     mediates the interaction of activated platelets with neutrophils
     and monocytes. Cell 59:305-312 (1989).
25. Sandberg H., Anderson L-O, Hoglund S. Isolation and
     characterization of lipid-protein particles containing platelet
     factor 3 released from human platelets. Biochem. J. 203:303-311
     (1982).
26. Kruth H.S. Platelet mediated cholesterol accumulation in cultured
     aortic smooth muscle cells. Science 227:1243-1245 (1985).
27. Kaplan K.L. Platelet granule proteins: localization and
     secretion. In Platelets in Biology and Pathology-2, J.L. Gordon
     (ed) Elsevier, Amsterdam p. 80-90 (1981).
28. Ross R. Platelet derived growth factor. Lancet 1:1179-1182 (1989).

29. Usuki K., Heldin N-E., Miyazono K., Ishikawa F., Takaku F., Westermark B., Heldin C-H. Production of platelet-derived endothelial cell growth factor by normal and transformed human cells in culture. Proc. Natl. Acad. Sci. USA 86:7427-7431 (1989).

30. Dyerberg J., Bang H.O. Hemostatic function and platelet polyunsaturated fatty acids in Eskimos. Lancet 2:433-435 (1979).

31. Dyerberg J., Bang H.O., Stoffersen E., Moncada S., Vane J.R. Eicosapentaenoic acid and prevention of thombosis and atherosclerosis. Lancet 2:117-119 (1978).

32. Siess W., Scherer B., Bohlig B., Roth P., Kurzmann I., Weber P.C. Platelet-membrane fatty acids, platelet aggregation, and thromboxane formation during a mackerel diet. Lancet 1:441-444 (1980).

33. Leaf A., Weber P.C. Cardiovascular effects of n-3 fatty acids. N. Engl. J. Med. 318:549-557 (318).

34. Kromhout D., Bosschieter E.B., Coulander C. The inverse relation between fish consumption and 20-year mortality from coronary heart disease. N. Engl. J. Med. 312:1205-1216 (1985).

35. Burr M.L., Gilbert J.F., Holiday R.M., Elwood P.C., Fehily A.M., Rogers S. Effects of changes in fat, fish, and fibre intakes on death and myocardial reinfarction: diet and reinfarction trial (DART). Lancet 2:758-761 (1989).

36. Weiner B.H., Ockene I.S., Levine P.H., Guenoud H., Fisher M., Johnson B.F. Inhibition of atherosclerosis by cod-liver oil in a hyperlipidemic swine model. N. Engl. J. Med. 315-841-846 (1986).

37. Davis H.R., Bridenstine R.T., Vesselinovitch D., Wissler R.W. Fish oil inhibits development of atherosclerosis in Rhesus monkeys. Arteriosclerosis 7:441-449 (1987).

38. Turpie A.G.G. Clinical studies: Evidence for intervention with specific antiplatelet drugs in arterial thromboembolism. Semin. Thromb. Hemostas. 14:41-49 (1988).

39. ISIS-2 collaborative group. Randomized trial of intravenous streptokinase, oral aspirin, both or neither among 17187 cases of suspected acute myocardial infarction: ISIS-2. Lancet II: 349 (1988).

40. The steering committee of the physicians health study research group. Preliminary report: findings from the aspirin component of the ongoing physicians health study. N. Engl. J. Med. 318:262- (1988).

SMOKING, PLATELET REACTIVITY AND FIBRINOGEN

Jack Kutti

Department of Medicine, Östra Hospital
University of Göteborg, Göteborg, Sweden

INTRODUCTION

It is universally accepted that smoking constitutes a major
independent risk factor for the development of atherosclerosis as well as
arterial thrombosis formation and its most serious consequences, i.e.,
coronary artery disease/acute myocardial infarction and stroke (A Report
of the Surgeon General, 1983). Likewise, it is well recognized that
platelets play an important role in the development of occlusive arterial
disease (Schafer and Handin, 1979); there is indeed substantial evidence
to suggest that abnormal platelet/vessel wall interaction is involved in
the pathogenesis of the atherosclerotic lesion itself (Weksler and
Nachman, 1981; Hoak, 1988). Also, a strong association between plasma
fibrinogen concentration and platelet aggregability was reported by Meade
et al. (1985). In atherosclerosis, lipids and fibrinogen accumulate in
the subintimal layers of the arterial wall. Fibrinogen is an acute phase
protein and plasma concentrations rise as a result of a wide range of
stimuli many of which are of a non-specific nature, e.g., chronic
inflammatory disorders, after surgery, in malignant disease and during
pregnancy. Indeed, atheroma displays some of the characteristics of an
inflammatory response. The present article attempts to review some of
the concepts as to how smoking relates to platelet reactivity and plasma
fibrinogen concentration, and thereby to arterial vascular disease.

PLATELET REACTIVITY

In their attempts to evaluate platelet reactivity in response to
smoking, clinical investigators have employed a variety of laboratory
tests thought to reflect in vivo platelet behaviour. Herein are
included, e.g., measurements of platelet survival, the bleeding time,
platelet aggregation in response to various agonists, platelet
aggregation ratio, plasma concentrations of platelet-specific proteins,
serum thromboxane B_2 (TXB_2) formation and serotonin (5-HT) release.
However, only measurement of the standardized template bleeding time can
be considered a true reflection of in vivo events. A carefully performed
bleeding time yields a valuable measure of platelet participation in
primary hemostasis with sufficient sensitivity and reliability to serve
as the best clinical screening test of platelet function (Harker and
Slichter, 1979).

Tobacco Smoking and Atherosclerosis
Edited by J. N. Diana
Plenum Press, New York, 1990

<u>Platelet survival studies</u>

The results of studies obtained using ^{51}Cr-labelled platelets have shown reduced platelet survival in habitual smokers (Fuster et al., 1981); further, significant lengthening of shortened survival toward normal occurred in smokers who discontinued smoking.

<u>Measurement of bleeding time</u>

The template bleeding time is a highly repeatable test when there is strict adherence to multiple technical points. The presence of a blood pressure cuff, which creates venostasis to insure adequate venous filling, and the direction of the incision are technical factors that influence the repeatability of the test (Mielke, 1982). The results may also be disturbed by the possible recent intake of aspirin or other non-steroidal anti-inflammatory drugs which necessitate additional caution whenever the test is performed.

Despite all these considerations having been taken into account, rather conflicting results emerged when the effect of acute smoking on the template bleeding time was investigated. Thus, one group of Danish workers demonstrated that the bleeding time became significantly shortened immediately after the smoking of two high-nicotine content cigarettes (Ring et al., 1983; Madsen and Dyerberg, 1984); no such shortening was recorded after smoking two nicotine-free cigarettes (Ring et al., 1983). However, in two other studies carried out on healthy habitual smokers, no such shortening in bleeding time was recorded after the acute smoking of two cigarettes (Davis and Davis, 1983; Kampman and Hornstra, 1988). Nor was there any difference as regards bleeding time in between steady state habitual smokers and non-smokers (Davis and Davis, 1983).

<u>Platelet aggregation studies</u>

Similarly, the results of in vitro platelet aggregation studies in response to various agonists have yielded conflicting results as regards the effect of smoking. Herein, it appears necessary to distinguish between results obtained from steady state chronic/habitual smokers and results obtained in the immediate response to acute smoking.

<u>Chronic smoking</u>. As regards in vitro platelet aggregation in response to a variety of agonists including adenosine diphosphate (ADP), epinephrine, collagen, thrombin and arachidonic acid (AA) as well as platelet aggregation ratio (Wu and Hoak, 1974) most authors appear to agree that platelet sensitivity does not differ between chronic smokers and non-smoking control subjects (Chao et al., 1982; Carlsson and Wennmalm, 1983; Dotevall et al., 1987; Rival et al., 1987; Berglund et al., 1988; Lassila and Laustiola, 1988). Indeed, in one study ADP-induced aggregability was even greater in non-smokers than smokers (Meade et al., 1985). More interestingly, it was demonstrated by Lassila and Laustiola (1988) that after exercise, smokers' platelets were desensitized to all doses of epinephrine and low doses of ADP and collagen and the levels of serum TXB$_2$ were lower than among non-smokers; this finding was supported by the decreased release of 5 HT and TXB$_2$ during aggregation induced with epinephrine. It was also shown that aggregation of platelets stimulated by epinephrine was significantly reduced in smokers after exercise and the refractoriness appeared to be maintained for 15 and 30 minutes afterwards (Lassila 1989).

In studies of chronic smokers at rest the plasma concentrations of α-granule proteins, beta-thromboglobulin (BTG) and platelet factor 4

(PF4), and serum TXB_2 production did not differ between smokers and
non-smokers (Dotevall et al., 1987; Rival et al., 1987, Lassila and
Laustiola, 1988). Recent work has demonstrated the presence of
significantly larger numbers of TXA_2 platelet receptors in smokers than
in non-smokers (Modesti et al., 1989).

<u>Acute smoking</u>. The effect of acute smoking on platelet reactivity
remains more controversial. Thus, some studies have demonstrated
augmented ADP-induced in vitro platelet aggregation (Levine, 1973; Belch
et al., 1984). It was, however, shown by Belch et al. (1984) that baseline platelet sensitivity to ADP in chronic smokers did not differ from
results obtained from a group consisting of age- and sex-matched non-
smokers; further, acute smoking did not affect the plasma levels of BTG.
Similarly, by using ^{111}In-labelled autologous platelets and gamma camera
imaging, Schmidt and Rasmussen (1984) demonstrated that shortly after
smoking there was a significant sequestration of radiolabelled platelets
in the spleen; concomitantly, the plasma levels for BTG and PF4 increased
significantly, and there was a fall in platelet aggregation ratio.

In the experience of most workers, however, acute smoking did not
sensitize the platelets to either ADP, epinephrine, collagen, AA or
ristocetin (Siess et al., 1982; László et al., 1983; Ring et al., 1983;
Pittilo et al., 1984; Ahlsten et al., 1986; Vicari et al., 1988).
Further, in none of these studies did acute smoking affect platelet
aggregation ratios, plasma BTG concentrations or serum TXB_2 formation
(Siess et al., 1982; László et al., 1983; Ring et al., 1983; Vicari et
al., 1988). However, in the work of Ahlsten et al. (1986) platelet
aggregation was elicited by lower concentrations of AA in male smokers
than in male non-smokers, but no such difference was found between female
smokers and non-smokers or between male and female smokers. Indeed,
there is evidence to suggest that cigarette smoke inhibits AA-induced
platelet aggregation as well as second phase of epinephrine-induced
aggregation (Mansouri and Perry, 1982); this aggregation inhibition was
shown to be due to the presence of carbon monoxide.

<u>Comments on platelet aggregation studies</u>. The above discrepancies as
to platelet activation in response to acute smoking are not easily
explained. As regards in vitro platelet aggregation it must be kept in
mind that such studies are carried out on anticoagulated blood/platelet-
rich plasma and measure the response to different agonists. The results
of these tests are therefore subject to a variety of artifacts produced
by e.g., the drawing of blood, presence of anticoagulant and centrifugation. Consequently, the results obtained will only be crude reflexions
of true in vivo platelet behaviour. In addition to methodological
differences between laboratories, there are differences as regards the
experimental designs. Further, age and sex, presence or absence of
concurrent disease or medication (other than aspirin and non-steroidal
anti-inflammatory drugs) are additional factors which might influence the
results of experimental work.

It has been convincingly demonstrated that ADP-induced aggregability
increases with age in both sexes and is greater in women than men (Meade
et al., 1985; Berglund et al., 1988; Vilén et al. (1989). In the study
of Vilén et al. (1989) comprising 120 healthy humans of both sexes, a
narrow range of final ADP concentrations (0.1–1.0 µM) was used for the
evaluation of primary and secondary aggregation. The rate of primary aggregation was shown to be significantly related to increasing age and this
was true for both sexes. As regards secondary aggregation, the same
pattern was seen in male subjects; however, no similar association was
detectable among the females. Consequently, in platelet aggregation studies the interpretation of data obtained invariably requires a careful

selection of age- as well as sex-matched control groups. These
considerations most likely also apply to other assays of platelet
activation such as release of α-granule proteins and serum TXB_2 formation.

PLASMA FIBRINOGEN

   Platelet aggregation requires the presence of divalent cations and
fibrinogen in the suspending medium; it appears that the platelet
fibrinogen receptor occupies a central role in this process regardless of
the agonist used. Activated platelets undergo a rapid transition in
shape from discoidal cells to spheres with pseudopods. During shape
change, hidden receptors for binding of fibrinogen are exposed on the
platelet surface; Ca^{2+}-dependent binding to these receptors is necessary
for platelet aggregation (Peerschke, 1985; Steen and Holmsen, 1987).

   In large prospective studies it has been convincingly shown that
elevated plasma fibrinogen constitutes an independent risk factor for the
development of myocardial infarction as well as stroke (Meade et al.,
1980; Wihelmsen et al., 1984). Indeed, there is little doubt that
smokers have significantly higher plasma fibrinogen values than
non-smokers (Belch et al., 1984; Dotevall et al., 1987; Kannel et al.,
1987; Berglund et al., 1988; Rosengren et al., 1989). Estimates to date
suggest that 25% to 50% of the relation of cigarette smoking to
occurrence of atherosclerotic cardiovascular disease is attributable to
the effect of smoking on fibrinogen levels, which in turn enhances
thrombotic tendencies leading to occlusive clinical events (Kannel et
al., 1987). However, among current smokers there does not appear to be
an obvious dose-response relationship between fibrinogen concentration
and the amount of tobacco smoked (Rosengren et al., 1989). Evidence also
exists that this elevation in plasma fibrinogen is reversible since
ex-smokers have values similar to non-smokers/subjects who never smoked
(Kannel et al., 1987; Berglund et al., 1988; Rosengren et al., 1989).

   It is well recognized that also peripheral arterial disease is
closely linked to smoking habits. In a recent study of patients with
femoropopliteal vein grafts it was shown that two principle factors,
continued smoking and plasma fibrinogen concentration, adversely affected
graft patency (Wiseman et al., 1989). In contrast, rather unexpectedly,
increased serum cholesterol and plasma low density lipoprotein
cholesterol concentrations were associated with improved patency.
Indeed, plasma fibrinogen concentration was the most important variable
predicting graft occlusion, followed by smoking markers; patency in
patients with plasma fibrinogen above the median value was only 57%
compared with 90% in those with concentrations below the median. It was
emphasized that forceful approach is needed to stop patients smoking
(Wiseman et al., 1989).

   In addition to high plasma fibrinogen concentrations chronic smokers
demonstrate elevated levels of other acute phase proteins including
α_1-antitrypsin, haptoglobin, orosomucoid and properdin factor B (Chao et
al., 1982). It could therefore well be that all those variables reflect
a primary non-specific overall inflammatory injury to the vascular intima
induced by chronic exposure to tobacco smoke (Dotevall et al., 1987).
This hypothesis is supported by the fact that the effect of smoking on
plasma fibrinogen concentration appears to be direct and reversible in
case smoking is stopped.

ACKNOWLEDGEMENT

   This work was in part supported by a grant from the Swedish Cancer
Society) project 432-B90-09XB).

REFERENCES

A Report of the Surgeon General. The health consequences of smoking: cardiovascular disease. US Department of Health and Human Services, Rockville, MD. (1983).

Ahlsten, G., Ewald, U., and Tuvemo, T. Arachidonic-acid-induced platelet aggregation is increased in male but not in female smokers, Prostaglandins Leukotrienes Med, 21:149 (1986).

Belch, J. J. F., McArdle, B. M., Burns, P., Lowe, G. D. O., and Forbes, C. D. The effects of acute smoking on platelet behaviour, fibrinolysis and haemorheology in habitual smokers, Thromb Haemost, 51:6. (1984).

Berglund, U., Wallentin, L., and von Schenck, H. Platelet function and plasma fibrinogen and their relations to gender, smoking habits, obesity and beta-blocker treatment in young survivors of myocardial infarction, Thromb Haemost, 60:21 (1988).

Carlsson, I., and Wennmalm, Å. Platelet aggregability in smoking and non-smoking subjects, Clin Physiol, 3:565 (1983).

Chao, F. C., Tullis, J. L., Alper, C. A., Glynn, R. J., and Sibert, J. E. Alterations in plasma proteins and platelet functions with aging and cigarette smoking in healthy men, Thromb Haemost, 47:259 (1982).

Davis, R. F. and Davis, J. W. The effect of smoking on the stressed template bleeding time, Ann Clin Res, 15:131 (1983).

Dotevall, A., Kutti, J., Teger-Nilsson, A. C., Wadenvik, H., and Wilhelmsen L. Platelet reactivity, fibrinogen and smoking, Eur J Haematol, 38:55 (1987).

Fuster, V., Chesebro, J. H., Frye, R. L., and Elveback, L. R. Platelet survival and the development of coronary artery disease in the young adult: effects of cigarette smoking, strong family history and medical therapy, Circulation, 63:546 (1981).

Harker, L. A., and Slichter, S. J. The bleeding time as a screening test for evaluation of platelet function, N Engl J Med, 187:155 (1972).

Hoak, J. C. Platelets and atherosclerosis, Semin Thromb Hemost, 14:202 (1988).

Kampman, M. T., and Hornstra, G. No acute effect of cigarette smoking, and risk of cardiovascular disease: insights from the Framingham study, Am Heart J, 113:1006 (1987).

Lassila, R., and Laustiola, K. E. Physical exercise provokes platelet desensitization in men who smoke cigarettes - involvement of sympathoadrenergic mechanisms - a study of monozygotic twin pairs discordant for smoking, Thromb Res, 51:145 (1988).

Lassila, R. The platelet α_2-adrenoceptor and prostacyclin sensitivity are not altered by cigarette smoking - a study of monozygotic twin pairs discordant for smoking, Thromb Res , 54:339 (1989).

László, E., Káldi, N., and Kovács, L. Alterations in plasma proteins and platelet functions with aging and cigarette smoking in healthy men, Thromb Haemost, 49:150 (1983).

Levine, P. H. An acute effect of cigarette smoking on platelet function. A possible link between smoking and arterial thrombosis, Circulation, 48:619 (1973).

Madsen, H., and Dyerberg, J. Cigarette smoking and its effect on platelet-vessel wall interaction, Scand J Clin Lab Invest, 44:203 (1984).

Mansouri, A., and Perry, C. A. Alteration of platelet aggregation by cigarette smoke and carbon monoxide, Thromb Haemost, 48:286 (1982).

Meade, T. W., Chakrabarti, R., Haines, A. P., North, W. R. S., Stirling, Y., Thompson, S. G., and Brozovic, M. Haemostatic function and cardiovascular death: early results of a prospective study, Lancet, 1:1050 (1980).

Meade, T. W., Vickers, M. V., Thompson, S. G., Stirling, Y., Haines,
    A. P., and Miller, G. J. Epidemiological characteristics of
    platelet aggregability, <u>Br Med J</u>, 290:428 (1985).
Mielke, C. H. Aspirin prolongation of the template bleeding time:
    influence of venostasis and direction of incision, <u>Blood</u>, 60:1139
    (1982).
Modesti, P. A. Abbate, R., Gensini, G. F., Colella, A., and Neri
    Serneri, G. G. Platelet thromboxane A_2 receptors in habitual
    smokers. <u>Thromb Res</u>, 55:195 (1989).
Peerschke, E. I. B. The platelet fibrinogen receptor, <u>Semin Hematol</u>,
    22:241 (1985).
Pittilo, R. M., Clarke, J. M. F., Harris, D., Mackie, I. J., Rowles,
    P. M., Machin, S. J., and Woolf, N. Cigarette smoking and
    platelet adhesion, <u>Br J Haematol</u>, 58:627 (1984).
Ring, T., Kristensen, S. D., Jensen, P. N., Mourits-Andersen, T.,
    Madsen, H., and Dyerberg, J. Cigarette smoking shortens the
    bleeding time, <u>Thromb Res</u>, 32:531 (1983).
Rival, J., Riddle, J. M., and Stein, P. D. Effects of chronic smoking
    on platelet function, <u>Thromb Res</u>, 45:75 (1987).
Rosengren, A., Wilhelmsen, L., Welin, L., Teger-Nilsson, A. C., and
    Tsipogianni, A. Social influences and cardiovascular risk factors
    as determinants of plasma fibrinogen in a general population
    sample of middle-aged men (submitted for publication) (1989).
Schafer, A. I., and Handin, R. I. The role of platelets in thrombotic
    and vascular disease, <u>Prog Cardiovasc Dis</u>, 22:31 (1979).
Schmidt, K. G., and Rasmussen, J. W. Acute platelet activation induced
    by smoking. In vivo and ex vivo studies in humans, <u>Thromb
    Haemost</u>, 51:279 (1984).
Siess, W., Lorenz, R., Roth, P., and Weber, P. C. Plasma catecholamines
    platelet aggregation and associated thromboxane formation after
    physical exercise, smoking or norepinephrine infusion,
    <u>Circulation</u>, 66:44 (1982).
Steen, V. M., and Holmsen, H. Current aspects on human platelet
    activation, <u>Eur J Haematol</u>, 38:383 (1987).
Vicari, A. M., Margonato, A., Macagni, A., Luoni, R., Seveso, M. P.,
    Vicedomini, G., and Pozza, G. Effects of acute smoking on the
    hemostatic system in humans, <u>Clin Cardiol</u>, 11:538 (1988).
Vilen, L., Jacobsson, S., Wadenvik, H., and Kutti, J. ADP-induced plate-
    let aggregation as a function of age in healthy humans, <u>Thromb
    Haemost</u>, 61:490 (1989).
Weksler, B. B., and Nachman, R. L. Platelets and atherosclerosis, <u>Am
    J Med</u>, 71:331 (1981).
Wilhelmsen, L., Svärdsudd, K., Korsan-Bengtsen, K., Larsson, B., Welin,
    L., and Tibblin, G. Fibrinogen as a risk factor for stroke and
    myocardial infarction, <u>N Engl J Med</u>, 311:501 (1984).
Wiseman, S., Kenchington, G., Dain, R., Marshall, C. E., McCollum, C. N.,
    Greenhalgh, R. M., and Powell, J. T. Influence of smoking and
    plasma factors on patency of femoropopliteal vein grafts, <u>Br Med
    J</u>, 299:643 (1989).
Wu, K. K., and Hoak, J. C. A new method for the quantitative detection
    of platelet aggregation in patients with arterial insufficiency,
    <u>Lancet</u>, 2:924 (1974).

VASCULAR AND PLATELET EICOSANOIDS, SMOKING AND ATHEROSCLEROSIS

J.Y. Jeremy and D.P. Mikhailidis

Department of Chemical Pathology and Human Metabolism
Royal Free Hospital and School of Medicine
University of London
Pond Street, London NW3 2QG, UK

INTRODUCTION

Vascular tissue synthesizes prostanoids (PGs) that modulate
contractility, myocyte and fibroblast proliferation, platelet aggregation
and leucocyte function, all of which are key components in the
pathophysiology of atherogenesis. Furthermore, platelet and leucocyte
release substances have been shown to stimulate the synthesis and release
of PGs from vascular cells. It is possible, therefore, that vascular PGs
are part of a protective response against atherogenic events.
Consequently, any disruption of PG synthesis (viz. as a result of
smoking) may accelerate the atherogenic state. Although smoking is a
major risk factor in the development of atherosclerosis, the combination
of smoking with other risk factor (e.g., diabetes, hypertension, hypercholesterolaemia) markedly increases the likelihood of death from
atherosclerotic disease. The present paper therefore discusses the
following:

1) How PGs relate to pathophysiology of atherosclerosis
2) Experimental findings on the effects of cigarette smoking
   on vascular and platelet prostanoid synthesis.
3) How other risk factors may interact with smoking to
   influence vascular permeability
4) Future directions and emphases on research into eicosanoids
   and smoking.

Before discussing the effects of cigarette smoking on vascular PG
synthesis, it is relevant to briefly review current concepts on the
pathogenesis of atherosclerosis and how they may relate to PGs.

The key event in the initiation and perpetuation of the fibrous
plaque, the pathognomonic lesion of atherosclerosis, is the proliferation
of arterial smooth muscle cells, followed by deposition of lipid and the
accumulation of collagen, elastic fibres and proteoglycans (1). Although
the trigger for these events is unknown, it has been proposed that
atherogenesis is initiated by endothelial cell injury leading to
adherence of platelets resulting in local release of platelet
constituents including TXA_2, platelet derived growth factor (PDGF),
platelet activating factor (PAF), serotonin (5HT) and histamine (2).

Tobacco Smoking and Atherosclerosis
Edited by J. N. Diana
Plenum Press, New York, 1990

Since this proposal, evidence for the platelet being the progenitor of
the atheroma has not been forthcoming. However, there is increasing
evidence that the macrophage is in fact the progenitor of the
atherosclerotic plaque (3). Studies on laboratory animals and
microscopic observations of human lesions revealed that monocyte adhesion
to vascular endothelium and subsequent migration of macrophages
(transformed monocytes) into the vascular intima is an early event in
fatty streak formation (4). Furthermore, it is now well established that
foam cells (the principal cell type in small lesions found at the edges
of most advanced plaques) are macrophages (5). Activated macrophages
synthesize and release a plethora of potent bioactive substances that are
mitogenic, chemoattractant, cytotoxic, vasoconstrictive, and increase
vascular permeability. These release substances include histamine,
serotonin, leukotrienes, hydroxy and hydroperoxy fatty acids,
interleukins (IL), tumor necrosing factor (TNF), TXA_2, PDGF-like
substance, toxic free radicals, and neutral proteases (elastase and
collagenase). The release of these substances is likely to elicit the
following consequences: 1) proliferation of intimal smooth muscle cells,
2) production of ceroid (cholesterol-containing lipid) within the
atherosclerotic lesion, 3) necrosis within and around the lesion
resulting in increased vascular permeability to lipoproteins and
cholesterol, and 4) the attraction of monocytes and other leukocytes
resulting in an amplification cascade of the above deleterious actions.

<u>Distribution, Control of Synthesis, and Properties of PGs in Blood Vessels</u>

Human and animal blood vessels possess the capacity to synthesize a
wide range of PGs, principally (and in this general quantitative rank
order): prostacyclin (PGI_2), PGE_2, PGD_2, $PGF_{2\alpha}$ and thromboxane A_2 (TXA_2)
(6,7). Although it is largely thought that the endothelium is the
principal source of PGI_2, smooth muscle cells (SMC) and fibroblasts also
possess the capacity to synthesize PGs (8,9). Furthermore, in terms of
the PG synthesizing capacity of the whole vessel the endothelium actually
contributes negligible quantities of PGs relative to that produced by the
media part of the vessel (10).

PGs are derived from certain essential fatty acids, principally
arachidonic acid. Fatty acid precursors are liberated from phospholipid
stores in plasma membranes by the hydrolytic action of phospholipase A_2
(PLA_2; activated by calcium). The fatty acids are then converted to PGs
by the sequential action of cyclooxygenase and synthases and/or
isomerases. Recent studies have established that excitatory receptor
activation (α-adrenergic (11), serotonergic (12), histaminergic (13),
protein kinase C^{14} (PKC) and G protein activators (15) all stimulate
vascular PG synthesis (for review see Ref. 16). In turn, receptor linked
PKC and G protein stimulated PG synthesis is inhibited by calcium channel
blockers (11,14,15,16), demonstrating that calcium channel activation is
a key event in the stimulation of vascular PG synthesis (for hypothetical
working model see Fig. 1).

It is well established that a major role of vascular PGs (viz., PGI_2)
is that of prevention platelet adhesion and aggregation at the endothelium
(17). Apart from the control of platelet adhesion and aggregation PGs
also possess properties that are directly related to the pathogenesis of
atherosclerosis. It has long been known that PGs inhibit the proliferation of smooth muscle cells in culture (18,19).

PGI_2 has been shown to inhibit the proliferation of arterial smooth
muscle cells in vivo (20). PGE_1 causes a dose dependent rise in cAMP
which is followed by a dose dependent increase in glycosaminoglycan (GAG)
secretion and reduced proliferation (21). In a recent elegant study

Libby et al. (22) demonstrated that IL-1 alpha and beta (a potent
stimulator of PG release by smooth muscle tissue and fibroblasts)
stimulated human (saphenous vein) vascular smooth muscle cells
proliferation only in the presence of cyclooxygenase inhibitors. They
further found that addition of PGE_1, PGE_2 (but not of PGI_2, $PGF_{2\alpha}$, or
$PGF_{1\alpha}$), in the presence of cyclooxygenase inhibitors, blocked IL-1
stimulated proliferation. This study therefore indicated that PGs of the
E series may play a key role in controlling vascular cell proliferation.
(For a recent comprehensive recent review on this topic see Ref. 23).

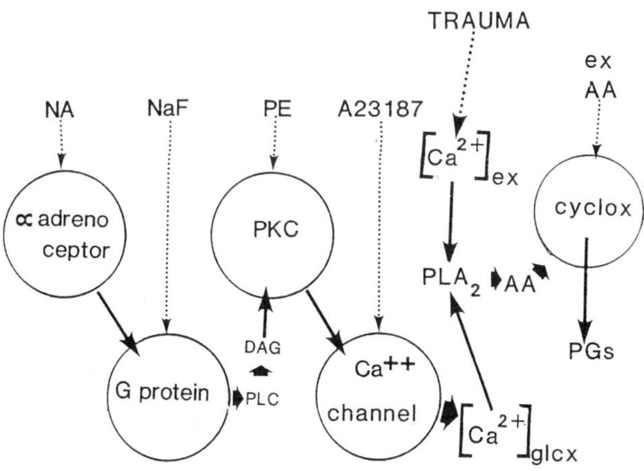

Fig. 1. Scheme for the intracellular control and site of action of
stimulators of PGI_2 synthesis by isolated vascular smooth muscle.
Excitatory agonists (e.g., noradrenaline) stimulates PGI_2 synthesis via
activation of alpha adrenoceptors. NaF stimulates PGI_2 synthesis through
G protein activation which in turn activates phospholipase C, thereby
generating diacyl glycerol (DAG), a protein kinase C activator. Phorbol
ester dibutyrate (PDBU: a DAG mimetic) stimulates PGI_2 through activation
of PKC possibly via cell surface calcium (channel) mobilization. A23187
stimulates PGI_2 synthesis through creation of artificial calcium
channels, which in turn activates phospholipase A_2 (PLA_2) which generates
arachidonate from phospholipid stores. Trauma (cutting, sonication,
freeze fracturing) stimulates PGI_2 synthesis by disrupting the cell
membrane, thereby stimulating PLA_2 through massive, non-calcium channel
mediated influx of extracellular calcium. Arachidonate (AA) stimulates
PGI_2 through direct conversion by the cyclooxygenase/synthetase/isomerase
system.

     Surprisingly little is known of the effect of PGs on leucocyte
function. PGI_2 has been shown to inhibit chemotaxis of polymorphonuclear
leucocytes PMNs (24) and PG_2 enhances by almost 200% the chemotactic
responsiveness of human monocytes to complement activated human serum

(25). The E-series PGs increase the levels of cAMP in leucocytes and
reduce extrusion of lysosomal enzymes (26). However, it is clear that
more work on PG-leucocyte interactions is required.

Interactions of PGs with Platelet and Leucocyte Release Substances

The following platelet release substances have all been shown to
stimulate vascular (endothelial and smooth muscle) PGI_2 synthesis: TXA_2
(27) ADP (28), 5HT (12), histamine (13), platelet activating factor (29)
(PAF) and PDGF (30). It was proposed that the release of PGI_2 in
response to these platelet-derived factors constitutes the mechanism by
which PGI_2 prevents the adhesion of platelets to vascular endothelium and
possibly also the elongation of thrombus formation (27). Leucocyte
release substances have likewise been shown to stimulate PG synthesis by
smooth muscle cells and fibroblasts: interleukins (31,32), leukotrienes
(33), histamine (12), serotonin (13). The release of vascular (both
smooth muscle and endothelial) PGs in response to proatherogenic platelet
and leucocyte release substances may constitute a fundamental protection
against atherosclerotic events. Given that PGs (viz., PGI_2 and the
E-series PGs) inhibit the known pathophysiological sequelae of
atherosclerosis (platelet adhesion, monocyte adhesion and migration,
smooth muscle cell proliferation) it is reasonable to predict that any
disruption of vascular PG synthesis (viz; through smoking) may contribute
to the atherogenesis.

Cigarette Smoking and Vascular PG Synthesis

Cigarette smoking causes morphological changes in the vascular
endothelium in laboratory animals and in umbilical and uterine vessels
obtained from smoking mothers (34,35). Furthermore, these morphological
changes appear to be associated with inhibition of PGI_2 synthesis (34)
(effects of smoking on the endothelium are reviewed in another chapter of
these proceedings). It has also been shown that cigarette smoking
significantly reduces urinary excretion of 6-oxo-$PGF_{1\alpha}$ secretion in man
(36) and umbilical vessels from smoking mothers produce less PGI_2 (in
vitro) than vessels obtained from non-smoking mothers (37). More
recently it has been shown that cigarette smoking reduces PGI_2 synthesis
by vascular endothelial cells (38).

Of the components of cigarette smoke, nicotine has attracted the
widest attention. Nicotine has been shown to cause morphological changes
of the vascular endothelium (39) and to inhibit in vitro PGI_2 synthesis
in vascular tissue (40,41). However, other groups and ourselves have
failed to demonstrate an inhibition of vascular prostanoid synthesis by
nicotine (42,43).

Since we were unable to elicit inhibition of vascular PGI_2 with
nicotine, we felt that it was possible that cigarette smoke was
diminishing vascular PG synthesis through indirect mechanisms (e.g.,
increased circulating NEFAs which in turn inhibit PG synthesis).
However, in order to rule out direct effects, we developed a simple means
of testing effects of cigarette smoke on PGI_2 release. Briefly, the
filter end of a cigarette (middle tar) was inserted into a 10-cm length
of tubing. This in turn was attached to tubing submerged in 10 ml of
Kreb's Ringer bicarbonate buffer (KRB). Vacuum was then applied to the
flask through another tube so that the ignited cigarette was completely
consumed over a 5-min period, the smoke bubbling through the KRB. Rat
aortae, rabbit aortae, and human umbilical arteries and veins were cut
into 2-mm rings and incubated in the KRB-containing cigarette smoke
extracts. It should be noted here that this form of stimulating PGs

bypass receptor linked calcium channels as described above (see Fig. 1
for explanatory notes). Released PGs were measured by specific
radioimmunoassays. We also investigated the effect of cigarette smoke
extracts on the conversion of [^{14}C]arachidonic acid to [^{14}C] 6-oxo-PGF$_{1\alpha}$
(the stable spontaneous hydrolysate of PGI$_2$). These experiments
demonstrated that cigarette smoke extracts inhibited both the spontaneous
release of PGI^2 as well as conversion of [^{14}C]-AA to [^{14}C]-oxo-PGF$_{1\alpha}$. We
concluded therefore that cigarette smoking directly inhibited PGI$_2$
synthesis at the cyclooxygenase level and that this effect may contribute
to the pathophysiology of smoking-related vascular disease via disruption
of the vessel-platelet interaction (viz., the PGI$_2$-TXA$_2$ balance).

Subsequent to these studies we concentrated our research on receptor
linked PG synthesis in vascular and other smooth muscle tissues and the
intracellular control of these mechanisms. The key to this form of study
was that we first allowed PG release from vessels elicited by preparative
handling (trauma) to subside (see Fig. 1 for explanatory notes). In one
particular study we investigated neurotransmitter control of PG synthesis
by the isolated rat trachea and found parasympathomimetics to be potent
stimulators of PG synthesis in that tissue (44). Since cigarette smoking
is of particular importance in the pathogenesis of airways disease (e.g.,
chronic bronchitis) we looked at the effect of cigarette smoke extracts
on de novo and carbachol-stimulated PG release by this tissue.
Surprisingly, we found that cigarette smoke extracts showed a biphasic
effect on the release of PG by this tissue: stimulation at low
concentrations and inhibition at high concentrations of extracts (44).
We reasoned that since cigarette smoke extracts contain an enormous
number of chemical components (> 2,000), this may constitute an
activation by certain components which was overridden (i.e., inhibited)
by other components at higher concentrations of cigarette smoke (44).

Having observed this phenomenon in the trachea model, we investigated
the effect of cigarette smoke extracts on de novo and stimulated PG
release by isolated rat aorta. Here again we found cigarette smoke
extracts to have a biphasic effect on both PGI$_2$ and PGE$_2$ synthesis:
i.e., stimulatory at low concentrations and inhibitory at high
concentrations (Fig. 2). Furthermore, cigarette smoke extracts
potentiated the stimulatory Fig. 2 effects of noradrenaline, phorbol
ester and calcium ionophore A23187 on aortic PGI$_2$ synthesis (Fig. 2).
These latter synergisms indicate that cigarette smoke extracts are acting
through systems common to noradrenaline, phorbol ester, and calcium
ionophore. Since stimulation of vascular PG synthesis by these compounds
is mediated by calcium mobilization, we investigated the effect of
calcium channel blockers on cigarette smoke extract-stimulated PG
synthesis. Here we found that nifedipine, diethylstilbestrol, and
verapamil all inhibited cigarette smoke extract-stimulated PGI$_2$ synthesis
at concentrations similar to those required to inhibit the aforementioned
stimulators of PGI$_2$ synthesis (Fig. 3). Thus it appears that cigarette
smoke contains a component(s) that actually stimulates vascular PGI$_2$
synthesis via calcium mobilization.

These results would seem to mitigate against prostanoids playing an
aetiological role in atherosclerosis since it has been postulated that
reduction in vascular PG synthesis would tend toward an atherogenic
state. It should be noted that the present vascular model does not
reflect possible effects on endothelial PG production since we have
recently shown that removal of endothelium does not significantly alter
PG output by the rat aorta when stimulated by agents such as
noradrenaline and phorbol ester (10). Thus, we do not preclude that
cigarette smoke inhibits endothelial PG synthesis.

In the context of the present data, a recent study has demonstrated that proliferating cells synthesize more PGI_2 than quiescent cells (45). We have also shown that mitogens (viz., phorbol esters) are potent stimulators of PGI_2 synthesis. The possibility arises, therefore, that the stimulatory action of cigarette smoke extracts on PGI_2 may reflect an activation of proliferation of vascular cells. Given the key role of SMC and fibroblast proliferation in atherogenesis, the possibility that cigarette smoke is directly eliciting cell proliferation warrants further investigation.

It is also of interest that epidemiological studies have shown that blood pressure tends to be lower in smokers than in non-smokers (46). This finding is surprising because smoking raises blood pressure acutely (46). This acute effect may be mediated by the release of catecholamines, but may also be the result of direct activation of vascular smooth muscle which is consistent with the stimulatory effect of cigarette smoke extracts on PGI_2 release reported here, since there is a direct relationship between vasoconstriction and the release of this eicosanoid (16).

Given the potential role of platelets in atherosclerosis, we also assessed the effect of cigarette smoke extracts on platelet thromboxane A_2 synthesis. Briefly, platelets were obtained from non-smoking volunteers. The platelets were washed free of plasma and resuspended in buffer. Following incubation in the presence of varying concentrations of cigarette smoke extracts, TXA_2 synthesis was stimulated with collagen ($20\mu g.ml^{-1}$). TXA_2 was measured as its stable spontaneous hydrolysate (TXB_2) by radioimmunoassay. As can be seen from Fig. 4 platelet TXA_2 was markedly inhibited by extremely low concentrations of cigarette smoke extracts.

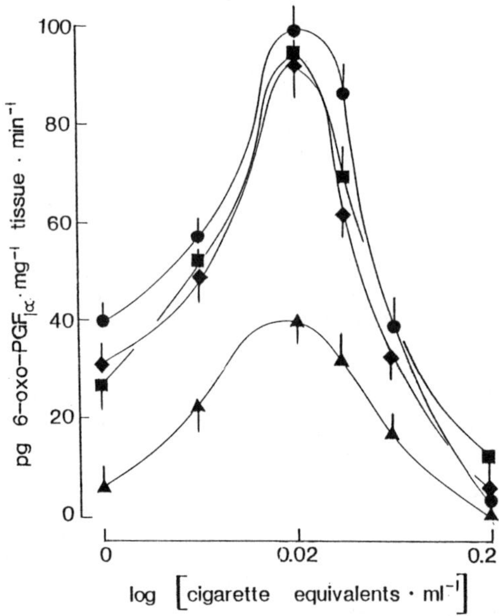

Fig. 2. Effect on in vitro rat aortic PGI_2 synthesis of cigarette smoke extract alone (▲) and in combination with $1\mu M$ noradrenaline (♦), 250nM phorbol ester dibutyrate (■) and $10\mu M$ calcium ionophore A23187 (●). Each point represents the mean of 6 determinations (± S.E.M.)

Data on effects of cigarette smoking on platelet function are
equivocal. However, our results are consistent with a large study on
platelet aggregation which showed that ADP-induced aggregation (ex vivo)
was diminished in long term smokers when compared with controls (47). It
is also of interest that smoking has been reported to be associated with
a decrease in the incidence of deep vein thrombosis (DVT) despite its
deleterious effects on the vasculature and hemostasis (48). There is,
therefore, evidence that, at least in some circumstances, the incidence
of vascular events is diminished in smokers. The exact mechanisms
involved in this phenomenon remain to be established.

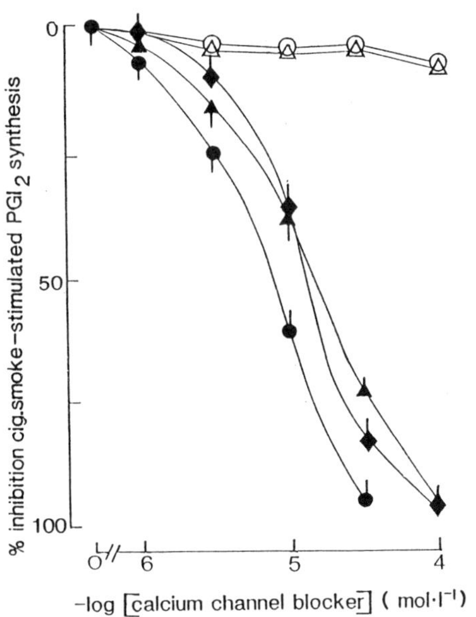

Fig. 3. Effect of calcium channel blockers: nifedipine (●);
diethylstilbestrol (▲) and verapamil (♦) on cigarette smoke extract
stimulated (0.02 equivalents per ml) PGI$_2$ synthesis by rat aorta. Each
point represents the mean of 6 determinations (± S.E.M.)

<u>Interrelationship between Smoking and other Atherosclerotic Risk Factors</u>

Epidemiologically, it has become increasingly apparent the
combination of one or more of the known risk factors in atherosclerosis
markedly enhances mortality due to vascular disease. For example, a
recent epidemiological study has shown that in diabetic patients with
high cholesterol and who smoke, there is a profound increase in the
probability of dying through myocardial infarction (49). In this
context, major risk factors have all been shown to result in diminished
vascular PG synthesis (diabetes mellitus; hyperlipidaemia, hypercholesterolaemia). Furthermore, the combinations of risk factors have
been shown to have additive inhibitory effects on vascular PG synthesis
(e.g., cigarette smoke and diabetes in the rat penis (50); diabetes and
hypercholesterolaemia in the rat aorta (51).

It is also important to consider that smoking also results in other
changes that are likely to increase the risk of thrombosis. For example,
smokers have significantly higher plasma viscosity (52), packed cell
volume (52,53) and plasma fibrinogen concentration (54,55) (an
established risk factor for IHD (54) than non-smokers. These factors
will disturb blood flow (52-56) and fibriniogen has also been shown to
activate platelets (57). Smoking is also known to adversely affect the
lipid profile (58), a major predictor of IHD. For example, smokers have
lower serum high density lipoprotein (HDL) concentrations than
non-smokers. This observation is relevant because it is well established
that HDL is inversely related to the risk of IHD. Furthermore, this
lipoprotein may exert beneficial effects on platelets and PGI_2 synthesis
by blood vessels (59).

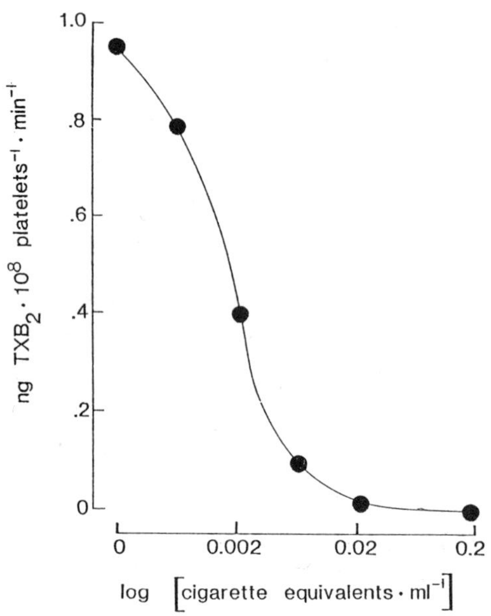

Fig. 4 Effect of cigarette smoke extracts on TXA_2 (as TXB_2) synthesis by
washed human platelets. Each point represents the mean of three
determinations.

We should also consider that smoking results in lipolysis and
consequently an increase in plasma NEFA concentration; this effect is
mediated by the release of catecholamines (60). These changes are of
possible relevance because of a multitude of reasons. Thus, elevated
plasma catecholamine concentrations have been associated with platelet
activation (61). Lipolysis is associated with endothelial damage
(34,39), decreased PGI_2 bioavailability (62) and possible diminished
synthesis (63) of this prostanoid. On the other hand the increased
availability of NEFA may lead to increased substrate (arachidonic acid)

142

availability and consequently, enhanced PG synthesis in both blood
vessels and platelets.

<u>Concluding Remarks</u>

Cigarette smoke contains potent bioactive substances that appear to
elicit, at least in vitro;
1) the stimulation or inhibition of vascular PG synthesis depending
   on concentration of smoke extract;
2) the inhibition of platelet TXA_2 synthesis.
These data would seem to indicate that atherogenesis associated with
smoking may not necessarily involve a disruption of the $TXA_2:PGI_2$
balance. However, these components of cigarette smoke may directly
elicit vascular cell proliferation and influence leucocyte function. The
identification and further investigation into the nature of these
compounds should be one of the main objectives of cigarette smoke-related
research.

REFERENCES

1. M.D. Haust, R.H. More and H.Z. Movat. The role of smooth muscle
    cells in the fibrogenesis of arteriosclerosis. Am. J. Pathol. 37:
    377 (1960).
2. R. Ross, J. Glomset and L. Harker. Response to injury and
    atherogenesis. Am. J. Pathol. 86: 675 (1977).
3. R. Ross and J.A. Glomset. The pathogenesis of atherosclerosis:
    an update. N. Engl. J. Med. 314: 488 (1977).
4. R.G. Gerrity. The role of monocytes in atherosclerosis. I
    Transition of blood borne monocytes into foam cells in fatty
    lesions. Am. J. Pathol. 103: 181 (1981).
5. M.J. Mitchinson and R.Y. Ball. Macrophages and atherogenesis.
    Lancet 2: 146 (1987).
6. P. Hadhazy, B. Malomvolgyi and K. Magyar. Endogenous prostanoids
    and arterial contractility. Prostagl. Leuk. Essential Fatty Acids
    Revs. 32: 175 (1988).
7. V. Tsang, J.Y. Jeremy, D.P. Mikhailidis, R.K. Walesby, J.V. Wright
    and P. Dandona. The release of prostacyclin by the human aorta.
    Cardiovasc. Res. 22: 489 (1988).
8. S. Moncada, A. Herman, E.A. Higgs and J.R. Vane. Differential
    formation of prostacyclin (PGX or PGI_2) by layers of the arterial
    wall. An explanation for the antithrombotic properties of vascular
    endothelium. Thromb. Res. 11: 323 (1977).
9. N.L. Baenzinger, M.J. Dillender and P.W. Majerus. Cultured human
    fibroblasts and arterial cells produce a labile platelet-inhibitory
    prostaglandin. Biochem. Biophys. Res. Commun. 78: 294 (1977).
10. J.Y. Jeremy and P. Dandona. Effect of endothelium removal on
    stimulatory and inhibitory modulation of vascular prostacyclin
    synthesis. Br. J. Pharmacol. 96: 243 (1989).
11. J.Y. Jeremy, D.P. Mikhailidis and P. Dandona. Adrenergic modulation
    of vascular prostacyclin synthesis. Eur. J. Pharmacol. 114: 133
    (1985).
12. S.R. Coughlin, M.A. Moskowitz, H.N. Antoniades and L. Levine.
    Serotonin receptor-mediated stimulation of bovine smooth muscle
    cell prostacyclin synthesis and its modulation by platelet derived
    growth factor. Proc. Natl. Acad. Sci. 78: 7134 (1981).
13. N.L. Baezinger, F.J. Fogerty, L.F. Mertz and L.F. Chernuta.
    Regulation of histamine-mediated prostacyclin synthesis in cultured
    human vascular endothelial cells. Cell. 24: 915 (1981).

14. J.Y. Jeremy and P. Dandona. The role of the diacycl glycerol protein kinase C system in mediating adrenoceptor-prostacyclin synthesis coupling in the rat aorta. Eur. J. Pharmacol. 136: 311 (1987).
15. J.Y. Jeremy and P. Dandona. Fluoride stimulates vascular prostacyclin synthesis: interrelationship of G proteins and protein kinase C. Eur. J. Pharmacol. 146: 279 (1988).
16. J.Y. Jeremy, D.P. Mikhailidis and P. Dandona. Excitatory receptor-prostanoid synthesis coupling in smooth muscle: mediation by calcium protein kinase C and G proteins. Prostagl. Leuk. Essential Fatty Acids Revs. 34: 215 (1988).
17. S. Moncada and J.R. Vane. Prostacyclin formation and effects. In Chemistry, Biology and Pharmacological activity of prostanoids. (S.M. Roberts and F. Scheinmann, eds.). Pergamon Press. p. 258 (1979).
18. J.J. Huttner, E.T. Gwebu, R.V. Panganamala, G.E. Milo and D.B. Cornwell. Fatty acids and their prostaglandin derivatives: inhibitors of proliferation in aortic smooth muscle cells. Science. 197: 289 (1977).
19. H.D. Peters, B.A. Peskar and P.S. Schonhofer. Influence of prostaglandins on connective tissue cell growth and function. Naunyn Schmiedeberg's Arch Pharmacol. 297: 587 (1977).
20. N. Owen. Prostacyclin can inhibit DNA synthesis in vascular smooth muscle cells. In Prostaglandins, leukotrienes and lipoxins. Biochemistry mechanisms of action and clinical applications. New York: Plenum Press. P. 193 (1985).
21. J. Nilsson and A.G. Olsson. Prostaglandin E_1 inhibits DNA synthesis in arterial smooth muscle cells stimulated with platelet derived growth factor. Atherosclerosis 53: 77 (1984).
22. P. Libby, S.J.C. Warner and G.B. Friedman. Interleukin 1: a mitogen for human vascular smooth muscle cells that induces the release of growth inhibitory prostanoids. J. Clin. Invest. 81: 487 (1988).
23. K.B. Pomerantz and D.P. Hajjar. Eicosanoids in regulation of arterial smooth muscle cell phenotype, proliferative capacity and cholesterol metabolism. Arteriosclerosis. 9: 413 (1989).
24. B.B. Weksler, J.M. Knapp and E.A. Jaffe. Prostacyclin (PGI2) synthesized by cultured endothelial cells modulates polymorphonuclear leucocyte functions. Blood (Suppl 1): 287 (1977).
25. W. McClatchney and R. Snyderman. Prostaglandin and inflammation: enhancement of monocyte chemotactic responsiveness by prostaglandin E_2. Prostaglandins 12: 415 (1976).
26. R.B. Zurier. Prostaglandins and inflammation. In Prostaglandins. Biology and Chemistry of Prostaglandins and Related Eicosanoids. (P.B. Curtis, ed.) Churchill Livingstone, Edinburgh. p. 595 (1988).
27. J.Y. Jeremy, D.P. Mikhailidis, P. Dandona. The thromboxane A_2 analogue U46619 stimulates vascular prostacyclin synthesis. Eur. J. Pharmacol. 107: 259 (1985).
28. A. Van Coevorden and J.M. Boeynaems. Physiological concentrations of ADP stimulate the release of prostacyclin from bovine endothelial cells. Prostaglandins 27: 615 (1984).
29. S.T. Test and N.U. Bang. Platelet activating factor stimulates prostacyclin synthesis by cultured human endothelial cells. Thromb. Haemostas. 46: 269 (1981).
30. S.R. Coughlin, M.A. Moskowits, B.R. Zetter, H.N. Antoniades and L. Levine. Platelet dependent stimulation of prostacyclin synthesis by platelet derived growth factor. Nature 288: 600 (1980).
31. V.R. Rossi, P. Breviario, E. Ghezzi, E. Dejana and A. Mantovani. Prostacyclin synthesis induced in vascular cells by interleukin-1. Science. 299: 174 (1985).

32. C.R. Albrightson, N.L. Baenziger, and P. Needleman. Exaggerated human vascular cell prostaglandin biosynthesis mediated by monocytes: role of monokines and interleukin 1. J. Immunol. 135: 1872 (1985).

33. C.W. Benjamin, N.K. Hopkins, T.D. Oglesby and R.R. Gorman. Agonist specific desensitisation of leukotriene C4-stimulated PGI_2 biosynthesis in human endothelial cells. Biochem. Biophys. Res. Comm. 117: 780 (1983).

34. Pittilo R.M., Machie I.J., Rowles P.M., MacLinn S.J. and Woolf N. Effects of cigarette smoking on the ultrastructure of rat thoracic aorta and its ability to produce prostacyclin. Thromb. Haemostas. 48: 173 (1982).

35. K. Asmussen and K. Kjeldsen. Intimal ultrastructure of human umbilical arteries. Observations on arteries from newborn children of smoking and non-smoking mothers. Circ. Res. 36: 570 (1975).

36. J.L. Nadler, J.S. Velasco and R. Horton. Cigarette smoking inhibits prostacyclin formation. Lancet 1: 1248 (1983).

37. C.H. Dadak, C.H. Leithner, H. Sinzinger and H. Silbergauer. Diminished prostacyclin formation in umbilical arteries born to women who smoke. Lancet 1: 94 (1981).

38. J. Reinders, H. Brinkman, J. Van Mourik and P. De Groot. Cigarette smoke impairs endothelial cell prostacyclin production. Arteriosclerosis. 6: 15 (1986).

39. H.A. Bull, R.M. Pittilo, N. Woolf and S.J. Machin. The effect of nicotine on human endothelial cell release of prostaglandins and ultrastructure. Br. J. Exp. Pathol. 69: 413 (1988).

40. A. Wennmalm and P. Alster. Nicotine inhibits vascular prostacyclin but not platelet thromboxane A_2 synthesis. Gen. Pharmacol. 14: 189 (1983).

41. T. Sonnefeld and A. Wennmalm. Inhibition by nicotine of the foundation of prostacyclin-like activity in rabbit and human vascular tissue. Br. J. Pharmacol. 71: 609 (1980).

42. J.Y. Jeremy, D.P. Mikhailidis and P. Dandona. Cigarette smoke extracts but not nicotine inhibit prostacyclin (PGI_2) synthesis in human, rabbit and rat vascular tissue. Prostagl. Leuk. Med. 19: 261 (1985).

43. O. Ylikkorkola, L. Vinikka and P. Lehtovirta. Effects of nicotine on fetal prostacyclin and thromboxane in humans. Obstet. Gynecol. 66: 102 (1985).

44. J.Y. Jeremy, D.P. Mikhailidis and P. Dandona. Muscarinic stimulation of rat tracheal prostanoid synthesis: studies on the effects of calcium corticosteroids and cigarette smoke. Eur. J. Pharmacol. 162: 117 (1989).

45. J. Larruye, D. Daret, J. Demond-Henri, C. Allieres and H. Bricaud. Prostacyclin synthesis in proliferative aortic smooth muscle cells. A kinetic in vivo and in vitro study. Atherosclerosis. 50: 63 (1984).

46. L. Wilhelmsen. Coronary heart disease: epidemiology of smoking. Am. Heart J. 115: 242 (1985).

47. T.W. Meade, M.V. Vickers, S.G. Thompson, Y. Stirling, A.P. Haines and G.J. Miller. Epidemiological characteristics of platelet aggregability. Br. Med. J. 290: 428 (1985).

48. R.M. Jones. Smoking before surgery: the case for stopping. Br. Med. J. 290: 1763 (1985).

49. A. Rosengren, L. Welin, A. Tsipogianni and L. Wilhelmsen. Impact of cardiovascular risk factors on coronary heart disease and mortality among middle aged diabetic men: a general population study. Br. Med. J. 299: 1127 (1989).

50. J.Y. Jeremy, C.S. Thompson, D.P. Mikhailidis and P. Dandona. The effect of cigarette smoke and diabetes mellitus on muscarinic stimulation of prostacyclin synthesis by the rat penis. Diab. Res. 3: 467 (1986).

51. H. Wey and M. Subbiah. 6-keto-$PGF_{1\alpha}$ synthesis in diabetic rat aorta: effect of substrate concentration and cholesterol feeding. Proc. Soc. Exp. Biol. Med. 171: 251 (1982).

52. E. Ernst, W. Koenig, A. Matrai, B. Filipiak, and J. Stieber. Blood rheology in healthy cigarette smokers. Results from the MONICA project, Augsburg. Arteriosclerosis 8: 385 (1988).

53. R. Aitchinson and N. Russell. Smoking-a major cause of polycthaemia. J. Roy. Soc. Med. 81: 89 (1988).

54. T.W. Meade, J. Imeson and Y. Sterling. Effects of changes in smoking and other characteristics on clotting factors and the risk of IHD. Lancet 2: 986 (1987).

55. G. Galea and R.J.L. Davidson. Haematological and haemorrheological changes associated with cigarette smoking. J. Clin. Pathol. 38: 978 (1985).

56. C.G. Caro, M.J. Lever, K.H. Parker and P.J. Fish. Effects of cigarette smoking on the pattern of arterial blood flow: possible insight into mechanisms underlying the development of arteriosclerosis. Lancet 2: 11 (1987).

57. D.P. Mikhailidis, M.A. Barradas, M.A. Maris, J.Y. Jeremy and P. Dandona. Fibrinogen mediated activation of platelet aggregation and thromboxane A_2 release: pathological implications in vascular disease. J. Clin. Pathol. 38: 1166 (1985).

58. C.S. Brischetto, W.E. Connor, S.L. Sonnor and J.D. Matarazzol. Plasma lipid and lipoprotein profiles of cigarette smokers from randomly selected families: enhancement of hyperlipidaemia and depression of high density lipoprotein. Am. J. Cardiol. 52: 675 (1983).

59. D.P. Mikhailidis and M.A. Barradas. Haemostatic effects of lipid-lowering drugs. J. Drug Develop. 2: 69 (1989).

60. A. Kershbaum, R. Khorsandian, R.F. Caplan, S. Bellet and C.J. Feinberg. The role of catecholamine in the free fatty acid response to cigarette smoking. Circulation. 28: 52 (1983).

61. H. Takeda. H. Kishikawa, M. Shinoharta. Effect of $alpha_2$ adrenoceptor antagonist on platelet activation during insulin-induced hypoglycaemia in type-2 (non-insulin dependent) diabetes mellitus. Diabetologia 31: 657 (1988).

62. D.P. Mikhailidis, A.M. Mikhailidis, M.A. Barradas and P. Dandona. Effect of non-esterified essential fatty acids on the stability of prostacyclin activity. Metabolism 32: 717 (1983).

63. J.Y. Jeremy, D.P. Mikhailidis and P. Dandona. Simulating the diabetic environment modifies in vitro vascular prostacyclin synthesis. Diabetes 32: 217 (1983).

MACROPHAGE INFLUENCE ON SMOOTH MUSCLE PHENOTYPE IN ATHEROGENESIS

Gordon R. Campbell and Julie H. Campbell

Department of Anatomy
University of Melbourne
Parkville, Victoria, 3052, Australia

SMOOTH MUSCLE CELLS OF AN ALTERED PHENOTYPE ARE PRESENT IN ATHEROMA

There is now a large body of evidence indicating that smooth muscle cells in the intima of human arteries involved in atherogenesis are phenotypically different from those of the underlying media. The differences include cell shape (Orekhov et al., 1984, 1986), actin isoform (Gabbiani et al., 1984), myosin isoform (Benzonana et al., 1988), tropomyosin content (Kocher and Gabbiani, 1986), intermediate filaments (Osborn et al., 1987), caldesmon (Glukhova et al., 1988), meta-vinculin (Glukhova et al., 1988), cyclic nucleotides (Tertov et al., 1987), PDGF (Wilcox et al., 1988), fibronectin (Glukhova et al., 1989), and expression of major histocompatibility complexes (Hansson et al., 1986).

We have used stereological techniques to analyse the volume fraction of myofilaments ($V_V$myo) in smooth muscle cells in the intima and media of human carotid artery sampled from atherosclerotic plaques. In atherosclerosis-free areas the $V_V$myo in smooth muscle cells of the intima is not significantly different from that of the underlying media, but adjacent to atheromatous plaques the $V_V$myo of the intimal cells is significantly lower than that of the subjacent medial cells (Mosse et al., 1985; 1986).

CHANGE IN SMOOTH MUSCLE PHENOTYPE ALTERS VASCULAR REACTIVITY

A number of studies suggest that the presence of intimal thickening and atherosclerotic plaques contribute to the development of vasospasm (Yokoyama et al., 1983; Heistad et al., 1987; Yamamoto et al., 1987; Lopez et al., 1989). In patients with coronary artery spasm there is close correlation between areas of spasm induced by ergonovine and the sites of atherosclerotic lesions (Schroeder et al., 1988; MacAlpin, 1980). While the exact mechanisms underlying vasospasm are unknown, vessels with atherosclerotic plaques have increased sensitivity to various vasoconstrictors such as serotonin and histamine but not to noradrenaline (Henry and Yokoyama, 1980; Ginsburg et al., 1984; Heistad et al., 1987; Lopez et al., 1989).

To study the effect of the development of intimal lesions upon vascular reactivity we examined the intimal thickening which develops in the rabbit carotid artery following endothelial denudation by a balloon catheter. After injury muscle cells migrated from the media to the intima where their subsequent proliferation formed a thickened intima, or neo-intima of longitudinally oriented smooth muscle cells (Fishman et al., 1975; Clowes et al., 1983). Using stereology we showed that the cells in both the neo-intima and the underlying media underwent a reversible change in phenotype (Manderson et al., 1989). At two weeks the $V_V$myo of the intimal cells was about 50% less than that of cells in the media of uninjured vessels, indicating the highly "synthetic" nature of the intimal cells. By six weeks the $V_V$myo had increased significantly and at 18 weeks the $V_V$myo was equivalent to that of control medial cells (Manderson et al., 1989).

To determine the changes in reactivity of the arteries following development of the experimental intimal thickening we studied the responses of ring segments to vasoactive drugs _in vitro_. Two weeks after endothelial denudation the ballooned arteries developed significantly less contractile force (E_{max}) than the control arteries to both KCl and the selective α_1-adrenoreceptor agonist methoxamine (Manderson et al., 1989b). The experimental arteries were also markedly less sensitive to methoxamine, as evidenced by the increases in the EC_{50} values, at both 2 weeks (approximately seven-fold) and six weeks (four-fold). In contrast, there was no significant alteration in the E_{max} in response to either serotonin or the thromboxane A2-mimetic U46619. However, the experimental arteries did show a slight, but significant increase in the sensitivity (i.e. lower EC_{50} values) to both these drugs. Despite our previous finding that regenerated endothelial cells produce endothelium-derived relaxing factor (EDRF), which is able to permeate the intimal thickening to produce vasorelaxation (Cocks et al., 1987), there was no apparent difference in reactivity between areas of the artery with regenerated endothelium and those lined by modified smooth muscle cells (Manderson et al., 1989). These results suggest, therefore, that following balloon catheter injury there is an apparent selective enhancement in reactiveness to the platelet-derived substances serotonin and the thromboxane A_2-mimetic U46619, a finding also reported in atherosclerotic vessels (Lopez et al., 1989; Wines et al., 1989).

We believe that these changes in reactivity may be related to changes in smooth muscle phenotype in the media and intimal thickening. Between two and six weeks (when there is a significant increase in $V_V$myo) the E_{max} to methoxamine increased significantly and the sensitivity to serotonin became more pronounced. Concomitant with these changes in phenotype there may be alterations in the density and/or affinity of receptors to various vasoconstrictors, similar to that suggested by preliminary reports in atherosclerotic arteries (Lucas et al., 1981).

PRIMARY CULTURE AS A MODEL FOR CHANGE IN PHENOTYPE (PHENOTYPIC MODULATION)

Smooth muscle cells from the aortic media which have been dispersed into single cells by the action of collagenase and elastase and seeded into primary culture, respond to appropriate stimuli by slow contraction (Chamley-Campbell et al., 1979). Under phase-contrast microscopy these cells appear ribbon- or spindle-shaped with the nucleus located centrally. The cytoplasm is phase-dense with an accumulation of cytoplasmic organelles in the perinuclear region. When the cells are seeded at less than 5×10^5 cells/mL, they spontaneously and gradually undergo a change in phenotype. The cytoplasm appears more abundant and less phase-dense as the cells become broader and flatter. Granularity in the perinuclear region becomes more apparent.

Using morphometry we have found that smooth muscle cells in the aorta of 9-week-old rabbits have a $V_V$myo of 39.5% ± 1.2%. When the aorta is dispersed into single cells then pelleted by centrifugation, the $V_V$myo is 35.4% ± 1.5%. Following plating in culture dishes at 3×10^5 cells/mL ($\approx 3 \times 10^5$/cm^2), the $V_V$myo on day 1 is 34.6% ± 2.3% and on day 2, 30.3% ± 0.8%. There is a significant decrease on day 3 (26.9% ± 2.8%) and on day 4 (22.5% ± 2.3%), falling sharply to 11.5% ± 1.6% on day 5 (Campbell et al., 1989). The cells remain relatively quiescent until this time, then commence logarithmic growth that continues at a uniform rate until confluence is achieved at day 9. By day 9 the $V_V$myo is 21.9% ± 1.2%. Two days after confluence, however, this has increased to 30.6% ± 0.6% (day 11) and is 32.0% ± 2.5% on day 12, indicating that the cells have almost returned to their original phenotypic state (Campbell, et al. 1989). Thus smooth muscle cells in primary culture when seeded at a specific density are capable of undergoing a reversible change in phenotype with time.

Smooth muscle cells from different species respond differently in primary cell culture (Campbell et al., 1989). For example, adult rat aortic smooth muscle cells undergo a more rapid change in phenotype and begin to proliferate earlier in culture than adult rabbit or pig cells. The degree of maturation of the animal (age), and hence the initial phenotype ($V_V$myo) of the cells, also affects the time in primary culture before isolated smooth muscle cells begin to proliferate. The initial seeding density in primary culture also affects the cells subsequent behaviour. For example, aortic smooth muscle cells from the 12 week old rabbit aortic media seeded at confluent density (6 x 10^5 cells/ml) have a $V_V$myo of 50.3 $\pm$ 2.4% which does not decrease significantly with time in culture. That is, cells seeded at confluent density do not undergo a change in phenotype.

WHAT CONTROLS SMOOTH MUSCLE PHENOTYPE?

Studies by Bissell et al., (1982, 1987) have stressed the importance of the extracellular matrix surrounding each cell as an integral part of the cellular functional unit. The cell, dependent on its phenotypic state, produces a particular matrix that in turn affects the cell, influencing many morphological and functional properties via interactions with membrane receptors and cytoskeletal components, which in turn affect gene expression (Bissell et al., 1982, 1987). Thus the pericellular matrix may be considered as an extension of the cell itself. Studies with many cell systems including fibroblasts, smooth muscle cells , and endothelial and epithelial cells have shown that matrix components regulate functions such as adhesion, shape, migration, proliferation, biosynthetic and degradative processes, morphogenesis, and differentiation (Hay, 1981; Hook et al., 1984; Hadley et al., 1985; Parry et al., 1987).

Smooth muscle cells in culture are known to synthesize matrix components that are both incorporated into the cell layer and secreted into the medium. Glycosaminoglycans are associated with the cell surface and specific interactions between matrix components occur and provide a means for linking the pericellular and interstitial matrix (Burke and Ross, 1979). In primary cell culture, a crude extract of glycosaminoglycans from the aortic intima plus inner media maintains sparsely-seeded smooth muscle cells with a high $V_V$myo. Treatment of this aortic extract with heparinase from *Flavobacterium heparinum* destroys the active factor, indicating that glycosaminoglycans of the heparan sulphate species are responsible; and indeed, addition of the closely related glycosaminoglycan heparin maintains sparsely-seeded smooth muscle in a high $V_V$myo (contractile) state (Chamley-Campbell and Campbell, 1981). This has recently been confirmed by Herbert et al. (1988) who also found a similar effect with pentosan polysulphate, a semi-synthetic sulphated polysaccharide. Smooth muscle cells in primary culture can also be maintained in the contractile state by seeding the freshly-dispersed cells at confluent density, or by placing sparsely-seeded cells with a spatially separated feeder layer of confluent contractile smooth muscle cells or endothelial cells (Campbell and Campbell, 1984) which produce large amounts of an antiproliferative heparan sulphate species (Fritze et al., 1985). It thus appears that the presence of heparin-like glycosaminoglycans is an important determinant of the phenotype that smooth muscle expresses, and that high levels of this glycosaminoglycan species maintain the cells in a high $V_V$myo state. Any factor which removes this substance (e.g. enzyme dispersion and dilution through sparse-seeding), may therefore induce the cells to undergo a change in phenotype to a lower $V_V$myo.

MACROPHAGES AND ATHEROGENESIS

The atherosclerotic plaque has a variable cellular composition depending on its stage of development, but the macrophage is always present in significant numbers (Aqel et al., 1985; Jonasson et al., 1986). These cells function primarily as scavengers, accumulating large amounts of lipid to become the classic foam cell (Fowler et al., 1979). However, in culture the macrophage also produces a potent stimulator of cellular proliferation (Martin et al., 1981). This factor resembles the platelet-derived growth factor (PDGF) in that it competes with ^{125}I-PDGF for binding to receptors and is mitogenic (Shimokado et al., 1985). Macrophages also secrete ß-transforming growth factor (ß-TGF) (Sporn and Roberts, 1986), as well as vasoactive agents and matrix-degrading enzymes (Page et al., 1978; Nathan et al., 1980).

Thus the macrophage may play a multiplicity of roles in the initiation and development of the atheromatous plaque. We have recently demonstrated that peritoneal macrophages grown in co-culture with a confluent monolayer of smooth muscle cells induce a significant decrease in the $V_V$myo of the smooth muscle cells after three days as compared with both the freshly isolated smooth muscle cells and those grown for three days in the absence of macrophages (Rennick et al., 1988). Since invasion of the artery wall by monocytes (which then differentiate into macrophages) is one of the earliest cellular events in experimental atherogenesis (Joris et al., 1983) we have postulated that it may be through the influence of these cells that smooth muscle phenotypic change is induced in the initial stages of atherogenesis (Campbell and Campbell, 1989).

MACROPHAGE INFLUENCE UPON SMOOTH MUSCLE PHENOTYPE IN PRIMARY CELL CULTURE

Macrophages are prominent cells both in the host defense mechanisms characteristic of chronic inflammatory responses as well as in atheroma. These mononuclear phagocytes produce a vast array of secretory products, some of which enable the macrophage to play important roles in remodelling extracellular matrix; for instance, neutral proteases degrade the matrix macromolecules elastin, collagen and glycoproteins. Of particular interest, however, is that endoglycosidic activity, specific for heparan sulphate, been described in macrophages (Savion et al., 1984). We were therefore interested in determining whether macrophages could initiate a change in smooth muscle phenotypic expression through degradation of heparan sulphate proteoglycans on the smooth muscle cell surface.

An assay for heparan sulphate-degrading enzymes was set up as described by Savion et al. (1984). Bovine aortic endothelial cells were seeded sparsely into 30 mm plastic culture dishes in the presence of 30 μCi/ml $Na_2[^{35}S]O_4$. Five to 7 days after the cells had achieved confluency, the endothelial cells were removed by Triton X-100, leaving a layer of ^{35}S-labelled extracellular matrix on the bottom of the dish. About 70% of the incorporated ^{35}S was in the form of heparan sulphate, and 30% in chondroitin sulphate and dermatan sulphate. The factor to be tested for heparan sulphate-degrading activity was added to the washed matrix for 24 hours at $37^{\circ}C$ at neutral pH, then an aliquot of the supernatant passed through a Sepharose 6B column and the fractions counted by liquid scintillation.

Firstly, as control, medium 199 without serum (1.5 ml, pH 7.0) was incubated with the thoroughly washed ^{35}S-labelled extracellular matrix. After gel filtration of the supernatant, all the radioactivity co-migrated with dextran blue (M_r = 2 x 10^6 daltons) to elute at the void volume (V_0), indicating that only high molecular weight undegraded proteoglycans had been released into the medium (Fig. 1a). Free $[^{35}S]O_4^{2-}$ applied directly to the column eluted with a K_{av} of 0.86. After incubation of the matrix with living mouse peritoneal macrophages or J774 macrophages (1 x 10^6 cells/ml), the peak at V_0 was small, and instead more than 80% of the radioactivity eluted with a K_{av} of 0.84, indicating most of the ^{35}S-labelled glycosaminoglycan chains of the proteoglycans had been completed degraded (Fig. 1b). When 10 μg/ml sodium heparin was added to the incubation with living macrophages, the size of the K_{av} 0.84 peak was decreased, and the peak at V_0 increased. As well, a number of small peaks appeared at K_{av} 0.5 or less.

When macrophages were lysed and a 100 μl aliquot incubated in 1.4 ml medium 199 without serum with the ^{35}S-labelled matrix, most of the degradation products eluted with a K_{av} of 0.60, with smaller peaks at V_0 and at K_{av} of 0.84 (Fig. 2a). Based on a comparison of the experimental K_{av} values on Sepharose 6B columns with those determined for chondroitin sulphate fractions of known molecular weight, it was estimated that the degradation fragments at K_{av} 0.60 had a molecular weight of about 10,000 daltons. When fresh macrophage-conditioned medium (4 x 10^6 cells/ml serum-free medium for 24 hours at $37^{\circ}C$) was incubated with ^{35}S-labelled matrix, most of the released radioactivity eluted at V_0, similar to that observed with fresh medium 199 without serum, with minor peaks at K_{av} 0.60 and 0.84 (Fig. 2b). Addition of an equal volume of the macrophage-conditioned medium and fresh medium to confluent 30 mm dishes of high $V_V$myo rabbit aortic smooth muscle, 8 days in primary culture, had no effect on smooth muscle phenotype after 4 days incubation. However, macrophage-conditioned medium was concentrated 10 fold with an Amicon concentrator, then increasing concentrations of ammonium sulphate added together with bovine serum albumin. Precipitates were obtained in the presence of 0-25%, 25-50%, 50-75% and 75-100% ammonium sulphate, then each of the four precipitates dissolved in 1 ml medium 199 without serum, and 100 μl aliquots added to confluent 30 mm dishes of rabbit aortic smooth muscle. After 4 days, the $V_V$myo of smooth muscle cells in fractions which precipitated with 0-25%, 25-50% and 75-100% ammonium sulphate were not significantly different from controls (p>0.5), but the $V_V$myo of 12 week old rabbit aortic smooth muscle cells in the presence of fraction 50-75% had a $V_V$myo of 42.7 $\pm$ 2.9%, which was significantly lower than that of control cells (49.2 $\pm$ 1.4%, p<0.15). Each of the precipitates was also tested for matrix-degrading activity. The 0-25%, 25-50% and 75-100% ammonium sulphate-precipitated fractions all produced large peaks at V_0 and another broad peak with a K_{av} of 0.37 - 0.42 (M_r = 30,000-40,000 daltons). Fraction 50-75% had a small peak at V_0 and a distinct, considerably larger peak at K_{av} 0.65 (Fig. 2c). That is, the fraction that degraded the ^{35}S-labelled proteoglycans into fractions of about 10,000 dalton molecular weight, was the same that induced a decrease in $V_V$myo of the rabbit aortic smooth muscle cells.

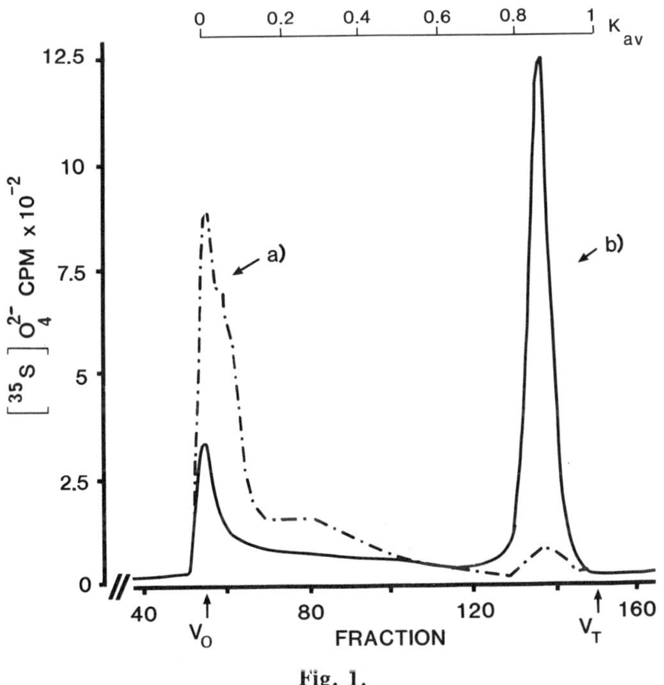

Fig. 1.

a)   *Medium 199 without serum (1.5 ml, pH 7.0) was incubated with ^{35}S-labelled matrix for 24 hours at 37°C and 0.5 ml of the supernatant applied for gel filtration through Sepharose 6B. Elution profile showed that all the radioactivity eluted at V_0, indicating that only undegraded proteoglycans had been released into the medium.*

b)   *Living macrophages (mouse peritoneal) at 1 x 10[6] cells/ml were included in the incubation. Elution profile showed that more than 80% of the radioactivity eluted with a K_{av} of 0.84 indicating that most of the ^{35}S-proteoglycans had been completely degraded to free (^{35}S)O4 2-.*

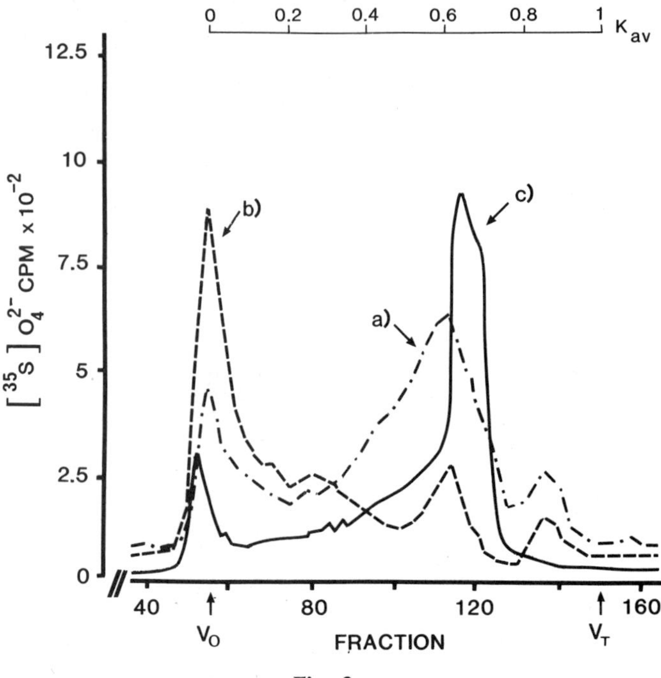

Fig. 2.

a)   Lysate of whole macrophages (mouse peritoneal) included in the incubation. Elution profile showed a peak at K_{av} = 0.60 with smaller peaks at V_O and K_{av} = 0.84.

b)   Macrophage-conditioned medium (without serum) was incubated with ^{35}S-labelled matrix. Elution profile showed a large peak at V_O (undegraded proteoglycan), and small peaks at K_{av} = 0.60 and 0.84.

c)   A fraction prepared from concentrated macrophage-conditioned medium by ammonium sulphate precipitation (50-75%), incubated with labelled matrix. Elution profile showed a large peak at K_{av} = 0.65.

Thus it was clear that the matrix-degrading activity of the macrophages is stored within the cells and that only small amounts are released into the incubation medium. To identify the subcellular location of the stored enzymes that degrade the ^{35}S-labelled proteoglycans, macrophage lysosomes were prepared by subcellular fractionation (DeDuve et al., 1955) and 100 μl of lysate added to confluent 30 mm dishes of rabbit aortic smooth muscle, 8 days in culture. After 4 days in the presence of the lysosomal lysate, the $V_V$myo of the cells was 38.2 $\pm$ 2.6%, which was significantly lower (p<0.1) than that of control cells (48.2 $\pm$ 0.5%). Incubation of the ^{35}S-labelled matrix with 100 μl of the macrophage lysosomal lysate at neutral pH produced a degradation peak with K_{av} 0.63 that was totally inhibited when 10 μg/ml heparin was included in the incubation (Fig. 3a,b). When the lysosomal lysate was incubated with the ^{335}S-labelled matrix at acid pH (pH 6.5 or less), the peak at K_{av} 0.63 was smaller and a second large peak at K_{av} 0.84 appeared (Fig. 3c).

Having established that the K_{av} 0.63 peak eluted in fractions 100 to 125, a second aliquot of the lysate was run on the same column and fractions 100 to 125 collected (5 ml). An aliquot (0.5 ml) of these pooled fractions was treated with low pH nitrous acid which selectively results in deamination of heparan sulphate chains to fractions less than M_r 1,000. When this mixture was passed through Sepharose 6B, no

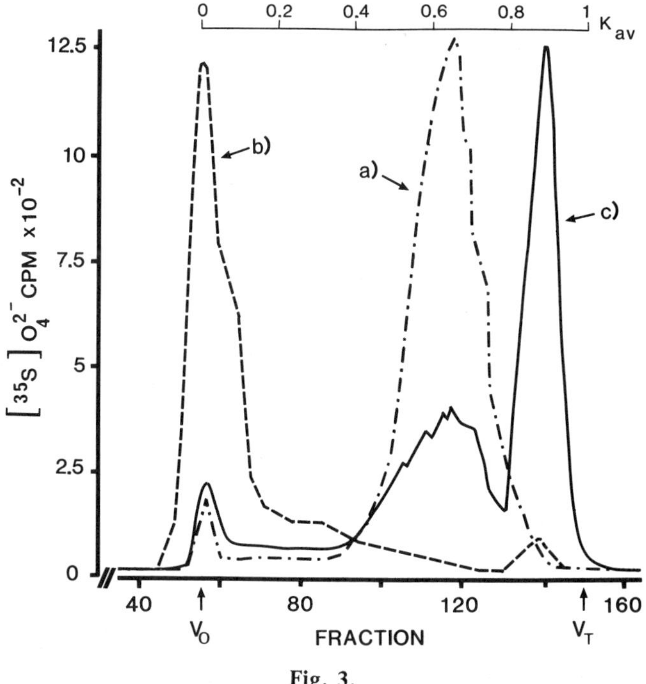

Fig. 3.

a)   *Incubation of ^{35}S-labelled matrix with macrophage lysosomal lysate at pH 7.0 for 24 hours at 37°C produced an elution profile with a large peak at K_{av} = 0.63.*

b)   *As above, but in the presence of 10 µg/ml sodium heparin. Peak at K_{av} = 0.63 does not occur and nearly all radioactivity eluted at V_o.*

c)   *Same preparation of macrophage lysosomal lysate as a), but incubation with ^{35}S-labelled matrix carried out at pH 6.1. A large peak eluted at K_{av} = 0.84, with a smaller peak at 0.63.*

peak at K_{av} 0.63 occurred, but instead a peak at K_{av} 0.84 appeared (Fig. 4a). In contrast, treatment of another 0.5 ml aliquot of pooled fractions 100 - 125 (peak K_{av} 0.63) with 10 units/ml chondroitin ABC lyase, which degrades chondroitin sulphate, resulted in only a very small peak at K_{av} 0.84, with most of the radioactivity continuing to elute with a K_{av} 0.63 (Fig. 4b).

Addition of 10 units/ml heparinase to the extracellular matrix consistently resulted in a large peak eluting at K_{av} 0.63, identical to that obtained in the presence of macrophage lysosomal lysate at neutral pH (see Fig. 3a). Again, this peak did not occur if 10 μg/ml heparin was included in the incubation, and also like the K_{av} 0.63 peak produced by lysosomal lysate, was degraded to smaller fragments and free $[^{35}S]O_4^{2-}$ (K_{av} = 0.84 to 0.87) after treatment with nitrous acid, but was unaffected by chondroitin ABC lyase (see Figs. 3b, 4a,b). Addition of 10 units/ml heparinase to confluent aortic smooth muscle cultures induced a significantly (p<0.1) lower $V_V$myo ($42.8 \pm 1.2\%$) compared with control cells ($51.0 \pm 3.2\%$) after 4 days.

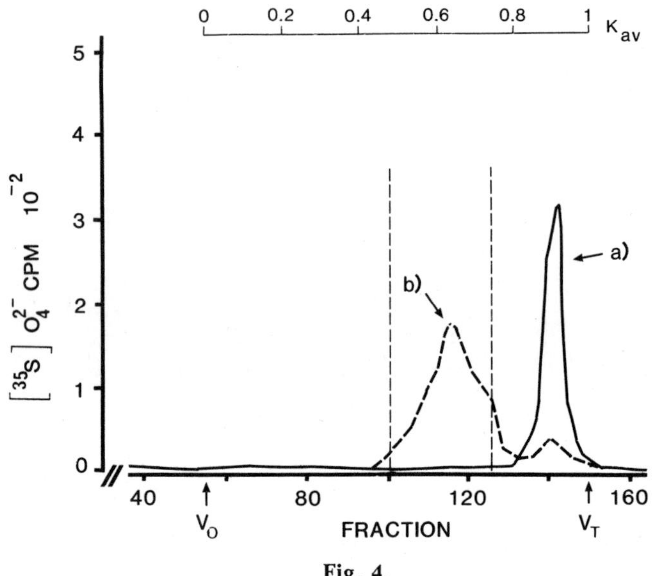

Fig. 4.

a)   An aliquot from fractions 100-125 from macrophage lysosomal lysate pH 7.0 was treated with low pH nitrous acid. Peak at K_{av} = 0.63 completely disappeared, and instead at peak a K_{av} = 0.84 appeared.

b)   As for a), but aliquot treated with chondroitin ABC lyase. The peak at K_{av} = 0.63 was mostly unaffected.

These results indicate that the degradation peak which elutes at $K_{av} \approx 0.63$ consists almost exclusively of heparan sulphate fragments, and suggest that the macrophages used in the present study have in their lysosomes a heparan sulphate degrading endoglycosidase which cleaves internal glycosidic bonds, and whose action on smooth muscle cells is sufficient to induce a change in phenotypic expression. The results further suggest that macrophages also possess sulphatases and/or exoglycosidases which sequentially release inorganic sulphates and monosaccharide residues from the non-reducing ends of the heparan sulphate fragments released by the endoglycosidase. The subcellular localisation of the exoglycosidases and/or sulphatases is also lysosomal, but unlike the heparinase, they are only active at acid pH, such as occurs in intact lysosomes.

Plasma membrane preparations of macrophages incubated with ^{35}S-labelled extracellular matrix produced a large, broad elution peak which began at V_O, peaked at K_{av} 0.24, then gradually tailed off, indicating that they possess enzyme activity, probably proteolytic, which releases high molecular weight ^{35}S-labelled species from matrix, but little or no activity to degrade these substances further. A similar elution profile occurred when 10 μg/ml trypsin was incubated with the ^{35}S-labelled matrix. One peak occurred at V_O (undegraded proteoglycan) and another broader peak at K_{av} of about 0.20.

Cell-associated heparan sulphate proteoglycans occur as membrane-intercalated glycoproteins where the core protein is anchored in the lipid interior of the plasma membrane, and the heparan sulphate chains bind to specific sites on collagen, laminin and fibronectin (Saunders and Bernfield, 1988). The function of the proteoglycan-mediated interaction is to promote the organisation of actin filaments in the attaching cell which also has the effect of stabilising cell morphology; thus removal and destruction of cell-surface heparan sulphate at these sites may initiate a change in smooth muscle phenotype through disorganisation of actin filaments with subsequent influences on gene expression (Bissell and Barcellos-Hoff, 1987). However, the observation that trypsin (which releases the heparan sulphate proteoglycans from the cell surface) does not by itself induce a change in smooth muscle phenotype, suggests that the heparan chains must be completely destroyed or otherwise removed from the vicinity of the cell for this to occur. The ability of free heparin to prevent a change in phenotype of those smooth muscle cells whose extracellular matrix and basal lamina have been degraded and removed during enzymatic isolation, supports this view.

Heparan sulphate proteoglycans and free glycosaminoglycan chains, including heparin, bind to specific proteins at the cell surface, are internalised with a half-life of 4-6 hours and degraded intracellularly (Bienkowski and Conrad, 1984). A small fraction of the heparan sulphate enriched in the rare 2-O-sulphate glucuronate units is transported to the cell nucleus where it has been implicated in cell growth control (Fedarko and Conrad, 1986), but whether internalised heparan sulphate/heparin affects smooth muscle phenotypic expression by the same or a similar mechanisms is unknown.

SIGNIFICANCE OF SMOOTH MUSCLE PHENOTYPIC MODULATION INDUCED BY MACROPHAGES IN ATHEROMA

Based on the studies reported here, we suggest the following scenario in relation to the genesis of atherosclerosis:

In hyperlipidemia, large numbers of monocytes enter the subendothelium, differentiate into macrophages and, in addition to releasing mitogens, oxidizing and taking up lipoproteins etc., release heparan sulphate proteoglycans from the surface of the smooth muscle cells by the action of released proteases or those present on the plasma membrane. The proteoglycans are phagocytosed by the macrophages and the heparan sulphate chains completely degraded in the lysosomes. This temporarily removes all heparan sulphate from the surface of the smooth muscle cells, and this, by an unknown mechanism, initiates the process of phenotypic modulation.

Our cell culture studies further indicate that change in phenotype confers upon the cells distinct alterations in biology. Prior to change in phenotype, while the cells have a high $V_V$myo, they do not proliferate in response to mitogens, synthesise minimal collagen, and accumulate little lipid even after exposure to high concentrations of ß-very low density lipoprotein (ß-VLDL). In contrast, smooth muscle cells of low $V_V$myo proliferate logarithmically in response to mitogens, synthesise greatly increased amounts of collagen (particularly type I) compared to cells with a high $V_V$myo, and accumulate considerably more lipid on exposure to ß-VLDL (Campbell and Campbell, 1985; Campbell et al., 1983, 1985; Chamley-Campbell et al., 1979; Ang et al., 1990). Since a low $V_V$myo, proliferation, synthesis of extracellular matrix (particularly collagen type I) and accumulation of lipid are all characteristic features of smooth muscle cells in atheroma, change in phenotype of smooth muscle cells may be an important initial event in the development of this disease.

Finally, we have shown that alterations in smooth muscle phenotype can alter the reactivity of arteries, and hence explains the observed hyper-reactivity to certain vasoconstrictors and the predisposition of sites of atherosclerotic lesions to vasospasm.

REFERENCES

Ang, A.H., Tachas, G., Campbell, J.H., Bateman, J.F. and Campbell, G.R., 1990, Collagen synthesis by cultured rabbit aortic smooth muscle cells: Alteration with phenotype, Biochem. J. 265:(in press).

Aqel, N.M., Ball, R.Y., Waldman, H. and Mitchinson, M.J., 1985, Identification of macrophages and smooth muscle cells in human atherosclerosis using monoclonal antibodies. J. Pathol. 146:197-204.

Benzonana, G., Skalli, O. and Gabbiani, G., 1988, Correlation between the distribution of smooth muscle or non-muscle myosins and α-smooth muscle actin in normal and pathological soft tissues. Cell Motil. Cytoskeleton. 11:260-274.

Bienkowski, J.H. and Conrad, H.E., 1984, Kinetics of proteoheparan sulphate synthesis, secretion, endocytosis, and catabolism by a hepatocyte cell line. J. Biol. Chem. 259:12989-12996.

Bissell, M.J. and Barcellos-Hoff, M.H., 1987, The influence of extracellular matrix on gene expression: Is structure the message? J. Cell Sci. 8(Suppl):327-343.

Bissell, M.J., Hall, H.G. and Parry, G., 1982, How does the extracellular matrix direct gene expression? J. Theoret. Biol. 99:31-68.

Burke, J.M. and Ross, R., 1979, Synthesis of connective tissue macromolecules by smooth muscle. Int. Rev. Connect. Tissue 8:119-157.

Campbell, G.R. and Campbell, J.H., 1985, Smooth muscle phenotypic changes in arterial wall homeostasis: Implications for the pathogenesis of atherosclerosis. Exp. Mol. Pathol. 42:139-162.

Campbell, G.R., and Campbell, J.H., 1987, Phenotypic modulation of smooth muscle cells in primary culture, in: "Vascular smooth muscle in culture," J.H. Campbell, G.R. Campbell, eds. Boca Raton. FL: CRC Press. pp. 39-56.

Campbell, J.H. and Campbell, G.R., 1984, Cellular interactions in the artery wall. in: "The Peripheral Circulation," S. Hunyor, J. Ludbrook, J. Shaw, M. McGrath, (eds)., Elsevier, New York, pp. 33-39.

Campbell, J.H. and Campbell, G.R., 1989, Potential role of heparanase in atherosclerosis. News In Physiol. Sci. 4:9-12.

Campbell, J.H., Kocher, O., Skalli, O., Gabbiani, G. and Campbell, G.R., 1989, Cytodifferentiation and expression of α smooth muscle actin mRNA and protein during primary culture of aortic smooth muscle cells. Correlation with cell density and proliferative state. Arteriosclerosis. 9:633-643.

Campbell, J.H., Popadynec, L., Nestel, P.J. and Campbell, G.R., 1983, Lipid accumulation in arterial smooth muscle cells. Depdenence on phenotype. Atherosclerosis 47:279-295.

Campbell, J.H., Reardon, M.F., Campbell, G.R. and Nestel, P.J., 1985, Metabolism of atherogenic lipoproteins by smooth muscle cells of different phenotype in culture. Arteriosclerosis 5:318-328.

Chamley-Campbell, J.H. and Campbell, G.R., 1981, What controls smooth muscle phenotype? Atherosclerosis 40:347-357.

Chamley-Campbell, J.H., Campbell, G.R. and Ross, R., 1981, Phenotype-independent response of cultured aortic smooth muscle to serum mitogens. J. Cell Biol. 89:379-383.

Clowes, A.W., Reidy, M.A. and Clowes, M.M., 1983, Kinetics of cellular proliferation after arterial injury I. Smooth muscle cell growth in the absence of endothelium. Lab. Invest. 49:327-333.

Cocks, T.M., Manderson, J.A., Mosse, P.R.L., Campbell, G.R. and Angus, J.A., 1987, Development of a large fibromuscular intimal thickening does not impair endothelium-dependent relaxation in the rabbit carotid artery. Blood Vessels 24:192-200.

DeDuve, C., Pressman, B.C., Gianetto, R., Wattiaux, R. and Applemans, F., 1955, Tissue fractionation studies. 6. Intracellular distribution patterns of enzymes in rat-liver tissue. Biochem. J. 60:604-617.

Fedarko, N.S. and Conrad, H.E., 1986, A unique heparan sulfate in the nuclei of hepatocytes: Structural changes with the growth state of the cells. J. Cell Biol. 102:587-599.

Fishman, J.A., Ryan, G.B., Karnovsky, M.J., 1975, Endothelial regeneration in the rat carotid artery and the significance of endothelial denudation in the pathogenesis of myointimal thickening. Lab. Invest. 32:339-351.

Fowler, S., Shio, H. and Haley, N.J., 1979, Characterization of lipid-laden aortic cells from cholesterol-fed rabbits. IV. Investigation of macrophage-like properties of aortic cell populations. Lab. Invest. 41:372-378.

Fritze, L.M., Reilly, C.F. and Rosenberg, R.D., 1985 An antiproliferative heparan sulphate species produced by post-confluence smooth muscle cells. J. Cell Biol. 100:1041-1049.

Gabbiani, G., Kocher, O., Bloom, W.S., Vanderkerkhove, J. and Weber, K., 1984, Actin expression in smooth muscle cells of rat aortic intimal thickening, human atheromatous plaque and cultured rat aortic media. J. Clin. Invest. 73:148-152.

Ginsburg, R., Bristow, M.R, Davis, K., Dibiase, A., Billingham, M.E., 1984, Quantiative pharmacologic responses of normal and atherosclerotic isolated human epicardial coronary arteries. Circulation 69:430-440.

Glukhova, M.A., Frid, M.G., Shekhonin, B.V., Vasilevskaya, T.D., Grünwald, J., Saginati, M. and Koteliansky, V.E., 1989, Expression of extra domain A fibronectin sequence in vascular smooth muscle cells is phenotype dependent. J. Cell Biol. 109:357-366.

Glukhova, M.A., Kabakov, A.E., Frid, M.G., Ornatsky, O.I., Belkin, A.M., Mukhin, D.N., Orekhov, A.N., Koteliansky, V.E. and Smirnov, V.N., 1988, Modulation of human aorta smooth muscle cell phenotype: A study of muscle-specific variants of vinculin, caldesmon, and actin expression. Proc. Natl. Acad. Sci. USA, 85:9542-9546.

Hadley, M.A., Byers, S.W., Suarex-Quian, C.A., Kleinman, H.K. and Dym, M., 1985, Extracellular matrix regulates Sertoli cell differentiation, testicular cord formation and germ cell development. J. Cell Biol. 101:1511-1522.

Hansson, G.K., Jonasson, L., Holm, K. and Claesson-Welsh, L., 1986, Class II MHC antigen expression in the atherosclerotic plaque: Smooth muscle cells express HLA-DR, HLA-DQ and the invariant gamma chain. Clin. Exp. Immunol. 64:261-268.

Hay, E.D. ed., 1981, Cell Biology of the extracellular matrix. New York, London: Plenum Press.

Heistad, D.D., Mark, A.L., Marcus, M.L., Piegors, D.J. and Armstrong, M.L., 1987, Dietary treatment of atherosclerosis abolishes hyperresponsiveness to serotonin: Implications for vasospasm. Circulation Res. 61:346-351.

Henry, P.D. and Yokoyama, M., 1980, Supersensitivity of atherosclerotic rabbit aorta to ergonovine. Mediation by a serotonergic mechanism. J. Clin. Invest. 66:306-313.

Herbert, J.M., Nuti, D., Paul, R. and Maffrand, J.P., 1988, *In vitro* and *ex vivo* regulation of vascular smooth muscle cell growth and phenotypic modulation by sulphated polysaccharides. Artery 16:1-14.

Hook, M., Kjellen, L., Johnson, S. and Robinson, J., 1984, Cell surface glycosaminoglycans. Annu. Rev. Biochem. 53:847-869.

Jonasson, L., Holm, J., Skalli, O., Bondjers, G. and Hansson, G.K., 1986, The human atherosclerotic plaque: Regional accumulations of T cells, macrophages and smooth muscle cells. Arteriosclerosis 6:131-138.

Joris, I., Zand, T., Nunnari, J.J., Krolikowski, F.J. and Majno, G., 1983, Studies on the pathogenesis of atherosclerosis. I. Adhesion and emigration of mononuclear cells in the aortic of hypercholesterolemic rats. Am. J. Pathol. 113:341-358.

Kocher, O. and Gabbiani, G., 1986, Cytoskeletal features of normal and atheromatous human arterial smooth muscle cells. Human Pathol. 17:875-880.

Lopez, J.A.G. Armstrong, M.L., Piegors, D.J. and Heistad, D.D., 1989, The effect of early and advanced atherosclerosis on vascular responses to serotonin, thromboxane A_2 and ADP. Circulation 79:698-705.

Lucas, C., Saffitz, J.E. and Henry, P.D., 1981, Monoaminergic receptor shift in atherosclerotic rabbit aorta. Circulation 64:Suppl.IV:286.

MacAlpin, R.N., 1980, Contribution of dynamic vascular wall thickening to luminal narrowing during coronary arterial constriction. Circulation 60:296-301.

Manderson, J.A., Cocks, T.M. and Campbell, G.R., 1989, Balloon catheter injury to rabbit carotid artery. II. Selective increase in reactivity to some vasoconstrictor drugs. Arteriosclerosis 9:299-307.

Manderson, J.A., Mosse, P.R.L., Safstrom, J.A., Young, S.B. and Campbell, G.R., 1989, Balloon catheter injury to rabbit carotid artery. I. Changes in smooth muscle phenotype. Arteriosclerosis 9:289-298.

Martin, B.M., Gimborne, M.A. Jr., Unanue, E.R. and Cotran, R.S., 1981, Stimulation of nonlymphoid mesenchymal cell proliferation by a macrophage-derived growth factor. J. Immunol. 126:1510-1515.

Mosse, P.R.L., Campbell, G.R. and Campbell, J.H., 1986, Smooth muscle phenotypic expression in human carotid arteries. II. Comparison of cells from areas of atherosclerosis-free diffuse intimal thickenings with those of the media. Arteriosclerosis 6:664-670.

Mosse, P.R.L., Campbell, G.R., Wang, Z-L. and Campbell, J.H., 1985, Smooth muscle phenotypic expression in human carotid arteries. I. Comparison of cells from diffuse intimal thickenings adjacent to atheromatous plaques with those of the media. Lab. Invest. 53:555-562.

Nathan, C.F., Murray, H.W. and Cohn, Z.A., 1980, The macrophage as an effector cell. N. Eng. J. Med. 303:622-626.

Orekhov, A.M., Karpova, I.I., Tertov, V.V., Rudchenko, S.A., Andreeva, E.R., Krushinsky, A.V. and Smirnov, V.N., 1984, Cellular composition of atherosclerotic and uninvolved human aortic subendothelial intima. Light microscopic study of dissociated aortic cells. Am. J. Pathol. 115:17-24.

Orekhov, A.N., Andreeva, E.R., Krushinsky, A.V., Novikov, I.D., Tertov, V.V., Nestaiko, G.V., Khashimov, Kh.A., Repin, V.S. and Smirnov, V.N., 1986, Intimal cells and atherosclerosis. Relationship between the number of intimal cells and major manifestations of atherosclerosis in the human aorta. Am. J. Pathol. 125:402-415.

Osborn, M., Caselitz, K., Püschel and Weber, K., 1987, Intermediate filament expression in human vascular smooth muscle and in arteriosclerotic plaques. Virchows Arch. A. 411:449-458.

Page, R.C., Davies, P. and Allison, A.C., 1978, The macrophage as a secretory cell. Int. Rev. Cytol. 52:119-157.

Parry, G., Cullen, B., Kaetzel, C.S., Kramer, R. and Moss, L., 1987, Regulation of differentiation and polarized secretion in mammary epithelial cells maintained in culture: Extracellular matrix and membrane polarity influences. J. Cell Biol. 105:2043-2051.

Rennick, R.E., Campbell, J.H. and Campbell, G.R., 1988, Vascular smooth muscle phenotype and growth behaviour can be influence by macrophages *in vitro*, Atherosclerosis 71:35-43.

Saunders, S. and Bernfield, M., 1988, Cell surface proteoglycan binds mouse mammary epithelial cells to fibronectin and behaves as a receptor for interstitial matrix. J. Cell Biol. 106:423-430.

Savion, N., Vlodavsky, I. and Fuks, Z., 1984, Interactions of T lymphocytes and macrophages with cultured vascular endothelial cells: Attachment, invasion and subsequent degradation of the subendothelial extracellular matrix. J. Cell Physiol. 118:169-178.

Schroeder, J.S., Bolen, J.L., Quint, R.A., Clarke, D.A., Hayden, W.G., Higgins, C.B. and Wexler, L., 1977, Provocation of coronary spasm with ergonovine maleate. New test with results in 57 patients undergoing coronary arteriography. Am. J. Cardiol. 40:487-491.

Shimokado, K., Raines, E.W., Madtes, D.K., Barnett, T.B., Benditt, E.R. and Ross, R., 1985, A significant part of macrophage-derived growth factor consists of at least two forms of PDGF. Cell 43:277-286.

Sporn, M.B. and Roberts, A.B., 1986, Peptide growth factors and inflammation, tissue repair and cancer. J. Clin. Invest. 78:329-332.

Tertov, V.V., Orekhov, A.N., Grigorian, G.Yu., Kurennaya, G.S., Kudryashov, S.A., Tkachuk, V.A. and Smirnov, V.N., 1987, Disorders in the system of cyclic nucleotides in atherosclerosis: Cyclic AMP and cyclic GMP content and activity of related enzymes in human aorta. Tissue and Cell 19:21-28.

Wilcox, J.N., Smith, K.M., Williams, L.T., Schwartz, S.M. and Gordon, D., 1988, Platelet-derived growth factor mRNA detected in human atherosclerotic plaques by in situ hybridisation. J. Clin. Invest. 82:1134-1143.

Wines, P.A., Schmitz, J.M., Pfister, S.L., Clubb Jr., F.J., Buja, L.M., Willerson, J.T. and Campbell, W.B., 1989, Augmented vasoconstrictor responses to serotonin precede development of atherosclerosis in aorta of WHHL rabbit. <u>Arteriosclerosis</u> 9:195-202.

Yamamoto, Y., Tomoike, H., Egashira, K. and Nakamura, M., 1987, Attenuation of endothelium-related relaxation and enhanced responsiveness of vascular smooth muscle to histamine in spastic coronary arterial segments from minature pigs. <u>Circulation Res.</u> 61:772-778.

Yokoyama, J., Akita, H., Mizutani, T., Fukuzaki, H. and Watanabe, Y., 1983, Hyperreactivity of coronary arterial smooth muscles in response to ergonovine from rabbits with hereditory hyperlipidemia. <u>Circulation Res</u>. 53:63-71.

CIGARETTE SMOKING AND PLATELET FUNCTION: RELATION TO NICOTINE, CARBON

MONOXIDE AND SATURATED FAT

Serge Renaud

INSERM, Unit 63
22 av. Doyen Lepine, Case 18
69675 Bron Cedex (France)

INTRODUCTION

Smoking has been associated with myocardial infarction, independently
of its association with atherosclerotic lesions (1). Since coronary
thrombosis is the proximate cause of acute myocardial infarction (2) as
confirmed recently by the success of thrombolysis to reopen coronary
arteries (3), smoking could act directly on the predisposition to
thrombosis. Platelet reactivity seems to play a key role in the
thrombotic tendency since, in a recent prospective study on 2500 subjects
in Wales (5), aspirin significantly reduced acute coronary events (4) and
platelet hyperaggregability to ADP was significantly related to the
prevalent cases of myocardial infarction. Thus, it is important to
determine whether smoking increases platelet reactivity and additionally
which smoke components could be involved in this effect. Finally, in the
Seven Country Study (6), smoking was less of a risk factor for heart
attacks in Southern Europe and Japan than in the USA and northern
Europe. Since the consumption of saturated fats was much lower in Japan
(3% of calories) and southern Europe (7-10%) than in the USA (18%) and
northern Europe (19-22%), the effect of smoking on platelets could be
additive to that of saturated fats which are known to markedly increase
their reactivity (7). Results of our studies on the effect of smoking,
smoke components and saturated fats on platelet function tests are
briefly reviewed hereafter.

Platelet Reactivity in Smokers and Non-Smokers

Platelet reactivity in relation to dietary habits and various other
parameters such as smoking was determined in 9 groups (20 to 49 subjects
per group) of male farmers from different regions of France and Great
Britain. Almost half of these farmers (113 vs 147) were smokers (7). In
multivariate analysis, smoking was significantly related to platelet
reactivity and positively related to clotting activity of platelets,
while aggregation to ADP, epinephrine and collagen was inversely related
to smoking. These relationships are expressed in a simple manner in Fig.
1. The mean platelet reactivity of the non-smokers is compared to that
of smokers and of heavy smokers (more than 20 cigarettes/day). While
clotting activity in heavy smokers is significantly accelerated, platelet
aggregation to ADP and epinephrine is significantly reduced in smokers
and heavy smokers.

Tobacco Smoking and Atherosclerosis
Edited by J. N. Diana
Plenum Press, New York, 1990

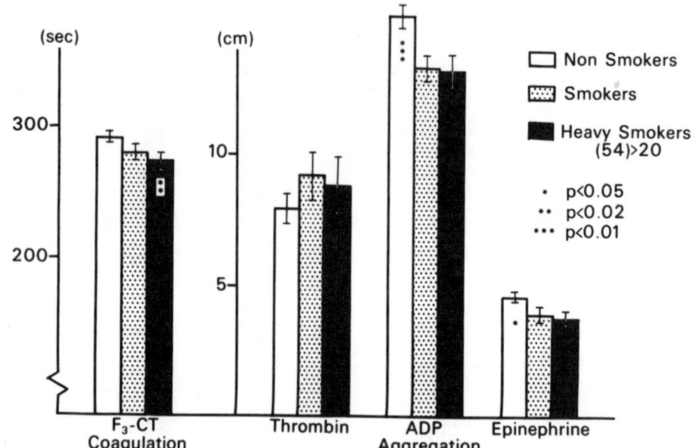

Fig. 1. F$_3$-CT = clotting activity of platelets
Non-smokers = 147   Smokers = 113
Heavy smokers (more than 20 cigarettes/day) = 54
Results: Mean ± S.E.

It has to be emphasized that venesection was performed in fasted
subjects deprived of smoking for at least 8 hours. Levine (8) has shown
that the effect of cigarette smoking to potentiate ADP induced
aggregation is an acute effect observed within 10 minutes and lost in 30
to 60 minutes. Thus, this last observation explains why 8 hours after
cigarette usage, there is no longer an increase in platelet reactivity to
various agonists. It has even been reported that by 2 hours after
smoking, platelet aggregability tends to be reduced (9).

Acute Effects of Cigarettes at Different Levels of Nicotine and Tar

In his studies, Levine (8) noted that the increase in platelet
aggregability was specific for tobacco smoking (could not be reproduced
by lettuce leaf cigarettes) provided the subjects were inhaling. In
preliminary studies, we were able to confirm that the increase in
platelet reactivity could be induced only in smokers who inhale. Thus in
all our studies (10,11), subjects were these who inhaled smoke and who
were deprived of smoking for 10 hours.

The specificity of tobacco smoking for increasing platelet reactivity
in smokers was confirmed in the following way. We compared, in 6
smokers, the clotting time and platelet aggregation to thrombin and
epinephrine before and 15 minutes after either no smoking, or smoking a
cigarette delivering 0.07mg nicotine (0.7mg tar) or 0.58 mg nicotine
(16.9mg tar). The results, shown in figure 2, indicate clearly that only
the high-nicotine, high-tar cigarette increased the clotting activity and
platelet aggregability. The extremely low-tar, low-nicotine cigarette
exerted no effect at all on the parameters examined.

To determine the extent to which the platelet response to smoking was
related to the level of nicotine delivered by the cigarette, 4 different
cigarettes were tested in 10 smokers under the conditions briefly
described in figure 2. The characteristics of those 4 cigarettes are
shown in Table 1.

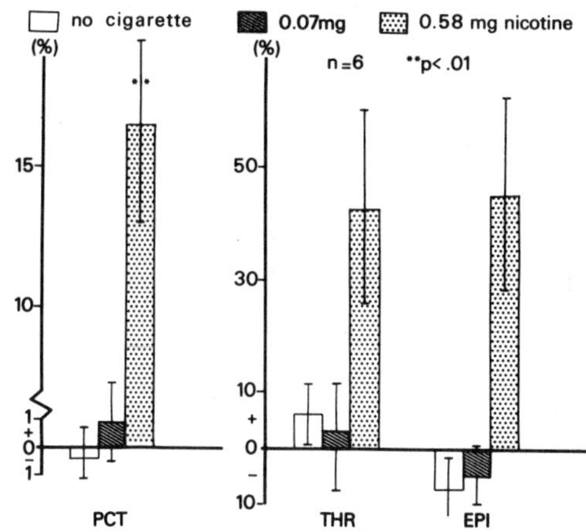

Fig. 2. Increase (%) in the response of the platelet
rich plasma clotting time (PCT) and platelet
aggregation to thrombin (THR) and epinephrine
(EPI) after smoking a cigarette in 6 smokers
fasted and deprived of smoking for 10 hours.
Each subject had 3 tests performed over a
period of approximately 3 weeks. The order
of the tests was randomly selected. For each
test, blood was removed twice at about 15
minute intervals. For one test, there was no
cigarette smoked. For the two other tests, a
cigarette (0.07) or 0.58 mg nicotine) was
smoked immediately after the first blood
removal and the second removal was performed
10 minutes after the end of smoking. (Adapted
from ref. 10). (An increase in the response of
clotting is in fact a shortening of the clotting
activity of platelets, 12).

With all the tests (coagulation, aggregation) the results were
somewhat similar in that cigarettes 1 and 2, with the lowest yields of
nicotine, were also associated with the lower effects on platelets.
Cigarettes 3 and 4, delivering the highest levels of nicotine, had the
most impact on platelets. Nevertheless, except perhaps for collagen
aggregation, parallelism between nicotine yield and platelet response was
far from convincing, especially for cigarette 3. It has been reported
that there is no relationship between blood nicotine and the nicotine
yield of cigarettes (13).

Table 1. Delivery of Nicotine, Tar, and Carbon
Monoxide of the 4 Cigarettes

Cigarette	1	2	3	4
Nicotine (mg)	0.18	0.40	0.58	1.44
Tar (mg)	1.8	9.8	16.9	23.8
Carbon monoxide (mg)	1.4	13.4	11.8	15.7

The effect of these 4 cigarettes on coagulation,
platelet aggregation, is shown in figure 3.

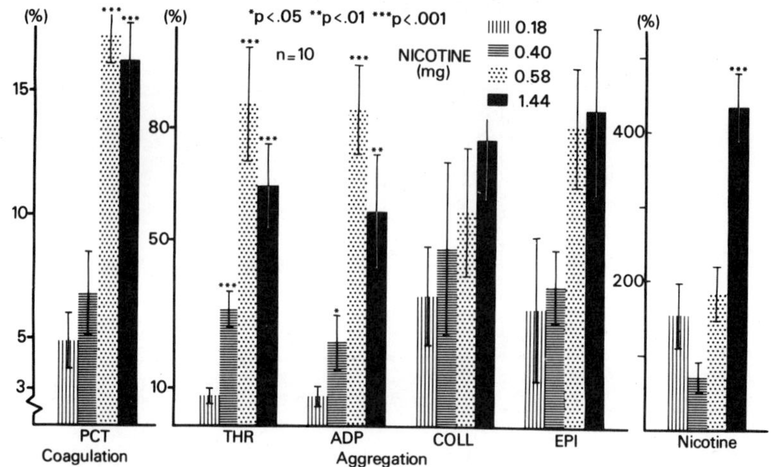

Fig. 3.   Increase (%) in coagulation (PCT), platelet aggregation response and blood nicotine 10 minutes after the 4 different cigarettes in 10 smokers (inhalers). Each subject (fasted and deprived of smoking for 10 hours) had 4 tests performed at approximately one-week intervals under the conditions described in figure 2. Coagulation and aggregation tests as in figure 2, in addition to platelet aggregation induced by ADP (adenosine diphosphate) and Coll (collagen). (Reproduced from ref. 10 with permission of Clin. Pharmacol. Ther.).

For that reason, in these experiments, we have determined the blood level of nicotine (as described elsewhere, 14). The results, also presented in figure 3, do not indicate a striking parallelism between the blood level of nicotine determined at 10 minutes and either the nicotine yield of those cigarettes or the response of platelets to clotting and

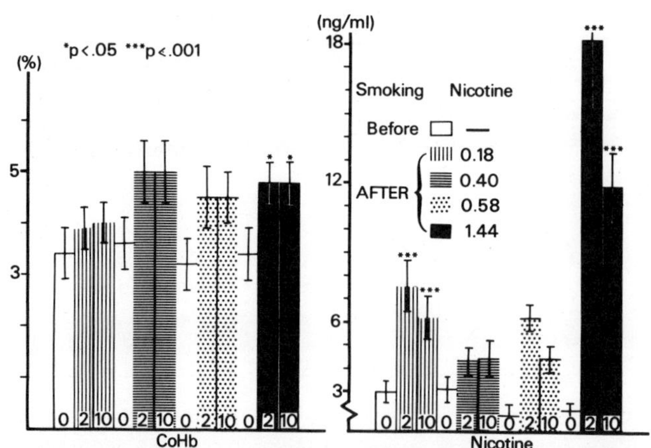

Fig. 4.   Carboxyhemoglobin and blood nicotine levels. Blood was removed at 0, 2 and 10 minutes after smoking a cigarette delivering 0.18 (1), 0.40 (2), 0.58 (3) or 1.44 mg (4) nicotine as in figure 3. (Adapted from ref. 10).

164

aggregation. It is known that nicotine has a short half-life in blood
after smoking (13,15). Thus in these studies, blood nicotine and
carboxyhemoglobin (by spectrometry according to a standard technique)
were also determined before and 2 minutes after smoking.

The level of carboxyhemoglobin was similar at 2 and 10 minutes, but
the blood level of nicotine was, with at least 3 cigarettes, slightly
higher at 2 minutes.

Further investigation (10) of the kinetics of blood nicotine
indicated that the highest blood levels were achieved at the end of
cigarette smoking, but these were quite similar to the levels at 2 and 10
minutes. Thus, none of the direct determinations performed in these
studies indicate a close relationship between platelet reactivity and the
level of either carboxyhemoglobin or nicotine (with the 4 cigarettes
utilized).

Since the blood level of nicotine was surprisingly different from
that which might have been expected from the nicotine yield of the 4
cigarettes, the blood level of nicotine was calculated utilizing the
carboxyhemoglobin level as an indication of the amount of smoke inhaled,
according to the following formula:

$$\frac{100 \times \% \text{ increase COHb (\%)} \times A}{B}$$

With A being nicotine yield and B being carbon monoxide.

From a similar equation of the % increase in the blood level of tar
can be calculated

$$\frac{\% \text{ increase COHb} \times D}{B}$$

With D being the tar delivered by the cigarette. The results
obtained from these calculations are reported in figure 5. The close
parallelism between the observed and predicted (calculated) level of
blood nicotine with a highly significant ($p<.001$) correlation coefficient
$r = .67$ for the 40 determinations completely validates the direct
determinations of blood nicotine. Another surprising result, also shown
in figure 5, is the parallelism between the predicted blood level of tar
and the response of platelets to aggregation, expressed as the mean of
the response to the 4 agonists utilized in the present study. That
comparison shows a parallelism which would be expected for a causal
relationship.

Influence of Nicotine on Platelet Reactivity In Vitro

To determine the extent to which nicotine per se could influence
platelet reactivity, nicotine was added, at levels found in smokers, to
platelet rich plasma from non smokers in vitro.

Nicotine at 10 ng had no effect on clotting but increased by 10% the
reactivity to thrombin and ADP. At 20 ng, nicotine enhanced the response
of coagulation by 11%; of aggregation to thrombin by 65% and aggregation
to ADP by 55%. In that connection it can be emphasized that the in vivo
studies had shown that the mean blood level of nicotine at 2 min. after
cigarette 1 was 16.0 ng/ml. The corresponding increase in the response
to coagulation was 16%; to thrombin, 65%; and to ADP, 58%. The mean
level of blood nicotine at 2 min with cigarette 4 was 8 ng; the increased
response to thrombin and ADP with this cigarette was 8%.

Consequently, nicotine in vitro at levels observed in smokers is able
to totally reproduce the effect on platelets which is induced in vivo by
smoking certain cigarettes.

Failure by some investigators (16) to observe the noxious effect of
nicotine on platelets appears to be due to the high concentration they
have utilized, as already indicated by Saba and Mason (17).
Nevertheless, for certain cigarettes (2 and 3, figure 3) neither the
nicotine yield nor the level of nicotine in blood could completely
explain the effect on platelets. Analysis of the alkaloid composition of
these cigarettes has shown that cigarettes 2 and 3 had higher levels of
nornicotine (2 to 3 fold) than the other cigarettes (10). Consequently,
nornicotine, which was not determined in the blood of our subjects,
probably contributed to the response of platelets observed with
cigarettes 2 and 3. Preliminary in vitro studies have shown that
nornicotine has a potentiating effect on platelet reactivity identical
and additive to that induced by nicotine.

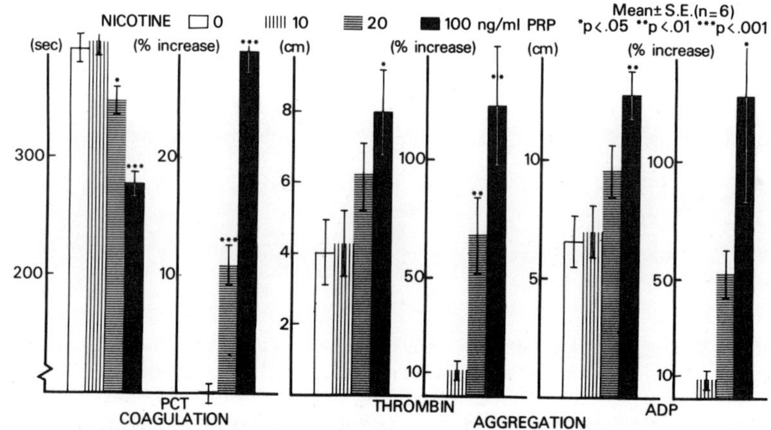

Fig. 5. Comparison between the observed level of blood
nicotine 2 minutes after the 4 different cigarettes
(experiment reported in figures 3 and 4) and the
level of nicotine which can be predicted from the
observed level of carboxyhemoglobin, as an indication
of the amount of smoke inhaled. Also reported in the
figure is the predicted amount of tar inhaled and the
response of platelets to aggregation (mean of the
response to thrombin + ADP + epinephrine + collagen).
(Adapted from ref. 10).

Thus nicotine (and probably nornicotine) having in vivo
characteristics in blood which show; 1) a peak preceding that of the
platelet reactivity, and 2) a short half-life comparable to that of
platelet hyperaggregability, can explain, at least in part, the effect of
smoking on platelets. By contrast, carboxyhemoglobin with a half-life of
2 to 3 hours (13) does not seem to offer an adequate explanation.
Nevertheless, further studies are certainly needed to determine whether
some substances contained in tar could be responsible for the as yet
unexplained effects of certain cigarettes on blood platelets.

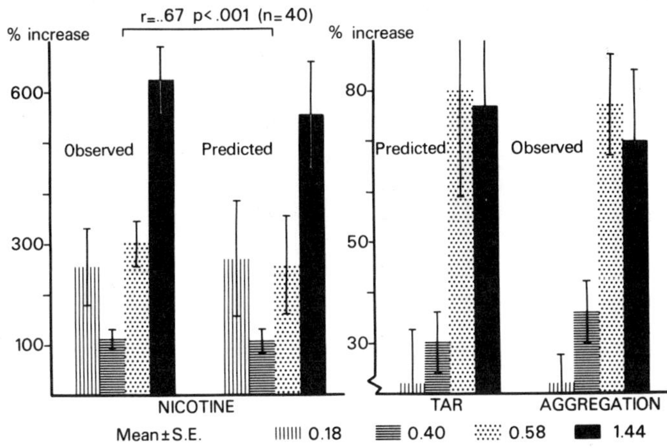

Fig. 6.   In vitro influence of nicotine on coagulation and
platelet aggregation. Nicotine diluted in saline
was added, at the concentration shown in the figure,
to platelet rich plasma (PRP), 2 minutes before
starting the test. During these 2 minutes, the
PRP was stirred at 1100 RPM for both coagulation
(PCT) and the aggregation to thrombin and ADP. The
PRP was obtained from 6 fasted non-smoker volunteers.
(Reproduced from ref. 10 with permission of Clin.
Pharmacol. Ther.)

Saturated Fats, Smoking and Platelet Reactivity

As already mentioned, smoking does not seem to predispose individuals
to myocardial infarction to the same extent; depending on countries or
regions. For example, in Southern Europe and Japan, smoking is less a
risk factor than in Northern Europe or in the USA (6). The
interpretation of this observation by A. Keys (6) is that the noxious
effect of cigarette smoking depends upon the level of serum cholesterol.
The effect could also depend directly on the intake of saturated fat,
usually greater in populations with higher serum cholesterol. One main
difference between saturated fat and serum cholesterol is that platelet
reactivity appears to be related only to the intake of saturated fat (7).

To be able to determine whether dietary habits would change the
platelet reactivity to smoking, we compared farmers from East and West
Scotland. These populations differed mainly in their intake of saturated
and polyunsaturated fats, but not in their level of serum lipids (11, 18)
and the subjects were all smokers (inhalers). Each subject was fasted
and deprived of smoking for at least 10 hours. After the first
venesection, subjects were asked to smoke the same high nicotine (3.5mg)
cigarette.

A second venesection was performed 10 minutes after smoking.

Results of the dietary survey (mean of weighing + recall) indicated
drastic differences in the intake of fats between the two regions
(Figure 7).

The hematologic results are shown in figure 7 and 8.

Before smoking, the clotting activity of platelets and the plasma
clotting time of platelet rich plasma (PCT) as well as the aggregation to

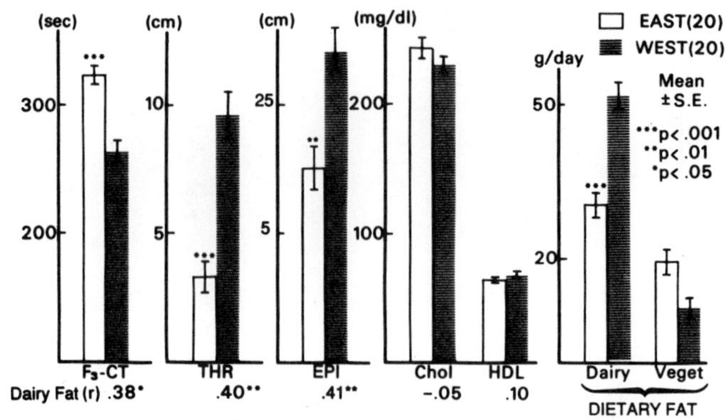

Fig. 7.   Blood parameters (before smoking) and dietary fat
          in farmers from East and West Scotland.
          F_3-CT = Clotting activity of platelets
          THR   = Thrombin aggregation in PRP (final
                    concentration 0.12 NIH Units per ml)
          ADP   = Adenosine diphosphate aggregation
                    (0.92×10^{-6} M in PRP)
          Chol = Total cholesterol, HDL = HDL cholesterol
          Veget = Vegetable fat
          r   = Correlation coefficients between intake
                    of dairy fat and blood parameters.

thrombin and primary aggregation to epinephrine were significantly
different between the East and West Scotland subjects. These tests were
all significantly correlated on an individual basis with the intake of
dairy fat (p from 0.05 to 0.01) (Fig. 7).

After cigarette smoking, as shown in figure 8, the response to all
the tests was significantly increased (shortened for the PCT). In West
Scotland subjects, the response to platelet aggregation (thrombin,
epinephrine) was more significantly increased than in East Scotland
subjects. Thus, the effect of smoking on platelet activity is enhanced
by the intake of saturated fat, but not by serum cholesterol which was
slightly lower in the subjects from West Scotland (11).

A striking result was the 4-fold increase in platelet aggregability
to thrombin and epinephrine after smoking one cigarette in the West
Scotland group compared with the aggregability before smoking in the East
Scotland group. Although such results should be confirmed by studies in
other countries, they could explain at least partly the higher
susceptibility to myocardial infarction of smokers in populations with a
high intake of saturated fat.

CONCLUSIONS

Recent studies indicate that platelet reactivity and coronary
thrombosis have a more important role in myocardial infarction than
formerly suspected (2-5). Smoking is known to markedly increase the risk
of myocardial infarction (MI) (19,20,21), but is weakly correlated, if at
all, with angina pectoris (22). The apparent rapid drop in the incidence
of myocardial infarction when men (19) and women (21) stop smoking
suggests that smoking is acting mostly on mechanisms other than
atherogenesis. This was confirmed directly by coronary angiography in
patients with MI (1). There was still, however, a very high association

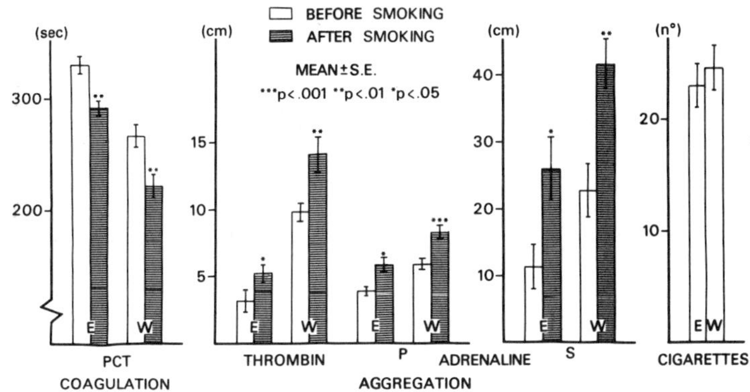

Fig. 8. Effects of smoking one high nicotine cigarette on
coagulation and platelet aggregation in male farmers
from East (E) and West (W) Scotland (20 smokers per
group, 40–45 years). Results compared before smoking
(deprived for the last 10 hours) and 10 minutes after
smoking.
PCT = recalcification plasma clotting time of platelet
rich plasma.
Ex vivo aggregation to thrombin and to adrenaline
(epinephrine).
P = primary; S = secondary
N° = number of cigarettes usually smoked.
(Adapted from ref. 11).

between smoking and MI after adjusting for occlusion. It was concluded
that smoking increased the risk of MI, primarily by mechanisms other than
coronary atherogenesis (1).

Thus a direct effect of smoking thrombogenesis can be postulated. It
is well documented that platelet reactivity plays a key role in coronary
thrombosis (4,5). Concerning the effect of smoking, several
investigators (8,23) have shown that cigarette smoking induces a marked,
transient increase in platelet aggregability, confirmed in our own
studies (10,11). We also observed that smoking increases the clotting
activity of platelets (10).

Even if the noxious effect of smoking one cigarette on platelets
lasts only 30 to 40 min, smoking one cigarette or more every hour would
probably result in an almost continuous increase in platelet reactivity
since the blood level of nicotine has been shown to rise to similar high
levels after each cigarette (15). Consequently, the effect of smoking on
blood platelets can explain, at least partly, the predisposition of
smokers for coronary thrombosis and MI. It also explains the varying
relationship between smoking and susceptibility to MI in different
countries, since the effect of smoking on platelets is additive to that
of saturated fats as shown by comparing smokers from East and West
Scotland (11). It may also explain the marked beneficial effect of the
combined response of a low saturated fat diet and reduction in cigarette
smoking to prevent myocardial infarction and sudden death (24).

Calculation of the level of blood nicotine based on the level of
carboxyhemoglobin as an indication of the smoke inhaled and on the
nicotine yield, produced identical results to those of the direct
determination of nicotine in blood. This indicates that, if the blood

nicotine level is not related to the nicotine yield as observed by
Russell et al. (13), and confirmed in our studies (10), it is because
smokers inhale different levels of smoke, depending on the cigarette used.

The agent responsible for the effect of cigarette smoke on platelet
reactivity, whether it is nicotine, tar, or carbon monoxide, is
apparently not completely elucidated. Carbon monoxide seems to be
excluded since its kinetics are markedly different from those of platelet
reactivity. Nicotine (and probably nornicotine) at levels found in
smokers can totally reproduce in vitro the effect observed after smoking
a cigarette. The kinetics of nicotine in blood are smoking a
high-nicotine cigarette. Nevertheless, it is still possible that
substances in the tar fraction could also play a role in the effect of
cigarette smoke on platelets.

REFERENCES

1. A.J. Hartz, P.N. Barboriak, A.J. Anderson, R.B. Hoffmann, and JJ.
    Barboriak. Smoking, coronary artery occlusion and nonfatal
    myocardial infarction, J.A.M.A. 246:851 (1981).
2. M.A. De Wood, J. Spores, R. Notske. Prevalence of total coronary
    occlusion during the early hours of transmural myocardial
    infarction, N. Engl. J. Med. 303:897 (1980).
3. AIMS TRIAL study group. Effect of intravenous APSAC on mortality
    after acute myocardial infarction: preliminary report of a
    placebo-controlled clinical trial, Lancet, 1:545 (1988).
4. V. Fuster, M. Cohen, J. Halpern. Aspirin in the prevention of
    coronary disease, N. Engl. J. Med. 321:183 (1989).
5. P.C. Elwood, S. Renaud, D.S. Sharp, A.O. Beswick, J.R. O'Brien and
    J.W.G. Yarnell. Ischaemic heart disease and platelet
    aggregation: the Caerphilly collaborative heart disease study
    (Submitted for publication).
6. A. Keys, Seven Countries. A multivariate analysis of death and
    coronary heart disease. Harvard University Press, Cambridge, MA,
    USA (1980).
7. S. Renaud, R. Morazain, F. Godsey, E. Dumont, C. Thevenon, J.L.
    Martin, F. Mendy. Nutrients, platelet function and composition in
    nine groups of French and British farmers, Atherosclerosis 60:37
    (1986).
8. P.H. Levine. An acute effect of cigarette smoking on platelet
    function, Circulation 48:619 (1973).
9. W. Siess, R. Lorenz, P. Roth, P.C. Weber, Plasma catecholamines,
    platelet aggregation and associated thromboxane formation after
    physical exercise, smoking or norepinephrine infusion,
    Circulation 66:44 (1983).
10. S. Renaud, D. Blache, E. Dumont, C. Thevenon, and T. Wissendanger.
     Platelet function after cigarette smoking in relation to nicotine
     and carbon monoxide. Clin. Pharmacol. Ther. 36:389 (1984).
11. S. Renaud, E. Dumont, F. Baudier, E. Ortchanian, and I.S. Symington.
     Effect of smoking and dietary saturated fats on platelet functions
     in Scottish farmers, Cardiovasc. Res. 19:155 (1985).
12. S. Renaud, and F. Lecompte. Hypercoagulability induced by hyper-
     lipemia in rat, rabbit and man. Role of platelet factor 3, Circ.
     Res. 27:1003 (1979).
13. M.A.H. Russell, M. Jarvis, R. Iyer, and C. Feyerabend. Relation of
     nicotine yield of cigarettes to blood nicotine concentrations in
     smokers, Br. Med. J. 1:972 (1980).
14. D. Blache, C. Thevenon, M. Ciavatti, and S. Renaud. A sensitive
     method for the routine determination of plasma nicotine by flame
     ionization gas-liquid chromatography, Analyt. Biochem. 143:316
     (1984).

15. M.E. McNabb, R.V. Ebert, and K. McCusker. Plasma nicotine levels produced by chewing nicotine gum, _J.A.M.A._ 248:865 (1982).

16. K. Brinson. Effect of nicotine on human blood platelet aggregation. _Atherosclerosis_ 20:127 (1975).

17. S.R. Saba, and R.G. Mason. Some effects of nicotine on platelets, _Thromb. Res_.

18. S. Renaud, R. Morazain, F. Godsey, E. Dumont, I.S. Symington, E.M. Gillanders, and J.R. O'Brien. Platelet functions in relation to diet and serum lipids in British farmers, _Brit. Heart J_. 46:562 (1981).

19. R. Doll, R. Peto. Mortality in relation to smoking: 20 years' observations on male British doctors, _Brit. Med. J_. 2:1526 (1976).

20. T. Gordon, N.B. Kannel, D. McGee, and T.R. Dawber. Death and coronary attacks in men after giving up cigarette smoking, _Lancet_ 2: 1345 (1974).

21. W.C. Willett, C.H. Hennekens, C. Bain, B. Rosner and F.E. Speizer. Cigarette smoking and non-fatal myocardial infarction in women, _Amer. J. Epidemiol_. 113:575 (1981).

22. J.T. Doyle, T.R. Dawber, W.B. Kannel, S.H. Kinch, and H.A. Kahn. The relationship of cigarette smoking to coronary heart disease, _J.A.M.A._ 190:886 (1964).

23. G. Grignani, G. Gamba and E. Ascari. Cigarette-smoking effect on platelet function, _Thromb. Haemost_. 37:442 (1977).

24. I. Hjermann, K. Velve Byre, I. Holme, and P. Leren. Effect of diet and smoking intervention on the incidence of coronary heart disease, _Lancet_ 2:1303 (1981).

ATHEROSCLEROSIS ALTERS THE RESPONSE
TO ACTIVATED PLATELETS AND LEUKOCYTES

Donald D. Heistad, J. Antonio Lopez, and Mark L. Armstrong

Departments of Internal Medicine and Pharmacology
Veterans Administration Medical Center and
University of Iowa College of Medicine
Iowa City, IA 52242

INTRODUCTION

This chapter addresses vasomotor consequences of atherosclerosis.
The major goal of the studies is to understand the pathophysiology of
spasm in atherosclerotic arteries.

First, we will provide evidence which suggests that vasoactive
products that are released by leukocytes may produce spasm of
atherosclerotic arteries. Second, the role of platelets in vasospasm
will be examined. Third, we will suggest that vasospasm may contribute
to several clinical syndromes. Finally, we will speculate that
vasoactive products that are released by platelets may affect
baroreceptor function.

We have studied effects of atherosclerosis in cynomolgus monkeys.
Monkeys develop atherosclerotic lesions that closely resemble the lesions
that occur in humans (1), and no other species (except humans) develop
lesions of comparable severity in the coronary, carotid, and limb
arteries, unless vascular injury is superimposed on an atherogenic diet.
Moderately severe atherosclerosis develops in monkeys that are fed an
atherogenic diet that contains about 40% fat and 0.7% cholesterol for 18
months to 5 years. Lesions are present in the epicardial coronary
arteries, proximal carotid arteries, and iliac arteries (Fig. 1), and the
distribution of the lesions closely resembles that which occurs in
humans.

ROLE OF LEUKOCYTES

Blood-borne monocytes frequently adhere to atherosclerotic lesions, and
the lesions usually are infiltrated by monocyte-macrophages. A major
hypothesis concerning the development and progression of atherosclerotic
lesions is that growth factors are released by monocyte-macrophages in
the lesions, and these factors stimulate cellular migration and
proliferation (2).

Tobacco Smoking and Atherosclerosis
Edited by J. N. Diana
Plenum Press, New York, 1990

We have proposed recently a new role for leukocytes in atherosclerotic lesions. Our hypothesis is that leukocytes may release vasoactive products that produce spasm of atherosclerotic arteries (3).The findings that led us to propose this hypothesis were that injections of a peptide, f-met-leu-phe, to activate leukocytes had little effect in normal monkeys and produced pronounced vasoconstriction in atherosclerotic monkeys (3). Other investigators have injected fMLP in normal animals (4) and after myocardial infarction (5), but our studies were the first in which fMLP was injected in atherosclerotic animals. We concluded that activation of leukocytes, with release of their vasoactive products, produces constriction or even spasm in atherosclerotic arteries.

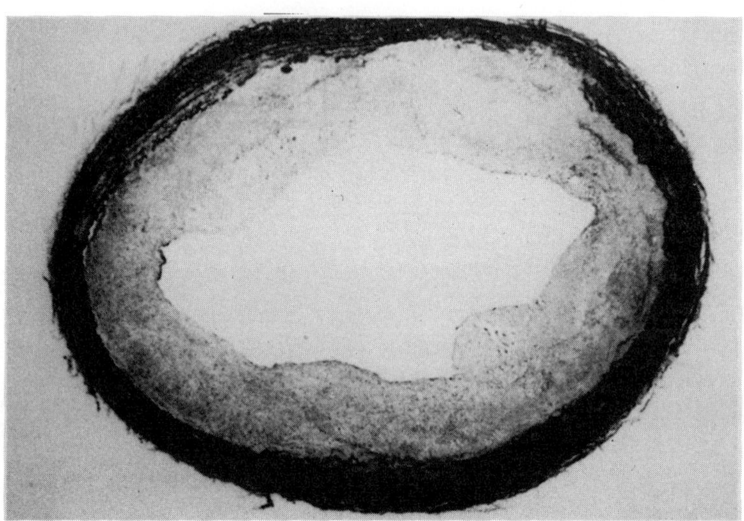

Figure 1. Carotid artery of a cynomolgus monkey that was fed atherogenic diet for about 18 months. The arteries typically develop pronounced intimal proliferation.

We have begun studies to determine which of the vasoactive substances that are released by leukocytes may account for the vasoconstrictor responses (Fig 2). Leukocytes, including monocyte-macrophages, release thromboxane, leukotrienes, prostaglandin E_2, and a variety of other potent vasoactive substances. Atherosclerosis potentiates vasoconstrictor responses to thromboxane (6), and we suggest that thromboxane may contribute to leukocyte-induced constriction of atherosclerotic arteries. A surprising finding was that prostaglandin E_2, which normally is a vasodilator, produces constriction of large arteries in atherosclerotic monkeys (3), and thus may also contribute to leukocyte-induced vasoconstriction. Our initial studies suggest that leukotrienes do not contribute to leukocyte-induced constriction of atherosclerotic arteries.

ROLE OF PLATELETS

The major hypothesis concerning the pathophysiology of vasospasm focuses on the role of platelets (7,8). Platelets may adhere to atherosclerotic plaques, aggregate, and release several potent vasoactive substances. Thromboxane is a potent vasoconstrictor that is released by platelets, but large amounts of ADP (which is a potent vasodilator) are also released by platelets and normally mask the effect of thromboxane.

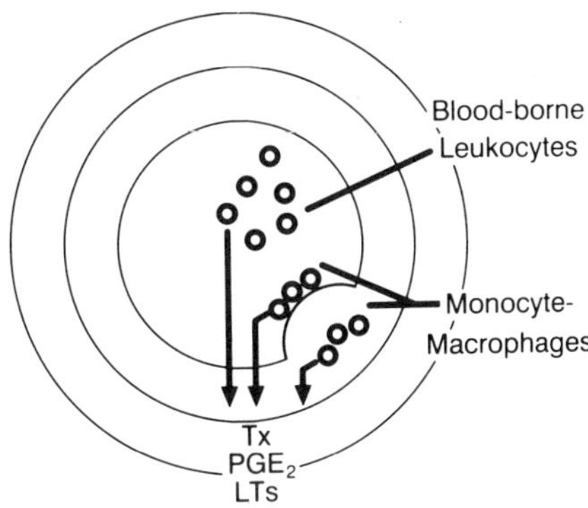

Figure 2. Activation of leukocytes by f-met-leu-phe produces vasoconstriction in atherosclerotic monkeys. This effect may be mediated by release of vasoactive substances from blood-borne leukocytes or from leukocytes that are on or in the vessel wall (monocyte-macrophages). Thromboxane (Tx) and prostaglandin E_2 (PGE_2) are released from leukocytes, and may contribute to leukocyte-induced vasoconstriction. Leukotrienes (LTs) do not appear to contribute to leukocyte-induced vasoconstriction.

Our studies suggest that serotonin, which also is released by platelets, produces profoundly different responses in normal and atherosclerotic monkeys. Serotonin has little effect, or produces vasodilatation, in normal monkeys, and produces pronounced vasoconstriction in atherosclerotic monkeys (9). We have suggested that potentiation of vasoconstrictor responses to serotonin (6,9) and thromboxane (6) in atherosclerotic arteries, with impaired vasodilator responses to ADP, may predispose to vasospasm when platelets release their vasoactive substances in atherosclerotic arteries.

To examine effects of vasoactive products that are released by platelets, we aggregated human platelets in vitro and injected the supernatant, which contained the vasoactive products that are released by platelets, in the perfused limb of monkeys (10). The supernatant was injected within 30 seconds after aggregation of platelets, so that thromboxane (which has a half-life of about 30 seconds) was only partially metabolized. Injection of the supernatant produced vasodilatation in normal monkeys, presumably because vasodilator effects of ADP are predominant. The vasodilator response to the supernatant was impaired in atherosclerotic monkeys, but not reversed to vasoconstriction.

Recently, we infused purified bovine collagen into the perfused limb of monkeys, to aggregate platelets in vivo (11). Infusion of collagen produced vasodilatation in normal monkeys, presumably because vasodilator effects of ADP are predominant. Infusion of collagen in atherosclerotic monkeys initially produced vasodilatation but, after several minutes, there was progressive and pronounced constriction of large arteries. These findings support the hypothesis that when platelets are activated in vivo, they may contribute to constriction or spasm of atherosclerotic arteries.

A major mechanism by which atherosclerosis potentiates vasoconstrictor responses and leads to vasospasm involves modulation of vascular tone by the endothelium. Aggregation of platelets in normal arteries in vitro produces relaxation (12). After removal of endothelium, aggregation of platelets produces contraction (12). Endothelium is present in atherosclerotic arteries, but endothelium-dependent relaxation is impaired (Fig. 3) (13-15). It is likely that endothelial dysfunction may lead to vasoconstrictor responses, instead of vasodilatation, to serotonin or aggregation of platelets in atherosclerotic arteries.

IMPLICATIONS FOR VASOSPASM

In every organ that we have studied, we have found that atherosclerosis alters vascular responses in a direction that would favor augmented vasoconstriction. There is modest potentiation of vasoconstrictor responses to a variety of agonists, but the potentiation is profound in response to serotonin and to activation of leukocytes. Based on these finding, we suggest that vasospasm may contribute to ischemia in several vascular beds, and that activation of leukocytes and/or platelets may play a pivotal role.

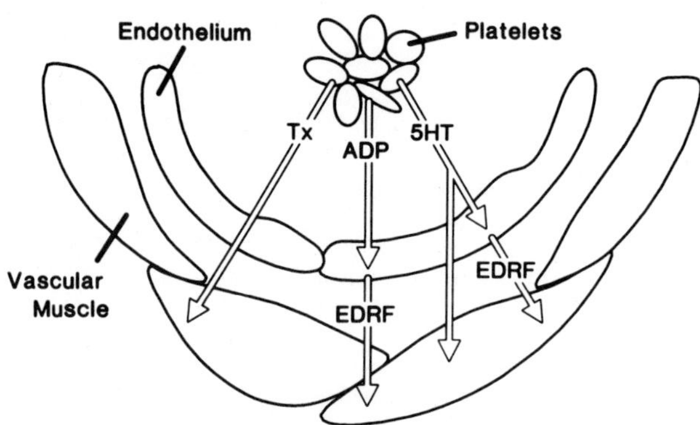

Figure 3. Aggregation of platelets releases thromboxane (Tx), ADP, and serotonin (5HT). Thromboxane acts directly on vascular muscle to produce constriction. ADP releases endothelium-derived relaxing factor (EDRF), which diffuses from endothelium to vascular muscle to produce relaxation. Serotonin has two opposing effects: it acts directly on vascular muscle to produce constriction, and it releases EDRF to produce relaxation. Endothelium is dysfunctional in atherosclerotic arteries, and release of EDRF is impaired, so the constrictor response to thromboxane and serotonin may predominate over the dilator response to ADP and serotonin. Thus, platelets produce relaxation in normal arteries, but they may produce contraction or spasm of atherosclerotic arteries.

Vasospasm clearly occurs in the coronary circulation and contributes to myocardial ischemia and angina (16). Recently, we have observed that coronary vasoconstrictor responses to serotonin are potentiated in atherosclerotic monkeys (17). Thus, altered vascular responsiveness may contribute to coronary vasospasm.

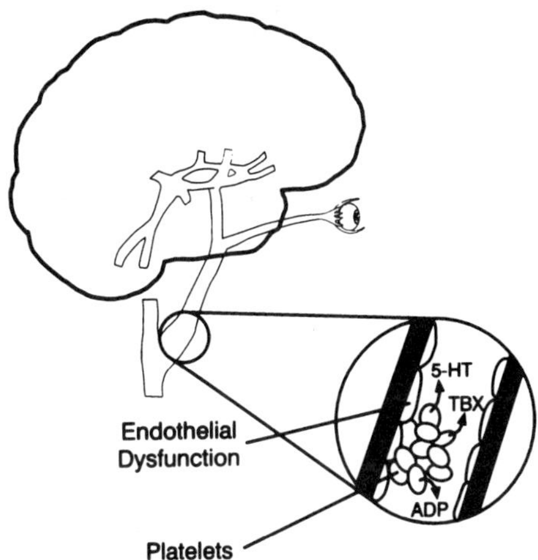

Figure 4. Transient ischemic attacks (TIAs) currently are attributed to adherence of platelets to atherosclerotic plaques in the carotid artery, with embolization to the eye and cerebrum. We have suggested that, when platelets adhere to the carotid arteries, aggregate, and release serotonin (5-HT), thromboxane (TBX), and ADP, endothelial dysfunction of the atherosclerotic carotid artery may lead to pronounced vasoconstrictor responses. Thus, we have proposed that vasospasm, as well as platelet emboli, may produce TIAs.

We have found that, in atherosclerotic but not normal monkeys, serotonin and thromboxane produce cerebral vasoconstriction and serotonin produces a profound reduction in retinal blood flow (18). Serotonin also produces reversible retinal ischemia and blindness in atherosclerotic monkeys. Based on these finding, we have proposed that alteration of vascular responses to products that are released by platelets may lead to vasospasm and to amaurosis fugax and transient ischemia attacks (TIAs) (Fig. 4) (18). This hypothesis contrasts with the current concept, that amaurosis fugax and TIAs can be attributed entirely to emboli. The current concept is that platelets adhere to atherosclerotic plaques in the carotid arteries and then embolize to produce retinal and cerebral ischemia. We are proposing that vasospasm may also occur in this setting.

PROLIFERATION OF VASA

Morphological studies indicate that vasa vasorum proliferate in atherosclerotic arteries (19). We have demonstrated a remarkable increase in blood flow through vasa in the atherosclerotic aorta, coronary arteries, and carotid arteries (20-22). Vasa in atherosclerotic arteries are hyperresponsive to serotonin (22,23).

We speculate that proliferation of vasa in the carotid artery may contribute to alterations of baroreceptor function by atherosclerosis. When platelets adhere to atherosclerotic plaques in the carotid arteries, serotonin and other platelet products may be delivered through vasa which are adjacent to baroreceptor endings in the carotid sinus (Fig. 5) (22). Serotonin affects discharge of baroreceptors (24) and, if a significant concentration of serotonin is achieved in vasa to the carotid sinus, platelet-derived serotonin might alter baroreceptor function.

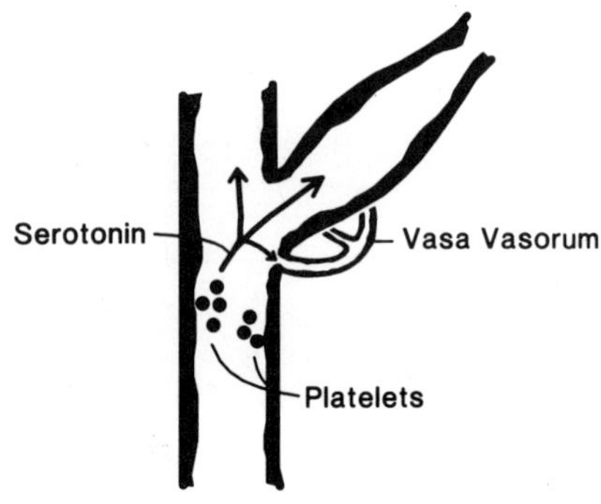

Figure 5. Atherosclerosis stimulates proliferation of vasa vasorum in the carotid sinus. We speculate that, when platelets adhere to atherosclerotic plaques in the carotid arteries, serotonin and other products that are released by platelets may be delivered through vasa to baroreceptor endings. These platelet products may produce dysfunction of baroreceptors.

Hyperresponsiveness to serotonin in vasa of atherosclerotic arteries (22,23) also may have functional implications. Serotonin produces pronounced constriction of atherosclerotic coronary and carotid arteries (17,25), and thus increases metabolic needs of the vessels. If serotonin compromises nourishment of the arteries by reducing blood flow through vasa (22,23), the combination of increased metabolic needs of the arterial wall and reduced flow through vasa vasorum may predispose the wall of the arteries to ischemia. Thus, we speculate that hyperresponsiveness of vasa to constrictor stimuli may play a role in the pathophysiology of vascular necrosis, which characterizes atherosclerotic arteries.

Acknowledgments

Original studies were supported by a Medical Investigatorship, Associate Investigatorship, and research funds from the Veterans Administration, and by NIH Grants HL 14230, HL 16066, NS 24621, and HL 14388.

References

1. M. Armstrong, and E. Warner, Morphology and distribution of diet-induced atherosclerosis in rhesus monkeys, Arch. Pathol. 92:395 (1971).

2. R. Gerrity, The role of the monocyte in atherogenesis. I. Transition of blood-borne monocytes into foam cells in fatty lesions, Am. J. Pathol. 103:181 (1981).

3. J. A. G. Lopez, M. Armstrong, D. Harrison, D. J. Piegors and D. D. Heistad, Vascular responses to leukocyte products in atherosclerotic primates, Circ. Res. 65:1078 (1989).

4. M. N. Gillespie, D. C. Booth, B. J. Friedman, M. R. Cunningham, M. Jay, and A. N. DeMaria, fMLP produces coronary vasoconstriction and myocardial ischemia in rabbits, Am. J. Physiol.(Heart Circ. Physiol) 23:254 (1988).

5. A. S. Evers, S. Murphree, J. E. Saffitz, B. A. Jakschik, and P. Needleman, Effects of endogenously produced leukotrienes, thromboxane, and prostaglandins on coronary vascular resistance in rabbit myocardial infarction, J. Clin. Invest. 75:992 (1985).

6. J. A. G. Lopez, M. Armstrong, D. Piegors and D. D. Heistad, Effect of early and advanced atherosclerosis on vascular responses to serotonin, thromboxane A_2, and ADP, Circulation 79:698 (1989).

7. P. Vanhoutte, and D. Houston, Platelets, endothelium and vasospasm, Circulation 72:728 (1985).

8. J. T. Willerson, L. D. Hillis, M. D. Winniford and L. M. Buja, Speculations regarding mechanisms responsible for acute ischemic heart disease syndrome, J. Am. Coll. Cardiol. 8:245 (1986).

9. D. D. Heistad, M. L. Armstrong, M. L. Marcus, D. J. Piegors and A. L. Mark, Augmented responses to vasoconstrictor stimuli in hypercholesterolemic and atherosclerotic monkeys, Circ. Res. 54:711 (1984).

10. J. A. G. Lopez, A. F. A. Brotherton, M. L. Armstrong, D. J. Piegors, and D. D. Heistad, Vascular responses to activated platelets in atherosclerotic primates, FASEB J. 2:A943 (1988).

11. J. A. G. Lopez, J. C. Hoak, M. L. Armstrong, D. J. Piegors, and D. D. Heistad, Effect of intravascular collagen in atherosclerotic primates, J. Am. Coll. Cardiol. (Abstract, in press).

12. D. S. Houston, J. T. Shepherd, and P. M. Vanhoutte, Aggregating human platelets cause direct contraction and endothelium dependent relaxation of isolated canine coronary arteries: role of serotonin, thromboxane A_2, and adenine nucleotides, J. Clin. Invest. 78:539 (1986).

13. P. C. Freiman, G. G. Mitchell, D. D. Heistad, M. L. Armstrong and D. G. Harrison, Atherosclerosis impairs endothelium-dependent vascular relaxation to acetylcholine and thrombin in primates, Circ. Res. 58:783 (1986).

14. T. J. Verbeuren, F. H. Jordaens, L. L. Zonnekeyn, C. E. Van Hove, M-C Coene, A. G. Herman, Effects of hypercholesterolemia on vascular reactivity in the rabbit, Circ. Res. 58:552 (1986).

15. J. B. Habib, C. Bossaller, S. Wells, C. Williams, J. D. Morrisett and P. D. Henry, Preservation of endothelium-dependent vascular relaxation in cholesterol-fed rabbit by treatment with the calcium blocker PN 200110, Circ. Res. 58:305 (1986).

16. A. Maseri, A. L'Abbate, G. Barold, S. Chierchia, M. Marzilli, A. M. Ballestra, S. Severij, O. Parodi, A. Distante and A. Pesula, Coronary vasospasm as a possible cause of myocardial infarction, N. Eng. J. Med. 299:1271 (1978).

17. W. M. Chilian, K. C. Dellsperger, S. M. Layne, C. L. Eastham, M. A. Armstrong, M. L. Marcus, and D. D. Heistad, Effects of atherosclerosis on the coronary circulation: potentiation of arterial and microvascular constrictor responses, Am. J. Physiol.:Heart (In Press).

18. J. K. Williams, G. L. Baumbach, M. L. Armstrong and D. D. Heistad, Hypothesis: vasoconstriction contributes to amaurosis fugax, J. Cereb. Blood Flow Metab. 9:111 (1989).

19. E. Geringer, Intimal vascularization and atheromatosis, J. Pathol.
     Bacteriol. 63:201 (1951).
20. D. D. Heistad, M. L. Armstrong, and M. L. Marcus, Hyperemia of the
     aortic wall in atherosclerotic monkeys, Circ. Res. 48:669 (1981).
21. D. D. Heistad, and M. L. Armstrong, Blood flow through vasa vasorum
     of coronary arteries in atherosclerotic monkeys, Arteriosclerosis
     6:326 (1986).
22. J. K. Williams, K. I. Orgren, M. L. Armstrong, and D. D. Heistad,
     Vasa vasorum in the carotid sinus of atherosclerotic monkeys:
     implications for baroreceptor function, Atherosclerosis 78:25
     (1989).
23. J. K. Williams, M. L. Armstrong, and D. D. Heistad, Vasa vasorum in
     atherosclerotic coronary arteries: responses to vasoactive
     stimuli and regression of atherosclerosis, Circ. Res. 62:515
     (1988).
24. K. Nishi, The action of 5-hydroxytryptamine on chemoreceptor
     discharges of the cat's carotid body, Br. J. Pharmacol. 55:27
     (1975).
25. D. D. Heistad, K. Breese, and M. L. Armstrong, Cerebral
     vasoconstrictor responses to serotonin after dietary treatment of
     atherosclerosis: implications for transient ischemic attacks,
     Stroke 18:1068 (1987).

PLATELETS IN CHRONIC SMOKERS SHOW A HYPERACTIVE

RESPONSE IN VITRO TO A FOREIGN SURFACE

Paul D. Stein, Jan Rival, and Jeanne M. Riddle

Henry Ford Hospital
Detroit, Michigan 48202

INTRODUCTION

A number of investigators have evaluated platelet function in
smokers. The results have been diverse. Part of the lack of clarity may
relate to the diversity of the characteristics of platelet function that
were evaluated, as well as the length of time following smoking during
which the test was performed. In this chapter, we will review the
literature in this regard, with particular emphasis upon the more
prolonged effects of smoking. In particular, we will review an
investigation of the effects upon platelets of chronic smoking that we
previously performed (1).

METHODS

We studied the effect of cigarette smoking upon platelets in 23
individuals who had never smoked (nonsmokers); and 20 persons who had
smoked various quantities of nicotine-containing cigarettes (smokers)
(1). None of the subjects had taken any antiplatelet or anticoagulant
drugs (aspirin, dipyridamole, allopurinol, sulfinpyrazone, propranolol,
heparin, coumadin) for at least one month prior to the investigation.
None of the smokers had smoked a cigarette for at least 4 hours prior to
any platelet studies. Among the smokers there were 10 men and 10 women.
Among the non-smokers there were 15 men and 8 women. The levels of
cholesterol, uric acid, triglycerides, glucose, and urea nitrogen (BUN)
were similar in both groups. The daily consumption of cigarettes among
smokers was 1.4 $\pm$ 0.5 packs, and the duration of smoking was 19 $\pm$ 12
years. Platelet function was measured both in the early phase of
activation (adhesion, formation of pseudopodia and spreading) and in the
later phases (aggregation as well as the release of platelet-specific
proteins) (1).

PLATELET ELECTRON MICROSCOPIC SURVEY

The platelet electron microscopic survey that we performed will be
described in somewhat more detail than the other investigations of
platelets, because the technique is less well known, although it has been
described in detail previously (1-3).

Tobacco Smoking and Atherosclerosis
Edited by J. N. Diana
Plenum Press, New York, 1990

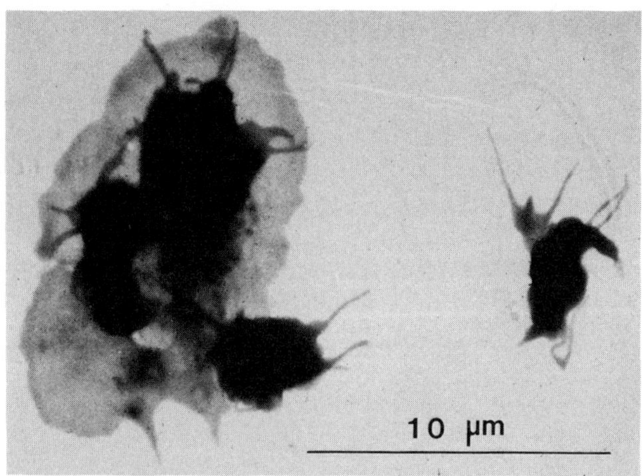

FIG. 1   Increased numbers of platelet aggregates (left) were present
after the platelets of smokers were exposed to an activating
surface.

The procedure is a standardized in vitro method that provides a
morphologic assessment of three separate phases of the platelet response
(adhesion, surface activation and aggregation). The major advantages of
this method, are that one can directly observe structural changes after
platelets contact an activating surface in vitro, and the test monitors
the early phase of the platelet response (Fig 1).

A 2 ml whole blood sample was withdrawn and discarded to eliminate
any procoagulant substances that might have been introduced during
venipuncture. Nine ml of whole blood was withdrawn into the 3.8% sodium
citrate, which was mixed briefly with the whole blood. The
anticoagulated blood specimen was transferred immediately to a 6 oz
polyethylene, square bottle with a wide mouth and a nonfrosted microscope
slide covered with a thin film of Formvar (1% Formvar dissolved in
ethylene dichloride) was introduced horizontally. The blood mixture
remained in contact with the Formvar film for exactly eight minutes.
During this incubation, blood platelets became attached to the surface of
the Formvar film. Afterward, the microscope slide was removed from the
bottle, drained briefly, and rinsed in Tyrode's solution. Rinsing
continued until erythrocytes were no longer grossly visible on the film
surface. The slide was then immersed in cacodylate buffered 3%
glutaraldehyde for 15 minutes at room temperature to stabilize the
structure of the adherent platelets. After fixation, the slide was
rinsed and air dried.

The edges of the slide were scraped with a scalpel blade to free the
Formvar film. Stainless steel, 200 mesh specimen grids were placed over
representative areas so that their convex surfaces contacted the Formvar
film. After mounting, platelet populations were viewed with a
transmission electron microscope operated at 50 KV.

In this procedure, the number of platelets present on the surface of
the Formvar film (a polyvinyl formal plastic film) provided a measure of
their adhesive capacity. Surface activation of single platelets was
demonstrated by the extrusion of pseudopodia and cytoplasmic spreading
between pseudopodia with reorganization of internal organelles.
Aggregation, indicative of the cohesive capacity of each population of

platelets, was directly observed by counting the number of platelet
aggregates seen during the classification of 100 single platelets.

Three distinct types of platelets were observed: round, dendritic,
and spread types. The round type was compact, had a smooth contour, was
uniformly electron-dense and corresponded to the disc-shaped circulating
platelet. The dendritic type was characterized by a compact, electron-
dense central area from which pseudopodia extruded. The spread type
showed varying degrees of cytoplasmic spreading between adjacent
pseudopodia, in addition to a relocation of its internal organelles.

A platelet differential count consisted of the enumeration of the
percent of round, dendritic and spread type platelets found on
examination of 100 consecutive single platelets. The number of platelet
aggregates seen during evaluation of the 100 platelets was also
recorded. At the time that the platelet electron microscopic survey was
performed, the number of circulating platelets per mm^3 was also counted.

Circulating Platelet Aggregates

The platelet aggregate ratio (index of circulating platelet
aggregates) was determined according to the method of Wu and Hoak (4).

Platelet Release Factors

Plasma levels of platelet factor 4 and beta-thromboglobulin (platelet
release factors) were measured using radioimmunoassay procedures (5).

RESULTS

Platelet aggregation after exposure to a foreign surface in vitro
(Formvar) was higher in smokers than in nonsmokers (80 $\pm$ 59 vs 43 $\pm$ 27
aggregates/100 platelets) (P<.01) (Fig. 2) (1). Studies which reflected
the in vivo status of platelets showed no difference among smokers versus
nonsmokers. These tests included the platelet count per mm^3 (310,000 $\pm$
82,000 vs 278,000 $\pm$ 78,000) (mean $\pm$ SD), platelet aggregate ratio (0.67 $\pm$
0.30 vs 0.86 $\pm$ 0.76), and levels of platelet-specific proteins released
in vivo including platelet factor 4 (ng/ml) (9.8 $\pm$ 5.2 vs 9.4 $\pm$ 5.3) and
beta-thromboglobulin (ng/ml) (53.9 $\pm$ 23.5 vs 49.1 $\pm$ 25.5). More platelet
aggregates on the electron microscopic differential count were found in
subjects who smoked 20 years or more than in subjects who smoked less
than 20 years (112 $\pm$ 60 vs 53 $\pm$ 45 platelet aggregates/100 platelets) (P
= 0.02) (Fig 2). None of the other platelet tests showed a difference in
relation to the duration of smoking. Platelet function among subjects
who smoked 0.5 - 1 pack daily compared to those who smoked more than 1
pack daily showed no significant difference.

The number of platelet aggregates determined by platelet differential
counts in vitro for smokers older than 50 years of age was increased in
comparison to smokers 50 years of age or younger (105 $\pm$ 54 vs 54 $\pm$ 55
aggregates/100 platelets) (P<.05). This increased tendency to form a
greater number of platelet aggregates with increasing age was not found
in nonsmokers.

The amount of surface activation (percent of the spread type platelet
and amount of aggregation), ratio of platelet aggregates, and levels of
platelet-specific proteins did not differ between males and females,
although the platelet count was higher in females irrespective of whether
they were smokers or nonsmokers.

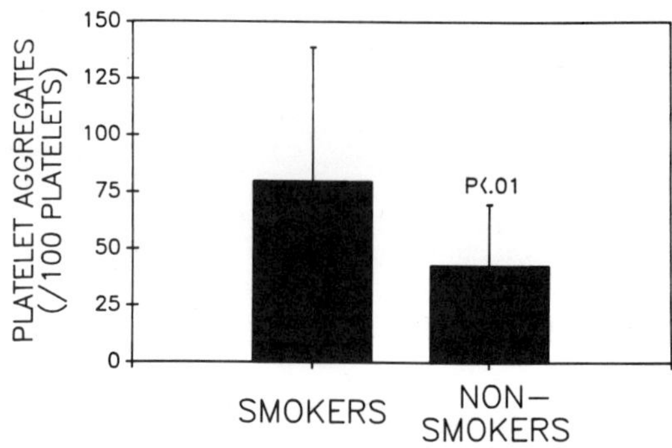

FIG. 2 Smokers showed more platelet aggregates in response to a
foreign surface in vitro than nonsmokers.

DISCUSSION

A number of investigators have evaluated the acute effects of smoking
upon platelet function. Some have shown a shortened bleeding time (6,7)
and others have shown no effect (8,9). Platelet stickiness seems to be
increased (10,11). Some have reported increased numbers of circulating
platelet aggregates (12-14); whereas others reported no change (9).
Regarding platelet aggregation, some investigators have shown an
increased response to various stimuli (15-18) whereas others showed no
effect (9,19,20). Thromboxane B-2 increased among smokers, but such
acute effects of smoking were not observed among habitual smokers (21).
Beta-thromboglobulin has been reported to increase with acute smoking
(13) and platelet factor 4 has been reported to increase (13,14) or
decrease (22). Thromboxane A_2 generation was increased with smoking
among nonsmoker volunteers (21) and ADP-induced thromboxane B_2 was also
increased with smoking in nonsmoker volunteers (21); although this effect
of smoking was not observed by others in nonsmoker volunteers (23).
Acute smoking among habitual smokers either diminished ADP-stimulated
thromboxane B_2 production (23) or had no effect (9). ADP-induced
electrophoretic mobility has been reported to be increased with smoking
(24). Observations of the effects of acute smoking upon platelets,
therefore, are quite inconsistent. It has been suggested that acute
smoking may produce an impaired metabolic capacity of the platelets,
rather than increasing platelet activation in vivo (23).

The more prolonged effects of smoking have been less extensively
studied than the acute effects. Regarding the platelet count, we and
others found no difference (1,21) although some investigators found it to
be elevated (25,26). Platelet survival, however, reportedly is shortened
in smokers, even during a period of abstinence from smoking (27). A
shorter coagulation of blood has been measured with thromboelastrography
(24). The tensile strength of blood clots is greater (24) and the number
and percent of adhesive platelets is also greater among smokers (26).

We and others found no significant difference of the number of
circulating platelet aggregates (1,12). Regarding release factors, in
our experience (1) and the experience of others (21) beta thromboglobulin
was not increased, nor was thromboxane B-2 (21,23), platelet factor 4
(1), platelet factor 3 (25) or platelet released serotonin (23,28).
Thromboxane B-2 release was diminished after stimulation with adenosine
diphosphate (ADP) (23), but no significant effect was seen with collagen
stimulation (23).

184

The electron microscopic evaluation of platelets showed that chronic
smokers formed significantly more platelet aggregates in response to a
foreign surface than nonsmokers. This increased cohesive capacity of
platelets from smokers was further enhanced if the subjects were older
than 50 years of age or had smoked cigarettes for more than 20 years.
Our electron microscopic study of the response of platelets to a foreign
surface is the only study of the early phases of platelet activation such
as the extension of pseudopodia, and cytoplasmic spreading with fusion
between adjacent pseudopodia. We found no significant difference in the
percentage of the spread type platelet on differential platelet counts
among chronic smokers and nonsmokers after exposure to an activating
surface (1).

The relative number of platelets on the surface of the Formvar film
reflects the capability of individual platelets to adhere to an
activating surface. During the differential counts of platelets obtained
from smokers, we uniformly saw numerous platelets adhered to the surface
of the Formvar film, qualitatively indicating at least a normal amount of
adhesiveness.

Our observations suggest that the circulating platelets of smokers
are hyperactive. Seemingly normal circulating platelets (based upon the
platelet count, aggregate ratio, and release factors in patients who have
not smoked for several hours) may in fact be primed, as shown by their
hyperactive response to a foreign surface in vitro. The mechanism by
which platelets are primed during smoking is unclear. Possible
mechanisms include increased levels of catecholamines (19,29), fatty
acids (30-32), nicotine (18), or carboxyhemoglobin (18).

REFERENCES

1. Rival, J., Riddle, J.M., Walz, D.A., Stein, P.D. Effects of
    chronic smoking on platelet function. Thrombosis Research 45,
    75-85 (1987).
2. Riddle, J.M. A method for evaluating the platelet surface response
    by using electron microscopy. Henry Ford Hosp. Med. J. 27,
    268-274 (1979).
3. Riddle, J.M. and Schatz, I.J. Platelet surface activation and
    inhibition during myocardial infarction. In: Platelet Adhesion
    and Aggregation in Thrombosis: Countermeasures. E.F. Mammen,
    G.F. Anderson and M.I. Barnhart (Eds.) New York, F.K. Schattauer
    Verlag (1970).
4. Wu, K.K. and Hoak, J.C. A new method for the quantitative detection
    of platelet aggregates in patients with arterial insufficiency.
    Lancet 2, 924-926 (1974).
5. Brown, T.R., Ho, T.T.S. and Walz, D.A. Improved radioimmunoassay
    of platelet factor 4 and beta-thromboglobulin in plasma. Clin.
    Chim. Acta. 101, 225-233 (1980).
6. Madsen, H. and Dyerberg, J. Cigarette smoking and its effects on
    the platelet-vessel wall interaction. Scand. J. Clin. Lab.
    Invest. 44, 203-206 (1984).
7. Ring, T., Kristensen, S.D., Jensen, P.N., Mourits-Andersen, T.,
    Madsen, H. and Dyerberg, J. Cigarette smoking shortens the
    bleeding time. Thromb. Res. 32, 531-536 (1983).
8. Kampman, M.T., Hornstra, G. No acute effect of cigarette smoking
    on bleeding time of habitual smokers. Thrombosis Research 52,
    287-294 (1988).
9. Hillbom, M., Muuronen, A., Lowbeer, C., Anggard, E., Beving, H.,
    Kangasaho, M. Platelet thromboxane formation and bleeding time is

influenced by ethanol withdrawal but not by cigarette smoking. <u>Thrombosis and Haemostasis</u> <u>52</u>, 419–422 (1985).

10. Ambrus, J.L. and Mink, I.B. Effect of cigarette smoking on blood coagulation. <u>Clin. Pharmacol. Ther.</u> <u>5</u>, 428–431 (1964).

11. Pittilo, R.M., Clarke, J.M.F., Harris, D., Mackie, I.J., Rowles, P.M., Machin, S.J., Woolf, N. Cigarette smoking and platelet adhesion. <u>British J. Haematology</u> <u>58</u>, 627–632 (1984).

12. Davis, J.W. and Davis, R.F. Acute effect of tobacco cigarette smoking on the platelet aggregate ratio. <u>Amer. J. Med. Sci.</u> <u>278</u>, 139–143 (1979).

13. Schmidt, K.G., Rasmussen, J.W. Acute platelet activation induced by smoking in vivo and ex vivo studies in humans. <u>Thromb Haemostas (Stuttgart)</u> <u>51</u>, 279–282 (1984).

14. Davis, J.W., Hartman, C.R., Lewis, H.D., Jr., Shelton, L., Eigenberg, D.A., Hassanein, K.M., Hignite, C.E., Ruttinger, H.A. Cigarette smoking-induced enhancement of platelet function: Lack of prevention by aspirin in men with coronary artery disease. <u>J. Lab. Clin. Med.</u> <u>105</u>, 479–483 (1985).

15. Levine, P.H. An acute effect of cigarette smoking on platelet function. <u>Circulation</u> <u>48</u>, 619–623 (1973).

16. Grignani, G., Gamba, G. and Ascari, E. Cigarette-smoking effect on platelet function. <u>Thrombos. Haem.</u> <u>37</u>, 423–428 (1977).

17. Bierenbaum, M.L. Fleischman, A.I., Stier, A., Somol, H. and Watson, P.B. Effect of cigarette smoking upon in vivo platelet function in man. <u>Thrombo. Res.</u> <u>12</u>, 1051–1057, (1978).

18. Renaud, S., Blache, D., Dumont, E., Thevenon, C. and Wissendanger, T. Platelet function after cigarette smoking in relation to nicotine and carbon monoxide. <u>Clin. Pharmacol. Ther.</u> <u>36</u>, 389–395 (1984).

19. Van der Giessen, W.J., Serruys, P.W., Stoel, I. Hugenholtz, P.G., De Leeuw, P.W., Van Vliet, H.H.D.M. Acute effect of cigarette smoking on cardiac prostaglandin synthesis and platelet behavior in patients with coronary heart disease. In: <u>Advances in Prostaglandin, Thromboxane and Leukotriene Research</u>. B. Samuelsson, R. Paoletti, and P.W. Ramwell (Eds.) New York, Raven Press, pp. 359–364 (1983).

20. Glynn, M.F., Mustard, J.F., Buchanan, M.R. and Murphy, E.A. Cigarette smoking and platelet aggregation. <u>Can. Med. Assn. J.</u> <u>95</u>, 549–553 (1966).

21. Mehta, P. and Mehta J. Effects of smoking on platelets and on plasma thromboxane-prostacyclin balance in man. <u>Prostaglandins, Leukotrienes and Medicine</u> <u>9</u>, 141–150 (1982).

22. Djaldetti, M., Agam, G. and Creter, D. Effect of cigarette smoking on platelet function. <u>Haemotologica</u> <u>62</u>, 575–580 (1977).

23. Marasini, B., Biondi, M.L., Barbesti, S., Zatta, G., Agostoni, A. Cigarette smoking and platelet function. <u>Thromboxis Research</u> <u>44</u>, 85–94 (1986).

24. Hawkins, R.I. Smoking, platelets and thrombosis. <u>Nature</u> 236, 450–542 (1972).

25. Chao, F.C., Tullis, J.L., Alper, C.A., Glynn, R.J. and Silbert, J.E. Alteration in plasma proteins and platelet functions with aging and cigarette smoking in healthy men. <u>Thrombosis. Haem.</u> <u>47</u>, 259–264 (1982).

26. Erikssen, J., Hellem, A. and Stormorken, H. Chronic effect of smoking on platelet count and "platelet adhesiveness" in presumably healthy middle-aged men (<u>Thrombos. Haem.</u> <u>38</u>, 606–611 (1977).

27. Mustard, J.F. and Murphy, E.A. Effect of smoking on blood coagulation and platelet survival in man. <u>Brit. Med. J.</u> <u>1</u>, 846–849 (1963).

28. Mansouri, A. and Perry, C.A. Inhibition of platelet ADP and
     serotonin release by carbon monoxide and in cigarette smokers.
     Experientia <u>40</u>, 515-517 (1984).
29. Seiss, W., Lorenz, R., Roth, P. and Weber, P.C. Plasma
     catecholamines, platelet aggregation and associated thromboxane
     formation after physical exercise, smoking or norepinephrine
     infusion. _Circulation_ <u>66</u>. 44-48 (1982).
30. Billimoria, J.D., Pozner, H., Metselaar, B., Best, R.W. and James,
     D.C.O. Effect of cigarette smoking on lipids, lipoproteins, blood
     coagulation, fibrinolysis and cellular components of human blood.
     Atherosclerosis <u>21</u>, 61-76 (1975).
31. Santos, M.T., Valles, J., Aznar, J., Beltran, M., Herraiz, M. Effect
     of smoking on plasma and platelet fatty acid composition in
     middle-aged men. _Atherosclerosis_ <u>50</u>, 53-62 (1984).
32. Murchison, L.E. and Fyfe, T. Effects of cigarette smoking on
     serum-lipids, blood-glucose, and platelet adhesiveness. _Lancet_ <u>1</u>,
     182-184 (1966).

PLATELET - VESSEL WALL INTERACTIONS IN INDIVIDUALS WHO SMOKE CIGARETTES

John J. Murray, Jacek Nowak, John A. Oates and Garret A. FitzGerald

Division of Clinical Pharmacology
Departments of Medicine and Pharmacology
Vanderbilt Medical Center
Nashville, TN 37232, USA

INTRODUCTION

Sudden cardiac death is the most frequent cause of mortality in industrialized societies accounting for the majority of the over 325,000 smoking-related deaths each year in the United States[1]. The strong link between cigarette smoking and the platelet-modulated syndrome of sudden cardiac death[2,3] implicates smoking as a cause of platelet activation. Consistent with this hypothesis of a direct activation of platelets is that platelet turnover appears to be accelerated in apparently healthy smokers[4,5]. However, an enhanced turnover of platelets may be the result of their increased clearance following membrane alterations after interaction with a damaged vasculature and conversely, thrombosis has been shown to occur *in vivo* without evidence of increased turnover[6-8]. Thus, an enhanced platelet turnover in smokers may not necessarily reflect a direct activation of platelets *in vivo*. In fact, *ex vivo* aggregation has been found to be either increased[9,10], unchanged[4,12-14] or significantly decreased[15,16] in chronic smokers.

Recently, it has been appreciated that the major urinary metabolites of thromboxane A_2, a product from arachidonic acid metabolism which causes platelet aggregation and smooth muscle contraction[17], are predominately derived from platelets under physiologic conditions[18]. Excretion of these metabolites, including 11-dehydro-thromboxane B_2 and 2,3-dinor-thromboxane B_2, is increased in syndromes of platelet activation *in vivo* such as during ischemic episodes in patients with unstable angina[19] and during experimentally induced thrombosis in the dog[20]. Thus, if the level of the urinary metabolites of thromboxane serves as an index of platelet activation *in vivo*, an increase in smokers would further substantiate the hypothesis that smoking causes platelet activation.

In addition to a possible direct effect of smoking on platelet activation, the vessel wall has also been suggested to be a target for the pathophysiologic consequences of cigarette smoking. The smoke products and specifically nicotine and carbon monoxide have been shown to cause pronounced intimal changes[21-23]. Furthermore, the production of prostacyclin, an endothelial cell metabolite from arachidonic acid possessing anti-aggregatory and vascular smooth muscle relaxing properties, has been shown to be reduced by cigarette smoke and nicotine *ex vivo*[24,25] and

Tobacco Smoking and Atherosclerosis
Edited by J. N. Diana
Plenum Press, New York, 1990

reported to be decreased *in vivo* in man[26]. A reduction in the biosynthesis of this eicosanoid would constrain the unexpected but normally enhanced response of the vessel wall to increase the generation of prostacyclin as occurs during vascular diseases involving enhanced platelet activation[19,27,28].

This brief overview will present a summary of our investigations into the mechanisms of smoking-induced platelet activation in man. Primarily by using the non-invasive indices of *in vivo* thromboxane A_2 and prostacyclin production, we suggest that cigarette smoking does cause platelet activation both by a direct effect as well as through a secondary effect of vascular damage.

EVIDENCE OF PLATELET ACTIVATION *IN VIVO* IN CHRONIC SMOKERS

Initially, we examined whether there was evidence of platelet activation *in vivo* in chronic smokers by determining their urinary levels of 2,3-dinor-thromboxane B_2 by gas chromatography-mass spectrometry[29]. Ten young (32±7 years), apparently healthy, male chronic smokers (>20 cigarettes per day for >5 years) collected daily urine specimens during a steady state period of smoking 20 cigarettes (1.3 mg nicotine) a day.

During the initial baseline period, the chronic smokers had significantly higher urinary levels of the thomboxane metabolite compared to age-matched, non-smoking controls[29,30]. When the smokers ceased smoking over the ensuing 3 weeks, the levels of 2,3-dinor-thromboxane B_2 fell by the first week. However, their 2,3-dinor-thomboxane B_2 levels remained significantly elevated above the non-smoking control levels and did not fall further over the ensuing two weeks. Compliance was assured by the absence of nicotine and cotinine in the urine and plasma as measured by high pressure liquid chromatography[31].

Following the three week abstinence period, smoking was reinstituted on a graduated protocol of 1, 2, and 3 cigarettes per hour, respectively, during an eight hour period on three consecutive days. The excretion of the thromboxane metabolite which had fallen to a new steady state level by the first week of abstinence increased back to the initial steady state level by the first day of resmoking. Thus, assuming that the increased biosynthesis of thromboxane is platelet-derived (as confirmed in the following study), these data support the fact that smoking increases platelet activation and that this activation persists despite a prolonged withdrawal from smoking. These findings would suggest that there is direct activation of platelets by smoking as reflected by the 2,3-dinor-thomboxane B_2 excretion and that this component of the activation is rapidly reversible. In addition, the stimulation of platelets persists significantly during the abstinence period which probably reflects an enhanced platelet-vascular interaction due to a long lasting injury to the vessel wall.

Ex vivo platelet aggregation studies were performed as well on these chronic smokers throughout the study[32]. Interestingly, despite the evidence *in vivo* from the urinary thromboxane metabolite for a decline in platelet activation, the *ex vivo* studies showed an enhanced aggregatory response of all of the chronic smokers' platelets to both ADP and epinephrine after the three weeks of withdrawal from smoking. Furthermore, the aggregatory response to ADP significantly decreased with the reinstitution of smoking, returning by the first day to the lower sensitivity observed during the initial baseline smoking period.

An additional finding was consistent with the paradoxical decrease in the *ex vivo* aggregation with smoking abstinence. The recovery of platelets in the platelet rich plasma when determined prior to normalization for aggregation studies (ie 300,000 $\mu l/\mu l$) was significantly decreased in all 10 chronic smokers during smoking withdrawal. Furthermore, following the reinstitution of smoking, the recovery of platelets increased again in all smokers. This change in the recovery of platelets in the platelet rich plasma occurred although their platelet counts in whole blood remained unchanged throughout the study. The higher recovery of platelets in the platelet rich plasma during smoking is consistent with less susceptibility to aggregation during the preparation of the platelets.

Such an artifactual selection of platelets *ex vivo* would result in only the less responsive platelets being sampled by phlebotomy during smoking. The more responsive platelets, activated by smoking, would form microaggregates or be adherent to the vessel wall *in vivo*. The decreased responsiveness of the platelets studied *ex vivo* may be secondary to their undergoing reversible aggregation and/or adhesion reactions *in vivo*, as has been described previously[6]. This explanation would account for the discrepancy between our *in vivo* data showing increased thromboxane A_2 production and our *ex vivo* data showing decreased aggregation. Interestingly, the largest population study of platelet aggregation was performed in chronic male smokers in whom the *ex vivo* aggregation response to ADP was decreased[16]. Correspondingly, our observations on thromboxane A_2 biosynthesis in chronic smokers has been confirmed by four separate groups (see Table 1).

Table1. Studies of platelet activation during chronic smoking by in vivo and ex vivo measures

Ex vivo (aggregation)	In vivo (thromboxane)
Increased	Increased
Ashby, et al.[9](1965)	Murray, et al.[29](1985)
Hawkins[10](1972)	Fischer, et al.[34](1986)
Unchanged	Nowak, et al.[33](1987)
Mustard and Murphy[4](1963)	Lassila, et al.[35](1988)
Murchison and Fyfe[12](1966)	Wennmalm [36](1988)
Rogers, et al.[14](1980)	Barrow, et al.[37](1989)
Siess, et al.[13](1982)	
Decreased	
Mansouri and Perry[15](1984)	
Murray, et al.[32](1985)	
Meade, et al.[16](1985)	
Nowak, et al.[33](1987)	

CELLULAR ORIGIN OF THE INCREASED *IN VIVO* THROMBOXANE A_2

Thromboxane A_2 can be synthesized by cells other than the platelet including the pulmonary macrophage. The extent to which the elevated 2,3-dinor-thomboxane B_2 metabolite was derived from the platelets was assessed next in a study of 5 chronic smokers (20 cigarettes/day, 1.3 mg nicotine) and 5 non-smoking, age-matched males by the administration of 20 mg of aspirin twice a day for 10 days. This low dose of aspirin was chosen because it is relatively selective for the platelet, having little effect on other tissues such as vascular endothelium[38,39]. Importantly, the time course of recovery for aspirin induced inhibition of thromboxane A_2 production is much faster in the macrophage (hours) than in the platelet (up to 10 days).

This study confirmed the finding in the initial study of the significantly elevated urinary 2,3-dinor-thromboxane B_2 levels in the smokers compared to controls[33]. Prior to the aspirin treatment, the *ex vivo* serum generation of thromboxane in smokers was not different from non-smokers. This result suggested that the increased excretion of 2,3-dinor-thromboxane B_2 was not due to an increased capacity of the smokers' platelets to generate thromboxane, although the capacity of platelets to form thromboxane greatly exceeds actual biosynthetic rates *in vivo*. Consistent with a platelet origin, however, low dose aspirin which inhibited the serum thromboxane generation >90% in both groups abolished the increment in the urinary thromboxane metabolite excretion in the smokers. Furthermore, the rate of full recovery of both the serum thromboxane and urinary thromboxane metabolite excretion in both groups returned to their pre-aspirin levels in 8 to 10 days after stopping aspirin indicative of a platelet derivation. Once again, 10 days after the aspirin treatment the increment in thromboxane biosynthesis in the smokers was apparent.

Prostacyclin production, as measured by its urinary metabolite 2,3-dinor-6-keto-$PGF_{1\alpha}$ with the sensitive and specific negative ion mass spectrometric method[40], was found to be significantly increased in the smokers compared to their matched non-smoking controls as has been observed recently by others[36]. During the period of aspirin treatment, the 2,3-dinor-6-keto-$PGF_{1\alpha}$ levels in the non-smokers were unchanged, implying a low systemic bioavailability of aspirin. The levels of this metabolite however fell slightly in the smokers following the aspirin treatment but remained consistently above the non-smokers' level. Thus, with the removal of the direct stimulus of increased platelet reactivity in the smokers by the aspirin treatment, the increased levels of prostacyclin fell to within normal levels but were still consistently elevated above the non-smokers' levels. Interestingly, the prostacyclin metabolite levels were not further suppressed as might be suggested from previous *in vivo* (measured by another metabolite of prostacyclin, 6-keto-$PGF_{1\alpha}$ using a radioimmunoassay) and *in vitro* studies showing a possible direct effect of nicotine on prostacyclin production[25]. Our finding would indicate a persistent stimulatory effect for the production of prostacyclin after the removal by aspirin of enhanced platelet-vascular interaction and may reflect a persistent intrinsic vascular injury related to the smoking.

Despite the evidence again in this study of increased platelet aggregation *in vivo* by the urinary thromboxane metabolites, this study, like the first, showed a significant decrease in platelet responsiveness to agonists *ex vivo*.

EFFECTS OF ACUTE SMOKING ON PLATELET ACTIVATION IN CHRONIC SMOKERS

The mechanisms by which cigarette smoking induces platelet activation is unclear. Smoking may activate platelets directly and/or via the smoking-induced activation of the sympathoadrenal pathway[41]. In addition, as suggested from the above studies platelets may also be activated secondary to smoking induced vascular damage[21-23,42]. To address the first possibility of direct platelet activation, we examined the effects of acute smoking on the production of the 11-dehydro metabolite of thromboxane A_2. This more recently developed stable isotope dilutional assay by gas chromatography-mass spectrometry in the negative ion-chemical ionization mode is a more sensitive urinary indicator of thromboxane A_2 generation than the less abundant 2,3-dinor metabolite[18]. This thromboxane metabolite has been shown to be elevated like the 2,3-dinor metabolite in chronic smokers[35,37].

For this study, 7 apparently healthy, young male chronic smokers were studied on 4 separate mornings at weekly intervals following an overnight (10 hour) abstinence from smoking. On the mornings of study following a two hour baseline urine collection period, they were randomized to smoking two tobacco containing cigarettes (1.3 mg nicotine) or two herbal cigarettes (no nicotine), sham smoking by puffing on a 6 inch straw or they were infused with epinephrine such as to simulate the plasma concentrations of epinephrine obtained after acute smoking[43]. Urine was collected in 2 hour intervals before, beginning with the experimental intervention, and for an additional two hours.

During the baseline period, despite the abstinence, both metabolites of thromboxane A_2 in the chronic smokers were elevated (195 ± 30 pg/mg creatinine for the 2,3-dinor metabolite and 849 ± 84 pg/mg creatinine for the 11-dehydro metabolite) which was an 84% and 7% increase, respectively, compared to non-smoking control values previously reported[18]. During the experimental period of smoking the tobacco containing cigarette, the levels of the 11-dehydro metabolite increased significantly by nearly two-fold and decreased towards the pre-smoking baseline level during the two hour recovery period (Figure 1). The other experimental conditions including smoking the herbal cigarettes, sham smoking through a straw or the

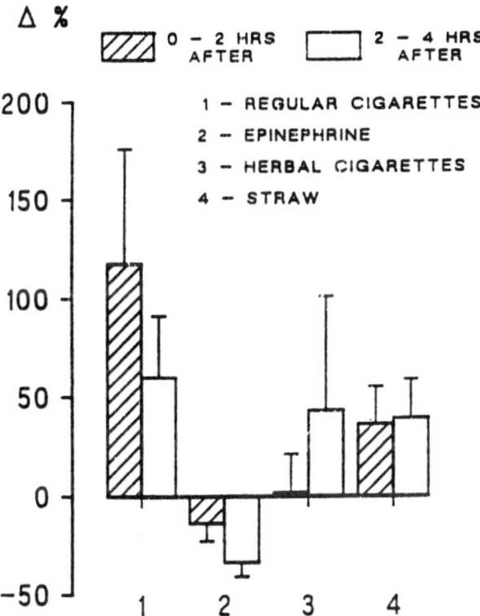

Figure 1.   Effects of acute interventions on 11-dehydro-thromboxane B_2 excretion. The excretion of the 11-dehydro metabolite is expressed as the percent of the baseline 2 hour collection and shown for the experimental period as well as for the following 2 hours. The four experimental conditions include smoking 2 tobacco (1.3 mg nicotine) cigarettes, 2 herbal cigarettes, sham smoking, and an infusion of epinephrine. Only smoking the tobacco cigarettes produced a significant increase ($p<0.05$) in the excretion of the 11-dehydro metabolite above the baseline period. The baseline levels were not different prior to the various treatments (849 ± 84 pg/mg creatinine).

epinephrine infusion did not produce any change in the excretion of the 11-dehydro metabolite. Concurrent determination of the urinary 2,3-dinor-metabolite of thromboxane A_2 as well as the 2,3-dinor-6-keto-$PGF_{1\alpha}$ did not show a change during any time period with any experimental manipulation including the smoking of tobacco cigarettes.

The data from this present study are consistent with the hypothesis of an indirect effect secondary to smoking-induced vascular damage contributing to the persistently increased excretion of thromboxane metabolites in chronic smokers despite a period of abstinence. The superimposed effects of acute smoking on platelet activation were detected only with the more abundant metabolite of thromboxane A_2, the 11-dehydro-thromboxane B_2. This agrees with evidence for such an acute effect using other approaches to monitor platelet function[9-11,44,45] and the rapid fall and increase of the 2,3-dinor-thromboxane B_2 excretion during the period of withdrawal and re-institution of smoking in our initial study. These acute effects result from a direct effect of smoking on platelets which is unrelated to the attendant increase in circulating epinephrine. Also consistent with the hypothesis of both direct and indirect mechanisms being involved in the activation of platelets with smoking is the finding of no further increase following acute smoking in the already elevated 2,3-dinor-6-keto-$PGF_{1\alpha}$ level in the chronic smokers.

SUMMARY

Our studies have shown that there an increased excretion of urinary metabolites of thromboxane A_2 in healthy, young male chronic smokers. This arachidonic acid metabolite from platelets reflects evidence of increased activation *in vivo*. These data contrast with the *ex vivo* study of platelets in chronic smokers and point out the fact that selection of cells for *ex vivo* study may not appropriately reflect the *in vivo* pathophysiologic situation. The platelet activation related to chronic smoking appears to result from both a direct, non-sympathoadrenally mediated activation which is rapidly inducible and reversible as well as a more persistent activation which long outlasts the smoke exposure. This latter mechanism appears to result from persistent vascular damage as reflected by the enhanced prostacyclin metabolite excretion. The acute, direct effect of smoking on the platelet appears to be a minor component of the altered platelet function. This latter inference may account for the inability in some studies to observe a small incremental, acute change superimposed on the persistently increased platelet reactivity secondary to the enhanced interactions with a damaged vasculature[37].

CONCLUSION

From the studies described in this review as well as previous work, a scheme of proposed contributing factors to sudden cardiac death produced by smoking is shown in Figure 2. A central factor in this scheme appears to be the enhanced platelet-vessel wall interaction.

The differential resolution of the platelet and vascular lesions may explain the biphasic decline in cardiovascular risk. This has been observed upon stopping smoking in otherwise apparently healthy individuals[46-48]. There is also a rapid decrease in mortality in quitters compared to persistent smokers following a myocardial infarction. In the latter group, mortality rates slowly normalize towards those of non-smokers[49].

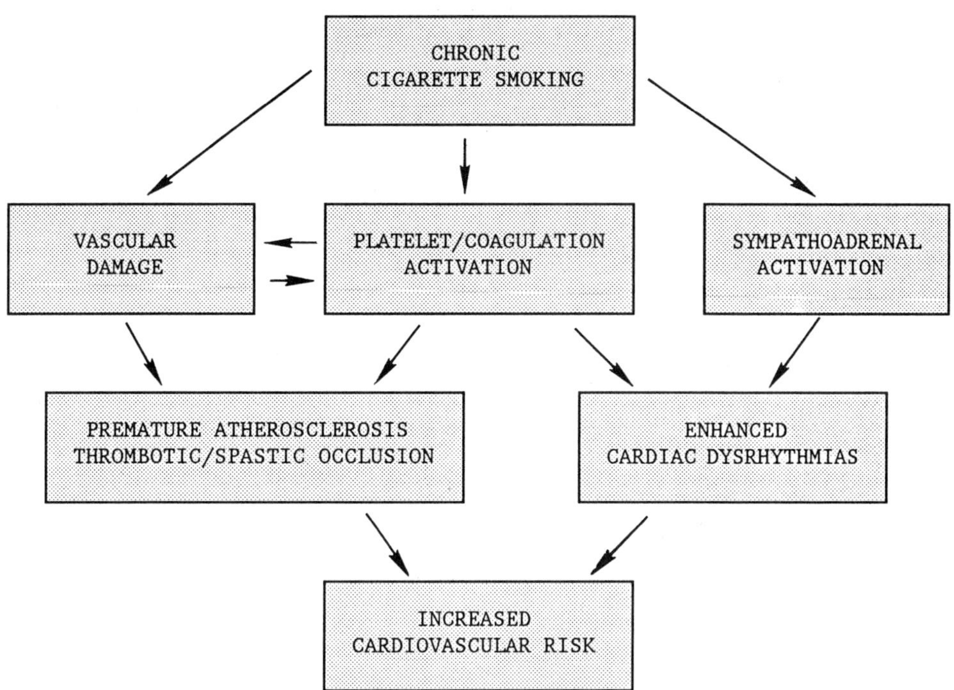

Figure 2. Proposed pivotal role of platelet-vascular interactions in smoking-induced cardiovascular disease.

ACKNOWLEDGEMENTS

This research was supported by grants from American Heart Association and NIH (GM 15431). JJM is an RJR Nabisco Scholar and GAF is an Established Investigator of the American Heart Association and the William Stokes Professor of Experimental Therapeutics. JN was supported in part by the E. and F. Fernstrom and Tore Nilson Foundations and the Swedish National Association against Chest and Heart Disease.

REFERENCES

1. W. B. Kannel. Update on the role of cigarette smoking in coronary artery disease. <u>Am. Heart J.</u> 101:319 (1981).
2. M. J. Davies and A. Thomas. Thrombosis and acute coronary artery lesions in sudden cardiac ischemic death. <u>N. Engl. J. Med.</u> 310:1137 (1984).
3. E. Falk. Unstable angina with fatal outcome: dynamic coronary thrombosis leading to infarction and/or sudden death. <u>Circulation</u> 71:699 (1985).
4. J. F. Mustard and E. A. Murphy. Effect of smoking on blood coagulation and platelet survival in man. <u>Br. Med. J.</u> 1:846 (1963).
5. V. Fuster, J. H. Cheseboro, R. L. Frye, and L. R. Elveback. Platelet survival and the development of coronary artery disease in the young adult: Effects of cigarette smoking, strong family history and medical therapy. <u>Circulation</u> 63:546 (1981).
6. B. L. Kinlough-Rathbone, M. A. Packam and J. F. Mustard. Vessel injury, platelet-adherence and platelet survival. <u>Arteriosclerosis</u> 3:529 (1983).

7.   P. D. Winocour, M. Cattaneo, D. Somers, M. Richardson, R. L. Kinlough-Rathbone and J. F. Mustard. Platelet survival and thrombosis. <u>Arteriosclerosis</u> 2:458 (1982).

8.   H. M. Graves, R. L. Kinlough-Rathbone, M. Richardson, L. Jorgensen, S. Moore and J. F. Mustard. Thrombin generation and fibrosis formation following injury to rabbit neointima: Studies of vessel wall reactivity and platelet survival. <u>Lab. Invest.</u> 46:605 (1982).

9.   P. Ashby, A. M. Dalby and J. H. Millar. Smoking and platelet stickiness. <u>Lancet</u> 2:158 (1965).

10. R. I. Hawkins. Smoking, platelets and thrombosis. Nature 236:450 (1972).

11. K. G. Schmidt and J. W. Rasmussen. Acute platelet activation-induced by smoking: *in vivo* and *ex vivo* studies in man. <u>Throm. Haemost.</u> 51:279 (1984).

12. L. E. Murchison and T. Fyfe. Effects of cigarette smoking on serum-lipids, blood glucose and platelet adhesiveness. <u>Lancet</u> 2:182 (1966).

13. W. Siess, R. Lorenz, P. Roth and P. C. Weber. Plasma catecholamines, platelet aggregation and associated thromboxane formation after physical exercise, smoking and norepinephrine infusion. <u>Circulation</u> 66:41 (1982).

14. W. R. Rogers, R. L. Bass III, D. E. Johnson, A. W. Kruski, C. A. McMahon, M. M. Montiel, G. E. Mott, R. L. Wilbur and H. C. McGill, Jr. Atherosclerosis-related responses to cigarette smoking in the baboon. <u>Circulation</u> 61:1188 (1980).

15. A. Mansouri and C. A. Perry. Inhibition of platelet ADP and serotonin release by carbon monoxide. <u>Experientia</u> 40:515 (1984).

16. T. W. Meade, M. V. Vicekers, S. G. Thompson, Y. Styirling, A. P. Haines and G. J. Miller. Epidemiological characteristics of platelet aggregability. <u>Br. Med. J.</u> 290:428 (1985).

17. B. Samuelsson. The role of prostaglandin endoperoxides and thromboxanes as bioregulators. <u>in</u>: "Proceedings of the Intrascience Symposium on the New Biochemistry of Prostaglandins and Thromboxanes." N. Kharsch and J. Fried, eds., Publisher New York Academic Press, Santa Monica, CA (1977).

18. F. Catella and G. A. FitzGerald. Paired analysis of urinary thromboxane B_2 metabolites in humans. <u>Throm. Research</u> 47:647 (1987).

19. D. J. Fitzgerald, F. Catella, L. Roy and G. A. Fitzgerald. Platelet activation in unstable coronary disease. <u>N. Engl. J. Med.</u> 315:983 (1986).

20. D. J. Fitzgerald, J. Doran, E. K. Jackson and G. A. FitzGerald. Coronary vascular occlusion mediated through thromboxane A_2-endoperoxide receptor activation in vivo. <u>J. Clin. Invest.</u> 77:496 (1986).

21. I. Asmussen and K. Kjeldsen. Intimal ultrastructure of human umbilical arteries. Observations on arteries from newborn children of smoking and nonsmoking mothers. <u>Circ. Res.</u> 36:579 (1975).

22. H. K. Thomsen. Carbon monoxide-induced atherosclerosis in primates. <u>Atherosclerosis</u> 20:233 (1974).

23. P. Ribeiro, R. Walesby, S. Edmonson, A. V. Jadhar, I. Traynor, C. M. Oakley and G. R. Thompson. Collagen content of atherosclerotic arteries is higher in smokers than in nonsmokers. <u>Lancet</u> 2:107 (1983).

24. J. H. Reinders, H-J. M. Brinkman, J. A. van Mourik and P. G. deGroot. Cigarette smoke impairs endothelial cell prostacyclin production. <u>Arteriosclerosis</u> 6:15 (1986).

25. T. Sonnenfeld and Å. Wennmalm. Inhibition by nicotine of the formation of prostacyclin-like activity in rabbit and human vascular tissue. <u>Br. J. Pharmacol.</u> 71:609 (1980).

26. J. L. Nadler, J. S. Velasco and R. Horton. Cigarette smoking inhibits prostacyclin formation. Lancet 1:248 (1983).

27. G. A. FitzGerald, B. Smith, A. K. Pedersen and A. R. Brash. Increased prostacyclin biosynthesis in patients with severe atherosclerosis and platelet activation. N. Engl. J. Med. 310:1065 (1984).

28. I. A. G. Reilly, L. Roy and G. A. FitzGerald. Thromboxane biosynthesis is increased in systemic sclerosis with Raynaud's phenomenon. Br.Med.J. 292:1087 (1986).

29. J. J. Murray, J. Nowak, J. A. Oates and G. A. FitzGerald. Biosynthesis of thromboxane A_2 and prostacyclin during chronic smoking and withdrawal in man. Clin. Res. 33:521A (1985).

30. I. A. G. Reilly and G. A. FitzGerald. Eicosanoid biosynthesis and platelet function with advancing age. Thrombosis Res. 41:545 (1986).

31. G. A. Kyerematen, M. D. Damiano, B. H. Dvorchik and E. S. Vesell. Smoking-induced changes in nicotine disposition: Application of a new HPLC assay for nicotine and its metabolites. Clin. Pharmacol. Ther. 32:769 (1982).

32. J. J. Murray, J. Nowak, J. A. Oates and G. A. FitzGerald. Platelet function during chronic smoking and withdrawal. Clin. Res. 33:350A (1985).

33. J. Nowak, J. J. Murray, J. A. Oates and G. A. Fitzgerald. Biochemical evidence of a chronic abnormality in platelet and vascular function in apparently healthy chronic cigarette smokers. Circulation 76:6 (1987).

34. S. Fischer, C. Bernutz, H. Meier and P. C. Weber. Formation of prostacyclin and thromboxane in man as measured by the main urinary metabolites. Biochim. Biophys. Acta 876:194 (1986).

35. R. Lassila, H. W. Seyberth, A. Haspanen, H. W. Schweer, M. Koskenvuo and K. E. Laustiola. Vasoactive and atherogenic effects of cigarette smoking- a study of monozygotic twins discordant for smoking. Br. Med. J. 297:955 (1988).

36. Å. Wennmalm. Smoking induced alterations in thromboxane biosynthesis: a population study. Presented at the Winter Prostaglandin Conference. Keystone, CO (1988).

37. S. E. Barrow, P. S. Ward. M. A. Sleightholm, J. M. Ritter and C. T. Dollery. Cigarette smoking: profiles of thromboxane- and prostacyclin-derived products in urine. Biochim. Biophys. Acta 993:121 (1989).

38. G. A. FitzGerald, J. A. Oates, J. Hawiger, R. L. Maas, L. J. Roberts II, J. A. Lawson and A. R. Brash. Endogenous biosynthesis of prostacyclin and thromboxane and platelet function during administration of aspirin in man. J. Clin. Invest. 71:676 (1983).

39. A. K. Pedersen, J. Nowak and G. A. FitzGerald. Slow administration of low dose aspirin:Enhanced inhibition of platelet cyclooxygenase. Circulation 72:772A (1985).

40. G. A. FitzGerald, A. R. Brash, I. A. Blair and J. Lawson. Analysis of urinary metabolites of thromboxane and prostacyclin by negative ion-chemical ionization gas chromatography-mass spectrometry. Adv. Prostglandin Thromboxane Res. 15:87 (1985).

41. P. Hill and E. L. Wyner. Smoking and cardiovascular disease- effect of nicotine on the serum epinephrine and corticoids. Am. Heart J. 87:491 (1974).

42. R. M. Pittilo, J. M. Clarke, D. Harris, I. J. Mackie, P. M. Rowles, S. J. Machin and N. Woolf. Cigarette smoking and platelet adhesion. Br. J. Haematol. 58:627 (1984).

43. P. E. Cryer, M. W. Haymond, J. V. Santiago and S. D. Shah. Norepinephrine and epinephrine release and adrenergic mediation of smoking-associated hemodynamic and metabolic events. N. Engl. J. Med. 295:573 (1976).

44. P. H. Levine. An acute effect of cigarette smoking on platelet function. A possible link between smoking and arterial thrombosis. <u>Circulation</u> 48:619 (1973).

45. J. J. Belch, B. M. McArdle, P. Burns, G. D. Lowe and C. D. Forbes. The effects of acute smoking on platelet behaviour, fibrinolysis haemarheology in habitual smokers. <u>Thromb. Haemostasis</u> 51:6 (1984).

46. T. Gordon, W. B. Kannel, D. McGee and T. R. Dawber. Death and coronary attacks in men giving up smoking. A report of the Framingham study. <u>Lancet</u> ii:1345 (1974).

47. R. Doll and A. B. Hill. Mortality in relation to smoking: Ten years' observation of British doctors. <u>Br. Med. J.</u> 1:1406 (1964).

48. L. Wilhelmsen. Coronary artery disease: Epidemiology of smoking and intervention studies of smoking. <u>Am. Heart J.</u> 115:242 (1988).

49. L. Rosenberg, D. W. Kaufman, S. P. Helmrich and S. Shapiro. Risk of myocardial infarction after quitting smoking in men under 55 years of age. <u>N. Engl. J. Med.</u> 313:1511 (1985).

ACUTE PLATELET ACTIVATION INDUCED BY SMOKING CIGARETTES:

IN VIVO AND EX VIVO STUDIES IN HUMANS

Kai G. Schmidt, Jens W. Rasmussen and Vagn Bonnevie-Nielsen

Department of Nuclear Medicine and Department of Clinical
Chemistry, Odense University Hospital, Denmark

The epidemiological and pathological evidence for a relationship
between cigarette smoking and atherosclerosis is considerable (1). In
particular, a strong correlation between cigarette smoking and acute
cardiac events in those with an already compromised coronary circulation
is evident (2). As platelets seem to play a central role for the
development of atherosclerosis and its thromboembolic complications (3),
it is natural that a vivid interest has been taken in the effects of
cigarette smoking on platelet function.

The immediate and chronic effects of smoking cigarettes on platelet
function and the platelet-vessel wall relationship have been investigated
in a large number of studies. The platelet functions studied include
platelet adhesion to foreign surfaces or endothelium, platelet
aggregability in response to various agonists (in plasma or whole blood),
extent of thromboxane B_2 formation, platelet release of the α-granule
constituents, beta-thromboglobulin and platelet factor 4, and estimation
of spontaneously formed platelet aggregates in vivo or during blood
sampling and processing. The platelet-vessel wall interaction has been
evaluated by bleeding time studies, and by investigating vessel wall
prostacyclin (PGI_2) synthesis, in vivo survival of PGI_2, platelet binding
of PGI_2, and platelet sensitivity to PGI_2. Suffice it to say that
inconsistent and conflicting results have been obtained. This is in fact
what would be expected considering the immense number of factors that may
affect the outcome of such studies. These include sex, age, vessel
status, smoking behaviour, and status of the sympathoadrenal system of
the subjects studied. The latter can be affected by the time of the
study, by the posture of the subjects, and by emotional stress factors.
In addition, numerous technical factors may affect the outcome of
platelet function studies, platelet aggregation in particular. It is
beyond the scope of the present paper to survey previously published
results and the inherent difficulties of interpretation. These subjects
have been reviewed recently (4,5).

Because in vitro platelet function tests are difficult to interpret
and often do not reflect the in vivo situation satisfactorily, we decided
to study whether changes in distribution in vivo of autologous
Indium-^{111}-labelled platelets occurred in response to smoking (6). In
addition, we sought circulating platelet aggregates (applying the
platelet aggregate ratio (PAR) test) and signs of release of the

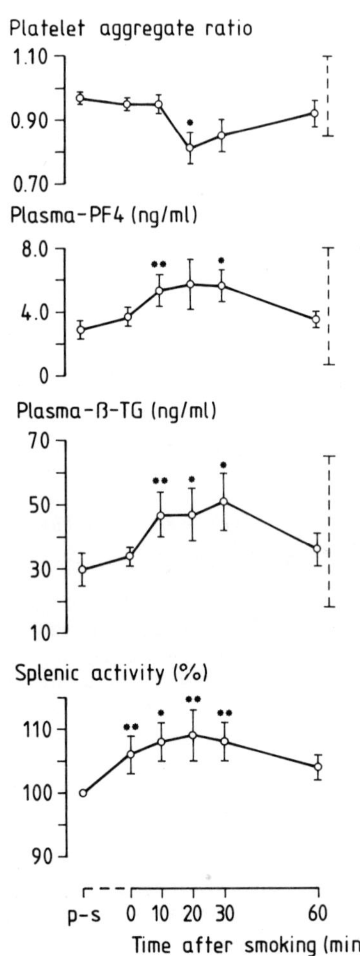

Figure 1. Platelet aggregate ratios, plasma PF4 and beta-TG values, and splenic activities prior to (p-s) and after smoking. Mean values and standard errors of the means are shown. Reference intervals are included. Asterisks indicate that the means differ significantly (* : p<0.01, ** : 0.01<p<0.05; Wilcoxon's rank sum test for paired data). From ref. 6; with permission, F.K. Schattauer Verlagsgesellschaft, mbH, Stuttgart.

α-granule proteins, beta-thromboglobulin (beta-TG) and platelet factor 4 (PF4). Ten healthy subjects (2 women, 8 men; 9 smokers and one non-smoker), who fasted and refrained from smoking for at least 12 hours, were studied. The smoking experiments were carried out 2 days after the injection of In-^{111} labeled autologous platelets. Computerized gamma camera images of the spleen, liver and lungs were obtained and the corresponding counts calculated, prior to smoking, immediately after smoking 2 medium tar cigarettes (each yielding 21 mg tar and 1.8 mg nicotine), and 10, 20, 30 and 60 min. after smoking. Concurrent blood samples were obtained for the PAR, beta-TG and PF4 determinations.

Figure 1 shows the results of the splenic imaging and the PAR, beta-TG and PF4 studies. It can be seen that, on the whole, the PAR declined in the post-smoking period, concomitantly with an increase in beta-TG and PF4 values and splenic activity. A characteristic feature was that the most pronounced signs of platelet activation as well as a concomitant increase in splenic activity were most frequently seen 20-30 min. after smoking. The courses in Fig. 1 cover a considerable interindividual variation. Thus the subjects could be roughly divided into 3 groups, corresponding to strong, moderate and weak or absent signs of platelet activation (3, 4 and 3 subjects, respectively). Two of the 3 subjects with strong and one subject with moderate biochemical signs of platelet activation showed a rather pronounced post-smoking increase in splenic activity (120-140% of the pre-smoking values). These 3 subjects and the third person showing biochemical signs of platelet activation showed a decline in hepatic activity (reaching 88-92% of the pre-smoking values) in the 10-30 min. post-smoking period. We did not observe any signs of pulmonary platelet trapping in response to smoking. Some subjects showed slight increments (amounting on the average to 2%) in circulating platelet-bound activity and platelet concentration.

The lowest post-smoking PAR values and the corresponding splenic activities were inversely correlated, i.e., signs of circulating platelet aggregates were associated with signs of increased platelet trapping in the spleen. The PAR values and the beta-TG and PF4 values did not correlate significantly. Figure 2 shows the relationship between PF4 and beta-TG, and the splenic activity. The latter was positively correlated to the concentration of both platelet proteins.

With the view of throwing some light on the intraindividual variability in smoking-induced platelet activation, six of the subjects already described took part in a repeat smoking experiment 6-12 months later. Except for the omission of the In-^{111} platelet study the procedure of the repeat investigation was identical to that of the first study. Some correlation between the results of study 1 and study 2 seemed to exist, inasmuch as the corresponding overall signs of platelet activation in the 6 subjects could be graded as S, S, M, M, M and W in the first study, and S, S, W, M S, M and W in the repeat study (S = strong, M = moderate, W = weak). However, as shown in Fig. 3, the separate results of the PAR, beta-TG and PF4 determinations in the two studies did not correlate significantly. Neither did the areas under the post-smoking PAR, beta-TG and PF4 curves.

We were somewhat intrigued by the post-smoking changes in splenic activity observed in some subjects in the original study. In view of the stimulated sympathoadrenal system in the post-smoking period (7) this was in fact not to be expected, inasmuch as one would predict an epinephrine-induced decrease in splenic blood flow (8) and a concomitant reduction in the size of the splenic platelet pool (9). As a possible explanation of our findings could be a smoking induced prolongation of the intrasplenic transit time of (activated?) platelets, we have

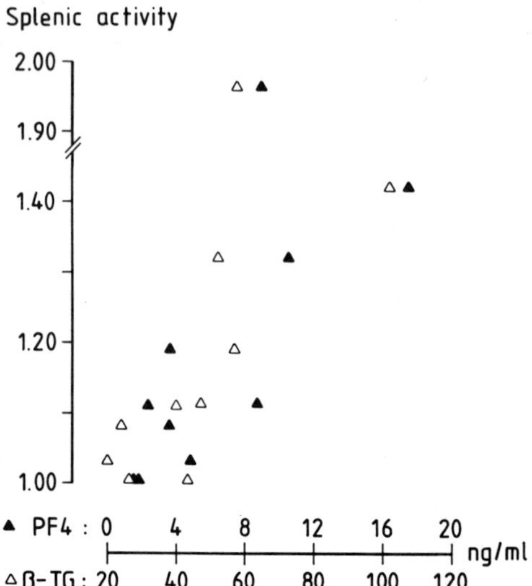

Figure 2. Plasma PF4 and beta-TG values measured concomitantly with the lowest platelet aggregate ratios, and the corresponding splenic activities (expressed as fractions of the pre-smoking exchangeable platelet pools). From ref. 6; with permission, F.K. Schattauer Verlagsgesellschaft, mbH, Stuttgart.

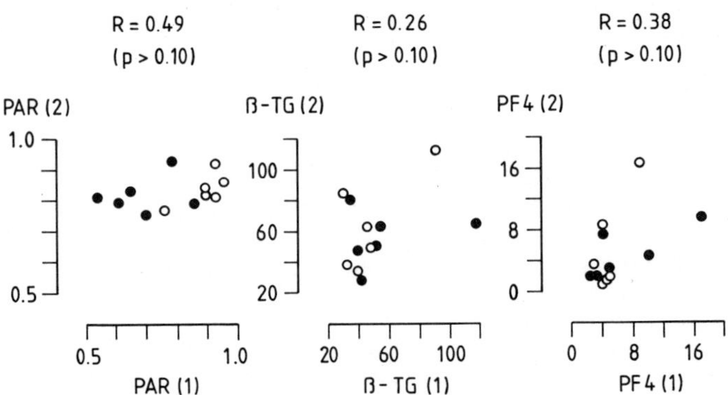

Figure 3. Corresponding results of PAR, beta-TG and PF4 determinations (fraction, ng/ml, and ng/ml, respectively) in 6 subjects studied on two occasions. Abscissa: study 1; ordinate: study 2. Open circles: results obtained in the early (0-10 min.) post-smoking period; closed circles: results from the late (20-60 min.) post-smoking period. Spearman's rank sum test was used for calculation of the R values.

initiated a study of initial platelet kinetcs in the early post-smoking
period. By application of the closed two-compartmental model of platelet
distribution between blood and spleen as proposed by Peters et al. (9),
the splenic blood flow (SBF) and mean platelet intrasplenic transit time
($\bar{t}$) can be computed as shown in Fig. 4. The calculations are based on
the fact that the changes in blood and splenic activities from 5 min.

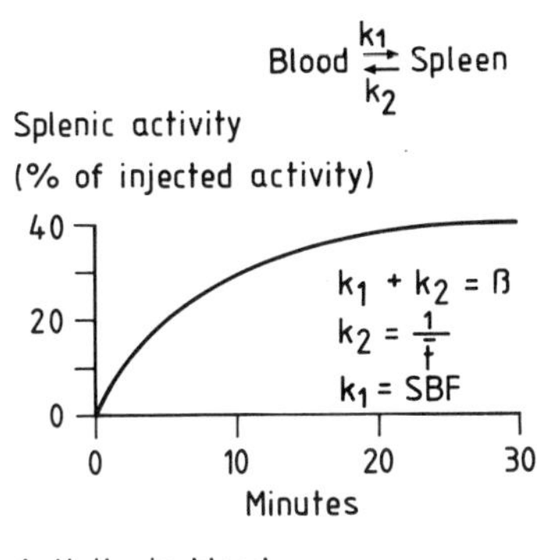

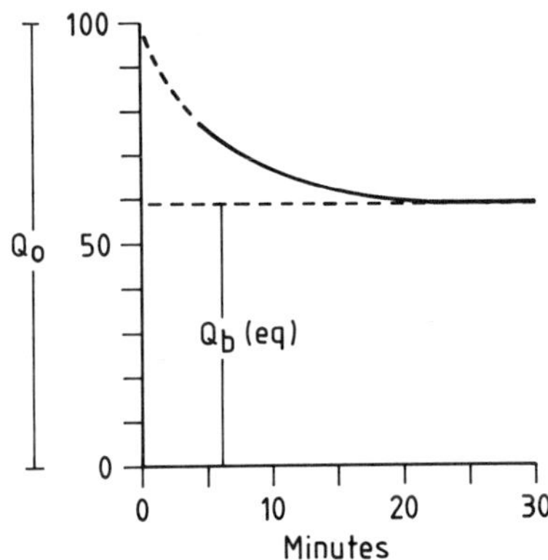

$$\frac{k_1}{k_2} = \frac{Q_0 - Q_b(eq)}{Q_b(eq)} \qquad \frac{Q_s(eq)}{Q_s(5)} = \frac{Q_0 - Q_b(eq)}{Q_0 - Q_b(5)}$$

Knowing $k_1 + k_2$ and $\dfrac{k_1}{k_2}$, SBF and $\bar{t}$ can be computed

<u>Figure 4</u>. Method of splenic blood flow (SBF) and mean intrasplenic
platelet transit time ($\bar{t}$) calculations. Q_s (5) denotes splenic activity
5 min. post-injection, Q_b (5) the corresponding circulating
platelet-bound activity. See text for further details.

post-injection to equilibrium (usually 30 min. post-injection) are
reciprocal. The activity in blood at time zero (Q_0) can be calculated as
shown, thus disregarding the initial transient platelet trapping in the
liver. Knowing the ratio between the rate constants k_1 and k_2 (which
equals the ratio between the partition coefficients at equilibrium) and
their sum (which is identical to the rate constant of the splenic uptake
curve), k_1 (= SBF) and k_2 (= the reciprocal of $\bar{t}$) can be computed.

In a previously published study (10) we have observed almost identical initial Chromium-51- and In-^{111}-platelet kinetics (Fig. 5). Consequently it would be possible, by application of a double isotope technique, to study the initial platelet kinetics and compute SBF and $\bar{t}$ prior to and immediately after smoking. In the present preliminary study 10 healthy male smokers were randomly assigned to smoking or sham smoking (5 in each group). Platelets were isolated from 100 ml samples of ACD-anticoagulated blood and two identical platelet suspensions labeled in a buffer medium with Cr-51 and In-^{111}, respectively, as previously described (10). With the subjects placed in the supine position, NaI crystal scintillation detectors were placed over the spleen, liver and the large vessels of the left upper hemithorax. Cr-51-labeled platelets (median dose 1.82 MBq, range 1.30-2.60 MBq) were injected prior to (sham) smoking, In-^{111}-labeled platelets (median dose 1.69 MBq, range 1.16-1.98 MBq) were injected immediately after (sham) smoking (2 cigarettes of the same brand as originally used was either kept unlighted between the lips or inhaled during a 12-15 min. period). Venous blood samples for determination of circulating platelet-bound activity were obtained through an indwelling catheter 2, 5, 10, 20 and 30 min. after the injection of labeled platelets. From the opposite arm blood samples were obtained by separate venipunctures 5 and 15 min. post-injection for PAR, beta-TG and PF4 determinations.

Figure 6 shows a Cr-51- and an In-^{111}-platelet splenic uptake curve from a sham-smoking subject. The curves have been smoothed by computerized fitting to an exponential function, which also yields the rate constant.

Figure 7 depicts the results of the SBF and $\bar{t}$ determinations. The smokers and sham smokers do not seem to differ in their responses. However, two smokers show a considerable prolongation of $\bar{t}$ concomitantly with an increase in SBF, which is the reverse of what would be expected if a change in SBF is the primary event. One of the sham smokers shows a similar but less pronounced pattern.

Figure 8 shows the corresponding relative changes in SBF and $\bar{t}$ in the two groups of subjects (In-^{111}-platelet results expressed as fractions of the Cr-51-platelet results). The shaded areas correspond to disproportionate prolongation of $\bar{t}$ compared to the changes in SBF (suggesting inverse proportionality between SBF and $\bar{t}$). The shade is graded according to the degree of disproportion. The two smokers mentioned above are represented by the two closed top symbols. The median time for the hepatic uptake curve to reach a maximum changed (Cr-51-platelet vs. In-^{111}-platelet studies) from 6.6 to 6.4 min. in the sham smokers, and from 6.1 to 6.6 min. in the smokers. The encircled symbols in Fig. 7 represent the subjects in whom the liver maximum was delayed following (sham) smoking. Interestingly, such a delay was only seen in the subjects showing a disproportionate prolongation of the intrasplenic platelet transit time.

Results of PAR determinations are available in 8 subjects (4 smokers and 4 sham smokers). A small decline in the PAR of the 15 min. sample following smoking was observed in one subject. In the remaining subjects the PAR did not change.

DISCUSSION

Even if the hitherto published results are inconsistent, there is firm evidence that platelets can be activated immediately after cigarette

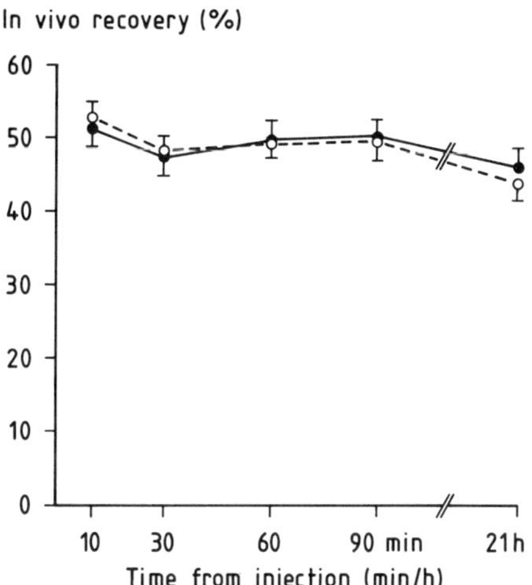

Figure 5. Courses of platelet-bound activity in the early post-injection period in 40 subjects studied after simultaneous injection of Cr-51-labeled (open circles) and In-111-labeled (closed circles) autologous platelets. The mean values and standard errors of the means are shown. From ref. 10; with permission, Munksgaard International Publishers Ltd., Copenhagen.

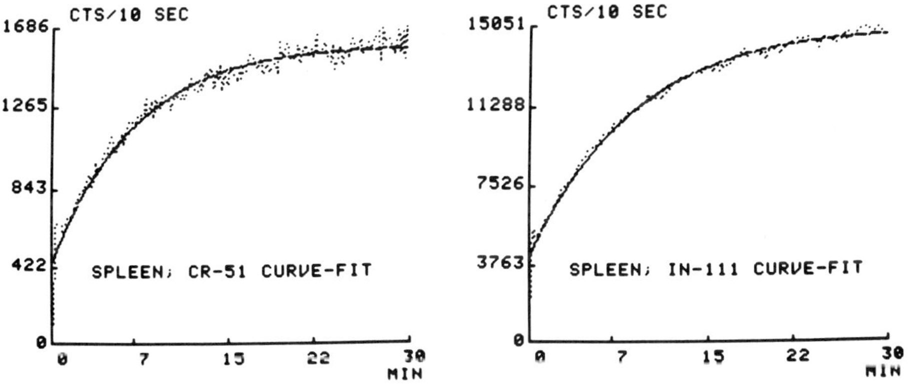

Figure 6. Splenic uptake curves following injection of Cr-51-platelets (left) and In-111-platelets (right) in a sham smoking person. The curves have been subjected to computerized smoothing. The dotted curves represent the original counts.

smoking. Our initial results definitely support this appraisal, although
it is peculiar that we found the most definite signs of platelet activation some time after the cessation of smoking. Levels of epinephrine
attainable under physiological conditions are able to activate platelets
or potentiate the activities of other platelet activating substances (11,
12). It is of interest in this context that Cryer et al. (7) found the
highest catecholamine values between 10 and 20 minutes after smoking.
Contributing to the inconsistent results published may be platelet
desensitization in some subjects, smokers in particular, due to a reduction in platelet α_2-receptor affinity induced by endogenous
catecholamines (13).

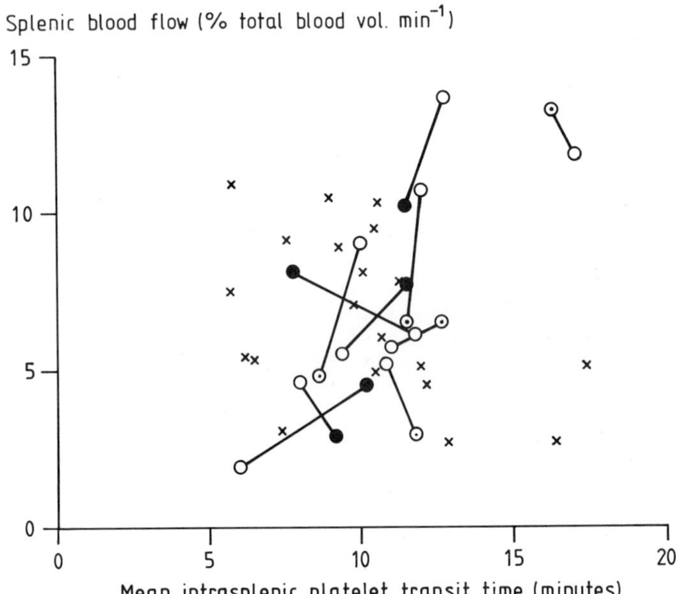

Figure 7. Corresponding results of mean platelet
intrasplenic transit time and splenic blood flow
determinations in 20 healthy subjects (x) and in
the 10 double isotope experiments. Results prior
to smoking (open circles) are connected with the
corresponding post-smoking results (closed circles:
circles with dots: sham smokers).

The discrepancy between our initially and recently performed PAR
tests seems difficult to explain. Even if platelet activation
undoubtedly takes place during the collection and processing of the blood
samples (14), and the number of blood samples collected in our first
study (6) was large, thereby increasing the risk of inadvertent platelet
activation, a more likely explanation is that platelets were counted in a
counting chamber in the first study (and the repeat study as well),
thereby excluding small aggregates, whereas an electronic particle counter

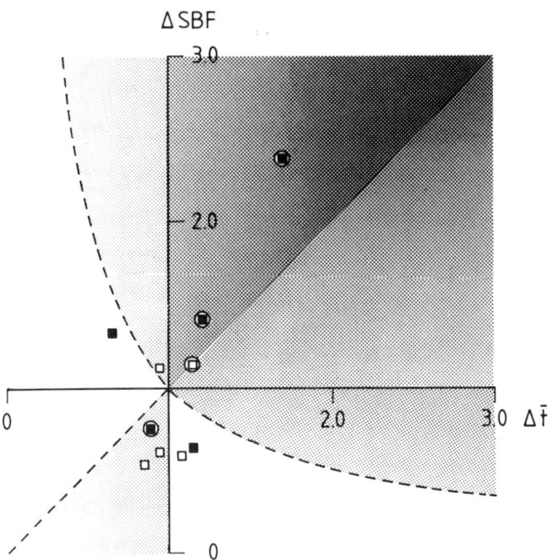

Figure 8. Corresponding relative changes in splenic blood flow (SBF) and mean platelet intrasplenic transit time (t̄) occurring during (sham) smoking, expressed as fractions of the pre-sham smoking (open symbols) and pre-smoking (closed symbols) values. The stippled lines correspond to corresponding values of SBF and t̄ expected to follow changes in SBF, presupposing an inverse proportionality between SBT and t̄. The shaded areas correspond to a disproportionate prolongation of t̄ (see text for details). Encircled symbols correspond to a delayed maximum of hepatic activtiy after (sham) smoking (see text).

was used in our recent study. That artefactual in vitro release of beta-TG and PF4 may have occurred in our studies is likely, considering the rather high PF4/beta-TG ratio in some of the subjects (15). We did not include platelet activation inhibitors like PGE$_1$ in our anticoagulants, which undoubtedly made our test system rather sensitive to platelet activation during blood sampling. Whatever the mechanisms, signs of platelet activation following smoking were evident.

In the light of the concurrent biochemical signs of platelet activation the delayed accumulation of platelets in the spleen in some subjects following smoking most likely reflects in vivo platelet activation. Besides, as we did not observe any signs of a change in spleen size in response to smoking, and an increased epinephrine level in the blood would tend to reduce the splenic perfusion (8), thereby reducing the size of the splenic platelet pool, our findings most likely reflect an increased platelet transit time through the spleen (9). In the immediate post-smoking period catecholamine-induced reduction in splenic blood flow and prolonged transit through the spleen of activated platelets could counteract each other, resulting in an unaltered level of splenic activity. Our recently performed studies of initial platelet kinetics do not lend firm support to these suggestions, however, even though two smokers showed signs of a disproportionate prolongation of the intrasplenic platelet transit time. The changes in splenic blood flow in response to smoking were inconsistent. This may reflect an catecholamine release insufficient for splenic blood flow reduction or adrenergic receptor down-regulation in smokers.

The initial post-injection trapping in the liver is presumed to be caused by a reversible activation or damage inflicted on the platelets during the isolation and labeling procedure (16). Factors generated in vivo (in response to smoking, for example) may contribute to this phenomenon, judged by the relationship between a delayed hepatic maximum and signs of a disproportionate prolongation of the intrasplenic platelet transit time in our recent study. Conversely, in our initial study (6) we observed an inexplicable post-smoking decline in hepatic activity in the subjects who, on the whole, exhibited the most pronounced signs of platelet activation, including an increase in splenic activity.

There is reason to believe that continuation and extension of the intrasplenic platelet kinetic studies will throw further light on the acute effects of smoking on platelet in vivo behavior. In particular, comparative studies of intrasplenic platelet transit and biochemical parameters reflecting platelet activation would be of interest. The role of the temporal relationship between the injection of labeled platelets and smoking for the evaluation of intrasplenic platelet kinetics should also be studied further.

There are weighty reasons to suggest that smoking in particular results in an augmented early platelet response, i.e., increases their adhesivity (17,18), and there is firm evidence of a disturbed platelet-vessel wall interaction after smoking (4,19,20). Considering the long intrasplenic transit time and the undoubtedly intimate platelet-endothelium relationship inside this organ, smoking would a priori seem capable of affecting intrasplenic platelet kinetics. If platelets are activated mainly outside the spleen, a transient hold-up (and deactivation?) in the spleen of activated platelets might damp the overall platelet reactivity. In accordance with this suggestion asplenic persons may show signs of increased platelet reactivity (21), and there is epidemiological evidence of an excess mortality from ischemic heart disease in splenectomized subjects (22).

ACKNOWLEDGEMENTS

This study was supported by grants from the Danish Medical Research Council (Grant No. 12-9305) and from the Danish Heart Foundation. The authors are grateful to Helle Mortensen and Ulla Mohr for skillful technical assistance and to Inge-Lise Neerholt for typing the manuscript.

REFERENCES

1. Hopkins PN, Williams RR. A survey of 246 suggested coronary risk factors. Atherosclerosis, 40: 1-52 (1981).
2. Kannel WB. Cigarettes, coronary occlusions and myocardial infarction. JAMA, 246: 871-2 (1981).
3. Schwartz CJ, Valente AJ, Kelley JL, Sprague EA, Edwards EH. Thrombosis and the development of atherosclerosis: Rokitansky revisited. Semin Thromb Hemostas, 14: 189-95 (1988).
4. Pittilo RM. Smoking and platelet-vessel wall interactions. In: Pittilo RM and Machin SJ (eds.). Platelet-vessel wall interactions. Springer-Verlag, pp 87-101 (1988).
5. FitzGerald GA, Oates JA, Nowak J. Cigarette smoking and hemostatic function. Am Heart J, 115: 267-71 (1988).
6. Schmidt KG, Rasmussen JW. Acute platelet activation induced by smoking. In vivo and ex vivo studies in humans. Thromb Haemostas, 51: 279-82 (1984).

7. Cryer PE, Haymond MW, Santiago JV, Shah SD. Norepinephrine and epinephrine release and adrenergic mediation of smoking-associated hemodynamic and metabolic events. N Engl J Med, 195: 573-7 (1976).

8. Wadenvik H, Kutti J. The effect of an adrenaline infusion on the splenic blood flow and intrasplenic platelet kinetics. Br J Haematol, 67: 187-92 (1987).

9. Peters AM, Lavender JP. Factors controlling the intrasplenic transit of platelets. Eur J Clin Invest, 12: 191-5 (1982).

10. Schmidt KG, Rasmussen JW, Rasmussen AD, Arendrup H. Comparative studies of the in vivo kinetics of simultaneously injected In-111- and Cr-51-labelled human platelets. Scand J Haematol, 30: 465-78 (1983).

11. Brezinski DA, Tofler GH, Muller JE, Pohjola-Sintonen S, Willich SN, Schafer AI, Czeisler C, Williams GH. Morning increase in platelet aggregability. Association with assumption of the upright posture. Circulation, 78: 35-40 (1988).

12. Ardlie NG, McGuiness JA, Garrett JJ. Effect on human platelets of catecholamines at levels achieved in the circulation. Atherosclerosis, 58: 251-9 (1985).

13. Hollister AS, FitzGerald GA, Nadeau JHJ, Robertson D. Acute reduction in human platelet α_2-adrenoreceptor affinity for agonist by endogenous and exogenous catecholamines. J Clin Invest, 72: 1498-1505 (1983).

14. Rohrer TF, Pfister B, Weber C, Imholf PR, Stucki P. Validity of the Wu-Hoak method for quantitative determination of platelet aggregation in vivo. Blut, 36: 15-20 (1978).

15. Kaplan KL, Owen J. Plasma levels of beta-thromboglobulin and platelet factor 4 as indices of platelet activation in vivo. Blood, 57: 199-202 (1981).

16. Goodwin DA, Bushberg Jt, Doherty PW, Lipton MJ, Conley FK, Dimanti CI, Meares CF. Indium-[111]-labeled autologous platelets for location of vascular thrombi in humans. J Nucl Med, 19: 626-34 (1978).

17. Pittilo RM, Clarke JMf, Harris D, Mackie IJ, Rowles PM, Machin SJ, Woolf N. Cigarette smoking and platelet adhesion. Br J Haematol, 58: 627-32 (1984).

18. Rival J, Riddle JM, Stein PD. Effects of chronic smoking on platelet function. Thromb Res, 45: 75-85 (1987).

19. Wennmalm Å. Interaction of nicotine and prostaglandins in the cardiovascular system. Prostagl, 23: 139-44 (1982).

20. Madsen H, Dyerberg J. Cigarette smoking and its effects on the platelet-vessel wall interaction. Scand J Clin Lab Invest, 44: 203-6 (1984).

21. Kenney MW, George AJ, Stuart J. Platelet hyperactivity in sickle-cell disease: a consequence of hyposplenism. J Clin Pathol, 33: 622-5 (1980).

22. Robinette CD, Fraumeni JF. Splenectomy and subsequent mortality in veterans of the 1939-45 war. Lancet, 2: 127-9 (1977).

ALTERATIONS OF ARACHIDONATE METABOLISM IN

CARDIOVASCULAR SYSTEM BY CIGARETTE SMOKING

Hsin-Hsiung Tai, Wen-Chang Chang, Ying Liu, and Shoshi Fukuda

Division of Medicinal Chemistry and Pharmaceutics
College of Pharmacy, University of Kentucky
Lexington, Kentucky 40536-0082

INTRODUCTION

Cigarette smoking is known to be a major risk factor for the development of a variety of cardiovascular diseases including myocardial infarction, thromboembolism and atherosclerosis(1). The underlying mechanisms by which cigarette smoking can contribute to these disorders are yet to be defined. Recently, two families of biologically potent fatty acids were discovered and found to have profound effects on the cardiopulmonary system(2,3). They may play dominant roles in the physiology of vascular homeostasis and in the pathophysiology of thrombosis, atherosclerosis and emphysema. Both families of fatty acids are derived from arachidonic acid by two separate pathways. The cyclooxygenase pathway leads to the formation of intermediary prostaglandin endoperoxides (PGG_2, PGH_2) followed by the synthesis of vasoconstrictive and proaggregatory thromboxane A_2 (TXA_2) in platelets(4) and that of vasodilatory and anti-aggregatory prostacyclin (PGI_2) in vascular wall(5). The opposing action of TXA_2 and PGI_2 provides a balanced interaction between platelets and vascular cells. TXA_2, PGI_2 and other prostaglandins (PG) are oxidized and inactivated by NAD^+-dependent 15-hydroxy prostaglandin dehydrogenase present in various tissues(6,7). The lipoxygenase pathway involves the formation of chemotactic 12-hydroxyeicosatetraenoic acid (12-HETE) in platelets(8) and possibly in vascular cells(9). This hydroxy fatty acid may control the migration and proliferation of smooth muscle cells in the development of atherosclerotic plaques(10).

We are interested if cigarette smoking may influence platelets and vascular cells interaction by altering arachidonate metabolic cascade. Our results indicated that thromboxane synthesis in platelets and prostacyclin production in aorta were not significantly changed in the first 8 weeks of smoke exposure in rats. However, 12-lipoxygenase pathway in platelets was significantly stimulated, and thromboxane and prostacyclin catabolic activity in lung was greatly decreased following 4 weeks of smoke exposure. Possible mechanisms that lead to decrease in pulmonary NAD^+-dependent 15-hydroxyprostaglandin dehydrogenase activity were also explored. Acrolein (CH_2=CHCHO) and oxidative species in the smoke may be contributing factors to the inactivation of prostaglandin catabolic enzyme in lung. Part of these findings has been described(11,12).

Tobacco Smoking and Atherosclerosis
Edited by J. N. Diana
Plenum Press, New York, 1990

METHODOLOGY

Materials

Arachidonic acid, glutathione (GSH), DL-dithiothreitol (DTT), human thrombin (3,000 units/mg), blue agarose, NAD^+, α-ketoglutarate, N-chloro-succinimide and bovine liver glutamate dehydrogenase (51 U/mg) were obtained from Sigma Chemical Co., St. Louis, MO. Acrolein, propionaldehyde, and chloramine T were supplied by Aldrich Chemical Co., Milwaukee, WI. Soluble calf skin collagen was purchased from Worthington Biochemical Corp., Freehold, NJ. $[1\text{-}^{14}C]$-Arachidonic acid (55 mCi/mmol) was obtained from Amersham, Arlington Heights, IL. PGE_2, TXB_2 and 15-keto-PGE_2 were kind gifts of the Upjohn Co., Kalamazoo, MI. 12-HETE was synthesized from arachidonic acid by human platelet suspension as described by McGuire et al(13). $15(S)\text{-}[15\text{-}^3H]$-$PGE_2$ was prepared according to Tai(14). Male Sprague Dawley rats of the same age were supplied by Harlan Sprague Dawley, Inc., Indianapolis, IN.

Smoke Inhalation

Rats were divided into two groups, shams and smoke exposed, following a 7-day acclimatization period. Smoked animals were exposed to fresh whole smoke for 10 minutes daily, 7 times a week for various length of time. Smoke exposure was accomplished by placing unanesthetized rats in a flexible restrainer allowing the nose to protrude into a moving column of air spiked every minute with fresh smoke generated from a University of Kentucky 2R1 reference cigarette. The bodies of the animals were never exposed to smoke. Sham animals were handled exactly as the smoked animals, but the moving air column contained only room air passed through the smoke generating equipment. Separate restrainers and smoke generators were used for the smoke and sham groups to prevent exposure of the shams to smoke exposure, 0.2 ml of blood was sampled randomly from the retro-orbital sinus of animals in each group immediately after smoke exposure and the % carboxyhemoglobin determined using a CO-oximeter (IL282, Instrumentation Laboratory Incorporated, Lexington, MA).

Preparation of Platelets and Different Tissues

Blood (10 ml) was withdrawn into one tenth volume of 3.8% sodium citrate from the abdominal aorta of rats under diethyl ether anesthesia. Platelet-rich plasma was prepared by centrifugation of citrated blood at 200 xg for 15 min at room temperature. Platelets were obtained by centrifuging the platelet-rich plasma at 1,000 xg for 10 min. The platelets were suspended in Ca^{+2}, Mg^{+2} free phosphate buffered saline containing 5.5 mM glucose. Platelet number was determined with a Model ZBI Coulter Counter.

Thoracic aortas were removed immediately after sacrifice. Aortas were then washed and suspended in 15 mM Tris-HCl buffer (pH 7.4) containing 140 mM NaCl and 5.5 mM glucose. Lung and kidney were removed immediately and frozen on dry ice block. Stomach was cut open to remove content and washed with distilled water before freezing on dry ice block. Lung, kidney and stomach were stored frozen at -80°C.

Preparation of Cytosolic Fraction from Platelets

Platelet pellets were suspended in 0.05 M Tris-HCl, pH 7.5 and freeze-thawed three times by using liquid N_2 and a water bath at 37°C. The suspension was centrifuged at 105,000 xg for 60 min. The supernatant designated as cytosolic fraction was used for 12-lipoxygenase assay.

Preparation of Crude Enzyme Extract from Different Tissues

Lung, kidney and stomach were homogenized in five volumes of 0.05 M
Tris-HCl, pH 7.5, using a modified Potter-Elvehjem homogenizer. The homoge-
nates were centrifuged at 27,000 xg for 20 min. The supernatant was used as
the source of NAD^+-dependent 15-hydroxyprostaglandin dehydrogenase.

Assay of Thromboxane Synthesis by Platelets

One ml of platelet suspension (6.5×10^8) incubated at 37°C was
challenged with arachidonic acid (3 µg/ml) or thrombin (1 U/ml) or collagen
(100 µg/ml) for 2, 10 or 20 min respectively. The reaction mixture was
acidified with 1N HCl and then neutralized with 1 M Tris-base before an
aliquot was diluted for radioimmunoassay of TXB_2. Radioimmunoassay was
carried out essentially as described by Tai and Yuan(15).

Assay of Prostacyclin Production by Aortas

Aortic rings weighing 6-8 mg were incubated in 0.5 ml of 0.05 M
Tris-HCl, pH 8.6 at 37° for 15 min. Buffer was acidified to pH 3.0 with 1N
HCl and then reneutralized to 7.4 with 1M Tris-base for radioimmunoassay of
6-keto-$PGF_{1\alpha}$. Radioimmunoassay was performed according to the method
described by Tai and Tai(16).

Enzyme Assay of 12-Lipoxygenase

The assay was done as described previously with some modifications(17).
The reaction mixture contained: [1-^{14}C]arachidonic acid (0.05 µCi, 0.82
nmol); GSH, 1 mM; and cytosolic fraction in a final volume of 1 ml of 0.05 M
Tris-HCl, pH 7.5. The reaction was performed at 37°C for 2 min and
terminated by acidification to pH 3.0. The reaction mixture was extracted
with 5 ml of ethylacetate. The organic phase was evaporated to dryness
under a stream of nitrogen. The residue taken up in ethanol was applied to
a silica gel G plate. The plate was developed in a solvent system of petro-
leum ether/ethyl ether/acetic acid (50:50:1). The substrate and products
were localized by autoradiography, and scraped separately into scintillation
vials and the radioactivity was determined by liquid scintillation counting.

Enzyme Assay of NAD^+-Dependent 15-Hydroxyprostaglandin Dehydrogenase

The different methods were employed:

a. Enzyme activity was determined by measuring the transfer of
tritium from 15(S)-[15-^{3}H]-PGE_2 to glutamate by coupling with glutamate
dehydrogenase as described previously(14). Briefly, the reaction mixture
contained: NH_4Cl, 5 µmoles; α-keto-glutarate, 1 µmole; NAD^+, 1 µmole;
15(S)-[15-^{3}H]-PGE_2, 1 nmole, 30,000 cpm; glutamate dehydrogenase, 100 µg and
crude enzyme extract in a final volume of 1 ml of 0.05 M Tris-HCl, pH 7.5.
The reaction was continued for 10 min at 37°C and terminated by the addition of 0.3 ml of 10% aqueous charcoal suspension. The radioactivity in the
supernatant after centrifugation (1,000 xg, 5 min) was determined by liquid
scintillation counting. Calculation of the amount of PGE_2 oxidized was
based on the assumption that no kinetic isotope effect was involved in the
oxidation of 15(S)-hydroxyl group of 15(S)-[15-^{3}H]-PGE_2 as a substrate.
The method was used for assaying crude enzyme extract.

b. Enzyme activity was measured by following the formation of NADH
fluorometrically. The reaction mixture contained 1.5 µmol NAD^+, 28 nmol
PGE_1, and enzyme in a final volume of 1.5 ml of 0.05 M Tris-HCl, pH 7.5.

The reaction was initiated by the addition of enzyme and allowed to proceed at room temperature. The rate of NADH formed was recorded by the increase in fluorescence at 460 nm with excitation at 340 nm, using an Aminoco-Bowman SPF coupled to a recorder. This instrument was standardized by known amount of NADH determined by direct measurement of absorbance at 340 nm using $\Sigma_M = 6.22 \times 10^3$ $M^{-1}cm^{-1}$. All the kinetic studies with inhibitors were carried out by this method.

<u>Purification of NAD^+-Dependent 15-Hydroxyprostaglandin Dehydrogenase from Swine Lung</u>

The enzyme was purified to the step of DEAE-Sephadex A-50 according to a previously described procedure(18). The specific activity of the purified enzyme was 0.12 U/mg protein.

RESULTS

Body weights and carboxyhemoglobin content after various lengths of cigarette smoke exposure are shown in Table I. Evidence that the smoked group actually inhaled cigarette smoke is demonstrated by the significant increase in % carboxyhemoglobin over shams immediately after a smoke exposure session.

Effect of cigarette smoke on thromboxane synthesis by platelets following 4 and 8 weeks of smoke exposure is shown in Table 2. Platelets were either incubated with exogenous arachidonic acid or stimulated with thrombin or collagen to release endogenous arachidonic acid. Thromboxane synthesis as determined by radioimmunoassay of TXB_2 was not significantly different between sham and smoked groups by any stimulation.

Effect of cigarette smoke on prostacyclin production by aortas following 4 and 8 weeks of smoke exposure is shown in Table 3. Prostacyclin production as measured by radioimmunoassay of stable product, $6\text{-keto-PGF}_{1\alpha}$, was found to exhibit no difference between sham and smoked groups in aortas which were allowed to produce prostacyclin from endogenously released arachidonate or stimulated with exogenous arachidonate.

Although synthesis of thromboxane and prostacyclin through the cyclooxygenase pathway was not significantly altered in platelets and aortas within 8 weeks of smoke exposure, production of 12-HETE through lipoxygenase pathway in platelets was found to increase significantly in smoked group when compared to sham group as shown in Fig. 1.

Catabolism of prostaglandins, thromboxane and prostacyclin as measured by the NAD^+-dependent 15-hydroxyprostaglandin dehydrogenase activity was

Table 1. Body Weight and Carboxyhemoglobin Content After 4 and 8 Weeks of Smoke Exposure

Group	0 Week	4 Weeks		8 Weeks	
	Wt (gm)*	Wt (gm)	COHb (%)+	Wt (gm)	COHb (%)+
Sham	290 + 2	314 + 4	1.1	347 + 9	0.5
Smoke	293 + 2	299 + 2	6.2	322 + 6	8..5

*Values for body weight represent the mean + S.E.M., n = 8.

+Values for COHb represent the mean % from two to three animals of that group.

Table 2. Effect of Cigarette Smoke on Thromboxane Biosynthesis by Platelets

Length of Smoke Exposure	Stimulators		TXB_2 (ng/6.5x10^8 Cells)		P-Value
			Sham	Smoke	
4 Weeks	Arachidonic Acid, 3 µg/ml[+]		150.3 $\pm$ 14.1[*] (100%)	161.2 $\pm$ 13.0 (107%)	NS[+]
	Thrombin	1 U/ml	24.0 $\pm$ 4.5 (100%)	20.4 $\pm$ 3.5 (85%)	NS
		2 U/ml	49.8 $\pm$ 5.8 (100%)	58.2 $\pm$ 6.8 (116%)	NS
	Collagen	100 µg/ml	15.3 $\pm$ 1.5 (100%)	14.6 $\pm$ 1.5 (95%)	NS
8 Weeks	Arachidonic Acid, 3µg/ml		395.8 $\pm$ 19.0 (100%)	435.7 $\pm$ 18.4 (110%)	NS
	Thrombin	1 µg/ml	58.7 $\pm$ 4.4 (100%)	65.0 $\pm$ 2.4 (112%)	NS
		2 U/ml	93.4 $\pm$ 3.4 (100%)	107.1 $\pm$ 4.2 (115%)	NS
	Collagen	100 µg/ml	49.2 $\pm$ 12.9 (100%)	41.2 $\pm$ 4.7 (84%)	NS

[*] Each value is the mean $\pm$ S.E.M. from 8 individual animals.

[+] Incubation of platelets with arachidonic acid, thrombin and collagen was performed for 2, 10 and 20 min, respectively.

[**] Values in the parenthesis are the percentage compared with their sham control.

[++] NS = Not significant

Table 3. Effect of Cigarette Smoke on Prostacyclin Production by Aortas

Length of Smoke Exposure	Stimulators	6-keto-PGF$_{1\alpha}$ (ng/mg Dry Aorta)		P-Value
		sham	smoke	
4 Weeks	None[**]	34.2 $\pm$ 5.0[*] (100%)[+]	33.3 $\pm$ 4.0 (100%)	NS[++]
	Arachidonate, 3 µg/ml	50.1 $\pm$ 6.7 (100%)	49.0 $\pm$ 5.0 (98%)	NS
8 Weeks	None	35.1 $\pm$ 3.4 (100%)	28.9 $\pm$ 2.0 (82%)	NS
	Arachidonate, 3µg/ml	45.2 $\pm$ 4.5 (100%)	38.9 $\pm$ 1.9 (86%)	NS

[*] Each value represents the mean $\pm$ S.E.M. from 8 individual animals.

[+] Values in the parentheses are the percentage compared with their sham control.

[**] Incubation in the absence or presence of arachidonate was performed for 15 min.

[++] NS = Not significant.

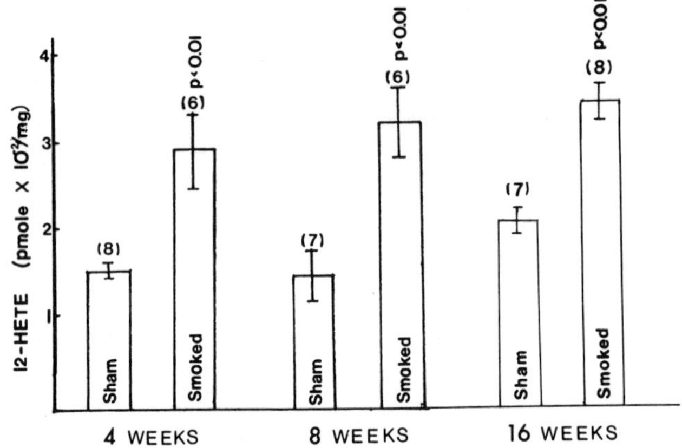

Fig. 1.   Effect of cigarette smoke exposure on platelet 12-lipoxy-
genase activity. Rats were exposed to cigarette smoke for 4,
8 and 16 weeks. Cytosolic fraction of platelets was assayed
for 12-lipoxygenase activity as described in the Methodology.

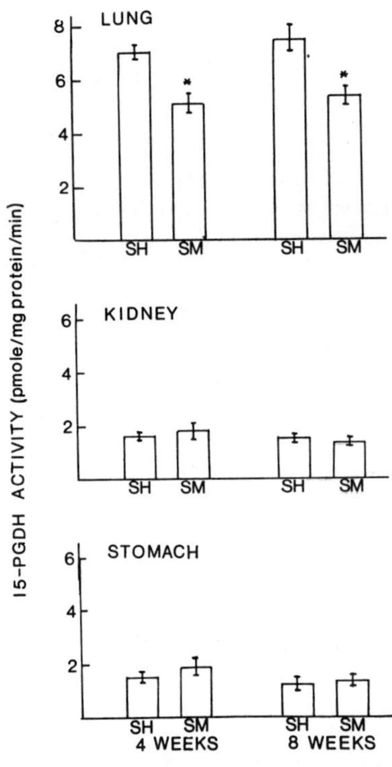

Fig. 2.   Pulmonary, renal and gastric NAD⁺-dependent 15-hydroxyprosta-
glandin dehydrogenase activities in sham and smoke exposed
rats. The length of smoke exposure was as indicated. Data
are means ± S.E.M. for n = 8. * indicates significantly
different from sham at p<0.05.

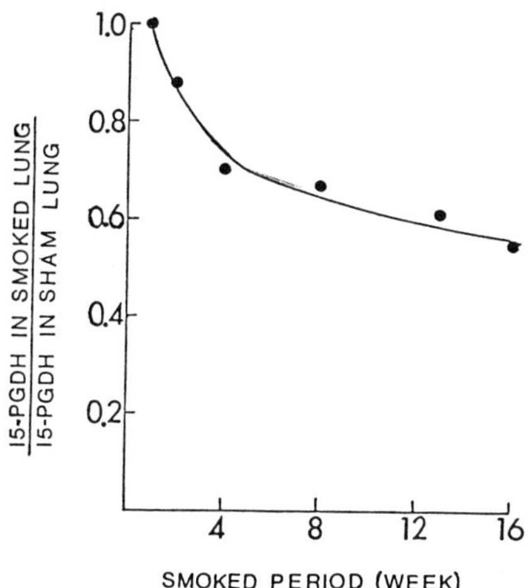

Fig. 3.   Effect of the length of smoke exposure in the pulmonary
NAD$^+$-dependent 15-hydroxyprostaglandin dehydrogenase
activity. The ratio of the enzyme activity in smoked lung to
that in sham lung was plotted against the length of smoke
exposure in week.

found to decrease significantly in lung but not in kidney or stomach following 4 and 8 weeks of smoke exposure when smoked group was compared to sham
group as shown in Fig. 2. Smoke induced decrease in pulmonary enzyme

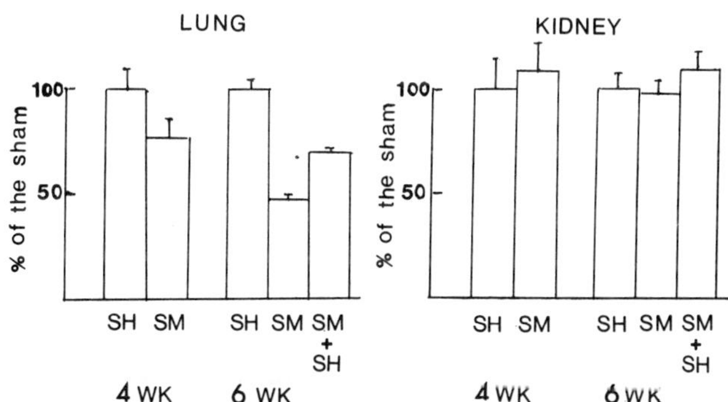

Fig. 4.   Reversibility of cigarette smoke induced inactivation of
pulmonary NAD$^+$-dependent 15-hydroxy prostaglandin dehydrogenase. Rats were smoked or shammed for 4 weeks and then the
smoked animals were divided into 2 groups (n = 6 for each
group). One group was allowed to smoke for two additional
weeks and the other was shammed for two more weeks. The
original shammed group was allowed to sham for a total of 4
and 6 weeks.

activity was very significant following 4 weeks of smoke exposure and
appeared to persist as long as 16 weeks as shown in Fig. 3. Inactivation of
the pulmonary enzyme appears to be irreversible since rats that were smoked
for 4 weeks and then shammed for 2 weeks showed about the same enzyme
activity as those that were smoked for 4 weeks. Apparently, cessation of
smoke exposure did not cause an immediate rebound of the enzyme activity
within the subsequent two weeks. Again, the renal enzyme activity was not
affected by cigarette smoking.

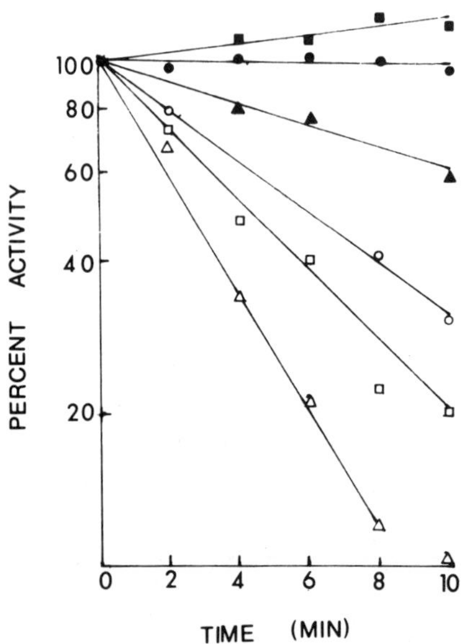

Fig. 5.   Inactivation of pulmonary NAD$^+$-dependent 15-hydroxyprosta-
glandin dehydrogenase by acrolein. Purified enzyme devoid of
DTT was preincubated with different concentrations of
acrolein or propionaldehyde for the indicated length of time
before an aliquot was removed for enzyme assay. -●-●-,
enzyme only; -■-■-, 1 mM propionaldehyde; -▲-▲-, enzyme plus
5 µM acrolein; -○-○-, enzyme plus 10 µM acrolein; -◻-◻-,
enzyme plus 20 µM acrolein; -△-△-, enzyme plus 50 µM
acrolein.

Possible mechanisms that may lead to decrease in pulmonary catabolic
enzyme activity were further explored. Acrolein, a α,β-unsaturated aldehyde
component in cigarette smoke, was found to cause a time dependent inactivation of a purified pulmonary NAD$^+$-dependent 15-hydroxyprostaglandin dehydrogenase, whereas propionaldehyde, a saturated aldehyde, showed no effect as
shown in Fig. 5. We employed swine lung enzyme preparation for studies
since rat lung was not available in good quantities for isolation and purification of the enzyme. The inactivation follows a pseudo - first order
kinetics with a limiting half-life of 1.4 min calculated from a T 1/2 vs
[acrolein]$^{-1}$ plot. (Data not shown). Inactivation by acrolein can not be
reversed by the addition of 1 mM GSH during assay but can be protected by

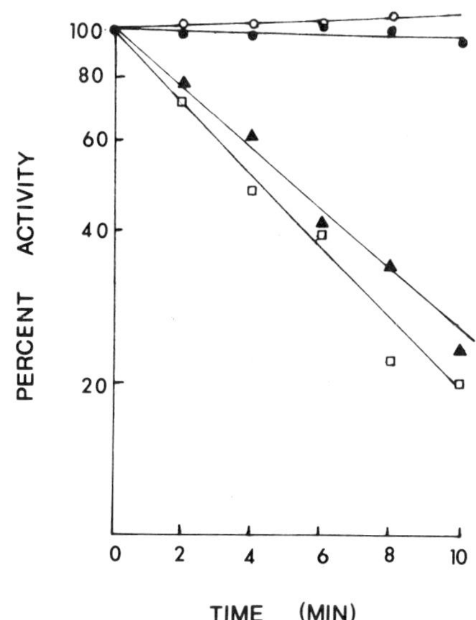

Fig. 6.   Effect of GSH on acrolein induced inactivation of pulmonary NAD$^+$-dependent 15-hydroxyprostaglandin dehydrogenase. Purified enzyme devoid of DTT was preincubated with or without 1 mM GSH for the indicated length of time before an aliquot was removed for enzyme assay. -●-● -, enzyme only; -o-o-, enzyme, 1 mM GSH and 20 μM acrolein during preincubation; -▲-▲-, enzyme plus 20 μM acrolein during preincubation and enzyme was then assayed in the presence of 1 mM GSH; -□-□-, enzyme plus 20 μM acrolein during preincubation.

prior incubation of the enzyme with GSH as shown in Fig. 6. Various oxidizing agents, namely, N-chlorosuccinimide, hydrogen peroxide and chloramine T, are also capable of irreversibly inactivating the enzyme as shown in Fig. 7.

DISCUSSION

Our data demonstrate that exposure of rats to cigarette smoke up to 8 weeks did not significantly alter the capability of platelets to metabolize either endogenous or exogenous arachidonate to thromboxane measured as TXB_2. This finding suggests that neither cyclooxygenase nor thromboxane synthetase activities were altered by smoking. Results from thrombin or collagen stimulation also suggest that receptor responsiveness of platelets to either agonists was not altered by smoking. Insignificant change of thromboxane synthesis in platelets has been reported for human smokers(9) and animals exposed to smoke(20). Therefore, a slightly increased plasma level of TXB_2 in human smokers reported by Mehta and Mehta(19) may be derived from sources other than platelets. Lubawy et al(20) recently demonstrated that lung microsomes from smoked rats did synthesize more TXB_2 from arachidonate than from shams. It is possible that the cyclooxygenase pathway in lung cells, notably fibroblasts which possess very active system synthesizing thromboxane(21), may be stimulated following chronic cigarette smoke exposure.

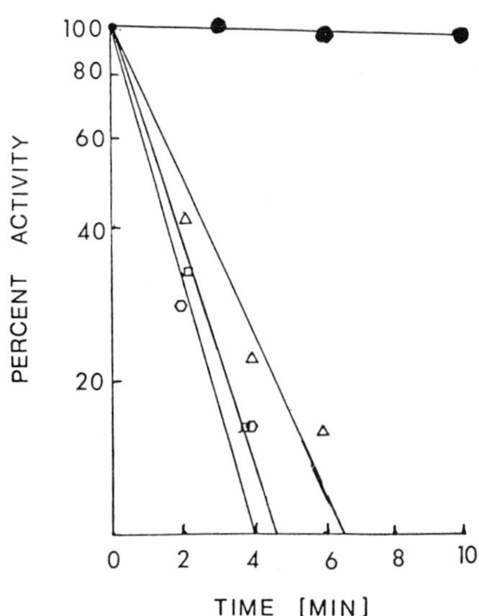

Fig. 7.   Inactivation of NAD^{+}-dependent 15-hydroxyprostaglandin
dehydrogenase activity by various oxidizing agents. Purified
enzyme devoid of DTT was preincubated with N-chlorosuccinimide (Δ; 50 μM), hydrogen peroxide (□, 10 mM) or chloramine
T (o, 50 μM) for the indicated length of time before an
aliquot was removed for enzyme assay.

Smoke exposure for 4 weeks also failed to change aortic prostacyclin
production from either endogenous or exogenous arachidonate. Although smoke
exposure for 8 weeks appeared to decrease aortic prostacyclin production
slightly, the large variation in sample values preclude any conclusion of
having statistical significance. Nevertheless, it seems to show a trend of
decreasing aortic production of prostacyclin as smoke exposure is prolonged.
Studies on the effect of longer smoke exposure on arachidonate metabolism in
cardiovascular system remain to be explored. Other laboratories have
reported that human smokers had significantly lower plasma levels of 6-keto-
PGF$_{1\alpha}$ (19,22). Although the source of circulating prostacyclin is not clear,
evidence for an effect of smoking on human vascular tissue has been
represented. Umblical arteries taken immediately after delivery from smoking
mothers produced only about half as much prostacyclin as arteries from non-
smokers(23). Mechanisms that lead to decreased synthesis of prostacyclin by
vascular tissues remain unclear. Although nicotine, a prominent component
in cigarette smoke, has been shown to decrease the efflux of 6-keto-PGF$_{1\alpha}$
following infusion of arachidonates in isolated rabbit heart(24) and in rat
aorta (25), it is doubtful that a sufficient amount of nicotine is present to
exhibit inhibition of prostacyclin synthesis from arachidonate in vascular
tissues excised from smokers or smoked animals. The effect of cigarette
smoke on vascular prostacyclin formation may be more complex than just
inhibition of synthesis.

Although cigarette smoke failed to show any significant effect on
cyclooxygenase pathway in platelets and aorta following 8 weeks of exposure,
it did stimulate significantly the lipoxygenase pathway in platelets
within 4 weeks of exposure. Increase in 12-lipoxygenase activity in

platelets may have several consequences. Firstly, 12-HETE is an
extremely potent agent to stimulate aortic smooth muscle cell migra-
tion(26). Migration of these cells from media to intima is considered an
early event following endothelial injury and platelet adhesion. Subsequent proliferation in intima and synthesis and release of connective
tissue proteins results in the formation of atheromatous plaques(27).
Significant stimulation of 12-lipoxygenase pathway may predispose
individuals to atherosclerosis commonly found in smokers. Secondly,
12-HETE, the immediate product of 12-lipoxygenase pathway may predispose
individuals to atherosclerosis commonly found in smokers. Secondly,
12-HETE, the immediate product of 12-lipoxygenase, modulates neutrophil
functions(28). It evokes neutrophil chemotactic and chemokinetic responses and augments lysosomal enzyme release initiated by chemotactic
fragment of C_5. Modulation of these neutrophil functions may be mediated
by LTB_4 since 12-HPETE was recently shown to stimulate leukotriene synthesis in neutrophils(29). Significant increase in 12-HPETE synthesis by
cigarette smoking may alter platelet-leukocyte interaction favoring the
formation of chemotactic, degranulating and vasoactive leukotrienes.
Thirdly, 12-HPETE like other hydroproxy fatty acids is a potent inhibitor
of prostacyclin synthetase(30) and may distort thromboxane/prostacyclin
ratio and disturb vascular homeostasis. In summary, the consequence of
increased activity of the platelet lipoxygenase pathway may contribute to
the development of atherosclerosis, emphysema and thromboembolism
commonly found in smokers.

Although synthesis of thromboxane and prostacyclin in the cardiovascular system did not appear to change significantly within the first 8
weeks of smoke exposure, catabolism of these vasoactive substances by
lung was decreased very significantly after 4 weeks of smoke exposure.
This finding is consistent with the observation that cigarette-smoke
exposure decreased pulmonary removal of PGE_2 in isolated rat lung(31).
Lung is considered the major site of inactivating most prostaglandins and
thromboxane except prostacyclin following their passage through this
organ(32,33). The consequence of decrease in catabolic enzyme activity
is the increase in circulating thromboxane/prostacyclin ratio despite no
change in their synthesis. Vascular homeostasis is thereby disturbed
favoring the development of thromboembolism commonly found in smokers.
The mechanism that leads to decreased dehydrogenase activity in lung
following cigarette smoke exposure is presently unknown. The fact that
lung but not kidney or stomach is affected by cigarette smoke suggests
that this organ which is at the first line of defense may be damaged
directly by some reactive components present in cigarette smoke.
Acrolein, a very reactive aldehyde found in smoke(34), can irreversibly
inactivate 15-hydroxyprostaglandin dehydrogenase. Inactivation by
acrolein can be prevented by prior incubation of the enzyme with GSH but
not by addition of GSH during assay. This finding is consistent with the
interpretation that acrolein is able to alkylate sulfhydryl group of
cysteine which is an essential amino acid residue for the cataltic
activity of 15-hydroxyprostaglandin dehydrogenase(7). Acrolein has been
shown to decrease the biosynthesis of PGE_2 but not that of TXB_2, from
arachidonate in rat alveolar macrophage(35). This was interpreted by the
GSH requiring sulfhydryl sensitive nature of PGE synthetase(36). The
sulfhydryl sensitive nature of 15-hydroxyprostaglandin dehydrogenase can
also account for the ready oxidation by N-chlorosuccinimide, chloramine T
and hydrogen peroxide as shown in this study. It is quite likely that
acrolein and oxidative species found in cigarette smoke may inactivate
the catabolic enzyme by alkylation of an essential cysteine residue or
oxidation of necessary cysteine or methionine residues. This oxidation
is reminiscent of the oxidation and inactivation of α_1-antitrypsin shown
previously(37,38).

In conclusion we have found that catabolism of prostaglandins and thromboxane but not their synthesis is affected in the early stage of smoke exposure. Increase in platelet activity of lipoxygenase pathway is another consequence of short term smoke exposure. The outcome of these biochemical changes may eventually contribute to the development of thromboembolism, atherosclerosis and pulmonary emphysema commonly found in smokers.

SUMMARY

Male rats were exposed to freshly generated cigarette smoke once daily for various lengths of time. Inhalation of smoke was verified by elevated levels of carboxyhemoglobin. Metabolism of arachidonate in the cardiovascular system to thromboxane and prostacyclin through the cyclooxygenase pathway and their further metabolism to 15-keto-derivates, and to 12-hydroxyeicosatetraenoic acid (12-HETE) through lipoxygenase pathway was investigated. Synthesis of thromboxane and prostacyclin in platelets and aortas respectively was not changed within 8 weeks of smoke exposure. However, formation of 12-HETE in platelets was significantly increased after 4 weeks of smoke exposure. Catabolism of thromboxane and prostacyclin as determined by NAD^+-dependent 15-hydroxyprostaglandin dehydrogenase activity was greatly decreased in lung but not in kidney and stomach following 4 weeks of smoke exposure. Increased 12-lipoxygenase activity in platelets may lead to stimulation of migration and proliferation of smooth muscle cells and to increased synthesis of leukotrienes in neutrophils. Decreased pulmonary prostaglandin catabolic activity may result in increase in circulating thromboxane/prostacyclin ratio and subsequently alteration of vascular homeostasis. The consequence of these biochemical changes may contribute to the development of atherosclerosis, thromboembolism and emphysema commonly found in smokers.

ACKNOWLEDGEMENTS

We are greatly indebted to Ms. Sharon Parker and Ms. Gwendolyn Cambron for their excellent technical assistance and to Ms. Jean Cavenee for typing this manuscript. This work was supported in part by a grant from the Kentucky Tobacco Research Board.

REFERENCES

1. Smoking and Health, a Report of the Surgeon General, DHEW publication 79-50066, p. 4-63 (1979).
2. B. Samuelsson, M. Goldyne, E. Granstrom, M. Hamberg, S. Hammarstrom, and C. Malmsten. Prostaglandins and Thromboxanes, Ann. Rev. Biochem. 47:997-1029 (1978).
3. S. Hammarstrom, S: Leukotrienes. Ann. Rev. Biochem. 52:355-377 (1983).
4. M. Hamberg, J. Svensson, and B. Samuelsson. Thromboxanes: a new group of biologically active compounds derived from prostaglandin endoperoxides, Proc. Nat. Acad. Sci. USA 72:2994-2998 (1975).
5. S. Moncada, R. Gryglewski, S. Bunting, and J.R. Vane. An enzyme isolated substance that inhibits platelet aggregation, Nature, 263:663-665 (1978).
6. E. Anggard, C. Larsson, and B. Samuelsson. The distribution of 15-hydroxyprostaglandin dehydrogenase and prostaglandin [13]-reductase in tissues of the swine, Acta. Physiol. Scand. 81:396-404 (1970).

7. D.T.Y. Kung-Chao, and H.H. Tai. NAD$^+$-dependent 15-hydroxyprostaglandin dehydrogenase from porcine kidney. I. purification and partial characterization, <u>Biochem. Biophys. Acta</u>. <u>614</u>:1-13, (1980).

8. M. Hamberg and B. Samuelsson. Prostaglandin endoperoxides. Novel transformation of arachidonic acid in human platelets, <u>Proc. Nat. Acad. Sci</u>. <u>71</u>:3400-3404 (1974).

9. J.E. Greenwald, J.R. Bianchine, and L.K. Wong. The production of the arachidonate metabolite HETE in vascular tissue, <u>Nature</u> <u>281</u>:588-589 (1979).

10. J. Nakao, H. Ito, W.C. Chang, Y. Koshihara, and S. Murota. Aortic smooth muscle cell migration caused by platelet-derived growth factor is mediated by lipoxygenase product(s) of arachidonic acid, <u>Biochem. Biophys. Res. Commun</u>. <u>112</u>:866-871 (1983).

11. W.C. Chang, S. Fukuda, and H.H. Tai. Cigarette smoking stimulates lipoxygenase but not cyclooxygenase pathway in platelets, <u>Biochem. Biophys. Res. Commun</u>. <u>115</u>:499-505 (1983).

12. W.C. Chang, S. Fukuda, and H.H. Tai. Pulmonary NAD$^+$-linked 15-hydroxyprostaglandin dehydrogenase activity is decreased by cigarette smoking, <u>Life Sci</u>. <u>34</u>:1261-1268 (1984).

13. J.C. McGuire, R.C. Kelly, R.R. Gorman, and F.F. Sun. Preparation and spectral properties of 12-hydroxyeicosatetraenoic acid (HETE), <u>Prep. Biochem</u>. <u>8</u>:147-153 (1978).

14. H.H. Tai. Enzymatic synthesis of 15(S)-[15-^{3}H]prostaglandins and their use in the development of a simple and sensitive assay for 15-hydroxyprostaglandin dehydrogenase, <u>Biochemistry</u> <u>15</u>:4586-4592 (1976).

15. H.H. Tai, and B. Yuan. Development of radioimmunoassay for thromboxane B$_2$, <u>Anal. Biochem</u>. <u>87</u>:343-349 (1978).

16. C.L. Tai and H.H. Tai. Radioimmunological assay of prostacyclin synthetase activity, <u>Prostaglandins Med</u>. <u>4</u>:399-408 (1980).

17. W.C. Chang and J. Nakao, H. Orimo and S. Murota. Effect of reduced glutathione on 12-lipoxygenase pathway in rat platelets, <u>Biochem. J</u>. <u>201</u>:771-775 (1982).

18. Y. Liu and H.H. Tai. Inactivation of pulmonary NAD$^+$-dependent 15-hydroxyprostaglandin dehydrogenase by acrolein, <u>Biochem. Pharmacol</u>. <u>34</u>:4275-4278 (1985).

19. P. Mehta and J. Mehta. Effect of Smoking on platelets and on plasma thromboxane-prostacyclin balance in man. <u>Prosta. Leuko. Med.</u> <u>9</u>:141-150 (1982).

20. W.C. Lubawy, M.A. Valentovic, J.E. Atkinson, and G.C. Gairola. Chronic cigarette smoke exposure adversely alters ^{14}C-arachidonic acid metabolism in rat lungs, aortas and platelets, <u>Life Sci</u>. <u>33</u>:577-584 (1983).

21. N.K. Hopkins, F.F. Sun, and R.R. Gorman. Thromboxane A$_2$ biosynthesis in human lung fibroblasts WI-38, <u>Biochem. Biophys. Res. Commun</u>. <u>85</u>:827-836 (1978).

22. M. Mascotti, L. Poggesi, G. Galanta, F. Trotta, and G. Serheria. Prostacyclin production in man, In: Clinical <u>Pharmacology of Prostacyclin</u>, edited by P.J. Lewis and J. O'Grady, Raven Press, New York, pp. 9-20 (1981).

23. C. Dadah, C. Leithner, H. Sinzinger, and K. Silberbauer, Diminished prostacyclin formation in umbilical arteries of babies born to women who smoke, <u>Lancet</u> <u>1</u>:94 (1981).

24. A. Wennmalm. Nicotine inhibits hydroxia and arachidonic acid-induced release of prostacyclin-like activity in rabbit heart, <u>Br. J. Pharmacol</u>. <u>69</u>:545-551 (1979).

25. F. Tan Hoor and J.F.A. Quadt. Effect of nicotine on prostacyclin production of the isolated pulsatingly perfused rat aorta, Fourth International Prostaglandin Conference, Abstracts, Washington, D.C., May 27-31, p. 52 (1979).

26. J. Nakao, T. Ooyama, H. Ito, W.C. Chang, and S. Murota. Comparative effect of lipoxygenase products of arachidonic acid on rat aortic smooth muscle cell migration, <u>Atherosclerosis</u> 44:339-342 (1982).

27. R. Ross and J.A. Glomset, Atherosclerosis and the arterial smooth muscle cell, <u>Science</u> 180:1332-1339 (1973).

28. E.J. Goetzl, H.R. Hill, and R.R. Gorman. Unique aspects of the modulation of human neutrophil function by 12L-hydroperoxy-5,8,10-14-eicosatetraenoic acid, <u>Prostaglandins</u> 19:71-85 (1980).

29. J. Maclouf, B. Fruteau de Laclos, and P. Borgeat. Stimulation of leukotriene biosynthesis in human blood leukocytes by platelet-derived 12-hydroperoxyeicosatetraenoic acid, <u>Proc. Natl. Acad. Sci</u>. 79:6042-6046 (1982).

30. J.A. Salman, D.R. Smith, R.J. Flower, S. Moncada and J.R. Vane. Further studies on the enzymatic conversion of prostaglandin endoperoxide into prostacyclin by porcine aorta microsomes, <u>Biochem. Biophys. Acta</u>. 523: 250-262 (1978).

31. Y.S. Bakhle, J. Hartiala, H. Toivonen, and P. Uotila. Effects of cigarette smoke on the metabolism of vasoactive hormones in rat isolated lungs, <u>Br. J. Pharmacol</u>. 65: 495-499 (1979).

32. P.J. Piper, J.R. Vane and J. H. Wyllie. Inactivation of prostaglandins by the lungs, <u>Nature</u> 225:600-604 (1970).

33. S. Moncada, R. Korbut, S. Bunting, and J.R. Vane, Prostacyclin is a circulating hormone, <u>Nature</u> 273:767-768 (1978).

34. A.D. Horton and M.R. Guerin. Determination of acetaldehyde and acrolein in the gas phase of cigarette smoke using cryothermal gas chromatography, <u>Tobacco Sci</u>. 18:19-22 (1974).

35. C.C. Grumdfest, J. Chang, and D. Newcombe, Aerolein: a potent modulator of lung macrophage arachidonic acid metabolism, <u>Biochem. Biophys. Acta</u>. 713:149-159 (1982).

36. N. Ogino, T. Miyamoto, S. Yamamoto, and O. Hayaishi, Prostaglandin endoperoxide E isomerase from bovine vesicular microsomes, a glutathione requiring enzyme, <u>J. Biol. Chem</u>. 251:890-895 (1977).

37. A. Janoff and H. Carp. Possible mechanisms of emphysema in smokers. In vitro supression of serum eleastase-inhibitory capacity by fresh cigarette smoke and its prevention by antioxidants, <u>Am. Rev. Respir. Dis</u>. 118:617-621 (1978).

38. D. Johnson and J. Travis. The oxidative inactivation of human α-1-proteinase inhibitor: Further evidence for methionine at the active center, <u>J. Biol. Chem</u>. 254:4022-4026 (1979).

EFFECT OF NICOTINE AND CARBON MONOXIDE ON PROSTACYCLIN

PRODUCTION BY THE RABBIT HEART

David James Effeney

University of Queensland
Princess Alexandra Hospital
Brisbane, Queensland, Australia

Tobacco smoking is an addictive disorder associated with excessive cardiovascular mortality (1-3). Cigarette smokers have twice the risk of death from coronary artery disease compared with their non-smoking contemporaries. Why cigarettes are addictive is not known with certainty. However, the most likely agent is nicotine (5). Bolus injections of the drug like single puffs of cigarettes result in a rapid rise in the blood concentration, a change in the sensorium and physiologic responses such as increased heart rate, a rise in the blood pressure and peripheral vasconstriction (6,7). Other compounds in this complex aerosol-carbon monoxide, tars, particulate matter, and cyanide-have potentially significantly deleterious effects on the heart and blood vessels as well as the organism as a whole.

Prostacyclin PGI_2 was first described in 1976 (8). This important molecule inhibits platelet aggregation, causes platelet disaggregation and the relaxation of smooth muscle (9). It has been postulated that inhibition of this potentially protective mechanism in blood vessel walls could be responsible in part for the initiation and progression of atherosclerotic plaque (9,10). Similarly, and perhaps more importantly, lack of prostacyclin production in response to the stimulus of ischaemia at the time of a major vessel occlusion could contribute to adverse outcome in acute coronary artery events. In a majority of studies nicotine has been shown to inhibit the synthesis and release of PGI_2 from vascular tissue and heart muscle in vitro although the in vivo data are conflicting (11). I am not aware of any studies of carbon monoxide on prostacyclin production.

This contribution to the meeting reports results of experiments designed to assess the effects of nicotine and carbon monoxide on the production of prostacyclin by the rabbit heart.

The aims of the four experiments were

Experiment 1: to show that nicotine was concentrated in
                vascular and cardiac tissue

Experiment 2: to determine the effect of nicotine on PGI_2
                production by the heart

Tobacco Smoking and Atherosclerosis
Edited by J. N. Diana
Plenum Press, New York, 1990

Experiment 3:   to determine the effect of CO on PGI$_2$ production
                by the heart

Experiment 4:   to determine the effects of nicotine + CO on PGI$_2$
                production by the heart

<u>Materials and Methods</u>

Much of the initial work on prostacyclin production and activity was
performed in rabbits and is well categorized (9,10). The animal has been
used as a model for the study of atherosclerosis; they are relatively
cheap, and easy to care for. We used 3kg male New Zealand while
rabbits. The "Principles of Laboratory Animal Care" and the "Guide for
the Care and the Use of Laboratory Animals" were followed.

The animals were anaesthetized with ketamine and xylazine (12). The
right jugular vein was exposed in the neck and a Silastic catheter (I.D.
0.020 in, Dow Corning, Midland, Michigan) was inserted through the
jugular vein into the superior vena cava. The catheter was tied in
position to the vein, and led from the lateral skin incision to a
separate stab wound on the back of the neck between the ears. A
stainless steel base plate was sutured to the nuchal muscles at the exit
site. Attached to this was a stainless steel spring shield, through
which the catheter was led. This arrangement allowed the animal freedom
of movement in the box while maintaining the catheter clear of the
animal. Rotational movements were absorbed by the spring and base plate
rather than the catheter.

The "Plexiglas" boxes were constructed to the standard size (27" x
27" x 16") for rabbits. Filtered fresh air (4.61/min), and in the CO
experiments, analytical grade CO (1.84ml/min), were introduced through
individual flow meters to a baffled mixing chamber in the box. The final
CO concentration was 400 parts per million. The box was exhausted into a
fume hood. Animals were exposed to CO for 7 to 10 days. Carboxyhaemoglobin was determined using the spectrophotometric method of Tietz and
Fiereck (13).

The catheter was attached to an infusion swivel (Instech
Laboratories, Philadelphia, PA) which was held in a rubber stopper in the
tower atop the boxes. Nicotine was infused by a syringe pump (Harvard
Apparatus, Southnatic, MA) as nicotine bitartrate in saline at
50ug/kg/hour at a flow rate of 2.7ml/hour for 7 to 10 days. The infusion
was final filtered through a Millipore filter.

At the time of sample collection, the animals were anaesthetized
breathing 100% O$_2$. Blood was sampled for estimation of plasma nicotine
and carboxyhaemoglobin. The heart was excised rapidly, washed thoroughly
in modified Krebs-Ringer solution saturated with oxygen pO$_2$ >300mmHg at
37° to clear the blood, and then immediately transferred to iced modified
Krebs-Ringer solution with oxygen. The Heart was then dissected into its
various chambers which were again washed with ice cold oxygenated buffer
and stored in the buffer on ice until used.

Platelet-rich plasma (PRP) was prepared from one human donor, who
took no antiplatelet drugs during the series of experiments. For each
experiment the aggregation of platelets was determined (14) for final
concentrations of ADP from 1.0 to 20 μM and a concentration was selected
that produced primary aggregation. This concentration, usually 2.5 or
5μM ADP, was then used for the rest of the assays performed that day.

Four rabbits were infused with nicotine for 7 days. The animals were anesthetized and blood was obtained for plasma nicotine determination. The heart and thoracic aorta were excised along with samples of most other organs. The aorta was opened, weighed, washed with normal saline to remove blood, and frozen in liquid nitrogen, then stored at -80°C. The heart was allowed to clear by beating in three washes of oxygenated saline at 37° then treated the same as the aorta. The tissue was subsequently dissolved in 5N NaOH and tissue nicotine concentrations determined.

Full thickness samples (wet weight approximately 100 mg) of walls of both ventricles, atria, and interventricular septum were washed in oxygenated Tris buffer (0.05M, pH 7.4) prior to incubation. The incubation solution was a modification of that of Ellis et al. (15), and consisted of 0.05M Tris and 16.7 μM sodium arachidonate. The volume was adjusted to 1000μl per 100mg tissue. All incubations were carried out at 23°C and the incubation solution was stirred by a Teflon coated stir bar placed so as not to perturb the tissue sample. Time of tissue incubation was 3 minutes. A 50μl aliquot was taken for the biological assay and the remainder of the solution was frozen and stored at -70°C for radioimmunoassay.

Two assays were performed to assess prostacyclin production.

1. Radioimmunoassay for 6-keto-PGF$_{1a}$, the stable metabolite of PGI$_2$ (New England Nuclear, Boston, MA).
2. Determination of inhibition of ADP induced platelet aggregation.

The samples of biologic fluids and infusates were collected and stored in nitric acid washed glassware. Nicotine determinations for all samples were performed in duplicate by the gas chromatographic method of Jacob et al. (16).

The data were analyzed using non-parametric tests: the Mann Whitney U Test corrected for ties, and the Wilcoxon, Matched Pairs Signed Rank Test corrected for continuity.

<u>Results</u>

Four animals were studied to determine the tissue concentration of nicotine in the heart and aorta and to compare these to the blood concentration of the drug. These data are shown in Table 1.

<u>Table I</u>

Nicotine Concentrations in Blood, Heart and Aorta

Rabbit No	Weight Kg	Infusate µg/ml	Blood ng/ml	Heart ng/g	Thoracic Aorta ng/g
113	2.96	143	50	73.5	177.5
114	3.50	167	52	120.4	139.3
115	3.26	116	52	104.6	153.8
116	3.16	129	46	106.4	82.0
Mean	3.22	139	50	101.2	138.2
SD	0.23	22	2.8	19.8	40.6

In all animals, the heart and aorta concentrated nicotine significantly above the plasma concentration.

Thirty-eight animals were used to assess the effects of nicotine and carbon monoxide on the production of PGI_2. Eleven served as controls, 11 received nicotine, 10 were exposed to 400ppm CO, while six received both pretreatments. The carboxyhaemoglobin was greater than 12% in the 16 exposed animals. There was no change in the control or nicotine groups. The plasma concentrations of nicotine in the control and CO rabbits were less than the level of detection (2.2ng/ml) of the measurement method.

Table II summarizes the plasma nicotine concentrations at the time of tissue collection for assays of PGI_2.

Table II

**Plasma Nicotine Concentrations at Time of
Tissue Collections for PGI-2 Production**

Rabbit	Weight Kg	Infusate ng/ml	Plasma ng/ml
101	3.50	142.6	20.3
103	3.06	126.5	15.3
104	3.66	131.1	29.0
105	3.58	165.4	43.8
107	3.10	121.8	47.2
108	3.42	171.3	48.5
109	3.57	211.0	50.4
111	3.00	148.5	51.3
112	2.94	139.8	40.1

The results of the biological assay for PGI_2-like activity are shown in Table III.

Table III

**Biological Assay of PGI-2 Like Activity
Percent Inhibition of Control Aggregation**

Rabbit No. Control	L Vent	R Vent	I.V.S.	L. Atrium
33	35	51	58	100
34	100	100	40	100
35	41	100	40	100
36	57	71	80	100
37	67	86	68	100
Mean	60	82	57	100
S.D.	25	20	17	–
Nicotine				
101	45	100	35	100
103	49	54	12	100
104	61	27	21	100
105	21	56	40	100
107	39	100	34	100
108	28	50	34	100
109	54	100	100	100
110	33	100	100	100
111	31	100	26	100
112	42	100	21	100

Table III (continued)

Mean	40	79	42	100
S.D.	12	29	31	–
Carbon Monoxide				
201	55	97	53	100
202	34	92	16	100
205	28	75	26	100
206	42	95	39	100
Mean	40	90	34	100
S.D.	12	10	10	–

In all samples tested, inhibition of ADP induced platelet aggregation was detected. The inhibition ranged from 1 to 100% of the control. These data were used qualitatively to show that the heart was producing biologically active substances.

The results of the radioimmunoassays for PGI_2 production expressed as pg of 6-keto-PGF_{1a}/mg of tissue/min are presented in Table IV.

Table IV

PGI_2 Production by the Rabbit Heart
(pg 6-Keto PGF-1-alpha/mg/min)
Means and Standard Deviation

Tissue incubated without arachidonic acid

		L Vent	R Vent	IVS	Atrium
Normal	(11)	3.9±1.7	7.1±3.3	3.9±1.4	18.7± 5.8
Nicotine	(11)	2.6±1.3	5.4±2.6	5.7±5.2	16.3± 3.4
CO	(10)	6.9±4.4	6.9±1.9	5.4±3.3	36.6±12.8
Nic & CO	(06)	7.0±2.5	9.0±3.7	7.0±5.1	40.7±15.5

Tissue incubated with arachidonic acid

		L Vent	R Vent	IVS	Atrium
Normal	(11)	8.5±2.8	16.8±7.4	8.9±3.2	36.6± 15.8
Nicotine	(11)	5.0±1.9	8.5±3.2	5.2±3.0	13.3± 3.5
CO	(10)	15.1±7.7	26.0±14	15.1±9.7	48.8±20.7
Nic & CO	(06)	22.6± 15	43.9±18	16.4±3.9	124.7±16.2

Tables V and VI show the statistical treatment of these data, a p-value of less than 0.05 was accepted for rejecting the null hypothesis.

Table V

PGI-2 Production by the Heart
Tissue Incubated with Arachidonic Acid
Statistical Analysis of Differences Between Treatment Groups*

	Control v Nicotine		Control v CO		Control v Nicotine+CO	
	(11)	(11)	(11)	(10)	(11)	(6)
L Vent	U = 17.5		U = 24.0		U = 2	
	P <0.01		P = 0.05		P <0.001	
R Vent	U = 12.0		U = 30.5		U = 4	
	P <0.001		P = 0.05		P <0.001	

*Mann-Whitney U Test, corrected for ties.

Table V (continued)

	Control v Nicotine		Control v CO		Control v Nicotine+CO	
	(11)	(11)	(11)	(10)	(11)	(6)
IVS	U = 19.5		U = 36		U = 4	
	P <0.01		N/S		P <0.001	
Atrium	U = 21.0		U = 36		U = 0	
	P <0.01		N/S		P <0.001	

	Nicotine v CO		Nicotine V Nicotine+CO		CO v Nicotine+CO	
	(11)	(10)	(11)	(6)	(10)	(6)
L Vent	U = 6		U = 0		U = 16	
	P <0.001		P <0.001		P N/S	
R Vent	U = 6		U = 0		U = 9	
	P <0.001		P <0.001		P <0.025	
I.V.S.	U = 14		U = 0		U = 22	
	P < 0.01		P <0.001		N/s	
Atrium	U = 0		U = 0		U = 0	
	P <0.001		P <0.001		P <0.001	

*Mann-Whitney U Test, corrected for ties.
N/S = Not Significant

Table VI

PGI-2 Production by the Heart
Tissue Incubated with Arachidonic Acid
Statistical Analysis of Differences
Between Chambers within Treatment Groups

	Control (11)	Nicotine (11)	CO (10)	Nicotine & CO (6)
L Vent v R Vent	T = 1	T = 0	T = 0	T = 0
	P <0.001	P <0.001	P <0.001	P <0.001
L Vent v I.V.S.	T = 29	T = 29	T = 22	T = 5
	N/S	N/S	N/S	N/S
L Vent v Atrium	T = 0	T = 0	T = 0	T = 0
	P <0.001	P <0.001	P <0.001	P <0.001
R Vent v I.V.S.	T = 2	T = 3	T = 7	T = 0
	P <0.001	P <0.005	P <0.025	P <0.001
R Vent v Atrium	T = 0	T = 0	T = 0	T = 0
	P <0.001	P <0.001	P <0.01	P <0.001
I.V.S. v Atrium	T = 0	T = 0	T = 0	T = 0
	P <0.001	P <0.001	P <0.001	P <0.001

*Wilcoxon matched pairs rank test, corrected for continuity.
N/S = Not Significant.

All parts of the heart produced PGI_2 in the control animals. Both atria produced similar quantities of PGI_2, so for clarity only the left atrial valves are shown. The right ventricle produced significantly more PGI_2 than the left ventricle and interventricular septum, while the atria produced significantly more PGI_2 than the other walls of the heart. The left ventricle and interventricular septum behaved in a very similar fashion. Nicotine reduced PGI_2 production significantly in all tissues examined, with a pattern of production by the various segments similar to

the controls. Carbon monoxide exposure increased PGI_2 production in this
model, which was reached statistically different from control values in
the left and right ventricle. The difference for PGI_2 production between
the nicotine and carbon monoxide treated animals was statistically
significant for each part of the heart.

Exposure to CO and infusion of nicotine together caused a significant
increase in PGI_2 production from control in all chambers. There was a
significant difference between nicotine treated animals treated with
nicotine + CO. Although there was an increase in PGI_2 production in
animals treated with the combination compared to CO alone this reached
statistical significance only in the right ventricle and atrium.

Discussion

The discovery and description of PGI_2 gave considerable impetus to
study the interaction between the platelet and the vessel wall (8,9,10).
There is no storage site for prostaglandins and they must be synthesized
de novo from arachidonic acid, which is mobilized by lipase hydrolysis of
phospholipids. Cyclo oxygenase converts the fatty acid to unstable
intermediates PGG_2 and PGH_2, which, in the endocardial cells are
converted to PGI_2. Marcus et al. (17) have demonstrated that endothelial
cells also synthesize PGI_2 from platelet-derived endoperoxides. This may
be the more important pathway in normal individuals, because endothelial
cells lack the enzymes to synthesize arachidonic acid from an exogenous
source (18,19).

The cardiac synthesis of prostaglandins from arachidonic acid has
been studied in man after infusion of arachidonic acid into the aorta
root with simultaneous blood sampling from the coronary sinus (20).
6-keto-PGF_{1a}, was the main C-14 prostaglandin recovered and accounted for
approximately 23% of the total produced. Animal experiments have shown
that cardiac tissue converts arachidonic acid into PGE_2, PGF_2, as well as
PGI_2. It has also been shown that ischaemia and hypoxia are potent
stimuli for the production of PGI_2 by the heart (21).

Cardiac PGI_2 behaves like prostaglandins elsewhere, it is a local
hormone (22) and most of the PGI_2 is produced by the endothelial cells of
the coronary vasculature (20).

Gerritsen and Printz (23) demonstrated that the major source of PGI_2
was the larger coronary arteries and that very little is contributed from
the resistance elements of the myocardial circulation.

Nicotine has been shown to inhibit synthesis and release of PGI_2 from
vascular tissue and heart muscle in vitro (11). The current state of
knowledge of the interaction of nicotine and the prostaglandins in the
cardiovascular system was reviewed by Wennmalm in 1982 (25).

The data reported here are the result of experiments attempting to
analyze the effect of carbon monoxide and nicotine on PGI_2 production by
the heart. We used full thickness cardiac wall samples to produce the
PGI_2 and presented the tissue with exogenous arachidonic acid in an
oxygenated buffer system. This avoided the extraction step that would
have been required had the experiment been undertaken in plasma, and
ensured that sufficient arachidonic acid was available to cells (18).

The heart and as much of the thoracic aorta as could be obtained were
used to assess the distribution of nicotine in those tissues. The dosage
of nicotine was chosen to approximate the blood nicotine concentrations
seen in cigarette smokers using more than 20 cigarettes per day. All 4

animals concentrated nicotine in the tissues. In the heart a
concentration of twice, and in the thoracic aorta 2 1/2 times, the blood
concentration was obtained. These data represent the tissue
concentrations in the heart during the experiments to assess PGI_2
production.

Nicotine has been found in the ganglia, neuromuscular junctions, and
in the adrenal medulla, and nicotine receptors have been demonstrated in
these tissues (26). As well, nicotine is metabolized by microsomal
enzymes and has been shown to bind to cytochrome P450 in hepatic cells.
Thus the nicotine in the tissue in these studies may be bound in the
plasma membranes, in the microsomes, or in various enzyme systems.

Birnstingl (27) exposed rabbits to CO at a concentration of 400ppm
and examined the progression of atherosclerosis. There was marked
acceleration of atherosclerosis within 20 weeks after multiple
exposures. These workers and others (28,29,30) have examined the effects
of CO on vessel wall in rabbits and postulate a direct toxic effect on
the endothelial cells. High carboxyhaemoglobin values also have been
associated with a higher prevalence of atherosclerosis in human smokers
(33). In light of the published data it seemed reasonable to use the
dose that had been measured in cigarette smoke (400 ppm) and expose the
animals to that dose for the duration of the experiment. All animals
tested fell within the preset limits of 12–30% although the majority of
animals had carboxyhaemoglobin concentrations of about 20%. These were
equivalent to the carboxyhaemoglobin seen in patients who were heavy
smokers and who had extensive cardiovascular disease (34). This minimum
level of 12% carboxyhaemoglobin was based on the observations of Ayres et
al. (31) that there was a shift from aerobic to anaerobic metabolism in
the myocardium at these values and that at carboxyhaemoglobin values of
10–20% metabolic activities are markedly altered in organs such as the
liver (32).

The various parts of the heart produced PGI_2 in similar proportions
in the control and treated animals. Much more PGI_2 was produced on a
weight basis by the atria and right ventricle than by the left ventricle
or interventricular septum. This adds further support to the concept
that PGI_2 is produced by the endocardial cells as the production was
inversely proportional to the amount of muscle tissue in each specimen.
It also points out that the left ventricle and interventricular septum
may be more vulnerable to the effects of acute ischaemia because they
produce less PGI_2.

The reasons for the differences in production of PGI_2 between
treatment groups is not known. The effects of carbon monoxide may be
direct toxic action on the endocardial cells or, much more likely, may be
due to an intracellular hypoxia causing the cell to produce increased
quantities of PGI_2. CO reduces tissue oxygen consumption by decreasing
O_2 carrying capacity and by shifting the Hb-O_2 dissociation curve to the
left (35). However, in this study the mechanism of increased PGI_2
production <u>must be cellular</u> as the incubation was carried out in a buffer
solution without the presence of haemoglobin. Acute tissue hypoxia was
prevented by performing all manipulations in oxygenated buffer systems,
and we did not see any increased PGI_2 production in the control
specimens. Thus, I believe that the observed phenomena are real and are
due to the carbon monoxide exposure. We have previously shown that
endothelial cells increase production of PGI_2 in areas of arterial
stenoses using similar methodology (36).

The data support the observations (24,25,36,37) that nicotine reduces
PGI_2 production by inhibiting metabolic pathways in the endocardial

cells. Because exogenous arachidonic acid was supplied to the cells, in
some experiments, the site of inhibition in those is in the synthetic
pathway from arachidonic acid to PGI_2. As suggested by Wenmalm (25) the
inhibition may be at the cyclo-oxygenase step in the pathway. If his
initial observations that prostacyclin synthetase activity is not blocked
by nicotine are confirmed, this would be further evidence for this site
of action.

The production of PGI_2 by the heart in an incubate without
arachidonic acid is much less than when exogenous arachidonic acid is
supplied. These data can be used to interpret changes in the cleavage
step by the enzyme phospholipase, if it is assumed that all the
arachidonic acid presented to the cyclo-Ooxygenase system under these
conditions is metabolized. When the tissue was incubated with
arachidonic acid, an excess of the PUFA was used and the production of
PGI_2 under these conditions, should be a measure of the capacity of the
cyclo-oxygenase system.

The large increase in PGI_2 production seen with CO and nicotine + CO,
and the decrease seen with nicotine alone, were due mainly to changes in
the cyclo-oxygenase pathways. In order to answer the question as to
whether cyclo-oxygenase or PGI_2 synthetase are affected, a series of
experiments are planned using similar pretreatment groups and presenting
the heart with endoperoxides as the substrate.

The hearts from animals treated with nicotine and exposed to CO
demonstrate further increased production of PGI_2 when challenged with
arachidonic acid. The biologic significance of increased production of
PGI_2 after CO exposure is unknown, as is the time course of the effect.
It is possible that the 7-10 day exposure may be the zenith of the effect
of CO and that protracted exposures of weeks to months will result in
substrate depletion, enzyme failure, or some other change which will
cause a fall in PGI_2 production.

A possible alternative explanation for the observations is that the
events are related to changes at the surface of the cell. That is,
nicotine and CO are affecting the permeability of the endocardial cell to
arachidonic acid and the increase and decrease seen in the end production
of the microsomal enzyme system are related to availability of
substrate. In other systems, CO has been shown to change permeability
(38,39). However, there is no direct evidence for or against this
hypothesis.

The experiments are relevant to the planning of future studies, for
they have shown that nicotine and CO administered alone and in
combination, in vivo, exert significant effects on PGI_2 production. The
left ventricle and interventricular septum behave identically and
conclusions drawn from one are applicable to the other. The rabbit heart
has a separate septal artery and the isolated perfused septum (40) has
been used extensively to study cardiac physiology. We are currently
continuing this series of experiments using an isolated perfused rabbit
interventricular septal preparation to confirm our initial findings and
to investigate the pathways involved in this more physiological model.

Acknowledgement

I thank Miss Ann Selfe for assistance in the preparation of the
manuscript.

References

1. Knapp PA, Bliss CM, Wells H. Addictive aspects of heavy cigarette smoking. Am J Psychiatry, 119:966-972 (1973).
2. Larson PS, Silvette H. Tobacco-Experimental and clinical studies. A comprehensive account of the world literature. Supplement III. Baltimore, MD, The Williams and Wilkins Company (1975).
3. Klein LW. Cigarette smoking, atherosclerosis and the coronary hemodynamic response: a unifying hypothesis. J Am Coll Cardiol, 4:972-074 (1984).
4. Kannel WB, Schatzkin A. Risk factor analysis. Prog Cardiovasc Dis, 26:309-332 (1983).
5. Lader M. Nicotine and smoking behavior. Br J Clin Pharmacol, 5:289-292 (1978).
6. Armitage AK, Dollery Ct, George CF, Houseman TH, Lewis PG, Turner DM. Absorption and metabolism of nicotine from cigarettes. Br Med J, 4:313 (1975).
7. Rosenberg J, Benowitz NL, Jacob P, Wilson KM. Disposition kinetics and effects of intravenous nicotine. Clin Pharmacol Ther, 28:517-522 (1980).
8. Bunting S, Grygleski R, Moncada S, Vane JR. Arterial walls generate from prostaglandin endoperoxides a substance (prostaglandin C) which relaxes strips of mesenteric and coeliac arteries and inhibits platelet aggregation. Prostaglandins, 12:897-913 (1976).
9. Moncada S. Biological importance of prostacyclin. Br J Pharmacol, 76:13-31 (1982).
10. Moncada S. Prostacyclin and arterial wall biology. Arteriosclerosis 2:193-207 (1982).
11. Sonnenfeld T, Wennmalm A. Inhibition by nicotine of the formation of prostacyclin-like activity in rabbit and human vascular tissue. Br J Pharmacol, 7:609-613 (1980).
12. Sanford TD, Colby ED. Effect of zylazine and ketamine on blood pressure, heart rate and respiratory rate in rabbits. Laboratory Science, 30:519-523 (1980).
13. Tietz NW, Fiereck EA. The spectrophotometric measurement of carboxyhemoglobin. Ann Clin Lab Sci, 3:36-42 (1973).
14. Born GVR. Aggregation of blood platelets by adenosine diphosphate and its reversal. Nature, 194:927-929 (1962).
15. Ellis EF, Wright KF, Jones PS, Richardson DW, Ellis CK. Effect of oral aspirin dose on platelet aggregation and vascular prostacyclin (PGI$_2$) synthesis in humans and rabbits. J Cardiovasc Pharmacol, 2:387-397 (1980).
16. Jacob P, Wilson M, Benowitz NL. Improved gas chromatographic method for the determination of nicotine and cotinine in biological fluids. Am J Chromatogr, 222:61-70 (1981).
17. Marcus AJ, Weksler BB, Jaffe EA, Broekman MJ. Synthesis of prostacyclin from platelet derived endoperoxides by cultured human endothelial cells. J Clin Invest, 66:979-986 (1980).
18. Spector AA, Kaduce TZ, Hoak JC, Fry GL. Utilization of arachidonic and linoleic acids by cultured human endothelial cells. J Clin Invest, 68:1103-1011 (1981).
19. Fleischer LN, Tall AR, Witte LD, Miller RW, Cannon PJ. Stimulation of arterial endothelial cell prostacyclin synthesis by high density lipoproteins. J Biol Chem, 57:6653-6655 (1982).
20. Nowak J Kaijser L, Wennmalm A. Cardiac Synthesis of prostaglandins from arachidonic acid in man. Prostaglandins Med, 4:205-214 (1980).
21. Wennmalm A. Nicotine inhibits hypoxia and arachidate induced release of prostacyclin like activity in rabbit hearts. Br J Pharmacol, 69:545-549 (1980).

22. Blair IA, Barrow SE, Wadell KA, Lewis PJ, Dollery CT. L Prostacyclin is not a circulating hormone in man. Prostacyclin, 23:579-589 (1982).

23. Gerritsen ME, Printz MP. Sites of prostaglandin synthesis in the bovine heart and isolated bovine coronary micro vessels. Circ Res, 49:1152-1163 (1981).

24. Wennmalm A. Effects of nicotine on cardiac prostaglandin and platelet thromboxane synthesis. Br J Pharmacol, 64:559-563 (1978).

25. Wennmalm A. Interaction of nicotine and prostaglandins in the cardiovascular system. Prostaglandins, 23:139-144 (1982).

26. Jaffe JH. Drug addiction and drug abuse. The Pharmacological Basis of Therapeutics, Gilman AG, Goodman LS and Gilman A Eds. New York, McMillan Publishing Co., 535-584 (1980).

27. Birnstingle MA, Hawkins LA, McEwan T. Experimental atherosclerosis during chronic exposure to carbon monoxide. Eur Surg Res, 2:92-96 (1970).

28. Birnstingle MA, Brinson K, Chakrabarti BK. The effects of short term exposure to carbon monoxide on platelet stickiness. Br J. Surg, 58:837-839 (1971).

29. Bellet S, DeGuzman NT, Kostis JB, Roman L, Fleischmann D. The effect of inhalation of cigarette smoke on ventricular fibrillation threshold in normal dogs and dogs with acute myocardial infarction. Am Heart J, 83:67-76 (1972).

30. Goldsmith JR, Aronow WS. Carbon monoxide and coronary heart disease: a review. Environ Res, 0:236-248 (1975).

31. Ayres SM, Gianelli S, Mueller H. Effects of low concentrations of carbon monoxide Part IV: Myocardial and systemic responses to carboxyhaemoglobin. An NY Acad Sci, 174:268-293 (1970).

32. Topping DL. Acute effects of carbon monoxide on the metabolism of perfused rat liver. Biochem J, 152:425-427 (1975).

33. Wals M, Howard S, Smith PG, Kjeldsen K. Association between atherosclerotic disease and carboxyhaemoglobin levels in tobacco smokers. Br Med J, 1:761-765 (1973).

34. Hawkins LM, Cole PV, Harris JRW. Smoking habits and blood carbon monoxide levels. Environ Res, 11:310 (1976).

35. Roughton FJW, Darling RC. The effect of carbon monoxide on the oxyhaemoglobin dissociation curve. Am J Physiol, 141:295-307 (1944).

36. Qvardfordt PG, Reilly LM, Lusby RJ, Effeney DJ, Ferrell LD, Price DC, Fuller J, Ehrenfeld SK. Prostacyclin production in regions of arterial stenosis. Surgery, 98:484-491 (1985).

37. Sotel I, Van Der Giessen WJ, Zwolsman E, Quadt FJ, ten Hoor F, Verheugt FW, Hugenholtz PG. Direct effect of nicotine on prostacyclin production in human umbilical arteries. Acta therapeutica, 6:32-38 (1980).

38. Siggaard-Anderson J, Kjedlsen K, Bond E, Peterson F, Astrup P. A possible connection between carbon monoxide exposure, capillary filtration rate and atherosclerosis. Acta Med Scand, 182:397-399 (1967).

39. Parving HH. The effect of hypoxia and carbon monoxide on plasma volume and capillary permeability to albumin. Scand J Clin Lab Invest, 30:49-55 (1972).

40. Langer GA, Brady AJ. The effect of temperature upon contraction and ionic exchange in rabbit ventricular myocardium. J Gen Physiol, 52:682-713 (1968).

RELATIONS BETWEEN SMOKING, FOOD INTAKE AND PLASMA LIPOPROTEINS

Anders G. Olsson and Jörgen Mölgaard

Department of Medicine, Faculty of Health Sciences
Linköping University Hospital
S-581 85 Linköping, Sweden

INTRODUCTION

Smoking is regarded as one of the major risk factors for the
development of atherosclerosis together with hyperlipoproteinaemia and
hypertension. It has independent power as a risk factor.

There is strong support for the conclusion that smoking cessation
conveys marked reduction in coronary heart disease (1). Smoking is a
causative factor for atherosclerosis in its own right but the atherogenic
effect is by all probability multi-factorial as the smoke contains a
large number of potentially toxic substances like carbon monoxide,
nicotine and tar products.

However, smoking is not an isolated risk factor for the development
of atherosclerosis. Many interactions exist and the smoking habit
probably exerts its deleterious effect on the vessel wall partly through
its influence on other risk factors. Well-known correlations exist, for
example, between smoking and plasma lipoprotein concentrations, blood
pressure, platelet aggregation, eicosanoid metabolism, etc. Recent
studies also indicate a relation between smoking and nutrition.

The purpose of the present review is to point out some recently
recognized relations between the smoking habit and food intake. This
will be illustrated by the relation between smoking and plasma
lipoprotein concentrations and fatty acid composition of various lipids.

SMOKING AND PLASMA LIPOPROTEIN CONCENTRATIONS

Many studies have demonstrated that the plasma lipoprotein
concentrations in smokers differ from those in non-smokers. In general,
apolipoprotein B-containing lipoproteins (very low density, VLDL,
intermediate density lipoprotein, IDL, and low density (LDL)
lipoproteins) tend to be increased in smokers (2). In patients with
different types of hyperlipoproteinaemia, the smoking habit in man was
significantly correlated with increased concentrations of
triglyceride-rich lipoproteins (3). High density lipoprotein (HDL) is
generally found to be lower in smokers than in non-smokers (4). Some
studies show that the plasma HDL fraction in particular seems to be

Tobacco Smoking and Atherosclerosis
Edited by J. N. Diana
Plenum Press, New York, 1990

affected by smoking; one study, for example, showed a mean HDL cholesterol
level among smokers of only 0.82 mmol/l (5). Thus the plasma lipoprotein
pattern is more atherogenic in smokers. This might be one mechanism by
which smoking exerts its atherogenic effect.

The mechanism behind these alterations of plasma lipoprotein patterns
in smokers is not fully understood. One possible explanation could be a
decreased activity of lipoprotein lipase as a toxic effect induced by
smoking. Indeed, it has been demonstrated that post-heparin plasma
lipoprotein lipase activity increases after smoking cessation (6).

One argument supporting a direct toxic effect exerted by some
component in the blood, derived from the smoke, is the rapid increase in
HDL cholesterol following cessation of smoking. Two weeks after
cessation of smoking, HDL cholesterol increases by about 30 percent (6).
This is in striking contrast to the more sluggish increase in plasma HDL
levels seen when plasma lipid lowering treatment is started in type IV
hyperlipoproteinaemia, a condition also characterized by low HDL
cholesterol levels (7). Regardless of whether treatment was clofibrate,
nicotinic acid, or a combination of clofibrate and pyridylcarbinol, very
tiny increases in HDL cholesterol were seen during the first nine days of
treatment. After 23 days of treatment, however, the HDL cholesterol had
increased to the same magnitude as that observed after smoking
cessation. One explanation for the more rapid plasma HDL increase in
smoking cessation may be the elimination of a direct toxic effect.

However, a considerable part of the explanation seems to involve a
change in dietary intake by those who quit smoking. Figure 1 shows the
relation between change in fat consumption and plasma HDL cholesterol as
studied by Stubbe et al. (6). The higher the increase in fat consumption
upon smoking cessation, the greater the increase in plasma HDL
cholesterol. The authors concluded that the higher fat intake, resulting
in more chylomicronaemia, led to a compensatory increase in lipoprotein
lipase activity and secondarily, as more triglyceride-rich lipoproteins
were processed through the lipolytic system, to an increase in HDL
cholesterol. These authors also demonstrated that particularly the HDL
subfraction HDL_2 increased. As this HDL subfraction is an end product of
lipolytic activity, this finding supports the concept of increased
lipolytic activity as being at least partly responsible for the HDL
increase (6).

It is also clear from the regression of the line in Figure 1,
however, that in those cases where individuals did not increase their fat
consumption, there was an increase in HDL cholesterol concentration of
about 0.1 mmol/l. This increase may be interpreted as the disappearance
of a toxic effect of smoking on HDL, be it through lipoprotein lipase or
through some other mechanisms.

These authors continued their work on the importance of diet change
as an explanation of the HDL increase after smoking cessation (8). The
effects of smoking cessation as such were dissociated from those of
dietary alterations by monitoring plasma lipoprotein concentration after
cessation of smoking in 12 subjects in whom diet was kept constant during
an initial 2-week period and during 2 weeks following smoking cessation.
Under these conditions plasma HDL cholesterol did not increase
significantly. No significant alterations were recorded regarding other
lipoprotein concentrations or lipoprotein lipase activity. These results
suggest that the marked rise in HDL cholesterol concentrations after
smoking cessation is largely related to spontaneous changes in dietary
habits which occur upon cessation of smoking.

<u>Smoking and Fatty Acid Composition of Diet and Various Lipids</u>

It is well known that there is a positive relation between intake of saturated fatty acids and coronary heart disease (CHD) (9). Polyunsaturated fatty acids, on the other hand, have generally been regarded as beneficial in this context as they decrease atherogenic plasma lipoproteins. In a study which we made some years ago (10) we compared adipose tissue fatty acid composition in two male populations, Edinburgh and Stockholm, with Edinburgh having about 3 times the incidence of CHD at the particular age of 40 years. The adipose tissue content of linoleic acid (18:2) was very significantly lower in Edinburgh men. Also, Edinburgh men smoked considerably more cigarettes.

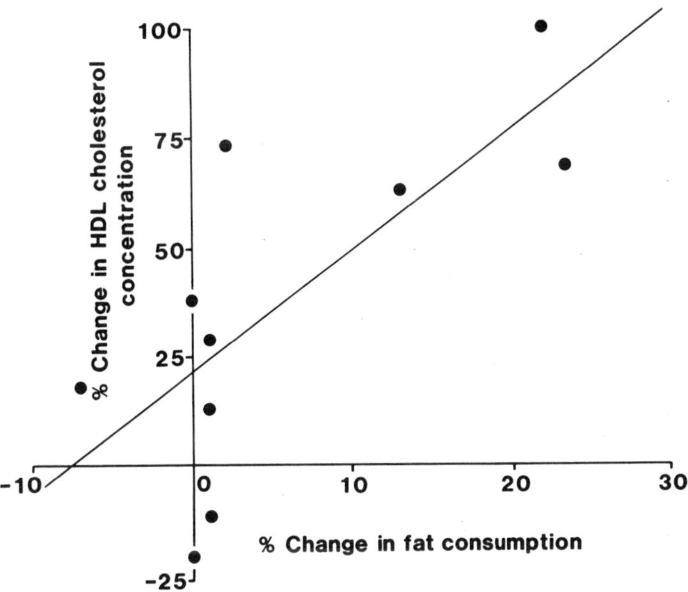

Fig. 1. Relation between changes in dietary fat intake and HDL cholesterol concentrations after cessation of smoking.

These observations were carried further in the Edinburgh group. The relation between fatty acid composition of adipose tissue and estimated risk for CHD was examined in a case control study of angina pectoris and acute myocardial infarction (11). Adipose tissue fatty acid composition reflects long-term dietary habits. It was found that an independent inverse relation exists between adipose tissue linoleic acid and the estimated relative risk of angina pectoris. A similar negative correlation was found for acute myocardial infarction, but this relationship was confounded by the smoking habit. The relation between the smoking habit and adipose tissue linoleic acid in controls is shown in Table 1. Adipose tissue linoleic acid was significantly lower in smokers when compared to non-smokers. The more cigarettes that were smoked the lower the percentage of adipose tissue linoleate.

Table 1. Smoking habit and mean (SEM) percentage adipose tissue
linoleic acid in controls. From ref. 11.

	Never smoked	Current cigarette smokers		
	n=13.7	1-10/day n=18	11-20/day n=44	$\geq$20/day n=50
Adipose tissue 18:2%	10.2(0.22)	11.9(1.1)	8.4(0.3)	8.5(0.3)

Dietary investigations using the technique of a 7-day prospective
weighed record showed that current cigarette smokers had a significantly
lower intake of linoleic acid than non-smokers and it corresponded with
the level of linoleic acid in adipose tissue of the same men. The low
intake by cigarette smokers was striking and significant whether
expressed as grams, percent of total fatty acids or percent of energy
(Table 2). Thus the low adipose tissue linoleic acid content most likely
reflects long-term dietary habits (11, 12). The intake of monounsaturated fatty acids was higher in smokers if expressed as percentage
of calories contributed (Table 2). However, oleic acid intake did not
differ between the groups when expressed as grams of fatty acid ingested
(12). Fiber intake was highest in the non-smoking group.

Other reported differences in food consumption in smokers vs
non-smokers are a lower fruit intake among smokers (12) and they are also
reported to have low blood ascorbic acid levels (13).

Why is this? One obvious explanation is that non-smokers are more
health conscious and have adopted healthier food habits. Surveys
indicate that smokers pursue a more unhealthy life style than non-smokers
(14).

In Naples a quite different pattern of fatty acid composition of food
and blood lipids in smokers and non-smokers is seen. Oleic acid was
lower in smokers compared to non-smokers in the plasma lipid
triglycerides, phospholipids and cholesteryl esters, all of which had
borderline significance, p < 0.1 (15).

The percentage contribution of different nutrients in the Naples men
is given in Table 3. While no differences are seen in the intakes of
saturated and polyunsaturated fatty acids between non-smokers and smokers
in Naples, non-smokers derive significantly more of their energy from
monounsaturated fatty acids than smokers. The Naples non-smokers and
smokers did not differ in the same way with respect to their food
preferences as did the Scottish men illustrated above. We therefore
think it is more probable that the differences seen between smokers and
non-smokers are more a result of cultural influences than of
physiological differences between the groups.

Another explanation would be that the smoking habit per se has an
influence on the choice of food, for example, through an effect on
taste. Fulton et al. found that smokers added more salt on their food
which may indicate an effect of smoking on taste (12). This, in turn,
might influence food preference.

Table 2. Dietary fat intake by smoking habit in Scottish men aged 45-54 years.

Mean daily intake	Never smoked (n=33)	Current cigarette smokers			P*
		$\leq$ 10 (n=11)	11-20 (n=35)	$\geq$21 (n=32)	
kcal	2814	2710	2657	2734	NS
Fatty acid (%)					
total kcal	39.1	36.1	37.8	37.8	NS
Dietary fatty acids					
(as % of all fatty					
acids) Saturated	46.9	46.3	47.2	48.5	NS
Monounsaturated	39.2	41.6	40.3	40.3	NS
Polyunsaturated	13.9	12.1	12.5	11.2	<0.01
Individual polyun-					
saturated fatty acids					
Linoleic acid	11.0	8.6	9.0	8.1	<0.001
Eicosapentaenoic acid	0.053	0.042	0.058	0.063	NS

*Wilcoxon rank sum: current cigarette smokers v those who never smoked, ex-smokers and pipe-cigar-smokers.

CONCLUSION

Obviously, a close and intricate relationship exists between smoking and dietary habits as exemplified in the short term by their effect on plasma lipoprotein concentrations at smoking cessation and in the long term by their effect on fatty acid composition of food and body storage sites. The low HDL concentrations among smokers and the low linoleic acid content in the smoker's adipose tissue are of potential harm in the process of atherogenesis. More research is needed to quantify more precisely the contribution of these effects on the development of CHD and other atherosclerotic diseases.

Table 3. Dietary intake in non-smoking and smoking Naples men from Rubba, unpublished.

	Non-smokers n=11	P	Smokers n-42
Total calories, kcal	2.085±234	n.s.	2.188±94
Protein, % energy	16.2±0.6	n.s.	15.0+0.5
Carbohydrates, % energy	49.8±1.7	n.s.	51.4±1.6
Fat, % energy	31.1±0.9	n.s.	28.4±1.6
Saturated	9.8±0.9	n.s.	8.9±0.6
Monounsaturated	18.0±0.9	<0.02	15.2±0.8
Polyunsaturated	3.4±0.9	n.s.	4.3±0.6

A primary goal should of course be to stop smoking or to decrease the
consumption of cigarettes. For those who cannot stop, it may be
advisable to take measures to increase HDL cholesterol and the intake of
polyunsaturated fatty acids.

From this short review it should be clear that, in addition to the
risk of tobacco smoking in the development of CHD in its own right, this
habit has harmful influences on dietary habits and lipoprotein
metabolism. These are only two areas in which smoking could have
impact. Other risks which the habit could influence are the adrenergic
system leading to elevated heart rate and blood pressure and increase in
plasma free fatty acids and in myocardial oxygen consumption. Smoking
increases fibrinogen and platelet adhesiveness and favours endothelial
injury.

The role of smoking in the pathogenesis of atherosclerosis is thus
extremely complex. Much more research is needed to fully understand
where and how smoking influences the clinical science and biology of
atherosclerosis.

REFERENCES

1. Yusuf, S. and Furberg, C.O. Single Factor Trials: Control
    through life style changes. In: Atherosclerosis, Biology and
    Clinical Science. Editor: Anders G. Olson. Churchill
    Livingstone, p.390-1 (1987).
2. Billimoria, J.S., Pozner, H., Metselaar, B., Best, F.W. and James,
    D.L.O. Effects of cigarette smoking on lipids, lipoproteins,
    blood coagulation, fibrinolysis and cellular components of human
    blood. Atherosclerosis, 21:61-76.
3. Olsson, A.G. Studies in asymptomatic primary hyperlipidaemia,
    Acta Med. Scand. Vol. 197, pp. 477-85 (1975).
4. Criqui, M.H., Wallace, R.B., Heiss, G., Mishkel, M., Schonfeld, G.
    and Jones, G.T.L. Cigarette smoking and plasma high density
    lipoprotein cholesterol, The Lipid Research Clinics Program
    Prevalence Study, Circulation 62, suppl IV, 70-76 (1980).
5. Stubbe, I., Eskilsson, J. and Nilsson-Ehle, P. High-density
    lipoprotein concentrations increase after stopping smoking,
    British Medical Journal, 284:1511-13 (1982).
6. Stubbe, I. and Nilsson-Ehle, P. Alterations in HDL subfractions and
    composition after smoking cessation. Relationships to lipoprotein
    lipase, hepatic lipase and LCAT activities. In Thesis: Coronary
    Risk Factor Modification by Ingo Stubbe, Studentilitteratur, Lund,
    pp. 155-84 (1982).
7. Ballantyne, D., Olsson, A.G. and Carlson, L.A. Acute effects of drug
    therapy in type IV hyperlipoproteinaemia, artery 4(2), 107-28
    (1978).
8. Quensel, M., Söderström, A., Agardh, C.D. and Nilsson-Ehle, P. High
    density lipoprotein concentrations after cessation of smoking:
    the importance of alterations in diet, Atherosclerosis, 75:189-93
    (1989).
9. Keys, A., Seven countries: a multivariate analysis of death and
    coronary heart disease. Cambridge, Massachusetts: Harvard Press
    (1980).
10. Logan, R.L., Thomson, M., Riemersma, R.A. Risk factors for ischaemic
    heart disease in normal men aged 40, Lancet I, 949-55 (1978).
11. Wood, D.A., Riemersma, R.A., Butler, S., Thomson, M., Macintyre, C.,
    Elton, R.A. and Oliver, M.F. Linoleic and eicosapentaenoic acids
    in adipose tissue and platelets and risk of coronary heart
    disease, The Lancet, 1:177-82.

12. Fulton, M., Thomson, M., Elton, R.A., Brown, S., Wood, D.A. and
    Oliver, M.F. Cigarette smoking, social class and nutrient
    intake: Relevance to coronary heart disease, European Journal of
    Clinical Nutrition, 42:797–803 (1988).
13. Riemersma, R. Unpublished.
14. Health and Lifestyle Survey, Preliminary report, p. 98, London:
    Health Promotion Research Trust (1987).
15. Rubba, P. Personal communication.

LIPOPROTEINS AND DIET IN THE

PATHOGENESIS OF ATHEROSCLEROSIS

James W. Anderson and
Tammy L. Floore

Metabolic Research Group
V.A. Medical Center
University of Kentucky
Lexington, Kentucky

INTRODUCTION

Atherosclerotic cardiovascular disease is a major cause of mortality
and morbidity in Western countries, leading to 759,000 deaths annually in
the United States[1]. The pathogenesis of atherosclerosis is complex,
multifactorial, and involves both environmental and genetic risk factors.
Steinberg[2] schematically illustrated the proposed interplay of blood factors
with arterial wall responses (Figure 1). Since many of these factors are
discussed elsewhere,[3] this review will focus on the central role of
lipoproteins in atherosclerosis. Specifically, this review will examine
the evidence linking dietary cholesterol and fat intake to serum
lipoproteins and coronary heart disease (CHD), the strength of the
relationships between atherosclerosis and serum lipoproteins, clinical
trials focusing on lowering serum lipoproteins to reduce risk for CHD,
animal and human regression studies, the effects of dietary fiber on serum
lipoproteins and CHD, and practical implications.

DIETARY FAT AND CHOLESTEROL AND CHD

The link between dietary cholesterol and fat intake and risk for
cardiovascular disease, especially CHD, is established by studies in
animals,[4] including non-human primates,[5,6] and epidemiologic studies in
humans[7,8]. For decades, dietary cholesterol has been the essential agent
for producing experimental atherosclerosis. Feeding laboratory animals
diets high in saturated fat and cholesterol results in hypercholesterolemia;
lipids accumulate in arterial vessels following relatively short periods of
high cholesterol intake[4,6]. Rhesus monkeys develop atherosclerosis
following ingestion of atherogenic diets[9,10]. The large number of such
studies and the consistency of their results provide strong evidence for a
causal role of cholesterol in CHD[6].

Numerous human experimental studies[11] establish that alterations in
dietary cholesterol intake result in changes in serum cholesterol.
Increased dietary cholesterol raises serum cholesterol, primarily through an
increase in low density lipoprotein (LDL) cholesterol particles[12].

Tobacco Smoking and Atherosclerosis
Edited by J. N. Diana
Plenum Press, New York, 1990

Stamler and Shekelle[13] reviewed the independent role of dietary cholesterol in CHD, suggesting that a 200 mg/1000 kcal lower cholesterol intake is associated with a 45% reduction in risk of death from CHD. Figure 2 shows the strong epidemiologic correlation between dietary cholesterol intake and risk for CHD[14].

Saturated fat intake is linearly related to the incidence of CHD,[13,15,16] independent of effects on serum lipids. Figure 3 illustrates the strong correlation between saturated fat intake and CHD mortality[16]. Saturated fat correlates more closely with CHD prevalence than any other variable[16].

SERUM LIPOPROTEINS AND CORONARY HEART DISEASE

High serum cholesterol levels constitute a known risk factor for CHD,[17] and the relationship between rising serum cholesterol levels and mortality and morbidity from CHD is well-known[18]. The 1984 National Institutes of Health Consensus Conference on Lowering Blood Cholesterol to Prevent Heart Disease concluded that serum cholesterol levels of the U.S. population were too high and strongly contributed to CHD[19]. Other epidemiological studies have related serum cholesterol to CHD incidence[20]. The Framingham study[21] and other prospective studies demonstrated a curvilinear relationship between serum cholesterol and CHD incidence, noting that a continuously-graded relationship exists between serum cholesterol and risk of CHD[18,22]. Other evidence supporting the cholesterol/CHD hypothesis comes from epidemiologic studies between and within populations. Aravanis[23] compared cholesterol levels of Southern Italians with those of Northern Europeans and Americans. Average serum cholesterol levels were lowest in Southern Italy, with correspondingly lower CHD death rates. These studies have firmly linked elevated serum cholesterol to increased heart disease mortality.

LDL-cholesterol is a major risk factor for CHD, and the strong relationship between total cholesterol and CHD is often correlated with elevated LDL-cholesterol. Of total serum cholesterol, LDL-cholesterol transports 70%, high density lipoprotein (HDL) cholesterol transports 20%, and other lipoproteins transport the remainder[24]. LDL-cholesterol can be taken up by cells in the arterial wall, which may help to explain the positive correlation between elevations of serum LDL-cholesterol and CHD risk[25,26].

In contrast to the positive relationship of total and LDL-cholesterol to CHD, serum HDL-cholesterol inversely correlates with CHD[27]. In the Framingham study,[28] measurable CHD risk reduction occurred with increased HDL-cholesterol levels. HDL-cholesterol levels showed the strongest relationship to risk for CHD in subjects older than 49 years.

Serum triglyceride elevations may also increase risk for CHD[29]. The Framingham data suggest that triglycerides may be a useful predictor of CHD development in women[30]. However, other investigators have questioned the independent impact of triglycerides on CHD development, except in persons with diabetes[31].

While total cholesterol and LDL-cholesterol are major contributors to the pathology of CHD, chylomicron remnants, intermediate-density lipoproteins, and other lipoprotein components may also be atherogenic[32,33].

Traditionally serum lipids are measured following a 10-16 hour fast, but Western individuals are predominantly in a postprandial state for most of the day. Zilversmit[32] suggested lipoproteins formed in the postprandial state may be active in the development of atherosclerosis.

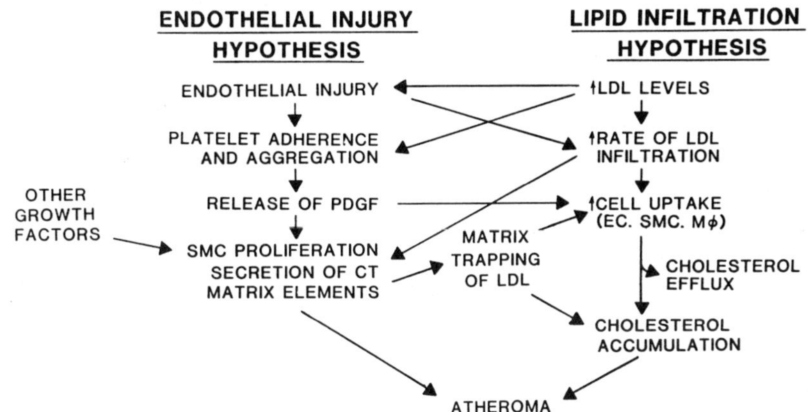

Figure 1. Proposed model of interaction of two theories of the pathogenesis of atherosclerosis. EC - endothelial cell, SMC - smooth
muscle cell, Mφ - macrophage, PDGF - platelet derived growth
factor, LDL - low density lipoprotein cholesterol. Redrawn
from D. Steinberg, _Arteriosclerosis_, 3:283, 1983.

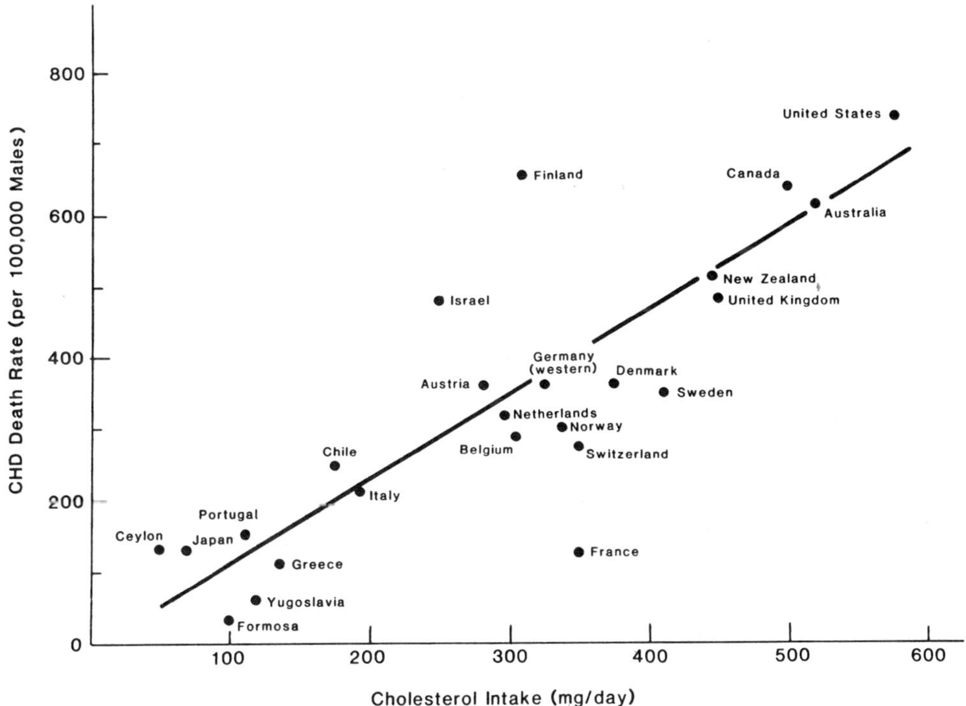

Figure 2. Correlation between dietary cholesterol intake and risk for
coronary heart disease. (Redrawn from W.E. Connor, et al.,
Prev. Med. 1:49:1972.)

Cohn and colleagues[34] reviewed studies of lipid, lipoprotein, and apolipoprotein responses to large amounts of fat. Other investigators[35,36,37] studied the postprandial lipid responses to varying fat loads, documenting the dynamic changes in postprandial lipid metabolism, including increases in triglyceride-rich lipoproteins and exchanges of cholesterol, cholesterol esters, and apolipoproteins between tissues and lipoproteins following fat ingestion. The mechanisms for the increased atherogenic potential of postprandial lipoproteins have not been described and require further study.

CLINICAL TRIALS - SERUM CHOLESTEROL AND CORONARY HEART DISEASE

Numerous clinical trials have evaluated the effects of reducing serum cholesterol levels on CHD morbidity and mortality. Whether the reduction is achieved by dietary or pharmacologic means, these studies indicate that reducing serum cholesterol reduces the incidence of CHD[38,39,40].

In a primary prevention trial conducted in Oslo, Norway, hypercholesterolemic male subjects reduced CHD risk with dietary changes to lower blood lipids and with reductions in cigarette smoking[38]. The reduction in the incidence of first myocardial infarction and sudden death correlated with reductions in serum total cholesterol and to a lesser extent, the reduction in smoking.

The Lipid Research Clinics Coronary Primary Prevention Trial[26,39] showed that lowering of serum total cholesterol and LDL-cholesterol by diet and cholestyramine significantly reduced incidence of CHD and death. This trial provided strong evidence for a causal role of total and LDL-cholesterol in CHD pathogenesis and showed that reducing cholesterol levels reduces risk of CHD in humans.

The Helsinki Heart Study[40] investigated the effect of gemfibrozil on CHD incidence. The gemfibrozil-treated group experienced marked increases in HDL-cholesterol and persistent reductions in levels of serum total and LDL-cholesterol, with an associated 34% reduction in the incidence of CHD.

The combined results of these and other trials[41,42] indicate that reduced CHD risk is related to the degree and duration of serum cholesterol lowering. Furthermore, numerous intervention studies indicate that lowering serum cholesterol levels reduces risk for atherosclerosis[26,39,40] and that regression of atherosclerotic lesions is possible through mobilization of arterial wall cholesterol, if a sizeable reduction in plasma cholesterol is achieved.

REGRESSION OF ATHEROSCLEROSIS

Experiments with hypercholesterolemic animals indicate that vigorously lowering serum cholesterol levels can delay progression or even cause regression[9]. Regression is studied experimentally in animals by inducing hypercholesterolemia with an atherosclerosis-inducing diet, examining a representative sample of animals through autopsy, then applying the regression treatment. After a treatment period ranging from weeks to years, the remaining animals are examined to determine the extent of regression.

Armstrong, et al.,[9] subjected rhesus monkeys to a high-fat, high-cholesterol diet for 17 months. Monkeys fed a cholesterol-free diet for an additional 40 months demonstrated 80% greater arterial lumen area than

Table 1. Serum lipid responses to different fibers. Values are expressed as median percentage change with number of studies in parentheses. (From Anderson, et al., <u>CRC Reviews</u>. In press.)

FIBER	TC	LDL	HDL	TG*
Guar	-11 (23)	-17 (10)	0 (14)	- 3 (18)
Pectin	-11 (16)	-	-2 (6)	+ 4 (10)
Psyllium	-14 (10)	-	-	-14 (4)
Oat bran	-16 (8)	-24 (6)	+2 (5)	- 8 (5)

*TC - total cholesterol; LDL - low-density lipoprotein cholesterol; HDL - high-density lipoprotein cholesterol; TG - triglyceride.

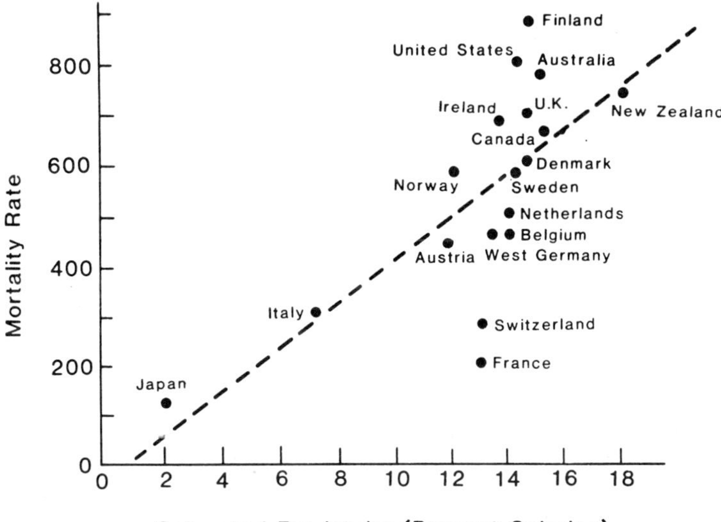

Figure 3. Correlation between saturated fat intake and coronary heart disease mortality. (Redrawn from D. Hegsted, et al., <u>J. Nutr.</u> 118:1184, 1988.)

monkeys with baseline atherosclerosis. These data suggest that regression
of atherosclerosis occurs in primates.

In humans, serial contrast angiography is generally used to evaluate
the extent of atherosclerosis[43,44]. Since direct visualization of human
lesions is not possible, investigators must extrapolate from animal studies
to interpret regression in humans. Limitations in quantitative and
qualitative assessment of lesion and lumen size in humans present the
greatest challenges in the evaluation of atherosclerosis regression. Given
these limitations, a growing body of research indicates that progression of
coronary atherosclerosis may be retarded when serum lipid levels are
significantly altered.

Both diet and drug regimens have been demonstrated to reduce serum
cholesterol and delay progression of CHD[43,45]. In a review of 10 studies,
Lavie and colleagues[46] identified the effects of lipid reduction on
progression and regression of CHD. The Cholesterol Lowering
Atherosclerosis Study[43] compared disease progression in non-smoking subjects
with CHD, some of whom had undergone coronary artery bypass graft surgery.
Subjects were randomly assigned to a group receiving Colestipol and niacin
along with a fat-restricted diet, or to a group receiving placebo medication
with a less restrictive diet. Angiography was performed at baseline and
again at 24 months to assess disease progression. In the drug-treated
group, total cholesterol decreased by 26%, LDL-cholesterol decreased by 43%,
and HDL-cholesterol increased by 37%. Disease progression was significantly
less after two years in the drug-treated group and regression of lesions was
found in 16% of these subjects compared to only 2% in the control group.

Brensike and colleagues[47] randomly assigned participants to a fat-
restricted diet with either cholestyramine or placebo. After five years of
follow-up, the drug-treated group demonstrated significantly less disease
progression than the placebo-treated group. Delayed progression was
associated with lower levels of LDL-cholesterol and higher levels of HDL-
cholesterol.

Another study of CHD progression was carried out by Arntzenius and
colleagues[8]. The Leiden intervention trial placed 39 subjects with CHD on a
strict vegetarian diet after baseline angiography. Repeat angiography after
two years showed that 18 of the 39 subjects experienced no coronary lesion
progression.

In summary, evidence suggests that regression of atherosclerotic
lesions occurs in animals and may occur in humans. Uncertainty remains
whether results of human trials with middle-aged hypercholesterolemic males
can be extrapolated to other groups.

DIETARY FIBER AND CORONARY HEART DISEASE

The positive effects of a fat- and cholesterol-restricted diet on serum
lipid levels are well known. However, dietary fiber intake is emerging as
the single dietary choice which may significantly affect all diet-related
risk factors for CHD, including hyperlipidemia, hypertension, diabetes, and
obesity.

For the past three decades, investigators have intensively examined the
cholesterol-lowering effects of dietary fibers in humans. Soluble fiber
sources such as pectin,[48] psyllium,[49,50] oats,[51] and guar[52] have been shown
to significantly lower serum cholesterol in humans. Table 1 summarizes
median percentage changes in serum lipids in previous studies evaluating
various fibers[3]. Diets high in complex carbohydrates and water-soluble

fibers reduce fasting serum cholesterol and triglyceride levels[53,54,55]. Other studies have reported the inverse relationship between dietary carbohydrate intake alone and CHD incidence[56,57,58]. Gordon, et al.[56] suggest that carbohydrate intake may be useful as a surrogate measure of dietary fiber.

In hypertensive individuals, a high-fiber diet has been shown to lower systolic blood pressure by 11% and diastolic blood pressure by 10%[59]. High-fiber intake lowers plasma glucose levels and insulin requirements in individuals with diabetes[60,61,62]. High-fiber intake also increases satiety, promoting weight loss or maintenance of normal body weight[63,64].

In addition to documented effects on CHD risk factors, dietary fiber may be independently protective against CHD[65,66,67]. Khaw and colleagues[68] reported that a 6 gram increase in daily dietary fiber intake was associated with a 25% lower mortality from CHD, independent of other dietary variables. The Baltimore Longitudinal Study[69] suggested that increasing fiber intake resulted in reduced CHD risk by reducing systolic and diastolic blood pressure, serum triglycerides and fasting plasma glucose. Other studies[66,67] have shown dietary fiber intakes to be inversely related to CHD mortality, independent of other risk factors[66,67,70]. Figure 4 illustrates the negative correlation between intake of dietary fiber and CHD incidence and mortality[65]. Collectively, these findings support the hypothesis that high dietary fiber intake may be independently protective of CHD mortality.

PRACTICAL IMPLICATIONS

Nutrition intervention is the treatment of choice for hyperlipidemia. The National Cholesterol Education Program (NCEP) Adult Treatment Panel[71] offers specific recommendations and guidelines for the detection and treatment of high-risk individuals. These recommendations are consistent with those of other groups[72,73].

Diet and exercise therapy should always be prescribed before and as an adjunct to drug treatment. The American Heart Association (AHA) diet program, which progressively decreases fat and cholesterol intake and increases carbohydrate intake, is often the diet of choice[74]. Moderate regular exercise helps prevent CHD,[75] and may exert protective effects by increasing HDL-cholesterol[76] and by altering other CHD risk factors such as blood pressure and body weight.

High-carbohydrate, high-fiber diets, originally developed for the treatment of diabetes mellitus, have also been used successfully in the treatment of hyperlipidemias[77]. The therapeutic high-fiber diet provides 70% of energy as carbohydrate, 18% as protein, 12% as fat, 50 mg cholesterol, and 60 g of dietary fiber daily. The high-fiber maintenance diet provides 55% of energy as carbohydrate, 25% as fat, 20% protein, 150 mg cholesterol, and 50 g of dietary fiber daily[78]. Individuals can use these diets at home, and they are suitable for the entire family. Table 2 contains fiber intake recommendations for the general public and for persons under supervision for metabolic conditions.

To meet the recommendations outlined above, individuals must usually decrease their intake of high-fat meats, dairy products, eggs, and refined sugars, while increasing intake of whole grains, legumes, fruits, and vegetables. Fruits, oats, barley, and beans are especially recommended for their soluble fiber content. Be Heart Smart[79] provides a detailed outline of these diet recommendations for the lay public.

Table 2.

FIBER RECOMMENDATIONS

	TOTAL FIBER	SOLUBLE FIBER
I. General Public (not under medical supervision)	20-35g/d 10-13g/1000 kcal	6-10.5g/d 3-4g/1000 kcal
Max	35g/d	

(Recommendations of many health organizations now form a consensus: the most effective diet for chronic disease prevention and treatment is liberal in complex carbohydrate and limited in fat. American Diabetes Association; Food and Nutrition Board; Committee on Diet, Nutrition and Cancer of the National Academy of Sciences; U.S. Dietary Guidelines for Americans; National Cancer Institute; American Cancer Society.)

	TOTAL FIBER	SOLUBLE FIBER
II. Metabolic Conditions (under medical super-vision; diabetes, obesity, hyperlipidemia)	35-50g/d 20-25g/1000 kcal	10.5-15g/d 6-7.5g/1000 kcal
Max	60g/d	18g/d

(consistent with American Diabetes Association recommendations)

*For all individuals with elevated cholesterol recommend an additional 5g/d soluble fiber from oats and beans.

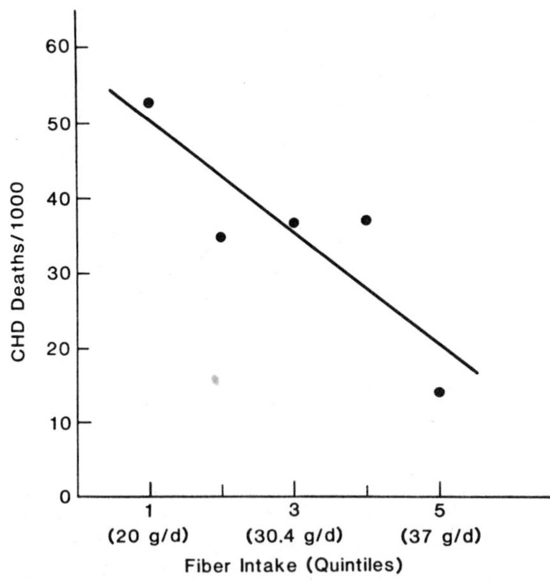

Figure 4. Illustration of inverse relationship between dietary fiber intake and coronary heart disease mortality. (Redrawn from D. Kromhout, et al., Lancet, 2:518, 1982.)

CONCLUSIONS

Atherosclerotic cardiovascular disease is a major cause of mortality and morbidity in the United States. Elevated serum cholesterol is firmly established as a risk factor for CHD. Dietary intake of saturated fat and cholesterol has been positively associated with CHD mortality, independent of effects on serum lipids. Numerous intervention trials have demonstrated significant reductions in CHD risk with cholesterol-lowering therapy. Aggressive nutrition therapy with resultant serum cholesterol reductions has been demonstrated to be effective in delaying progression of CHD and may lead to regression of atherosclerotic lesions.

Increasing dietary fiber intake is an important therapeutic measure for reducing CHD risk factors. Long-term use of high-fiber and low-fat diets can significantly decrease serum total and LDL-cholesterol. Nutrition management is the cornerstone of therapy for hyperlipidemias and also improves other CHD risk factors.

REFERENCES

1.   U.S. Department of Health and Human Services, "The Surgeon General's Report on Nutrition and Health," U.S. Government Printing Office, Washington (1988).

2.   D. Steinberg, Lipoproteins and atherosclerosis: a look back and a look ahead, Arteriosclerosis 3:283 (1983).

3.   J.W. Anderson, D.A. Deakins, T.L. Floore, B.M. Smith, and S.E. Whitis, Dietary fiber and coronary heart disease, CRC Reviews, in press.

4.   N. N. Anitschkow, A history of experimentation on arterial atherosclerosis in animals, in: "Cowdry's Arteriosclerosis: A Survey of the Problem," C.C. Thomas, Springfield (1967).

5.   M.G. Bond, B.C. Bullock, D.A. Bellinger, and T.E. Hamm, Myocardial infarction in a large colony on nonhuman primates with coronary artery atherosclerosis, Am. J. Pathol. 101:675 (1980).

6.   R.W. Wissler, Principles of the pathogenesis of atherosclerosis, in: "Heart Disease: A Textbook of Cardiovascular Medicine," E. Braunwald, ed., W.B. Saunders Company, Philadelphia (1984).

7.   A. Keys, "Seven Countries: A Multivariate Analysis of Death and Coronary Heart Disease," Harvard University Press, Cambridge (1980).

8.   A.C. Arntzenius, D. Kromhout, J.D. Barth, J.H.C. Reiber, A.V.G. Bruschke, B. Buis, C.M. van Gent, N.Kempen-Voogd, S. Strikwerda, and E.A. van der Velde, Diet, lipoproteins, and the progression of coronary atherosclerosis New Engl. J. Med. 312:805 (1985).

9.   M.L. Armstrong, E.D. Warner, and W. E. Connor, Regression of coronary atheromatosis in Rhesus monkeys, Circulation Res. 27:59 (1970).

10. T.B. Clarkson, M.G. Bond, B. C. Bullock, K.J. McLaughlin, and J.K. Sawyer, A study of atherosclerosis regression in Macaca mulatta, Exp. Mol. Pathol. 41:96 (1984).

11. C.J. Glueck, D.J. Gordon, J.J. Nelson, C.E. Davis, and H. A. Tyroler, Dietary and other correlates of changes in total and low density lipoprotein cholesterol in hypercholesterolemic men: the lipid

research clinics coronary primary prevention trial, <u>Am. J. Clin. Nutr.</u> 44:489 (1986).

12. F.M. Sacks, J. Salazar, L. Miller, J.M. Foster, M. Sutherland, K.W. Samonds, J.J. Albers, and E.H. Kass, Ingestion of egg raises plasma low density lipoproteins in free-living subjects, <u>Lancet</u> 1:647 (1984).

13. J. Stamler and R. Shekelle, Dietary cholesterol and human coronary heart disease: the epidemiologic evidence, <u>Arch. Pathol. Lab. Med.</u> 112:1032 (1988).

14. W.E. Connor, and S.L. Connor, The key role of nutritional factors in the prevention of coronary heart disease, <u>Prev. Med.</u> 1:49 (1972).

15. A. Keys, Diet, <u>in</u>: "Seven Countries: A Multivariate Analysis of Death and Coronary Heart Disease," Harvard University Press, Cambridge (1980).

16. D.M. Hegsted, and L.M. Ausman, Diet, alcohol and coronary heart disease in men, <u>J. Nutr.</u> 118:1184 (1988).

17. Inter-Society Commission for Heart Disease Resources, Optimal resources for primary prevention of atherosclerotic disease, <u>Circulation</u> 70:153A (1984).

18. J. Stamler, D. Wentworth, and J.D. Neaton, Is relationship between serum cholesterol and risk of premature death from coronary heart disease continuous and graded?, <u>JAMA</u> 256:2823 (1986).

19. Consensus Conference, Lowering blood cholesterol to prevent heart disease, <u>JAMA</u> 253:2080 (1985).

20. S.M. Grundy, Cholesterol and coronary heart disease: a new era, <u>JAMA</u> 256:2849 (1986).

21. W.B. Kannel, W.P. Castelli and T. Gordon, Cholesterol in the prediction of atherosclerotic disease, <u>Ann. Intern. Med.</u> 90:85 (1979).

22. The Pooling Project Research Group, Relationship of blood pressure, serum cholesterol, smoking habit, relative weight, and ECG abnormalities to incidence of major coronary events: Final report of the Pooling Project, <u>J. Chron. Dis.</u> 31:201 (1978).

23. C. Aravanis, The classic risk factors for coronary heart disease: experience in Europe, <u>Prev. Med.</u> 12:16 (1983).

24. N. Rifai, Lipoproteins and apolipoproteins: composition, metabolism and association with coronary heart disease. <u>Arch. Pathol. Lab. Med.</u> 110:694 (1986).

25. W. P. Castelli, J.T. Doyle, T. Gordon, C.G. Hames, M.C. Hjortland, S.B. Hulley, A. Kagan, and W.J. Zukel, HDL cholesterol and other lipids in coronary heart disease: The Cooperative Lipoprotein Phenotyping Study, <u>Circulation</u> 55:767 (1977).

26. The Lipid Research Clinics Program, The Lipid Research Clinics Coronary Primary Prevention Trial Results: reduction in incidence of coronary heart disease, <u>JAMA</u> 251:351 (1984).

27. D.J. Gordon, and B.M. Rifkind, High-density lipoprotein - the clinical implications of recent studies, <u>New Engl. J. Med.</u> 321:1311 (1989).

28. W.B. Kannel, High-density lipoproteins: epidemiologic profile and
     risks of coronary artery disease, <u>Am. J. Cardiol.</u> 52:9B (1983).

29. H. Aberg, H. Lithell, I. Selinus, and H. Hedstrand, Serum
     triglycerides are a risk factor for myocardial infarction but not for
     angina pectoris, <u>Atherosclerosis</u> 54:89 (1985).

30. T. Gordon, W.P. Castelli, M.C. Hjortland, W.B. Kannel, and T.R. Dawber,
     High density lipoprotein as a protective factor against coronary heart
     disease: the Framingham Study, <u>Am. J. Med</u>. 62:707 (1977).

31. K.M. West, M.M.S. Ahuja, P.H. Bennett, A. Czyzyk, O. Mateo De Acosta,
     J.H. Fuller, B. Grab, V. Grabauskas, R.J. Jarrett, K. Kosaka, H. Keen,
     A.S. Krolewski, E. Miki, V. Schliack, A. Teuscher, P.J. Watkins, and
     J.A. Stober, The role of circulating glucose and triglyceride
     concentrations and their interactions with other "risk factors" as
     determinants of arterial diseae in nine diabetic population samples
     from the WHO multinational study, <u>Diabetes Care</u> 6:361 (1983).

32. D.B. Zilversmit, Atherogenesis: a postprandial phenomenon,
     <u>Circulation</u> 60:473 (1979).

33. B.H. Chung, J.P. Segrest, K. Smith, F.M. Griffin, and C.G.
     Brouillette, Lipolytic surface remnants of triglyceride-rich
     lipoproteins are cytotoxic to macrophages but not in the presence of
     high density lipoprotein, <u>J. Clin. Invest</u>. 83:1363 (1989).

34. J.S. Cohn, J.R. McNamara, S.D. Cohn, J.M. Ordovas, and E.J. Schaefer,
     Postprandial plasma lipoprotein changes in human subjects of different
     ages, <u>J. Lipid Res</u>. 29:469 (1988).

35. R.J. Havel, Early effects of fat ingestion on lipids and lipoproteins
     of serum in man, <u>J. Clin. Invest</u>. 36:848 (1957).

36. T.G. Redgrave, and L.A. Carlson, Changes in plasma very low density and
     low density lipoprotein content, composition, and size after a fatty
     meal in normo- and hypertriglyceridemic man, <u>J. Lipid Res</u>. 20:217 (1979).

37. G.R. Castro, and C.J. Fielding, Effects of postprandial lipemia on
     plasma cholesterol metabolism, <u>J. Clin. Invest</u>. 75:874 (1985).

38. I. Hjermann, K. Velve Byre, I. Holmf, P. Leren, Effect of diet and
     smoking intervention on the incidence of coronary heart disease, <u>Lancet</u>
     2:1303 (1981).

39. Lipid Research Clinics Program, The Lipid Research Clinics Coronary
     Primary Prevention Trial Results: the relationship of reduction in
     incidence of coronary heart disease to cholesterol lowering, <u>JAMA</u>,
     251:365 (1984).

40. M.H. Frick, O. Elo, K. Haapa, O.P. Heinonen, P. Heinsalmi, P. Helo,
     J.K. Huttunen, P. Kaitaniemi, P. Koskinen, V. Manninen, H. Maenpaa, M.
     Malkonen, M. Manttari, S. Norola, A. Pasternack, J. Pikkarainen, M.
     Romo, T. Sjoblom, and E.A. Nikkila, Helsinki Heart Study: primary-
     prevention trial with gemfibrozil in middle-aged men with dyslipidemia,
     <u>N. Engl. J. Med</u>. 317:1237 (1987).

41. Committee of Principal Investigators, A co-operative trial in the
     primary prevention of ischaemic heart disease using clofibrate, <u>Br.
     Heart J</u>. 40:1069 (1978).

42. L. Wilhelmsen, G. Berglund, D. Elmfeldt, G. Tibblin, H. Wedel, K. Pennert, A. Vedin, C. Wilhelmsson, and L. Werko, The multifactor primary prevention trial in Goteborg, Sweden, _Eur. Heart J._ 7:279 (1986).

43. D.H. Blankenhorn, S.A. Nessim, R. L. Johnson, M.E. Sanmarco, S.P. Azen, and L. Cashin-Hemphill, Beneficial effects of combined Colestipol-Niacin therapy on coronary atherosclerosis and coronary venous bypass grafts, _JAMA_ 257:3233 (1987).

44. B.G. Brown, E.L. Bolson, C.D. Pierce, R.B. Peterson, and H.T. Dodge, Regression of atherosclerosis in man: current data and their methodological limitations, _in_: "Regression of Atherosclerotic Lesions - Experimental Studies and Observations in Humans," M. R. Malinow and V. H. Blaton, eds., Plenum Press, New York (1984).

45. D.H. Blankenhorn, R.L. Johnson, W.J. Mack, H.A. El Zein, and L. I. Vailas, The influence of diet on the appearance of new lesions in human coronary arteries, _JAMA_ 263:1646 (1990).

46. C.J. Lavie, G.T. Gau, R.W. Squires, and B. A. Kottke, Management of lipids in primary and secondary prevention of cardiovascular diseases. _Mayo Clin. Proc._ 63:605 (1988).

47. J.F.Brensike, R.I. Levy, S.F. Kelsey, E.R. Passamani, J.M. Richardson, I.K. Loh, N.J. Stone, R.F. Aldrich, J.W. Battaglini, D. J. Moriarty, M.R. Fisher, L. Friedman, W. Friedewald, K.M. Detre, and S.E. Epstein, Effects of therapy with cholestyramine on progression of coronary arteriosclerosis: results of the NHLBI Type II Coronary Intervention Study, _Circulation_ 69:313 (1984).

48. A. Keys, F. Grande, and J.T. Anderson, Fiber and pectin in the diet and serum cholesterol concentration in man, _Proc. Soc. Exp. Biol. Med._ 106:555 (1961).

49. J.E. Garvin, D.T. Forman, W.R. Eiseman, and C.R. Phillips, Lowering of human serum cholesterol by an oral hydrophylic colloid, _Proc. Soc. Exp. Biol. Med._ 120:744 (1965).

50. J.W. Anderson, N. Zettwoch, T. Feldman, J. Tietyen-Clark, P. Oeltgen, and C.W. Bishop, Cholesterol-lowering effects of psyllium hydrophilic mucilloid for hypercholesterolemic men, _Arch. Intern. Med._ 148:292 (1988).

51. A.P. DeGroot, R. Luyken, and N.A. Pikaar, Cholesterol-lowering effect of rolled oats, _Lancet_ 2:303 (1963).

52. D.J.A. Jenkins, D. Reynolds, B. Slavin, A.R. Leeds, A.L. Jenkins, and E.M. Jepson, Dietary fiber and blood lipids: treatment of hyper-cholesterolemia with guar crispbread, _Am. J. Clin. Nutr_ 33:575 (1980).

53. J.W. Anderson, and N. J. Gustafson, Alternative lipid-lowering diets, _Practical Cardiology_ 14:84 (1988).

54. J.W. Anderson, L. Story, B. Sieling, W.-J.L. Chen, M.S. Petro, and J. Story, Hypocholesterolemic effects of oat-bran or bean intake for hypercholesterolemic men, _Am. J. Clin. Nutr._ 40:1146 (1984).

55. J.W. Anderson, W.-J.L. Chen, and B. Sieling, Hypolipidemic effects of high-carbohydrate, high-fiber diets, _Metabolism_ 29:551 (1980).

56. T. Gordon, A. Kagan, M. Garcia-Palmieri, W.B. Kannel, W.J. Zukel, J. Tillotson, P. Sorlie, and M. Hjortland, Diet and its relation to coronary heart disease and death in three populations, _Circulation_ 63:500 (1981).

57. M.R. Garcia-Palmieri, P. Sorlie, J. Tillotson, R. Costas, E. Cordero, and M. Rodriguez, Relationship of dietary intake to subsequent coronary heart disease incidence: the Puerto Rico Heart Health Program, _Am. J. Clin. Nutr_. 33:1818 (1980).

58. D.L. McGee, D.M. Reed, K. Yano, A. Kagan, and J. Tillotson, Ten-year incidence of coronary heart disease in the Honolulu Heart Program, _Am. J. Epidemiol_. 119:667 (1984).

59. J.W. Anderson, Plant fiber and blood pressure, _Ann. Intern. Med_. 98:842 (1983).

60. J.W. Anderson and K. Ward, High-carbohydrate, high-fiber diets for insulin-treated men with diabetes mellitus, _Am. J. Clin. Nutr_. 32:2312 (1979).

61. H.C.R. Simpson, R.W. Simpson, S. Lousley, R.D. Carter, M. Geekie, T.D.R. Hockaday, and J.I. Mann, A high carbohydrate leguminous fibre diet improves all aspects of diabetic control, _Lancet_ 1:1 (1981).

62. L. Story, J.W. Anderson, W.-J.L. Chen, D. Karounos, and B. Jefferson, Adherence to high-carbohydrate, high-fiber diets: long-term studies of non-obese diabetic men, _J. Am. Diet. Assoc_. 85:1105 (1985).

63. M. Krotkiewski, Effect of guar gum on body-weight, hunger ratings and metabolism in obese subjects, _Br. J. Nutr_. 52:97 (1984).

64. J.W. Anderson, and C.A. Bryant, Dietary fiber: diabetes and obesity, _Am. J. Gastroenterol_. 81:898 (1986).

65. D. Kromhout, E.B. Bosschieter, and C.D.L. Coulander, Dietary fibre and 10-year mortality from coronary heart disease, cancer, and all causes, _Lancet_ 2:518 (1982).

66. J.N. Morris, J.W. Marr, and D.G. Clayton, Diet and heart: a postscript, _Br. Med. J_. 2:1307 (1977).

67. L.H. Kushi, R.A. Lew, F.J. Stare, C.R. Ellison, M.E. Lozy, G. Bourke, L. Daly, I. Graham, N. Hickey, R. Mulcahy, and J. Kevaney, Diet and 20-year mortality from coronary heart Disease. The Ireland-Boston Diet-Heart Study, _New Engl. J. Med_. 312:811 (1985).

68. K.-T. Khaw and E. Barrett-Connor, Dietary fiber and reduced ischemic heart disease mortality rates in men and women: a 12-year prospective study, _Am. J. Epidemiol_. 126:1093 (1987).

69. J. Hallfrisch, J. D. Tobin, D. C. Muller, and R. Andres, Fiber Intake, Age, and Other Coronary Risk Factors in Men of the Baltimore Longitudinal Study (1959-1975), _J. Gerontology: Med. Sci_. 43:M64 (1988).

70. D. Kromhout and C. de Lezenne Coulander, Diet, prevalence and 10-year mortality from coronary heart disease in 871 middle-aged men: the Zutphen Study, _Am. J. Epidemiol_. 119:733 (1984).

71. The Expert Panel, Report of the National Cholesterol Education Program Expert Panel on detection, evaluation, and treatment of high blood cholesterol in adults, <u>Arch. Intern. Med.</u> 148:36 (1988).

72. Study Group, European Atherosclerosis Society, Strategies for the prevention of coronary heart disease: a policy statement of the European Atherosclerosis Society, <u>Eur. Heart J</u>. 8:77 (1987).

73. The Canadian Consensus Conference on the Prevention of Heart and Vascular Disease by Altering Serum Cholesterol and Lipoprotein Risk Factors, Canadian Consensus Conference on Cholesterol: A final report, <u>Can. Med. Assoc. J.</u> 139:1 (1988).

74. A Joint Statement of the Nutrition Committee and the Council on Arteriosclerosis, Recommendations for treatment of hyperlipidemia in adults, <u>Circulation</u> 69:1065A (1984).

75. L.-G. Ekelund, W.L. Haskell, J.L. Johnson, F.S. Whaley, M.H. Criqui, and D.S. Sheps, Physical fitness as a predictor of cardiovascular mortality in asymptomatic North American men, <u>N.Engl. J. Med</u>. 319:1379 (1988).

76. P.D. Wood, W.L. Haskell, H. Klein, S. Lewis, M.P. Stern, and J.W. Farquhar, The distribution of plasma lipoproteins in middle-aged male runners, <u>Metabolism</u> 25:1249 (1976).

77. J.W. Anderson, W.-J.L. Chen, and B. Sieling, Hypolipidemic effects of high-carbohydrate, high-fiber diets, <u>Metabolism</u> 29:551 (1980).

78. J.W. Anderson, and N.J. Gustafson, Dietary fiber and heart disease: current management concepts and recommendations, <u>Top. Clin. Nutr</u>. 3:21 (1988).

79. J.W. Anderson, "Be Heart Smart", The HCF Nutrition Research Foundation, Inc., Lexington (1989).

APOLIPOPROTEIN PROFILES IN RANDOMLY SELECTED SMOKERS AND MATCHED CONTROLS

Jörgen Mölgaard and Anders G. Olsson

Faculty of Health Sciences, University Hospital
S-581 85, Linköping, Sweden

INTRODUCTION

Lipids are insoluble in water and have to be incorporated in
water-soluble lipid-protein complexes called lipoproteins for trans-
portation in the blood. The proteins in the lipoprotein complexes are
called apolipoproteins. Lipoproteins have a hydrophobic core containing
apolar lipids (triglycerides and cholesterylesters) and a hydrophilic
surface containing apolipo-proteins, unesterified cholesterol and
phospholipids. Apolipo-proteins are thus structural proteins in the
lipoprotein complexes, but they also have important functions in the
lipid metabolism.

Several different apolipoproteins are known. Apo B and A-1 are
important apolipoproteins with respect to atherosclerosis. Apo B is
found in the atherogenic lipoproteins very low density lipoprotein
(VLDL), intermediate density lipoprotein (IDL), and low density
lipoproteins (LDL). Apo B is a ligand for the LDL receptor and plays an
important role in the liver catabolism of LDL. Apo A-1 is the major
protein of high density lipoproteins (HDL) and has an "antiatherogenic
role." There is evidence that apo A-1 may function as a ligand for the
HDL in "reverse cholesterol transport," the process by which cholesterol
is transported from peripheral tissues (from the arterial wall) to the
liver. Apo B and apo A-1 are also found in the chylomicrons. As
apolipoproteins are part of the lipoproteins, changes in their concentrations also change the apolipoprotein levels. Roughly one can say
that changes in plasma concentrations of apo B are correlated to LDL
cholesterol concentrations and apo A-1 levels reflect HDL cholesterol
concentrations.

Several studies have shown that high levels of LDL cholesterol and
therefore apo B are associated with an increased risk for ischemic heart
disease (1), while the risk is decreased in people having high HDL and
apo A-1 concentrations (2,3). HDL can be divided into two main subfractions HDL_2 and HDL_3. HDL_2 cholesterol is thought to be the
"protective" part of HDL cholesterol. Apo A-1 is a structural protein of
both subclasses.

<u>Smoking, Lipids, and Lipoproteins</u>

Smoking and hyperlipidemia are well known major risk factors for atherosclerotic disease, e.g., myocardial infarction and intermittent claudication. Several studies have focused on the possibility that smoking may affect the concentrations of plasma lipids or lipoproteins in an unfavorable way (4). Some of the deleterious effects of smoking in atherosclerosis could theoretically be mediated through a more athero- genic lipid or lipoprotein profile in smokers. Epidemiological studies have shown an association between smoking and decreased levels of the "protective" HDL cholesterol (5). The effect of smoking on HDL seems to be a decrease in both the HDL_2 and HDL_3 subfractions (6). There have been some speculations that part of the atherosclerotic effect of smoking could be mediated through a reduction in HDL cholesterol (6). A possible cause for differences in HDL cholesterol between smokers and non-smokers could also, however, be due to differences in diet, alcohol intake, and physical exercise (7).

Ex-smokers have higher HDL cholesterol than smokers (8). The increase of HDL cholesterol in ex-smokers is probably due to changes in dietary habits. When diet is kept constant after smoke cessation, no significant changes are found in HDL cholesterol (9).

<u>Apolipoproteins in Smokers</u>

There are only a few studies on the effect of smoking on apolipo- proteins. In general they show a decrease in apo A-1. One study showed a reduction in apo A-1 and an increase in apo B. The effect was variable depending on the subject's age and number of cigarettes smoked (10). Another study also showed a decrease in apo A-1 and apo A-11 correlating with the degree of cigarette smoking (11). A German study has shown decreased levels of HDL apolipoprotein A-1 in smokers compared to non-smokers (12).

<u>Plasma Apolipoprotein Profiles in Randomly Selected Smokers and Matched Controls</u>

We wanted to study the possibility of an enhanced atherogenic effect of smoking on apolipoproteins, lipoproteins and plasma lipids. Our interest in this question is due to an ongoing study of intermittent claudication. Patients with intermittent claudication have in most cases a history of heavy cigarette smoking for many years.

From an epidemiological study of 15,253 middle-aged men (45-69 yrs.) we had detailed information about smoking habits in 87%. In a random sample consisting of 140 subjects - 50 current smokers, 20 ex-smokers, and 70 sex- and age-matched controls - we took fasting blood samples for determination of apolipoproteins A-1 and B and lipoprotein fractions VLDL, LDL, HDL, HDL_2 and HDL_3. The subjects were examined by a physi- cian, and appropriate laboratory tests were taken to exclude diseases giving secondary hyperlipidemia.

Apolipoproteins A-1 and B were determined by rocket electro- immunophoresis (13). Lipoproteins were determined by preparative ultracentrifugation and the cholesterol and triglyceride contents were analyzed by colorimetric enzymatic methods. HDL was divided into its two main fractions HDL_2 (d=1.063-1.125 kg/l) and HDL_3 (d=1.125-1.21 kg/l).

RESULTS

The results are given in Tables 1-4. Smokers had a significantly higher level of apo B in the VLDL fraction and this significance was due to the group of heavy smokers (Fig. 1). There were no differences in apo B in VLDL in moderate or ex-smokers compared to controls (Table 2). Compared to non-smokers, heavy smokers had significantly higher plasma total triglyceride, VLDL triglyceride and VLDL cholesterol concentrations (Table 3). Smokers also had a slightly increased plasma total cholesterol concentration. The difference, however, was not significant. When smokers are grouped into heavy ($\geq$ 15 cigarettes per day) and moderate smokers (< 15 cigarettes per day) all the differences in plasma lipids and lipoproteins could be explained by the group of heavy smokers. Interestingly enough moderate smokers had lower total and LDL cholesterol than their controls, though this difference was not significant. In ex-smokers there were higher plasma cholesterol levels than in controls. This was mainly due to higher HDL cholesterol levels. There was a significant difference in HDL_3 cholesterol and borderline significance differences in total plasma HDL and HDL_2 cholesterol. VLDL cholesterol was practically the same in ex-smokers and controls (Table 3).

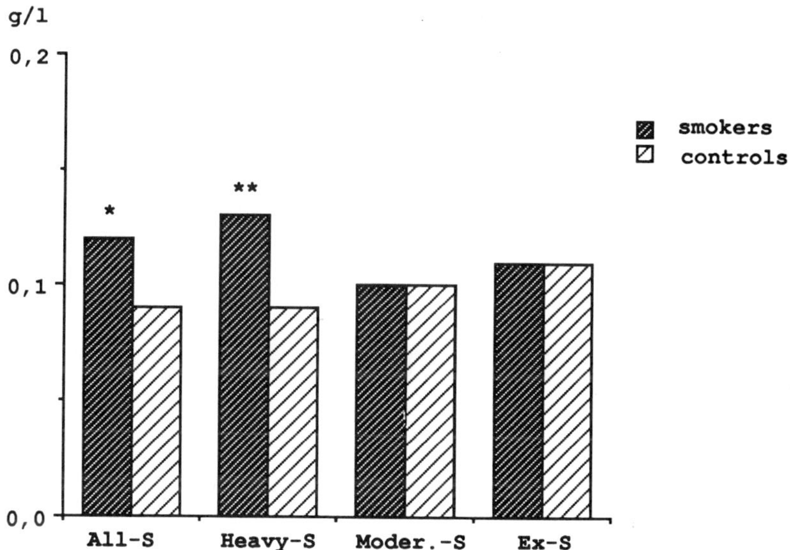

Fig. 1. Apolipoprotein B concentrations in very low density
         lipoproteins (VLDL) in smokers (S) and sex- and
         age-matched controls. Moder. = moderate

DISCUSSION

Our study shows that smoking increases apo B in the VLDL lipoprotein fraction. We did not see any effect of smoking on apo A-1 or HDL cholesterol in current smokers compared to controls as some studies have shown. The discrepancy in results may be due to our matching of smokers

TABLE 1   APOLIPOPROTEIN A-I IN PLASMA, HDL, HDL$_2$ AND HDL$_3$

Mean values ± SD (g/l)	Total Apo A-I	HDL A-I	HDL$_2$ A-I	HDL$_3$-A-I
all smokers (n=50)	1.32±0.19	1.24±0.20	0.35±0.14	0.89±0.13
controls (n=50)	1.34±0.17	1.26±0.17	0.37±0.11	0.90±0.10
heavy smokers (n=30)	1.32±0.18	1.25±0.18	0.34±0.11	0.91±0.14
controls (n=30)	1.33±0.16	1.25±0.16	0.35±0.10	0.90±0.10
moderate smokers (n=20)	1.31±0.22	1.23±0.24	0.37±0.18	0.86±0.11
controls (n=20)	1.35±0.18	1.29±0.19	0.39±0.12	0.89±0.11
ex-smokers (n=20)	1.33±0.13*	1.29±0.14*	0.39±0.10*	0.90±0.09
controls (n=20)	1.24±0.12	1.17±0.13	0.33±0.09	0.85±0.08

* = $p < 0.05$,
HDL A-I   = Concentration of APO A-I in HDL
HDL$_2$ A-I = Concentration of APO A-I in HDL$_2$
HDL$_3$ A-I = Concentration of APO A-I in HDL$_3$

TABLE 2   APOLIPOPROTEIN B IN PLASMA , LDL AND VLDL

Mean values ± SD (g/l)	Total APO B	LDL- APO B	VLDL-APO B
all smokers (n=50)	1.28±0.36	1.16±0.31	0.12±0.07*
controls (n=50)	1.21±0.27	1.11±0.25	0.09±0.05
heavy smokers (n=30)	1.35±0.36	1.22±0.32	0.13±0.08**
controls (n=30)	1.23±0.28	1.15±0.26	0.09±0.04
moderate smokers (n=20)	1.17±0.33	1.06±0.29	0.10±0.04
controls (n=20)	1.18±0.25	1.06±0.23	0.10±0.06
ex-smokers (n=20)	1.23±0.33	1.12±0.35	0.11±0.11
controls (n=20)	1.23±0.28	1.12±0.23	0.11±0.09

* = $p < 0.05$, ** = $p < 0.01$
LDL-APO B   = Concentration of APO B in low density lipoproteins
VLDL-APO B = Concentration of APO B in very low density lipoproteins

TABLE 3   PLASMA LIPIDS AND LIPOPROTEINS

Mean values ± SD (mmol/l)	Total Chol	LDL Chol	VLDL Chol	Total TG	VLDL TG	Quetelet [kg/m²]
all smokers (n=50)	5.45±0.96	3.85±0.86	0.41±0.29*	1.30±0.60*	0.90±0.54*	24.7±2.7
controls (n=50)	5.33±0.82	3.86±0.76	0.29±0.21	1.01±0.47	0.65±0.41	24.7±2.3
heavy smokers (n=30)	5.64±1.00	4.02±0.82	0.48±0.34*	1.45±0.69*	1.05±0.62*	25.4±2.
controls (n=30)	5.35±0.93	3.89±0.82	0.28±0.19	0.97±0.41	0.60±0.34	24.6±2.1
moderate smokers (n=20)	5.18±0.85	3.60±0.89	0.30±0.13	1.06±0.31	0.68±0.29	23.3±2.2*
controls (n=20)	5.28±0.62	3.81±0.69	0.31±0.25	1.09±0.55	0.73±0.50	24.8±2.5
ex-smokers (n=20)	5.60±1.10	3.99±12	0.46±0.96	1.26±1.51	0.89±1.49	26.0±3.2
controls (n=20)	5.33±0.89	3.88±0.78	0.43±0.46	1.32±1.03	0.96±1.05	25.3±2.3

* Significant differences compared to controls   p < 0.05,

Quetelet index = Weight(kg)/Height(m)²
Chol=Cholesterol, TG=Triglycerides
LDL=Low density lipoproteins, VLDL=Very low density lipoproteins

TABLE 4 HIGH DENSITY LIPOPROTEINS (HDL)

Mean values ± SD (mmol/l)	Tot. HDL Chol	HDL_2Chol	HDL_3 Chol	HDL TG
all smokers (n=50)	1.13±0.42	0.36±0.29	0.77±0.17	0.10±0.04
controls (n=50)	1.14±0.26	0.36±0.17	0.77±0.13	0.10±0.04
heavy smokers (n=30)	1.10±0.31	0.33±0.20	0.77±0.16	0.11±0.05
controls (n=30)	1.11±0.23	0.34±0.16	0.76±0.11	0.10±0.05
moderate smokers (n=20)	1.17±0.55	0.39±0.38	0.77±0.18	0.10±0.02
controls (n=20)	1.19±0.31	0.39±0.18	0.80±0.15	0.10±0.03
ex-smokers (n=20)	1.20±0.32	0.39±0.18	0.81±0.17*	0.10±0.04
controls (n=20)	1.01±0.25	0.28±0.11	0.70±0.10	0.10±0.03

* = p < 0.05,
Chol=Cholesterol, TG=Triglycerides, Tot.=Total
HDL_2 and HDL_3 are the two main subfractions of HDL

with controls of the same sex and age, factions that otherwise influence
the HDL levels. There were no significant differences in other confounding explanatory factors such as diet and alcohol when a random
subsample of smokers and controls was analyzed. The study also showed a
significant elevation of plasma triglycerides, VLDL triglycerides, and
VLDL cholesterol concentrations. The effect of smoking on triglycerides
has been found in other studies (14) and is probably due to an impaired
lipolysis in smokers. Some investigators have shown that after a fatty
meal the concentrations of triglycerides increase more and remain
increased for a longer time in smokers than in non-smokers (15).
Ex-smokers had significantly higher levels of apo A-1 in total plasma,
HDL, HDL_2, and HDL_3 and in general a light to moderate increase in plasma
lipids and lipoproteins. The explanation for these increases is most
likely due to changes in dietary habits after smoking cessation (9).

CONCLUSION

The atherogenic effects of smoking on apolipoproteins, lipoproteins,
and plasma lipids are seen in heavy smokers but not in moderate smokers.
Smoking increases apo B in VLDL as well as plasma triglycerides, VLDL
triglyceride and VLDL cholesterol concentrations. The light increase in
plasma cholesterol in heavy smokers is due to an increase in VLDL
cholesterol. Heavy smokers thus seem to have an impairment in their
triglyceride metabolism.

REFERENCES

1. Lipid Research Clinics Program. The Lipid Research Coronary Primary
      Prevention Trial. Results I. Reduction in incidence of coronary
      heart disease, JAMA, 251:351 (1984).
2. Wilson, P.W.F., Abbott, R.D., and Castelli, W.P. High density
      lipoprotein cholesterol and mortality. The Framingham Heart
      Study, Arteriosclerosis, 8:737 (1988).
3. Kottke, B.A., Zinsmeiste, A., Holmes, D.R., Kneller, R.W.,
      Hallaway, B.J., and Mao, S.J.T. Proceedings Mayo Clinic,
      Apolipoproteins and coronary heart disease, 61:313 (1986).
4. Craig, W.Y., Palomaki, G.E., and Haddow, J.E. Cigarette smoking and
      serum lipid and lipoprotein concentrations: an analysis of
      published data, Br. Med. J., 298:784 (1989).
5. Criqui, M.H., Wallace, R.B., Heiss G., Mishkel, M., Schonfeld, G.
      and Jones, G.T.L. Cigarette smoking and plasma high-density
      lipoprotein cholesterol. The Lipid Research Clinics Program
      Prevalence Study, Circulation, 62 (suppl. IV), IV-70 (1980).
6. Nesje, L.A., and Mjøs, O.D. Plasma HDL cholesterol and the
      subclasses HDL_2 and HDL_3 in smokers and non-smokers, Artery,
      13(1):7 (1985).
7. Heiss, G., Johnson, N.J., Reiland, S., Davis, C.E., and Tyroler, H.
      The epidemiology of plasma high density lipoprotein cholesterol
      levels. The Lipid Research Clinics Program Prevalence Study,
      Circulation, 62 (suppl. IV): IV-116 (1980).
8. Moffatt, R.J. Effects of cessation of smoking on serum lipids and
      high-density lipoprotein-cholesterol, Atherosclerosis, 74:85
      (1988).
9. Quensel, M., Soderstrom, A., Aagardh, C.D., and Nilsson-Ehle, P.
      High density lipoprotein concentrations after cessation of
      smoking: the importance of alterations in diet, Atherosclerosis,
      75:189 (1989).

10. Cuesta, C., Sánchez-Muniz, F.J., García-La Cuesta, A., Garrido, R., Castro, A., San-Felix, B., and Domingo, A. Effects of age and cigarette smoking on serum concentrations of lipids and apolipoproteins in a male military population, Atherosclerosis, 80:33 (1989).

11. Berg, K., Borresen, A-L, and Dahlen, G. Effect of smoking on serum levels of HDL apoproteins, Atherosclerosis, 34:339 (1979).

12. Assman, G. The effects of cigarette smoking on serum levels of HDL cholesterol and HDL apolipoprotein A-1, J. Clin. Chem. Clin. Biochem., 22:397 (1984).

13. Von Schenck, H., Mölgaard, J. Olsson, A.G. Electroimmunoassay and preliminary values for apolipoprotein Al and B, Lab Med., 11:194 (1987).

14. Brischetto, C.S., Connor, W.E., Connor, S.L. and Matarazzo, J.D. Plasma lipid and lipoprotein profiles of cigarette smokers from randomly selected families: Enhancement of hyperlipidemia and depression of high-density lipoprotein, Am. J. Cardiol., 52:675 (1983).

15. Richmond, W., Seviour, P.W., Teal, T.K., and Elkeles, R.S. Impaired intravascular lipolysis with changes in concentrations of high density lipoprotein subclasses in young smokers, Br. Med. J., 295:246 (1987).

NORMALIZATION OF HIGH DENSITY LIPOPROTEIN CHOLESTEROL

FOLLOWING CESSATION FROM CIGARETTE SMOKING

Robert J. Moffatt

Exercise Physiology Laboratory
Department of Nutrition, Food, and Movement Sciences
The Florida State University
Tallahassee, Florida 32306

INTRODUCTION

Coronary artery disease (CAD) is recognized as the chief cause of death in the industrialized world. Epidemiological data have revealed several factors which correlate highly with the incidence of this disease. Some of the most meaningful predictors of CAD are blood pressure, diabetes, obesity and serum cholesterol levels. Recent evidence suggests that in addition to the plasma concentration of cholesterol the manner in which cholesterol is distributed is of importance.

High density lipoprotein cholesterol (HDL-C) has been indicated as a positive factor in deterring the development of atherosclerosis (1-3). In fact, a close relationship exists between low levels of HDL-C and atherosclerotic CAD. Pearson et al. (4) report that as the level of HDL-C decreases, the number of diseased coronary arteries increases. Furthermore, a decrease in normal HDL-C levels of 10 mg/dl doubles the risk of CAD (1).

Cigarette smoking has also been implicated as a major contributor to risk of CAD. Although the mechanisms by which cigarette smoking exerts its influence have yet to be elucidated, several reports have demonstrated an inverse relationship between smoking and HDL-C (5-8). The relationship between cigarette smoking and HDL-C is of interest since elevated levels of HDL-C are purported to provide a protective factor against CAD.

Although information is available regarding cross sectional comparisons of smokers, non-smokers and ex-smokers (5-8), there are no controlled studies examining the effect of smoking or cessation from smoking on a longitudinal basis. Therefore, it was of interest to examine the effects of cessation from smoking and the resumption of smoking on serum lipids and HDL-C levels.

Tobacco Smoking and Atherosclerosis
Edited by J. N. Diana
Plenum Press, New York, 1990

METHODS

Subjects

Twenty-six smoking and ten non-smoking females volunteered to participate in this study. Subjects completed questionnaires regarding smoking status, medication and physical activity patterns. To qualify as a smoker, subjects must have smoked a minimum of 20 cigarettes per day for the past five consecutive years. Non-smoking subjects had never smoked. None of the subjects were using oral contraceptives or other lipid altering medications at the time of the study. All subjects were free of known heart disease and none were post-menopausal. Subjects were classified as sedentary and did not participate in aerobic activity. Written informed consent was provided in accordance with Human Subject Protocol.

Design

Smokers were classified into one of three groups. Group one (SMOKERS) consisted of eight smokers who continued the habit throughout the duration of the study. Of the 27 smokers who entered the smoking cessation program, 12 successfully completed the 60 day program and comprised the second group (EX-SMOKERS). Additionally, a third group of six smokers who were only successful in abstaining for a period of 30-36 days resumed smoking and were identified as RE-SMOKERS. The data from the remaining nine subjects who were unable to successfully abstain from smoking for a minimum of 30 days were not included in this study. A fourth group of 10 subjects served as non-smoking controls (NON-SMOKERS).

Food records and activity questionnaires were collected weekly throughout the study to determine nutrient composition, fat consumption, alcohol consumption and exercise patterns. A modified version of American Lung Association smoking cessation program was employed for a total of 10 weeks.

Lipid and Lipoprotein Assays

Blood samples for total cholesterol, HDL-C and triglycerides were drawn from an antecubital vein. Subjects were instructed to report to the laboratory for testing between the hours of 8:00 and 9:00 a.m. following a 12-hour overnight fast and abstinence from smoking. All samples were collected at the same time of day before cessation (Day 0) and 30 and 60 days after cessation from smoking. Total cholesterol content was determined by the method of Allain, et al. (9). High density lipoprotein was measured by precipitating all other cholesterol fractions with phosphotungstate magnesium (10) leaving a supernatant which was measured for cholesterol content. Triglyceride levels were quantitatively assayed by an enzymatic hydranalysis technique described by Beckman Instruments (11). Quality control samples were also assayed to determine precision. The coefficient of variation was determined to be less than 2.4% for HDL-C and total cholesterol and 3.9% for triglycerides.

Data Analysis

A repeated measures ANOVA was employed to evaluate the effects of cessation from smoking or resumption of smoking on dependent scores. Differences between treatment means were determined by the Scheffe analysis. The 0.05 level of confidence was pre-determined as the criterion for acceptance of statistical significance.

CIGARETTE SMOKING CAUSES ACUTE CHANGES IN ARTERIAL WALL MECHANICS AND THE
PATTERN OF ARTERIAL BLOOD FLOW IN HEALTHY SUBJECTS: POSSIBLE INSIGHT
INTO MECHANISMS OF ATHEROGENESIS

Colin G. Caro

Centre for Biological and Medical Systems
Imperial College of Science, Technology and Medicine
London, SW7 2AZ

Cigarette smoking is widely believed to increase the extent and
severity of atherosclerosis, but the underlying mechanisms have not been
delineated (1-5). There are changes in the plasma lipids and the
rheology of the blood in smokers, but these are unable fully to account
for atherosclerosis, in particular its focal distribution in the
circulation. Atherosclerosis principally affects the intima of
thick-walled arteries and within these vessels' regions of branching and
curvature. There is evidence consistent with the view that the mass
transport properties of the walls of blood vessels and the blood flow
pattern determine this distribution. We consider the mechanisms and
report acute changes we have found in arterial wall mechanics and
arterial blood flow in healthy human subjects after smoking cigarettes.
These changes may help explain the association between smoking and other
factors and atherosclerosis.

VESSEL WALL TRANSPORT AND ATHEROSCLEROSIS

There is normally outward movement of material from the lumen of
arteries across the wall to adventitial lymphatics and the medial layer
in thick-walled vessels may hinder this movement and thereby favor
accumulation in the intima (6-9). The movement is believed to be
influenced by among other factors arterial blood pressure and the tone of
the medial smooth muscle; noradrenaline-induced contraction of the smooth
muscle in isolated arteries reduces the porosity of the media and
relaxation of the smooth muscle with nitrates has the opposite effect
(10-12).

ARTERIAL BLOOD FLOW AND ATHEROSCLEROSIS

The blood flow in the arteries in human subjects is usually laminar,
but is nevertheless highly complicated. It is unsteady, varying during
the cardiac cycle and, indeed, reverses in direction during the cardiac
cycle in some arteries. In addition, the flow is three-dimensional at
sites of branching and curvature, which are preferred sites for the
occurrence of atherosclerosis.

Tobacco Smoking and Atherosclerosis
Edited by J. N. Diana
Plenum Press, New York, 1990

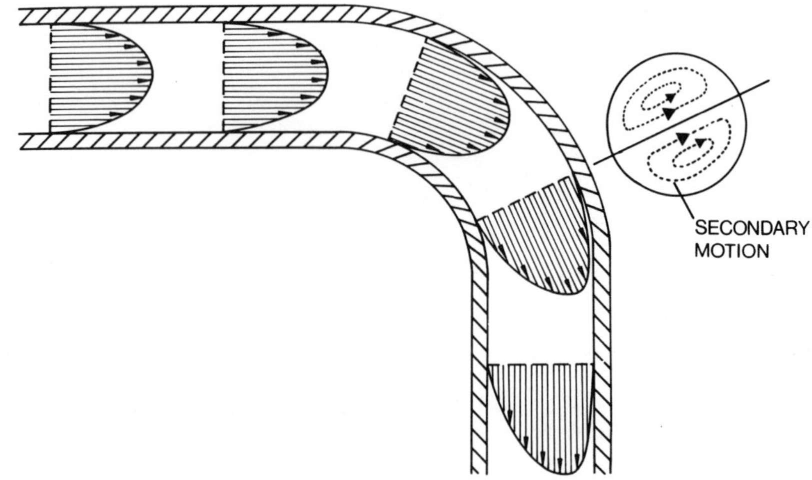

Fig. 1

Velocity profiles in a curved tube. The faster moving fluid is displaced
to the outer wall as a result of the secondary motions.

    Fig. 1 shows schematically steady flow in a curved tube, representing
in simplified form the flow in a curved artery. The axial velocities are
shown in the plane of curvature. Upstream, in the straight section of
the tube, the velocity profile is symmetrical whereas within the bend it
is skewed towards the outer wall.

    The change of direction causes a force to act on the fluid which
deflects the faster-moving fluid around the bend. The slower-moving
fluid near the wall has less inertia and is deflected to a lesser extent
and as a result transverse or secondary motions are set up. A
consequence is that the shear stress, which is the product of the
velocity gradient and the fluid viscosity, becomes higher at the outer
wall than at the inner wall (13).

    The details of steady flow in a bifurcation are more complicated but
can be understood qualitatively from similar arguments (Fig. 2). As the
faster-moving fluid in the centre of the tube approaches the flow
divider, forces are set up which deflect it into the daughter tubes. The
slower-moving fluid near the walls is subjected to the same forces and
transverse or secondary motions are established in each daughter tube.
In addition, new boundary layers develop on the flow divider similar to
those in the entrance region of a tube. As a result, the shear stress at
the flow divider can be very large compared to that at the opposite outer
wall.

    Downstream of both the bend and the bifurcation, the secondary
motions die out on a length scale which depends on the Reynolds number.

    If the fluid velocity is high enough and the curvature is great
enough, the force acting on the fluid may cause flow separation and flow
reversal at the outer wall of the bifurcation (Fig. 3). This will
greatly alter the distribution of the shear stress at the wall.

    There have been several studies of unsteady flow in curved and
branching tubes, but it is impossible to summarize the results in a
simple way. The velocity profiles differ from those of steady flow, much

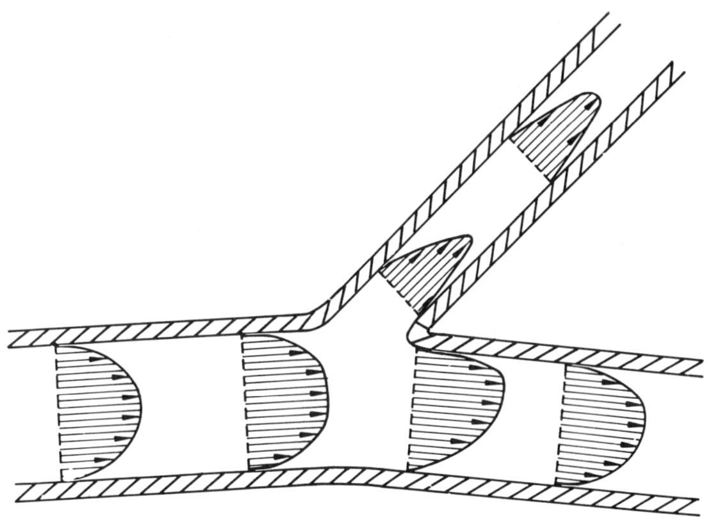

Fig. 2

Velocity profiles in a branching tube showing the highest shear occurs on
the flow divider while the shear is lowest on the opposite outer walls.

as do those in a straight tube (13), but the time relationships and the
secondary motions are very complicated. In addition, as in steady flow,
increasing velocity rather than increasing wall shear can cause flow
separation and reversal and a decrease in wall shear.

The flow near a vessel wall can have numerous effects influencing,
for example, the morphology and metabolism of the endothelium (20,31-35),
the movement of material into and across the wall (36-38), mixing
processes in the blood adjacent to the wall, and the interaction of blood
cells with the wall (39).

There is now wide agreement concerning the preferred sites of
occurrence of atherosclerosis in arteries. The disease in human subjects
exhibits a predilection for outer walls at branches and inner walls in
curved vessels, which are sites where the wall shear stress is on average
low and undergoes large oscillations of direction during the cardiac
cycle (14-20). The mechanisms which associate this flow pattern with the
development of atherosclerosis are, however, unknown.

ARTERIAL BLOOD FLOW AND CIGARETTE SMOKING

The unsteadiness of the flow at sites of predilection for
atherosclerosis has led to the suggestion that flow pulsatility plays a
part in the development of the disease (21,22). The pulsatility of
arterial flow is determined by many factors, including the blood
pressure, the stiffness of the vessels, the mean blood velocity, pulse
wave reflections, the heart rate and the ejection pattern of the heart.

The possibility that the arterial flow pattern is implicated in
atherosclerosis has encouraged us to carry out studies in healthy human
subjects on the effects of agents likely to influence the flow.
Cigarette smoking acutely increases the heart rate and arterial blood
pressure and can be expected therefore to alter arterial wall stiffness
and the pattern of arterial flow.

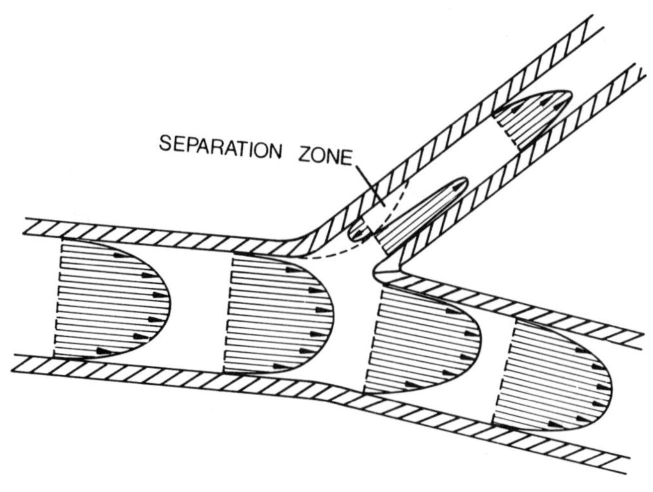

Fig. 3

Velocity profiles in a branching tube in the presence of flow separation

Subjects and Methods

Fifteen healthy subjects (13 men, 2 women, age range 20-61 yr, mean age 30.7 yr) were studied supine under controlled temperature conditions (23). Seven of the subjects were habitual smokers. Blood velocity was measured in the superficial femoral artery 10 cm distal to the bifurcation of the common femoral artery with a 4.8 MHz multi-channel Doppler ultrasound device (MAVIS, International Ltd.). Signals from up to 20 range gates yielded velocity profiles across the width of the artery at 25 ms intervals throughout the cardiac cycle. These were subjected to ensemble averaging and velocity waveforms and time-averaged velocity values were obtained. Vessel diameter was calculated from the velocity profile at the time of peak forward velocity. We also measured brachial artery blood pressure with a cuff device and a pulse wave transit time – the delay between the onset of forward flow in the common carotid artery (detected with an 8 MHz continuous wave Doppler ultrasound device) and in the superficial femoral artery.

The subjects rested for 30 min before a series of control measurements were made over a 30 min period. They then smoked two middle-tar cigarettes at a rate of one inhalation per min and further measurements were made immediately after the cessation of smoking and 10 and 20 min later. The significance of differences in changes in measurements between subjects was determined from a two-way analysis of variance.

The 10 min post-smoking values (Table 1) showed a rise of heart rate, an increase of mean arterial pressure without significant increase of artery diameter, a decrease in pulse wave transit time, no significant change in time-averaged mean blood velocity and a reduction of the pulsatility index (PI); the pulsatility index is defined as the difference between the maximum and minimum velocities during the cardiac cycle divided by the time-averaged mean velocity. The measurements tended to return to their control values by about 20 min after the cessation of smoking. There were no obvious differences between the responses of the habitual smokers and the non-smokers. Sham smoking (a similar pattern of inhalation with two unlit cigarettes) caused no significant departure from control values.

Table 1. Changes measured in 15 healthy subjects
after smoking two middle-tar cigarettes
(% change from pre-smoking)

	Average control value	SD*	Change (%) after smoking	P
Heart rate (min^{-1})	64.6	2.3	+19.1	<0.0005
Mean blood pressure (mm Hg)	82.8	3.6	+10.7	<0.0005
Vessel diameter (mm)	5.9	0.39	+ 0.5	NS
Pulse transit time (ms)	106.6	2.5	− 6.7	<0.0005
(Time-averaged mean velocity (cm s^{-1})	12.3	4.4	+ 8.2	NS
Pulsatility index	6.63	1.08	−15.8	<0.005

NS = not significant

*   Calculated from a two-way analysis of variance which
accounted for subject-to-subject variations.

The constancy of arterial diameter despite the increase of blood
pressure and the reduction of the pulse-wave transit time both imply that
cigarette smoking increases arterial wall stiffness.

An earlier study found that cigarette smoking caused a reduction in a
pulse-wave transit time in patients with occlusive vascular disease but
not in healthy subjects (24). The diminution of PI that we observed was

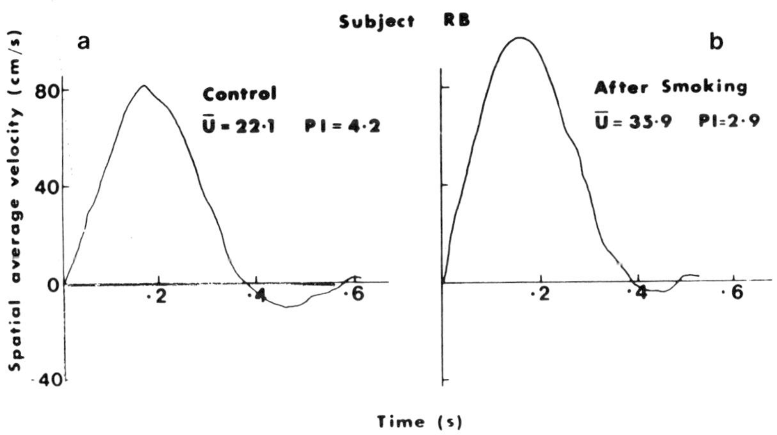

Fig. 4

Effects of smoking on spatial average blood velocity for a single subject

(a) before smoking: time-averaged velocity, $\bar{U}$ = 22.1 cm s^{-1}, PI = 4.2
(b) immediately after smoking two cigarettes: $\bar{U}$ = 35.9 cm s^{-1}, PI = 2.9

commonly associated with a decrease in reverse flow (Fig. 4). Such alterations of the flow pattern would strongly modify the distribution of velocities near the wall (17, 25-27) including that at sites of predilection
for atherosclerosis. Preliminary observations at sites of arterial
branching in healthy human subjects using a specially developed hybrid
echo-imaging: multichannel Doppler ultrasound device support that belief
(28).

The results with smoking can be compared with those we obtained in a
similar study in healthy subjects using the nitrate vasodilator,
isosorbide dinitrate (ISDN) (29). Twenty mg ISDN by mouth caused an
increase of the diameter of the superficial femoral artery despite a fall
of blood pressure and an increase of an apparent pulse-wave transit
time. Both findings are consistent with a decrease of arterial
stiffness. In addition, there was a fall in the time-averaged mean
arterial blood velocity and an increase in PI which was associated with
an increase in both the magnitude and duration of reverse flow. An
increase of PI with ISDN has been reported by others (30).

The findings that smoking in healthy subjects apparently increases
arterial wall stiffness - a change which may be associated with reduction
of medial porosity - and reduces flow pulsatility and that the
vasodilator, ISDN, causes the opposite effects may help explain why
smoking and other factors favor the development of atherosclerosis.

REFERENCES

1. Su, C. Actions of nicotine and smoking on circulation. Pharmacol.
    Ther. 17:129-141 (1982).
2. Surgeon-General. The health consequences of smoking: Cardiovascular
    disease. US Dept. of Health and Human Services Publication
    DHHS(PHS), 84-50204 (1983).
3. Auerbach, O., Hammond, E.C., Garfinkel, L. Smoking in relation to
    atherosclerosis of the coronary artery. N. Engl. J. Med.
    273:775-779 (1965).
4. Garrison, R. J., Kannel, W.B., Feinleib, M., Castelli, W.P.,
    McNamara, P.M., Padgett, S.J. Cigarette smoking and HDL
    cholesterol. The Framingham offspring study. Atherosclerosis
    30:17-25 (1978).
5. Holme, I., Enger, S.C., Helgeland, A. Risk factors and raised
    atherosclerotic lesions in coronary and cerebral arteries:
    Statistical analysis from the Oslo study. Arteriosclerosis
    1:250-256 (1981).
6. Caro, C.G., Lever, M.J., Laver-Rudich, Z. Net albumin transport
    across the wall of the rabbit common carotid artery perfused in
    situ. Atherosclerosis 37:497-511 (1980).
7. Smith, E.B., Staples, E.H. Intimal and medial plasma protein
    concentrations and endothelial function. Atherosclerosis
    41:295-308 (1982).
8. Fry, D.L. Mathematical models of arterial transmural transport.
    Am. J. Physiol. 248:H240-263 (1985).
9. Caro, C.G., Jay, M., Lever, M.J. Labelled albumin uptake by rabbit
    aorta, pulmonary artery and common carotid artery. J. Physiol.
    371:85P (1985).
10. Caro, C.G., Lever, M.J. Effect of vasoactive agents and applied
    stress on the albumin space of excised rabbit carotid arteries.
    Atherosclerosis 46:137-146 (1983).

11. Caro, C.G., Lever, M.J. Baldwin, A., Tedgui, A. Influence of con-
     vection and vasoactive agents on the mass transport properties of
     the arterial wall. In: Schettler, G., Nerem, R.M.,
     Schmid-Schonbein, H., Morl, H., Diehm, C., (eds.) "Fluid dynamics
     as a localizing factors for atherosclerosis." Berlin: Springer-
     Verlag, 129-134 (1983).
12. Lever, M.J. Effects of smooth muscle tone on interstitial transport.
     Int. J. Microcirc. Clin. Exp. 4:294 (1985).
13. Caro, C.G. and Parker, K.H. The effect of haemodynamic factors on
     the arterial wall. In: Atherosclerosis: Biology and Clinical
     Science. A.G. Olsson (Ed) Churchill Livingstone, Edinburgh (1987).
14. Caro, C.G., Fitz-Gerald, J.M. Schroter, R.C. Atheroma and arterial
     wall shear. Observation, correlation and proposal of a shear
     dependent mass transfer mechanism for atherogenesis. Proc. R.
     Soc. London B. 177:109-159 (1971).
15. Friedman, M.H., Hutchins, G.M. Bargeron, C.B., Deters, O.J., Mark.
     F.F. Correlation between intimal thickness and fluid shear in
     human arteries. Atherosclerosis 39:425-436 (1981).
16. Svindland, A.D., Walloe, L. Localization of early atherosclerotic
     lesions in carotid and coronary bifurcations in humans - a bifurc-
     ation of the high shear stress hypothesis. In: Schettler, G.,
     Nerem, R.M., Schmid-Schonbein H., Morl, H., Diehm, C., (eds.)
     "Fluid dynamics as a localizing factor for atherosclerosis."
     Berlin: Springer-Verlag, 212-215 (1983).
17. Giddens, D.P., Zarins, C.K., Glagov, S., Bharadvaj, B.K., Ku, D.N.
     Flow and atherogenesis in the human carotid bifurcation. In:
     Schettler, G., Nerem, R.M., Schmid-Schonbein, H., Morl, H., Diehm,
     C. (eds.) "Fluid dynamics as a localizing factor for athero-
     sclerosis, Berlin: Springer-Verlag, 38-45 (1983).
18. Sabbah, H.N., Khaja, F., Grymer, J.F., Hawkins, E.T., Stein, P.D.
     Blood velocity in the right coronary artery: relation to the dis-
     tribution of atherosclerotic lesions. Am. J. Cardiol. 53:1008-12
     (1984).
19. Sakata, N., Joshita, T., Ooneda, G. Topographical study on arterio-
     sclerotic lesions at the bifurcations of human cerebral arteries.
     Heart Vessels 1:70-73 (1985).
20. Yoshida, Y., Yamaguchi, T., Caro, C.G., Glagov, S. and Nerem, R.M.
     (eds) "Role of blood flow in atherogenesis, Springer-Verlag, Tokyo
     (1987).
21. Fry, D.L., Responses of the arterial wall to certain physical
     factors. In: Porter, R., Knight, J. "Atherogenesis: Initiating
     factors." Ciba Foundation Symposium 12 (new series). Elsevier,
     Amsterdam, 93-125 (1973).
22. Spence, J.D., Effect of antihypertensive drugs and blood velocity.
     In: Schettler, G., Nerem, R.M. Schmid-Schonbein, H., Morl, H.
     Diehm, C. (eds.) "Fluid dynamics as a localizing factor for
     atherosclerosis. Berlin: Springer-Verlag, 141-144 (1983).
23. Caro, C.G., Lever, M.J., Parker, K.H. and Fish, P.J. Effect of ciga-
     rette smoking on the pattern of arterial blood flow: possible
     insight into mechanisms underlying the development of arterio-
     sclerosis. The Lancet, 11-13 (July 4, 1987).
24. Lusby, R.N. Bauminger, B., Woodcock, J.P. Skidmore, R., Baird, R.N.
     Cigarette smoking: Acute main and small vessel haemodynamics
     responses in patients with arterial disease. Am. J. Surg.
     142:169-173 (1981).
25. Reneman, R.S., Van Merode, T., Hick, P., Hoeks, A.P.G. Flow velocity
     patterns in and distensibility of the carotid artery bulb in
     subjects of various ages. Circulation 71:500-509 (1985).

26. Batten, J.R., Nerem, R.M. Model study of flow in curved and planar arterial bifurcations. Cardiovasc. Res. 16:178-186 (1982).

27. Lutz, R.J., Hsu, L., Menawat, A., Zrubek, J., Edwards, K.H. Comparison of steady and pulsatile flow in a double branching arterial model. Biomechanics 16:753-766 (1983).

28. Caro, C.G., Fish, P.J., Parker, K.H., Watkins, N., Lever, M.J. and Light, H. Effects of vasoactive agents on arterial flow pattern - relevance to atheroma. In: "Biology of the Arterial Wall." Satellite Meeting of 8th Int. Symposium on Atherosclerosis, Siena, 199-208 (1988).

29. Caro, C.G., Fish, P.J., Goss, D.E. Effect of isosorbide dinitrate on arterial haemodynamics in man. J. Physiol. 365:93P (1985).

30. Mahler, F., Brunner, H.H., Bollinger, A., Casty, M., Anliker, M. Changes in phasic femoral artery flow induced by various stimuli: A study with percutaneous pulsed Doppler ultrasound. Cardiovasc. Res. 11:454-60 (1977).

31. Dewey, C.F. Effects of fluid flow on living vascular cells. J. Biomech. Eng. 106:31-35 (1984).

32. Frangos, J.A., Eskin, S.G., McIntyre, L.V., Ives, C.L. Flow effects on prostacyclin production by cultured human endothelial cells. Science 227:1477-79 (1985).

33. Vanhoutte, P.M. The end of the quest? Nature 327:459-60 (1987).

34. Lansman, J.B. Going with the flow. Nature 331:481-482 (1988).

35. Sprague, E.A., Steinbeck, B.L., Nerem, R.M. and Schwartz, C.J. Influence of a steady-state fluid-imposed wall shear stress on the binding, internalization and degradation of low-density lipoprotein by cultured arterial endothelium. Circulation 26, 648-656 (1987).

36. Fry, D.L. Certain histological and chemical responses of the vascular interface to acutely induced mechanical stress in the aorta of the dog. Circulation Res. 24:93-108 (1969).

37. Caro, C.G., Nerem, R.M. Transport of ^{14}C-4-cholesterol between serum and wall in perfused dog-common carotid artery. Circulation Res. 32: 189-205 (1973).

38. Caro, C.G., Lever, M.J., Tarbell, J.M. Effect of luminal flow rate on transmural fluid flux in the perfused rabbit common carotid artery. J. Physiol. 365:92P (1985).

39. Karino, T., Motomiya, M., Goldsmith, H.L. Flow patterns in model and natural branching vessels. In: Schettler, G., Nerem, R.M. Schmid-Schonbein H., Morl, H., Diehm, C. (eds.) "Fluid dynamics as a localizing factor for atherosclerosis." Berlin: Springer-Verlag, 60-70 (1983).

ACKNOWLEDGEMENTS

Support is acknowledged from: MRC, Wellcome Trust, National Heart Research Fund, Bayer UK Ltd., Schwarz Pharma, AB Hassle, BUPA Medical Foundation Ltd., British Telecom.

HOW DOES THE ARTERIAL ENDOTHELIUM SENSE FLOW?

HEMODYNAMIC FORCES AND SIGNAL TRANSDUCTION

Peter F. Davies and Randal O. Dull

Department of Pathology
Pritzker School of Medicine
University of Chicago

INTRODUCTION

The focal nature of atherosclerotic lesions is associated with
patterns of altered blood flow in the major arteries (1-3), although the
precise nature of the flow in such regions is unclear. At the interface
between flowing blood and the arterial wall, a confluent monolayer of
endothelial cells operates as a signal-transduction system for hemodynamic forces associated with flow. Investigations of the influence of
pressure, stretch and shear stress upon endothelial biology have therefore been conducted with a view to linking the precise flow profiles and
the vessel wall pathophysiology. It is now clear that early atherogenesis develops in the presence of an intact endothelial monolayer (4-6,
10) consistent with the pivotal role that the endothelium may play in
this disease process. The mechanisms by which physical forces influence
endothelial biology, however, have yet to be fully defined.

By the use of in vitro techniques, together with correlative in vivo
studies, a catalog of endothelial responses to defined flow forces has
been established. Some reactions occur over lengthy periods of time,
whereas others respond within milliseconds. Using newly developed
techniques in physiology and biochemistry, measurements of instantaneous
responses to flow have led to a better understanding of the interactions
between physical forces and the cell membrane. We will briefly review
the data obtained over the last two decades and attempt to organize them
from the perspective of a temporal sequence of responses to address the
question "how do endothelial cells detect and respond to hemodynamic
forces?"

High and Low Shear Stress Forces in Atherogenesis

Hemodynamic forces can be resolved into two principal components; (1)
pressure acting perpendicular to the endothelial surfaces and (2) shear
stress, the dragging, frictional force tangential to the cell surface in
the direction of the flow. While underlying smooth muscle and extracellular matrix in arteries together with the endothelium are subjected
to pressure, the endothelium is the major recipient of the wall shear
stress. Following Fry's observations (7, 8) of endothelial denudation in
arterial segments in vitro exposed to high shear stress forces of > 400
dynes/cm^2, investigations were designed to determine whether endothelial

damage leading to focal atherogenesis occurs as a result of locally
enhanced hemodynamic shear stress. Subsequent analysis of lesion
distribution and extensive arterial modeling in vitro has supported an
alternative point-of view; that lesions develop at sites which are
subjected to low shear stress forces (9, 11). Furthermore, shear stress
levels in vivo in excess of 400 dynes/cm^2 are extremely rare in the
arterial circulation where values typically range from zero to peaks of
approximately 100 dynes/cm^2 in pulsatile flow. We have suggested, based
upon in vitro measurements, that the magnitude of shear stress may be
only one factor in determining the responsiveness of endothelial cells to
these forces. Studies on cell turnover in complex chaotic flow conditions when compared with well defined laminar flow have demonstrated
that the flow characteristics rather than the magnitude of the shear
stress may play a prominent role in determining the cellular responses
(12). Disturbances in steady laminar flow such that the endothelium is
exposed to shear stress of changing frequency and direction may be as
important as the level of the shear strews. The development of more
sophisticated experimental flow systems should help resolve these
questions.

<u>Blood Flow Patterns in Arteries</u>

The measurement of blood flow in vivo at the resolution required to
determine shear stress levels acting at the endothelium is beyond the
scope of present technologies despite advances in anemometry, pulse
Doppler, and laser Doppler techniques. In addition, a description of the
complexities of flow in very small regions, where there are only a few
endothelial cells, may be required for an understanding of the focal
initiation of atherogenesis. There has therefore been a reliance upon
detailed modeling of the arterial circulation using a number of transparent models and vascular casts. These together with fixed human
arteries made transparent by organic solvents (13) allow flow patterns to
be observed and analyzed (14-18). From such data, a consensus has
emerged concerning the flow behavior associated with defined branch
points, such as the carotid artery in the human circulation. When the
flow analyses generated by groups such as that of Goldsmith and colleagues (13) are considered in relation to the detailed morphological
and histological studies of Glagov and colleagues (19, 20), valuable
insights have evolved concerning flow in relation to the development of
atherosclerotic lesions (21).

While blood flow in regions away from branches in unidirectional and
laminar with a shear stress range of approximately 40 dynes/cm^2 (14),
flow separation occurs at bifurcation, curvatures and branches near the
outer wall of the major vessel in early systole. Just after peak
systole, the separation point moves a short distance proximally, and
helical patterns of secondary laminar flow develop within the zone of
separation. This zone represents a region of disturbed flow. Before
diastole, the separation zone expands towards the inner wall where
vortices may form and in extreme cases some turbulent characteristics of
flow may occur transiently at the inner wall. Throughout the cardiac
cycle, helices of complex secondary laminar flow occur within the
separation zone. Particles, circulating cells and macromolecules may be
trapped by recirculation within the vortex before eventually rejoining
the main flow. Thus, particle (or cell) residence time in the
recirculation zone may be considerably extended, an effect that has
implications for the efficacy of chemotactic gradients or permeability
disturbances associated with early lesion development. When wall shear
stress is measured in these regions, a low average shear stress in the
region of flow separation adjacent to the outer wall is found which
correlates well with the regions of development of the atherosclerotic

lesion. As the cardiac cycle progresses, the zone of disturbed flow
moves back and forth over short distances, for example within the
boundaries of the carotid sinus. Thus, this region will contain areas of
endothelial surface which are exposed periodically to positive, zero and
negative shear stress forces throughout the cardiac cycle (22). Such
flow characteristics are probably involved in the localization of
atherosclerotic lesions.

Throughout this discussion, it should be borne in mind that the fluid
mechanics of blood flow is more complex than classic Poiseuille flow
which applies only to a rigid tube of circular cross section subjected to
steady flow. In the arterial circulation, blood flow is pulsatile, blood
is a non-Newtonian fluid, and the artery is a compliant, tortuous tube of
changing cross sectional shape and area with many branches. Recent
developments in non-Newtonian modeling may provide better estimates of
predicted forces in vivo (23).

Further considerations of the responses of endothelial cells to
defined shear stresses will be considered primarily in vitro. First, the
slower, more sustained responses of chronic exposure to shear stress will
be briefly summarized, and second, we will discuss the very fast responses
which may provide insights into the structures responsible for the detection of physical forces in these cells. Tables 1, 2 and 3 summarize the
responses of cultured endothelium.

Slow Responses of Endothelial Cells to Shear Stress In Vitro

Cell Shape Change. It has been recognized for many years that the
orientation of arterial endothelial cells closely follows the predicted
flow patterns in vivo. A classic experiment by Flaherty and colleagues
(24) demonstrated that when arterial tissue was surgically rotated 90°
such that the shear stresses were at right angles to their previous
direction, the cells realigned with the flow over a period of days.
Shear stress as a major determinant of endothelial cell shape was first
demonstrated in vitro when a confluent monolayer was exposed in a
rotating cone device to unidirectional fluid shear stress (25). The
cells changed from a polygonal to an ellipsoidal shape and aligned
uniformly with the direction of flow as shear was increased above 5
dyne/cm^2. Numerous subsequent studies have confirmed that cell
reorientation is time and shear dependent, reversible, and measurable by
axial ratio determinations (26). Flow mediated cell shape changes occur
over at least several hours (e.g., 80 dynes for 8 hours; 8 dynes for 48
hours) and are accompanied by reorganization of the cytoskeleton. It is
probable that the cytoskeletal organization determines the changes in the
shape of the cell as a response to the shear stress. The cytoskeleton
itself is a prime candidate for transduction of plasma membrane events to
the interior of the cell. However, the precise mechanical link and its
biochemical mediation (e.g., phosphorylation) is at present, unclear.

Membrane Deformation. The mechanical properties of the surface membrane
of the endothelial cell appears to change after exposure to steady shear
stress (27). Deformation of the cell surface was measured by direct
application of a suction pipet (micropipet aspiration). As the cell
changed from a polygonal to an elongated configuration, the membrane
stiffness increased 3 fold and this was accompanied by distinct cytoskeletal changes as detected by simultaneous immunodecoration. The
cytoskeletal properties in the cortical region immediately underlying the
plasma membrane are probably responsible for the observed quantitative
changes in membrane deformability. Again, the precise morphology and
mechanism of such changes are poorly understood.

TABLE 1. <u>SLOW RESPONSES</u>
[> 3h.]

<u>Effect:</u>

1. SHAPE CHANGE Laminar shear stress > 5 dynes/cm^2 > 24h — Reorientation & alignment with flow. Reversible.

Significance: Adaptation to minimize drag

2. CYTOSKELETAL REARRANGEMENT Laminar shear stress 6 – 26 dynes/cm^2 ; > 6h — redistribution of F-actin.

Significance: Probable mechanism of cell shape change

3. LDL METABOLISM Laminar shear stress 30 dynes/cm^2 ; 24h — LDL degradation stimulated.

Significance: Endothelial cholesterol balance

4. MECHANICAL STIFFNESS Laminar shear stress 10 – 85 dynes/cm^2 ; > 3h — Cell 'surface' deformability decreased.

Significance: Deformability of membrane cytoskeleton decrease associated with cell alignment

5. CELL CYCLE <u>Turbulent</u> shear stress 1.5 – 14.0 dynes/cm^2 ; > 3h — Commitment to cell cycle.

Significance: Loss of contact inhibition of growth in confluent monolayer; related to flow disturbance.

<u>Endocytosis</u>. Another endothelial membrane event is the dynamic and continuous formation of vesicles by invagination of the plasma membrane, the process known as endocytosis. The basal rate of fluid phase endocytosis (pinocytosis) which was measured in a confluent monolayer of endothelial cells increased up to 5 fold in a time- and force-dependent fashion by application of shear stress (28). However, when the flow was continued for longer periods (in excess of six hours), adaptation of the cells to the flow stimulus occurred, as reflected by alternate inhibition and stimulation of pinocytosis at decreasing amplitudes until the basal rate of pinocytosis was reattained. The removal of cells to static conditions after chronic exposure to steady flow also resulted in a transient increase in pinocytosis rate followed by an adaptation to the new conditions, suggesting that temporal fluctuations in shear stress may influence endothelial cell function. When cells were exposed to rapidly oscillating shear stress at 1 Hertz frequency in laminar flow, the pinocytosis rate remained at the basal level and never increased. To obtain a sustained elevation of pinocytosis, it was necessary to subject cells to fluctuations in shear stress amplitude with a cycle time of at least 15 minutes. These data indicate that whatever the sensing mechanism in the endothelial cell that allows it to adjust its rate of pinocytosis to a given shear stress, a sustained signal needs to be retained for longer periods than the rapid oscillations associated with pulsatile blood flow such as occur during diurnal cycles, exercise and smoking are candidates for regulation of such cellular changes. In related studies, elevated shear stress has been shown to increase low

$$\text{TABLE 2.}\quad \underline{\text{INTERMEDIATE RESPONSES}}$$
[mins – 2 h]

<u>Effect:</u>

1. PROSTACYCLIN
a) Steady shear stress
0.9, 14 dynes/cm^2–,
2 min.
Burst of PGI2 release

b) Pulsatile laminar shear stress
Mean 10 dynes/cm^2; < 1 min.
PGI2 release rate increased over steady flow

Significance: PGI2 regulation of vascular tone and anti-thrombotic properties.

2. PINOCYTOSIS
Steady and oscillating laminar shear stress;
> 5 dynes/cm^2, minutes.
Stimulation followed by adaptation

Significance: Transient elevation in the rate of formation of plasma membrane vesicles.

3. ENDOTHELIN
Steady shear stress in laminar flow.
5-10 dynes/cm^2, 1-2-h.
2-3 fold increase mRNA levels

4. TISSUE PLASMINOGEN ACTIVATOR (tPA)
Steady laminar shear stress
> 4 dynes/cm^2
2-3 fold increased secretion

density lipoprotein endocytosis (and associated cholesterol metabolism) via increased expression of cell surface lipoprotein receptors (29). Thus, shear stress modulates both the formation of the vesicular volume (pinocytosis), as well as discrete receptor mediated events associated with the forming vesicle, but regulated via quite different mechanisms.

<u>Cell Cycle Commitment (Cell Growth)</u>. In confluent arterial endothelial cells in vivo, ^{3}H-thymidine autoradiography of DNA reflects a low incidence of cell turnover, typically 0.1% labelled (30). A similar monolayer in vitro shows a higher labeling index in the range 1 to 6% labeled cells (31). Both circumstances reflect a low turnover in the confluent monolayer. When cells are induced to align with flow in vitro, the labeling index remains unchanged (12), reflecting a remarkable maintenance of cell-cell contact during the reorientation process. In support of the in vitro results, alterations of the flow pattern in the rabbit aorta by coarctation have resulted in realignment of the endothelium in vivo, a change which was not associated with the development of atherosclerotic lesions. In contrast, a focal increase in endothelial turnover has been reported in regions subjected to disturbed flow in vivo (32). In order to model disturbed flow in vitro turbulent (Chaotic)flow has been applied to endothelial monolayers and their cell cycle responses measured (12, 33). Turbulent flow induced large increases in endothelial turnover at average shear stresses as low as 1.5 dyne/cm^2 demonstrating that cell turnover is

more sensitive to relatively low shear stresses in turbulent flow than to much higher shear stresses applied in laminar flow. The results imply that endothelial cells are also susceptible to the characteristics of flow, as well as to the magnitude of the shear stresses alone. Our interpretation of these data is that the chaotic frequencies, directions and durations of shear stresses which occur in turbulence are responsible for the loss of contact inhibition of cell growth, although the mechanisms are unclear. It may be that a physical pulling apart of adjacent cells results in loss of contact inhibition; alternatively, the flow-sensing mechanisms of the cell may initiate a series of intracellular events resulting in loss of contact inhibition. The difference between these two concepts are that one is a purely mechanical result, whereas the other retains some elements of intracellular modulation.

TABLE 3.   <u>FAST RESPONSES</u>
[Millisec – secs]

		Effect:
1. ION CHANNELS	Laminar shear stress activates K current. 0.1–16.5 dynes/cm^2. Milliseconds.	Opens K selective ion channels; Hyperpolarization.
	Stretch activates a cation – selective channel.	Cation permeability. Depolarization.
2. INOSITOL PHOSPHOLIPIDS		
	Laminar shear stress	Generation of IP^3.
3. INTRACELLULAR CALCIUM		
	Laminar shear stress	Increase of cytosolic calcium from internal sources.
4. EDRF	Flow characteristics poorly defined. Release of **humoral** agent targeted to smooth muscle.	Vascular smooth muscle relaxation.

Other endothelial responses that occur over several hours include the stimulation of tissue plasmmogen activator (tPA) secretion in response to steady laminer shear stress of 15 and 25 dynes/cm^2 (34). Lower levels (4 dynes/cm^2) did not elicit this response. Simultaneous measurement of tissue tPA inhibitor secretion revealed no change. Higher flow rates in vivo would therefore favor fibrinolytic activity.

A recent publication (35) has described a modest stimulation of endothelin mRNA and increased release of endothelin peptide by low shear stress levels in vitro when compared with stationary control cells. Endothelin is a potent vasoconstrictor synthesized by endothelium which acts antogonistically to EDRF.

<u>Fast Responses of Endothelial Cells To Hemodynamic Forces - Are These the
Flow-Sensors?</u>

The transduction of information from the cell surface to appropriate
intracellular systems occurs following mechanical, humoral and electrical
stimuli. There are useful parallels between these multiple stimuli and
the translation of the response at the plasma membrane into an ionic
and/or biochemical signal. The signals are part of an information
processing system in the cell; furthermore, information may be transmitted beyond the single cell to its neighboring cells. Numerous
examples of specialized cells which detect mechanical stimuli can be
found in all vertebrates. For example, the mechanisms within the ear
allow airborne vibrations to be converted into electrical (ionic) signals
which are recognized as sound (see review, 36). Other examples include
specialized neurons within the skin, lungs and joints which detect touch,
stretch and movement respectively. Therefore, it is not surprising that
endothelial cells which are exposed to various mechanical forces may have
evolved mechanisms that allow the cell to detect mechanical forces and to
respond very rapidly. Blood flow is known to influence constriction and
relaxation of arteries, as well as being implicated in the localization
of atherosclerotic lesions, and therefore, the role of the endothelium as
a "Mechanosensor" of flow-related forces is of great interest. While the
nature of the flow sensor is unknown, recent investigations of very early
responses to flow have provided insight into plasma membrane events and
second messenger generation associated with the endothelial flow response.

<u>Could Endothelial Ion Channels be Flow Sensors?</u>

<u>Shear Stress-Activated K^+ Channels</u>. Whole cell recordings of single
endothelial cells using the patch clamp technique have demonstrated that
the transmembrane flux of K^+ is regulated as a function of the shear
stresses acting on the cell surface (figure 1; ref 37). A K^+ selective
ionic current ($I_{K.S}$) was identified which varies in magnitude and
duration as a function of shear stress with a half maximal effect at
approximately 1 dyne/cm^2. The ion channel activity desensitized slowly
and recovered rapidly and fully when flow was stopped. $I_{K.S}$ activity
represents the earliest and fastest stimulus-response coupling of
hemodynamic forces in endothelial cells reported to date. The effects of
$I_{K.S}$ are to hyperpolarize the endothelial cell. The flow characteristic
in these experiments was steady laminar over the range 0 to 17 dyne/cm^2.
The $I_{K.S}$-shear stress relationship saturated above 15 dyne/cm^2. The
current-voltage relation rectified sharply above the potassium holding
potential, displaying a negative slope in the depolarized range similar
to the potassium inward retifier currents reported in other cell types.
When reversal potential was plotted as a function of extracellular
potassium, the relationship had a slope consistent with an ion selective
membrane. $I_{K.S}$ was blocked by extracellular perfusion with a solution
containing Ba^{2+} and by internal and external Cs^+. When atrial cells were
exposed to flow under the same conditions as endothelial cells, they did
not exhibit $I_{K.S}$ indicating some specificity associated with the
endothelial cells.

<u>Stretch-Activated Channels in Endothelial Cells</u>. Lansman and colleagues
(38) have described stretch-activated ion channels in endothelial cells
and suggested that they could be involved in the response to mechanical
forces generated by blood flow. Unlike $I_{K.S}$, the channels are selective
to a range of cations (Ca^{++}, K^+, and Na^+, other cations) and open when
the cell membrane is stretched. Similar channels have been observed in
skeletal muscle; they conduct an inward depolarizing current under
physiological conditions, a response which is opposite to the effect of
$I_{K.S}$. Furthermore, the stretch-activated channels were not stimulated by

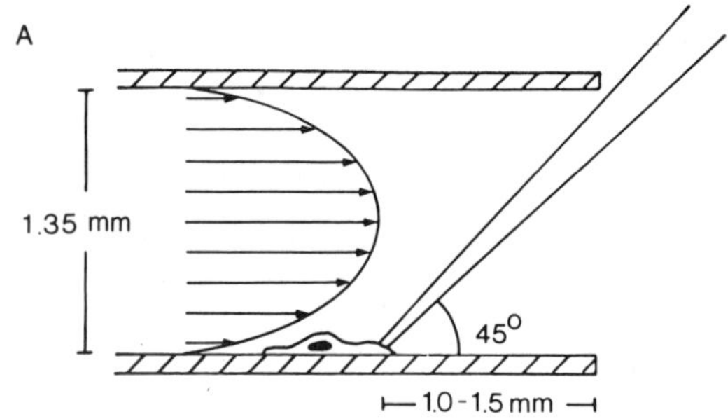

B

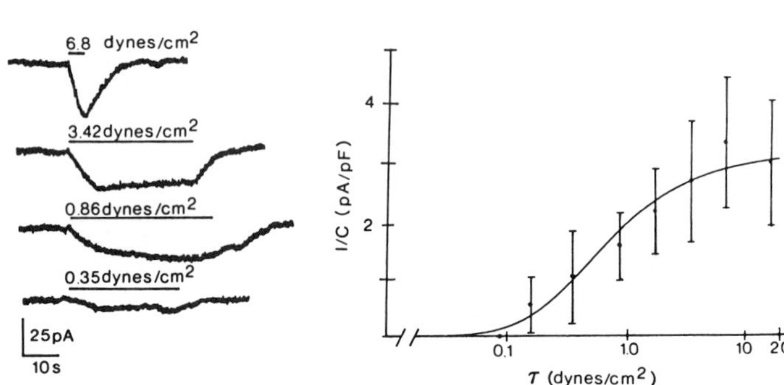

Figure 1.   A. Experimental design showing endothelial whole cell
               recording via a patch clamp pipette in a laminar flow
               tube. The tube was perfused using a syringe pump to create
               a laminar flow profile.
            B. Relationship between fluid shear stress amplitude at the
               cell surface and whole cell current. Tracings show the
               time course for whole cell currents from a single cell
               exposed to 4 different flow rates.
               The whole cell current normalized by cell capacitance is
               plotted as a function of the fluid shear stress from 7
               experiments. Maximum current was 3.2 pApF^{-1} and half
               maximal current was evoked by a shear stress of 0.7
               dynes/cm². (reprinted from ref. 37, with permission).

pure laminar flow-generated shear stress in the range 0-20 dynes/cm²
(39). These observations may reflect different mechanisms appropriate to
changing flow conditions in the arterial circulation. For example, the
stretch-activated channels may be more readily activated at locations
where the endothelial membrane is likely to be subjected to distension,
perhaps at flow dividers and bifurcations. Thus, two potential
candidates for flow sensing in endothelial cells have been identified,
one responds to shear stress and the other to pressure and stretch. As
recently summarized (37, 39) the membrane potential changes associated
with flow may regulate blood vessel tone via direct actions upon
underlying smooth muscle cells. Two general mechanisms of such activity
have been proposed. First, consistent with the existence of gap
junctions between endothelial and smooth muscle cells in the arterial
circulation, electrical coupling has been proposed as an efficient
mechanism for regulation of smooth muscle membrane potential (40).
Hyperpolarization associated with $I_{K.S}$ when transmitted directly to
adjacent smooth muscle cells will result in relaxation. Second,
endothelial cells release a potent vasorelaxant, endothelial-derived

relaxing factor(s) (EDRF). These are soluble nitro derivatives (41)
which are secreted and act on the smooth muscle cell to regulate
guanylate cyclase; elevation of cyclic GNP results in vasorelaxation.
Flow and agonist-induced EDRF release is associated with
hyperpolarization of the endothelial cell (42); thus endothelial membrane
potential is associated with vasoactive responses whatever mechanism
operates to transmit the signal to the smooth muscle cells.

<u>Arachidonic Acid Metabolism</u>. The release of arachidonic acid from
endothelial phospholipids either by the activity of a phospholipase A2 or
by hydrolysis of diacylglycerol following pholpholipase C cleavage of
inositol phospholipids, results primarily in the release of prostaglandin
I_2 (PGI2; prostacyclin) a potent inhibitor of platelet aggregation and a
vasorelaxing agent. Several reports have shown that PGI2 production by
cultured endothelial cells is stimulated following exposure to steady and
pulsatile flow (43, 44). Bursts of PGI2 release were observed which were
followed by a decline in PGI2 production over subsequent minutes. Longer
term stimulation of PGI2 production rate increased as a linear function
of shear stress up to 24 dynes/cm^2. Furthermore, when a pulsatile flow
of about 1 Hertz was applied to the cells, the PGI2 production rate
increased to a level 2.4 times that of the cells exposed to the same
average shear stress in steady flow. PGI2 production, although rapid, is
the end stage of a cascade of biochemical responses in the cell membrane
similar to that which occurs upon ligand-receptor stimulation. The
mechanical force may in some way bypass or mimic the agonist-receptor
coupling. The mobilization of intracellular calcium is intimately
associated with the same biochemical pathways as discussed below.

<u>Mobilization of Intracellular Calcium</u>. Single cell measurements of endo-
thelial intracellular free calcium concentrations are currently under investigation in a number of laboratories. Calcium is an important second
messenger within most cell types. Preliminary studies have demonstrated
that flow induces the release of internally-sequestered calcium in a
shear stress-dependent manner (45). The removal of external calcium did
not abolish the calcium increase, suggesting that mobilization occurs
from endoplasmic reticulum or other intracellular storage pools. The
shear stress-dependent calcium response appears to occur in approximately
80% of the cell population, but varies in magnitude from cell to cell.
This may reflect some form of heterogeneity in populations of cultured
endothelial cells. In other ongoing work, Frangos and colleagues
(personal communication) have demonstrated that shear stress stimulated
the formation of inositol-1, 4, 5-triphosphate (IP_3) the most widely
known mediator of intracellular calcium release (46). IP_3 is generated
as a product of phospholipase C action upon inositol phospholipid. Thus,
activation of inositol phospholipid plays a key role in the responses of
endothelial cells to shear stress via arachidonic acid metabolism and
calcium mobilization.

The analogy to receptor activation is provocative; there is a
threshold level, a dose-dependency and saturability, an increase in IP_3
formation and a mobilization of intracellular calcium, all responses
associated with agonist-receptor coupling. Neither cellular hyperpolarization nor depolarization alters resting calcium levels (Dull,
unpublished), thus excluding voltage-activated calcium channels in
regulating cytosolic calcium. It is unknown whether the shear-induced
calcium response and K^+ current are activated via the same mechanosensor
or occur independently. Since shear rates which stimulate the K^+ current
do not increase calcium, the K^+ current may not be a calcium activated K^+
channel, although other K^+ channels in endothelial cells are of this type.

We hypothesize that the mechanosensor or flow-receptor is a membrane
structure analogous to other membrane receptors. Because of its
structure, or perhaps interaction with cytoskeletal elements, the sensor
is sensitive to shear stress. Consistent with the receptor model for the
flow sensing mechanism, we further suggest that the mechanosensor is
linked with intracellular regulatory mechanisms. In general, receptor
activation stimulates phospholipase C which cleaves phosphatidyl
inositol-1, 4 bisphosphate into water soluble IP_3 and membrane bound
diacylglycerol. The diacylglycerol activates protein kinase C.
Diffusible IP_3 appears to induce the release of calcium from intracellular stores via specific, high affinity receptors. Once released,
cytosolic calcium may then participate in the modulation of other enzyme
activities such as associated with the calmodulin system as well as
feedback modulation of phospholipase C and protein kinase C. These
enzyme systems can initiate a cascade of events including the activation
of ion channels, enhanced stimulation-secretion coupling, cytoskeletal
rearrangement and the induction of transcription and translation of
specific proteins, including those associated with chronic exposure to
flow as discussed above.

In conclusion, studies of endothelial responses to flow embrace fluid
dynamics, histology, pathophysiology, electrophsiology, biochemistry, and
cell and molecular biology. A multidisciplinary approach is essential
for an understanding of the mechanisms involved and their relationship to
the physiology and pathology of blood vessels. The levels of complexity
reflect the importance of the biological implications of stiumulus-
response coupling at the level of mechanical forces in the circulation.

<u>Acknowledgement</u>

This work was supported by NIH grants HL 36028 and HL 36049.

REFERENCES

1. C.G. Caro, J.M. Fitzgerald, and R.C. Schroter, Atheroma and arterial
    wall shear, Observation, correlation and proposal of a shear
    dependent mass transfer mechanism for atherogenesis, <u>Proc R Soc
    Lond B Biol Sci</u> 177:109-159 (1971).
2. J.T. Flaherty, V.J. Ferrans, J.E. Pierce, T.E. Carew, III, and D.L.
    Fry, Localizing factors in experimental atherosclerosis. <u>In</u>:
    "Atherosclerosis and Coronary Heart Disease", pp. 40-83. W.
    Likoff, B.L. Segal, W. Insull, and S.J. Moyer, eds. Grune and
    Stratton. (1972).
3. J.F. Cornhill and M.R. Roach, A quantitative study of the
    localization of atherosclerotic lesions in the rabbit aorta,
    <u>Atherosclerosis</u> 23:489 (1976).
4. P.F. Davies, M.A. Reidy, T.B. Goode, and D.E. Bowyer, Scanning
    electron microscopy in the evaluation of endothelial integrity of
    the fatty streak lesion of atherosclerosis. <u>Atherosclerosis</u>
    25:125-130 (1976).
5. T.B. Goode, P.F. Davies, M.A. Reidy, and D.E. Bowyer, Aortic
    endothelial cell morphology observed in situ by scanning electron
    microscopy during atherogenesis in the rabbit, <u>Atherosclerosis</u>
    27:235-251 (1977).
6. A. Faggiotto, R. Ross, and L. Harker, Studies of hyper-
    cholesterolemia in the non-human primate, I. Changes that lead to
    fatty streak formation, <u>Arteriosclerosis</u> 4:323 (1984).
7. D.L. Fry, Acute vascular endothelial changes associated with
    increased blood velocity gradients, <u>Circ Res</u> 22:165-197 (1968).

8. D.L. Fry, Response of the arterial wall to certain physical factors, *Ciba Found Symp* 12:93-110 (1972).

9. C.K. Zarins, D.P. Giddens, B.K. Bharadvaj, V.S. Sottiurai, R.F. Mabon, and S. Glagov, Carotid bifurcation atherosclerosis. Quantitative correlation of plaque localization with flow velocity profiles and wall shear stress, *Circ Res* 53:502-514 (1983).

10. D.N. Ku, D.P. Giddens, C.K.Zarins, and S. Glagov, Pulsatile flow and atherosclerosis in the human carotid bifurcation. Positive correlation between plaque location and low and oscillating shear stress, *Arteriosclerosis* 5:293-301 (1985).

11. S. Glagov, C.K. Zarins, K.E. Taylor, R.A. Bomberger, and D.P. Giddens, Evidence that high flow velocity and endothelial disruption are not the prinicipal factors in experimental plaque localization. *In*: "Fluid Dynamics as a Localizing Factor for Atherosclerosis", pp. 208-211, G. Schettler, R.M. Nerem, H. Schmid-Schonbein and H. Morl, ed., Springer, Berlin (1983).

12. P.F. Davies, A. Remuzzi, E.J. Gordon, C.F. Dewey, and M.A. Gimbrone, Turbulent fluid shear stress induces vascular endothelial cell turnover in vitro, *Proc Natl Acad Sci* USA 83:2114-2117 (1986).

13. T. Karino, M. Motomiya, and H.L. Goldsmith, Flow Patterns in Model and Natural Branching Vessels, *in*: "Fluid Dynamics as a Localizing Factor for Atherosclerosis", pp. 60-70, G. Schettler, ed., Springer-Verlag, Heidelberg (1983).

14. C.F. Dewey, Jr, Dynamics of arterial flow, *Adv Exp Med Biol* 115:55-103 (1979).

15. M.H. Friedman, C.B. Bargeron, G.M. Hutchins, F.F. Mark, and O.J. Deters, Hemodynamic measurements in human arterial casts and their correlation with histology and luminal area, *J Biomech Eng* 102:247-251 (1980).

16. T. Karino and H.L. Goldsmith, Disturbed flow in models of branching vessels, *Trans Am Soc Artif Intern Organs* 26:500-505 (1980).

17. D.N. Ku and D.P. Giddens, Laser Doppler anemometer measurement of pulsatile flow in a model carotid bifurcation, *J Biomech* 20:407-421 (1987).

18. R.J. Lutz, J.N. Cannon, K.B. Bischoff, R.L. Dedrick, R.K. Stiles, and D.L. Fry, Wall shear stress distribution in a model canine artery during steady flow, *Circ Res* 41:391-399 (1977).

19. S. Glagov, E. Weisenberg, C.K. Zarins, M.P. Stankunavicius, and B.A. Kolettis, Compensatory enlargement of human atherosclerotic coronary arteries, *New Engl J Med* 316:1371-1375 (1987).

20. S. Glagov, Hemodynamic risk factors: mechanical stress, mural architecture, medial nutrition and the vulnerability of arteries to atherosclerosis, *in*: "The Pathogenesis of Atherosclerosis", pp. 164-199, R.W. Wissler and J.C. Geer, eds., Williams and Watkins, Baltimore, (1972).

21. C.K. Zarins, D.P. Giddens, B.K. Bharadvaj, V.S. Sottiurai, R.F. Mabon, and S. Glagov, Carotid bifurcation atherosclerosis. Quantitative correlation of plaque localization with flow velocity profiles and wall shear stress, *Circ Res* 53:502-514 (1983).

22. D.N. Ku, D.P. Giddnes, C.K. Zarins, and S. Glagov, Pulsatile flow and atherosclerosis in the human carotid bifurcation. Positive correlation between plaque location and low and oscillating shear stress, *Arteriosclerosis* 5:293-301 (1985).

23. D.N. Ku and D. Liepsch, The effects of non-newtonian viscoelasticity and wall elasticity on flow at a 90 bifurcation, *Biorheology* 23:359-370 (1986).

24. J.T. Flahery, J.E. Pierce, V.J. Ferrans, D.J. Patel, W.K. Tucker, and D.L. Fry, Endothelial nuclear patterns in the canine arterial tree with particular reference to hemodynamic events, *Circ Res* 30:23-32 (1972).

25. C.F. Dewey, S.R. Bussolari, M.A. Gimbrone, and P.F. Davies, The
    dynamic response of vascular endothelial cells to fluid shear
    stress, J Biomech Engin 103:177-185 (1981).
26. A. Remuzzi, C.F. Dewey, P.F. Davies, and M.A. Gimbrone, Orientation
    of endothelial cells in shear fields in vitro, Biorheology
    21:617-630 (1984).
27. M. Sato, M.J. Levesque, and R.M. Nerem, Micropipette aspiration of
    cultured bovine aortic endothelial cells exposed to shear stress,
    Arteriosclerosis 7:276-286 (1987).
28. P.F. Davies, C.F. Dewey, S.R. Bussolari, E.J. Gordon, and M.A.
    Gimbrone, Influence of hemodynamic forces on vascular endothelial
    function, J Clin Invest 73:1121-1129 (1984).
29. E.A. Sprague, V.L. Steinbach, R.M. Nerem, and C.J. Schwartz,
    Influence of a laminar steady-state fluid-imposed wall shear
    stress on the binding, internalization and degradation of low
    density lipoproteins by cultured arterial endothelium, Circulation
    76:648-656 (1987).
30. S.M. Schwartz, and Benditt, E.P. Benditt, Clustering of replicating
    cells in aortic endothelium, Proc Natl Acad Sci USA 73:651-653
    (1976).
31. S.M. Schwartz, Selection and characterization of bovine aortic
    endothelial cells, In Vitro 14:966 (1978).
32. B.A. Caplan, and C.J. Schwartz, Increased endothelial cell turnover
    in areas of in vivo Evans Blue uptake in the pig aorta,
    Atherosclerosis 17:401 (1973).
33. H.P. Sdougos, S.R. Bussolari, and C.F. Dewey, Secondary flow and
    turbulence in a cone-and-plate device, J Fluid Mech 138:379-404
    (1984).
34. S.L. Diamond, S.G. Eskin, and L.V. McIntire, Fluid flow stimulates
    tissue plasminogen activator secretion by cultured human
    endothelial cells, Science 243:1483-1485 (1989).
35. M. Yoshizumi, M. Murihara, T. Sugiyama, F. Takaku, M. Yanagisawa, T.
    Masaki, and Y. Yazaki, Hemodynamic shear stress stimulates
    endothelin production by cultured endothelial cells, Biochem
    Biophys Res Comm 161:859-864 (1989).
36. A.J. Hudspeth, How the ear's works work, Nature 341:397-404 (1989).
37. S.P. Olesen, D.E. Clapham, and P.F. Davies, Haemodynamic shear
    stress activates a K+ current in vascular endothelial cells,
    Nature 331:168-170 (1988).
38. J.B. Lansman, T.J. Hallam, and T.J. Rink, Single stretch-activated
    ion channels in vascular endothelial cells as mechanotransducers,
    Nature 325:811-813 (1987).
39. P.F. Davies, How do vascular endothelial cells respond to flow?,
    News in Physiol Sci (NIPS) 4:22-26 (1989).
40. P.F. Davies, S.P. Olesen, D.E. Clapham, E.E. Morrel, and
    F.J. Schoen, Endothelial Communication, Hypertension 11:563-572
    (1988).
41. P.M. VanHoutte, G.M. Rubanyi, V.M. Miller, and D.S. Houston,
    Endothelial relaxing factors, Ann Rev Physiol 48:307-320 (1986).
42. R. Busse, H. Fichtner, A. Luckhoff, and M. Kohlhardt, Hyper-
    polarization and increased free calcium in Ach stimulated
    endothelial cells, Am J Physiol 255:H965-H969.
43. E.F. Grabowski, E.A. Jaffe, and B.B. Weksler, Protascyclin
    production by cultured endothelial cell monolayers exposed to step
    increases in shear stress, J Lab Clin Med 105:36-43 (1985).
44. J.A. Frangos, S.G. Eskin, L.V. McIntire, and C.L. Ives, Flow Effects
    on Prostacyclin Production by Cultured Human Endothelial Cells,
    Science 227:1477-1479 (1985).
45. R.O. Dull, and P.F. Davies, Differential endothelial cytosolic
    calcium responses to hemodynamic shear stress in vitro,
    Circulation Suppl 2:481 (1989).

46. M.J. Berridge, and R.F. Irvine, Inositol Phosphates and cell signalling, _Nature_ 341:197-205.
47. B.L. Langille, M.A. Reidy, and R.L. Kline, Injury and repair of endothelium at sites of flow disturbances near abdominal aortic coarctations in rabbits, _Arteriosclerosis_ 6:146-154 (1986).

Edzard Ernst* and
Wolfgang Koenig°

*Dept. Physical Medicine and Rehabilitation Medical School
 3000 Hannover 61, PO-Box 610180, FRG
°Dept. Internal Medicine
 University of Ulm, FRG

INTRODUCTION

Smoking limits blood fluidity as measured by in vitro methods such as blood and plasma viscosities, hematocrit (for the present purpose interchangeable with hemoglobin), red cell aggregation and blood cell filtration (1). Blood rheology, in turn, can determine blood flow (2) and might play a role in the development of atherosclerosis (3). The present paper summarizes our work concerning the effects of smoking on blood rheology. Firstly, a hemorheological deficit is verified by means of an epidemiological study. Secondly, the dose dependency of the detrimental effects are analyzed in a clinical trial. Thirdly, an intervention study shows that most of the rheological effects are reversible upon cessation of smoking.

METHODS

Epidemiological Study

In the MONICA-project, Augsburg, a random sample (n=5312) of the population aged 25 to 64 years was drawn. Four-thousand twenty-two individuals were included in the present analysis. They were submitted to a standardized questionnaire and blood samples were drawn for plasma viscosity, total serum protein and hemoglobin. A detailed smoking history was recorded. Participants who, according to strict criteria, were not healthy, were excluded. The complete methodology of this trial is published elsewhere (4).

Dose-Effect Study

One-hundred fifty healthy, male blood donors were grouped according to smoking history (Table 1). They were analyzed for hematocrit, total white cell count, plasma viscosity, blood viscosity, red cell deformability (filterability), red cell aggregation and fibrinogen. Cross-sectional differences were evaluated. For methodological details, see reference 5.

Tobacco Smoking and Atherosclerosis
Edited by J. N. Diana
Plenum Press, New York, 1990

Table 1. Details on the various test groups

	n	Age	Definition
Non-smokers	30	42 ± 6	Never smoked regularly in their life
Ex-smokers	30	40 ± 4	Had smoked for at least 5 years and stopped 1-2 years before testing
Smokers I	33	38 ± 8	Smoking history >5 years, average cigarette consumption 10-20/day
Smokers II	28	39 ± 5	Smoking history >5 years, average cigarette consumption 21-40/day
Smokers III	29	43 ± 7	Smoking history >5 years, average cigarette consumption > 40/day

Abstention Study

Fourteen healthy male smokers (more than 20 cigarettes/day) with an average smoking history of 7 years participated in an abstention program of 8 weeks. The control group of this trial were those 17 individuals who relapsed into smoking. At fortnightly intervals the same hemorheological parameters as above were tested. For methodological details see reference 6.

RESULTS

Epidemiological Study

Table 2 and Table 3 show the mean values and standard errors of the mean (SEM) for hemoglobin, plasma viscosity, and total serum protein according to age for healthy non-smokers and healthy smokers. In men, there is no significant difference ($p > 0.05$) in hemoglobin between smokers and non-smokers. In female smokers, hemoglobin is significantly higher compared to female non-smokers (13.7 vs. 13.3 g/dl, $p > 0.02$).

For men there is a consistent relationship between age and plasma viscosity within smokers and a lack of one with non-smokers (interaction with $p = 0.002$). In the two youngest age groups, no smoking effect is noted. However, male smokers over 45 years old are associated with significantly elevated plasma viscosity values compared to non-smokers. In women, the effect of smoking is not modified by age (interaction with $p = 0.20$). There is, however, a non-significant trend for plasma viscosity to increase with age in female smokers similar to the one found in men. The mean values of plasma viscosity do not differ significantly between smokers and non-smokers (1.23 vs. 1.22 mPas; $p = 0.29$). No significant effect of age ($p = 0.41$) is noted.

For men there is a significant interaction between age and smoking ($p = 0.03$). Total serum protein declines with age in non-smoking men but is nearly constant in smokers. In smokers, the values of serum protein for the two lower age groups parallel those of non-smokers on a slightly lower level. In women, age ($p < 0.02$) and smoking ($p < 0.03$) are significantly related to total serum protein. Total serum protein in female non-smokers is higher than in smokers (71.4 vs. 70.7 g/l) (Tables 2 and 3). As with men, female non-smokers show a decrease of total serum protein with age, while smokers do not. The interaction is weaker than with men ($p = 0.06$).

Plasma viscosity in men shows a strong dependence on smoking habits modified by age. Because of the interaction between smoking and age,

Table 2. Hemoglobin, plasma viscosity, and protein in
non-smokers by age and sex

			n*	Hemoglobin (g/dl)	Plasma viscosity (mPas)	Protein (g/1)
Men			274	14.8(0.009)	1.23(0.003)	72.3(0.21)
25	to	34 yrs	103	14.9(0.15)	1.22(0.004)	73.2(0.31)
35	to	44 yrs	74	15.0(0.18)	1.24(0.006)	72.8(0.38)
45	to	54 yrs	66	14.4(0.14)	1.22(0.006)	71.1(0.44)
55	to	64 yrs	31	14.7(0.19)	1.23(0.009)	70.4(0.54)
Women			293	13.3(0.08)	1.23(0.003)	71.4(0.24)
25	to	34 yrs	79	13.4(0.17)	1.24(0.008)	72.8(0.42)
35	to	44 yrs	117	13.3(0.12)	1.22(0.005)	70.9(0.38)
45	to	54 yrs	73	13.2(0.15)	1.23(0.007)	70.6(0.47)
55	to	64 yrs	24	13.3(0.24)	1.26(0.010)	71.6(1.00)

Values are means ± SD   *n = number of participants

which is found mainly in the higher age group, the variables, "duration
of smoking" and "number of cigarettes," are included in a model for
smokers. This analysis is restricted to plasma viscosity in men. A weak
interaction between the duration of smoking and the number of cigarettes
smoked is found (p=0.07). Heavy smokers, who had smoked for 20 years or
longer, are associated with higher plasma viscosity than smokers with a
shorter smoking history. However, the number of cigarettes smoked per
day has a statistically significant main effect on plasma viscosity in
male smokers (p=0.03). The effect of age is also significant (p=0.02).

Dose-Effect Study

Compared with non-smokers, smokers consuming between 10 and 20
cigarettes show significantly elevated white blood cell counts plasma
viscosity and low and middle shear blood viscosity. As the number of
cigarettes smoked increases, additional abnormalities are detected: an
elevation of hematocrit and high shear blood viscosity (21-40 cig/day) as
well as a rise of fibrinogen and a decrease of blood cell filterability
(more than 40 cig/day). Hemorheological variables deteriorate in
parallel with a rise in cigarette consumption, most clearly demonstrable
for hematocrit, plasma viscosity and blood viscosity. Ex-smokers' blood

Table 3. Hemoglobin, plasma viscosity, and protein in
smokers by age and sex

Men			185	14.9(0.10)	1.24(0.004)	72.0(0.25)
25	to	34 yrs	88	15.0(0.15)	1.23(0.005)	72.2(0.36)
35	to	44 yrs	61	14.8(0.18)	1.23(0.007)	71.8(0.45)
45	to	54 yrs	27	14.9(0.23)	1.27(0.012)	71.9(0.65)
55	to	64 yrs	9	15.0(0.50)	1.27(0.012)	72.0(1.08)
Women			102	13.7(0.11)	1.22(0.006)	70.7(0.37)
25	to	34 yrs	56	13.7(0.16)	1.21(0.008)	70.6(0.51)
35	to	44 yrs	32	13.5(0.17)	1.23(0.011)	71.0(0.63)
45	to	54 yrs	11	13.9(0.40)	1.22(0.014)	69.9(1.14)
55	to	64 yrs	3	14.3(0.22)	1.31(0.009)	72.2(2.03

Table 4. Rheological variables in men with different smoking history (means and SD).

Parameter dimension	Hct %	WBC x10^9/l	PV mPas	BV 1 mPas	BV 2 mPas	BV 3 mPas	BCF units	BCA units	Fib g/l
Non-smokers	44.1	5.9	1.20	23.3	13.7	4.4	0.67	12.4	2.7
	±3.4	±2.1	±0.06	±8.0	±5.2	±0.6	±0.10	±3.2	±0.4
Ex-smokers	44.9	6.4	1.21	24.5	14.5	4.5	0.63	12.5	2.7
	±3.5	±1.9	±0.07	±9.2	±6.8	±0.7	±0.12	±4.3	±0.5
Smokers I	45.7	7.9	1.24	27.4	17.1	4.8	0.60	12.9	3.0
	±3.8	±2.0*	±0.06*	±8.1*	±7.7*	±0.9	±0.11	±4.1	±0.3
Smokers II	46.6	7.8	1.26	29.8	18.8	5.1	0.60	12.1	3.2
	±3.2*	±1.8*	±0.07*	±10.2*	±7.6*	±0.8*	±0.12	±4.4	±0.4
Smokers III	48.2	8.0	1.26	31.2	20.0	5.2	0.58	13.7	3.4
	±3.5*	±2.2*	±0.06*	±9.7*	±6.7*	±0.9*	±0.10*	±3.9	±0.5*

Abbreviations:

Hct = hematocrit, WBC = white blood cell count, PV = plasma viscosity, BV = blood viscosity; 1 = at shear rate 0.7 s-1, 2 = at shear rate 2.4 s-1, 3 = at shear rate 94.5 s-1; BCF = blood cell filterability, BCA = blood cell aggregation, Fib = Fibrinogen
* = p < 0.01 in comparison to non-smokers

rheology is not significantly altered compared with either non-smokers or
smokers consuming 10-20 cig/day (Table 4).

<u>Abstention Study</u>

Two weeks after abstention from smoking, low shear blood viscosity
has fallen significantly. Two weeks later, all viscosity values have
declined. Furthermore, the white blood cell count was reduced. This
holds true also for the 6- and 8-week observations where, in addition,
hematocrit, blood cell filterability and fibrinogen were also
significantly altered (Table 5). A progressive change is apparent for
most variables during abstention. No significant alterations of any
parameter occurred in the relapse group.

DISCUSSION

The results confirm a pronounced hemorheological deficit in smokers
(7). This is predominantly due to hemoconcentration,
hyperfibrinogenemia, and leukocytosis (4,5,6). New findings are,
firstly, the apparent sex difference in the smoking-induced rheological
abnormalities; secondly, a clear dose-effect dependency; and thirdly, the
reversibility of the rheological deficit upon cessation of smoking.

It is not clear why women yield different results than men. In women
the threshold for increased fibrinogen synthesis could be higher than in
men. If this were true, only excessive smoking would lead to a rise in
plasma viscosity, while the average rheology of the total female
population would remain statistically unchanged. In our analysis, female
smokers showed an increase in hemoglobin but their plasma viscosity
values were unaffected by smoking. This suggests an enhanced
erythropoiesis in female smokers. Hence, the threshold for fibrinogen
synthesis in women may be higher and the threshold for an erythropoietic
stimulus may be lower. Clearly these hypothesis need further
experimental testing.

The dose dependency of blood rheology and smoking might have been
expected. Relative hypoxia seems to be the main stimulus for the
rheological sequaele of smoking, and this is obviously related to the
number of cigarettes inhaled.

The reversibility of the rheological abnormalities finally is good
news for all who manage to quit. This process seems somewhat faster than
the change of other body functions. It could contribute to the
normalization of blood perfusion after giving up smoking which is
suggested by various lines of evidence (2).

The biological meaning of these hemorheological findings may be
twofold. First, the hemorheological deficit of smokers could be a marker
of an increase in arteriosclerotic risk. According to our working
hypothesis (3) both the early endothelial damage and the changes leading
to decreased blood fluidity are caused by similar surface adsorption
phenomena taking place both at the inner vessel lining and at the outer
surface of blood cells (in fact at all surfaces that come into contact
with blood plasma). This could explain the most striking association
between cardiovascular risk factors and pathological flow properties of
blood. Second, the viscosity of blood will increase the viscous
component of the total peripheral resistance (7) and could participate in
reducing blood flow in organs such as the heart or the brain (2,8). With
all other factors constant, perfusion is directly dependent on the
fluidity of blood (7). In situations where the vasomotor reserve is
limited, hemorheological mechanisms can lead to a substantial reduction

Table 5. Longitudinal study of hemorheological variables after abstention
of smoking (means and SD).

Parameter (dimension)	Baseline (n = 14)	2 weeks (n = 14)	4 weeks (n = 14)	6 weeks (n = 11)	8 weeks (n = 10)
BV 0.7 s-1 (mPas)	30.2±9.4	27.5±8.8*	26.01±8.2*	26.2±8.4*	24.1±7.5*
BV 2.4 s-1 (mPas)	18.6±5.4	17.9±5.1	15.5±5.0*	14.8±4.8*	14.5±3.9*
BV 94.5 s-1 (mPas)	5.2±5.4	5.2±0.8	4.9±0.8*	4.7±0.7*	4.6±0.7*
PV (mPas)	1.26±0.04	1.26±0.05	1.24±0.04*	1.22±0.05*	1.22±0.06*
Hct (%)	49.1±4.0	48.0±3.5	46.3±3.6	45.9±3.9*	45.7±3.5*
BCF (units)	0.52±0.10	0.51±0.11	0.57±0.10	0.56±0.09	0.60±0.10*
RCA (units)	14.4±3.9	14.0±4.2	14.3±4.0	13.8±3.8	13.2±3.9
Fib (g/l)	3.8±1.2	3.8±1.1	2.9±0.9	2.6±0.8	2.4±1.0*
WBC ($\times 10^6$/l)	7.4±2.0	7.1±2.1	7.2±2.0*	6.8±1.8*	6.2±1.7*

* $p < 0.01$, compared to baseline

Abbreviations in Tab. 4

in volume flow. On the microcirculatory level, rheological properties of
blood cells may interfere with vasomotion, leading to a significant
maldistribution of oxygen supply (8). Both these possibilities need
further, careful evaluation and experimental testing.

REFERENCES

1. G. Galea and R.J.L. Davidson. Haematological and haemorheological
    changes associated with cigarette smoking, J. Clin. Path. 38:978
    (1985).
2. R.L. Rogers, J.S. Meyer, B.W. Judd and K.F. Mortel. Abstention
    from cigarette smoking improves cerebral perfusion among elderly
    chronic smokers, J. Amer. Med. Ass. 253:2970 (1985).
3. E. Ernst, T. Weihmayr, M. Schmid, M. Baumann and A. Matrai. Cardio-
    vascular risk factors and hemorheology-physical fitness, stress
    and obesity, Atherosclerosis 59: 263 (1986).
4. E. Ernst, W. Koenig, A. Matrai, B. Filipiak and J. Stieber, Blood
    rheology in healthy cigarette smokers, Arteriosclerosis 8: 385
    (1988).
5. E. Ernst, A. Matrai, C. Schmölzl and I. Magyarosy, Dose-effect
    relationship between smoking and blood rheology. Brit. J.
    Haematol. 65: 485 (1987).
6. E. Ernst and A. Matrai. Abstention from chronic cigarette smoking
    normalizes blood rheology, Atherosclerosis 64: 75 (1987).
7. S. Chien, J. Dormandy, E. Ernst and A. Matrai, Clinical Hemorheology,
    M. Nijhoff, Dordrecht (1987).
8. E. Ernst, A. Matrai and J. Dormandy. Blood cells and the control
    of circulation, In: Cerebral ischemia and hemorheology, A.
    Hartmann, W. Kuschinsky, eds., Springer, Berlin (1987).

EFFECTS OF CIGARETTE SMOKING ON CORONARY VASCULAR DYNAMICS:

RELATIONSHIP TO CORONARY ATHEROSCLEROSIS

Lloyd W. Klein and Annabelle S. Volgman

Northwestern Memorial Hospital and
Northwestern University Medical School
Chicago, Illinois

Substantial evidence has been elucidated regarding the mechanisms
that link cigarette smoking with the development and accelerated
progression of atherosclerosis. Influences on platelet function, lipid
metabolism, coronary artery tone, the sympathetic nervous system, the
endocrine system, and oxygen transfer and dissociation from hemoglobin
have been identified. The purpose of this chapter is to review previous
work outlining the effect of smoking on coronary and systemic
hemodynamics, and to discuss how repetitive alterations in vascular tone
may be an integral part of the pathway by which smoking leads to an
increased risk of coronary atherosclerosis.

Systemic Hemodynamic Effects of Cigarette Smoke and Nicotine

The inhalation of cigarette smoke and the parenteral administration
of nicotine lead to a well-defined acute cardiovascular response. In
healthy humans and animal subjects, this includes increases in blood
pressure, heart rate, and cardiac output (1). Smoking either low or high
nicotine cigarettes produces similar changes in systemic hemodynamic
parameters (2). Although myocardial contractile force and velocity of
contraction also increase, changes in stroke volume are variable. Thus,
the increase in cardiac output is due mainly to the increase in heart
rate.

Smoking also increases both preload and contractility, which together
with the increased heart rate and systolic blood pressure, cause a
significant increase in myocardial demand within the first 2-3 minutes
(1,2). These same parameters return to baseline at a time interval of
about 15-20 minutes. In patients with significant coronary artery
disease, smoking causes acute left ventricular dysfunction, sometimes
associated with decreased stroke volume (3). This effect is partly due
to the induction of regional myocardial ischemia and partly a consequence
of the increased afterload resulting from increased peripheral resistance
caused by arteriolar vasoconstriction.

In subjects with coronary artery disease who are not restricted in
activity, i.e., not supine on a laboratory table, cigarette smoking
causes a significant rise in the arterial pressure five minutes after the
onset of smoking and occurs with all types of activity except when the
subject is lying in bed before sleep (4). Smoking causes an acute fall

in the blood pressure and heart rate for 8-10 beats after the first inhalation of tobacco smoke, perhaps involving a vasovagal reaction, but this is followed by a rebound rise in blood pressure to levels higher than the control value.

Mechanisms of Systemic Hemodynamic Effects

Nicotine stimulates all sympathetic ganglia, resulting in the release of norepinephrine from postganglionic adrenergic terminals. Peripheral arteriolar vasoconstriction is produced by direct ganglionic stimulation, by facilitation of the conduction of impulses through ganglia, and by chemoreceptor activation in the medulla and the aortic and carotid bodies. Arterial blood pressure rises consequent to the increased peripheral resistance (5). Heart rate increases as a result of both adrenal catecholamine release and direct chronotropic effects. Although the adrenal secretion of catecholamines increases acutely with cigarette smoking, the increases in heart rate and blood pressure occur prior to the rise in norepinephrine and epinephrine levels (6). Therefore, the initial hemodynamic changes are likely due to norepinephrine release from adrenergic axon terminals within the cardiovascular tissues, and not to augmented release of circulating catecholamines. There is no evidence that smoking affects the renin-angiotensin system. Richards et al. found no changes in these hormones measured in the venous circulation in six healthy volunteers after cigarette smoking (7).

Mechanisms of Increased Myocardial Demand

The ultimate significance of these diverse neurohumoral effects, from the cardiac viewpoint, is to increase myocardial oxygen demand and consumption (1). The increased demand is dictated by all of the observed changes in heart rate, contractility and loading conditions resulting from the systemic hemodynamic response to smoking. An alternative, though not necessarily contradictory, explanation is that smoking increases the myocardial production of adenosine, thereby directly increasing both cardiac contractility and oxygen consumption (8). The correlation between the increased demand due to smoking and cardiac adenosine appears to be independent of beta-adrenergic and muscarinic influences, suggesting the possibility of a powerful and important relationship.

Effects on Blood Oxygen and Relation to Myocardial Supply

The oxygen-carrying capacity of the blood is diminished in chronic smokers as a result of elevated carboxyhemoglobin levels (9). Carbon monoxide, a major component of cigarette smoke, is irreversibly bound to hemoglobin. This reduces oxygen exchange in the capillary beds and limits its availability for myocardial cellular aerobic metabolism. The resultant high concentration of serum carboxyhemoglobin, together with the induction of relative hypoxia, leads to polycythemia in some individuals.

Effects on Coronary Blood Flow

Several studies have delineated the acute coronary hemodynamic response to cigarette smoking, and the available data suggests that cigarette smoking causes important coronary vasomotor effects (10-14). In normal subjects, the acute coronary hemodynamic response to smoking is an increase in coronary blood flow with a slight decrease in coronary resistance (10-12,14). However, chronic smokers are less able to conduct an increased amount of blood through the coronary vasculature in response to a sudden increase in myocardial oxygen demand (decreased coronary flow

reserve); this may be a result of a chronic vasoconstrictive effect of
smoking on the coronary arteriolar bed (13,14), an effect that is more
pronounced in heavy smokers (more than one pack per day). Reduced
vasodilator reserve due to chronic smoking has also been shown in the
human hand, (15) providing evidence of a general microcirculatory
abnormality.

The presence of a significant coronary artery stenosis alters the
observed response. Typically, coronary flow remains the same or even
decreases, with a concomitant rise in coronary resistance (10). In
patients with proximal stenoses, and hence those with a large segment of
myocardium at risk, coronary resistance may rise by as much as 10% (10).
The precise site in the coronary vasculature at which these
vasoconstrictor effects take place – the stenosis itself or in the distal
vasculature – remains uncertain.

These results have been confirmed in studies by Martin et al. (11)
and Nicod et al. (12). Both studies compared coronary sinus blood flow
at comparable levels of myocardial oxygen demand before and during
cigarette smoking. By incrementally pacing the atrium, observed changes
in coronary sinus blood flow can be ascribed to the direct effects of
smoking, and not to a consequence of changes in the heart rate. Both
studies concluded that, in patients with coronary artery disease, smoking
decreases coronary sinus blood flow and increases coronary vascular
resistance independently of changes in heart rate and double product.

Clinical Sequelae

The net result of these effects is that in patients with coronary
artery disease, increased myocardial oxygen demand may not be
counter-balanced by increased coronary flow, which could potentially
precipitate myocardial ischemia. Smoking increases the coronary
arterio-venous oxygen difference in patients with coronary artery disease
equally to that of normal persons, reflecting the fact that myocardial
demand increases substantially (10,14). In this setting, cigarette smoke
or nicotine may elicit an increase in the metabolic requirements of the
myocardium that cannot be matched by adequate increases in supply. So as
a result, patients with symptomatic coronary disease, cigarette smoking
can lead to deleterious clinical effects, inducing angina pectoris at a
lower than usual workload (1) and of a more severe nature (16). The
elevated carboxyhemoglobin levels may further provoke angina, reducing
exercise capacity before the onset of symptoms and increasing their
severity (1).

Interestingly, electrocardiographic ST segment depression may occur
before the subjective complaint of pain (4). These findings were more
recently supported by Deanfield et al. (17) using positon emission
tomography. During exercise, eight of thirteen study patients had
angina, ECG changes and abnormal regional myocardial perfusion. Of these
eight patients, six had decreased rubidium uptake after smoking in the
same segment of myocardium affected during the exercise; however, during
smoking, these patients did not complain of angina. These data support
the concept that cigarette smoking may be a causative factor in producing
silent myocardial ischemia.

Possible Mechanisms of Effect on Coronary Vascular Tone

Four possible mechanisms have been proposed to explain the coronary
vasomotor effects of smoking: (1) rise in arginine vasopressin; (2)
inhibition of prostacyclin synthesis and imbalance of intracoronary

concentrations of prostacyclin and thromboxane; (3) alpha adrenergically
mediated vasoconstriction; and (4) coronary vasospasm.

The first suggested mechanism, the rise of arginine vasopressin
(18,19) in response to cigarette smoking, is not well supported by the
literature. It has been shown by Nicod et al. (20) that coronary flow
changes are unrelated to changes in plasma levels of arginine
vasopressin, which do not acutely change in response to smoking.

The second mechanism involves the complex interaction of platelets,
vascular endothelial tissue and prostaglandin metabolism. Prostacyclin
(PGI_2), which is generated by the endothelial cells, is a potent
vasodilator and inhibitor of platelet aggregation. Thromboxane A2 (TXA2)
is produced and released by platelets, mediating both platelet
aggregation and vasoconstriction. The local balance between these two
prostaglandins is widely believed to be a major controlling influence of
both vascular tone and thrombosis. Several studies have investigated the
effects of cigarette smoking on the relative rates of synthesis and
metabolism of these substances in an effort to draw conclusions about a
possible cause and effect relationship.

One line of investigation suggests that smoking and nicotine may
inhibit the synthesis of prostacyclin. In an in vitro study involving
human umbilical artery, rabbit aorta, rat aorta and rat lung tissue,
Jeremy et al. showed that cigarette smoke, although not nicotine
infusion, inhibits prostacyclin synthesis (21). Wennmalm and Alster
found that both smoking and oral indomethacin, a prostacyclin synthetase
inhibitor, prevents the normal reactive hyperemic response in human
arterial arm flow. Interestingly, smoking has no additive effect when
subjects are pre-treated with indomethacin, suggesting a common
mechanism. Further, both nicotine and indomethacin decrease prostacyclin
synthesis without a demonstrable effect on thromboxane synthesis in
rabbit or human platelet microsomes (22,23). Nadler et al. found that
chronic smokers have significantly reduced urinary prostacyclin
metabolites compared to non-smokers (24,25). Acute high-nicotine
cigarette smoking decreases urinary prostacyclin metabolite excretion in
both chronic smokers and non-smokers. Mikhailidis proposed that smoking
causes a rise in serum non-esterified fatty acids, which may cause
prostacyclin synthesis inhibition, accelerate the decay of prostacyclin,
and influence the renal clearance of prostacyclin metabolites (26).

Furthermore, it is well known that chronic smoking alters endothelial
structures, denuding the vessel surface and promoting platelet adhesion.
This alteration in vascular integrity may well explain decreased
prostacyclin synthesis in smokers (27) and these endothelial changes may
result from repetitive changes in coronary tone due to smoking. Hence,
smoking may damage coronary vascular endothelium via repeated hemodynamic
stress, leading to the first step in the induction of atherosclerosis
(28).

Several observations, however, do not support these findings. No
consistent steady state change in arterial concentrations of prostacyclin
in chronic smokers with coronary artery disease has been found (29).
Furthermore, chronic smokers may actually have increased urinary
prostacyclin metabolites (30). In addition, although smokers and
non-smokers have the same serum levels of thromboxane, smokers have a
significantly increased level of urinary thromboxane metabolites. The
administration of aspirin reverses this finding, and when the aspirin is
stopped, the level of urinary thromboxane metabolites are again increased
in the smokers after seven days. Neither Nicod (12) nor Martin (11) were
able to demonstrate significant changes in aortic or coronary sinus

prostacyclin with smoking. However, these negative results may have
technical explanations (28). Consequently, the literature on
prostaglandin metabolism and smoking must be regarded as conflicting at
this time.

The third postulated mechanism is coronary vasoconstriction, either
at the stenosis site or in the distal arterioles. It is thought that
increased catecholamines during smoking enhances alpha-adrenergic
stimulation without beta-adrenergic compensation (10). Recently, this
hypothesis has gained new support from the finding of beta-adrenergic
receptor down-regulation in chronic cigarette smokers. Laustiola et al.
recently demonstrated in a study of monozygotic twins discordant for
cigarette smoking that the smoking twins had decreased beta-adrenergic
receptors in their lymphocytes, (31) suggesting that the acute autonomic
neurohumoral response to smoking eventually leads to chronic changes.
This finding could well explain the diminished coronary flow reserve in
chronic smokers (13,14).

Laustiola's findings also support the observations of Winniford and
colleagues (32). They found that the administration of phentolamine, an
alpha-adrenergic blocker, during smoking abolished the vasoconstrictor
effects, while the concomitant administration of propranol, a
beta-adrenergic blocker, increased these effects. Although this suggests
that the smoking-induced flow changes are due to unopposed
alpha-adrenergic stimulation, Winniford's data may merely reflect the
combination of two competing pharmacologic responses rather than simple
receptor blockade.

Winniford et al. have further demonstrated that the administration of
verapamil, nifedipine, and nitroglycerin also prevent smoking-induced
coronary flow changes (33). It must be remembered, however, that the
therapeutic effects of these three drugs are multifactorial, and include
preventing coronary artery vasoconstriction or spasm, redistributing
regional flow and decreasing preload and afterload (34). Thus, while
these observations are important in a clinical sense, they do not provide
solid support for any single mechanism.

Although an epidemiologic association between smoking and coronary
spasm has been suggested (35,36), frank coronary spasm caused by smoking
rarely, if ever, occurs. Maomad (37) presented five patients with
variant angina in whom smoking appeared to cause diffuse changes in
coronary artery caliber, but only two had focal spasm induced by
smoking. These observations have not been confirmed in other
laboratories at this time.

A very interesting study in support of either smoking-induced
vasoconstriction or frank vasospasm was performed by Cox et al. (38).
The effect of two years of smoking on canine coronary arteries was
studied. The authors found that the maximum values of active force
development was increased in the animals exposed to smoking, as compared
to the control group. No significant differences were found in the
passive mechanical properties or collagen and elastin content of arteries
from the two groups. They suggested that chronic smoking may augment
coronary artery smooth muscle contractility.

Finally, there is some evidence for interaction of these mechanisms
in a study by Folts and Bonebrake (39). They studied the responses of
canine circumflex coronary arteries in which experimental 60-80%
occlusions were placed. Phentolamine prevented an acute occlusive
platelet thrombus in 18 of the 21 dogs studied. Catecholamines are known
to stimulate platelet aggregation and thrombus formation by an alpha-

adrenergic mechanism, an effect augmented by smoking. Thus, this study
provides evidence of the complexity of the relationship between all of
the proposed mechanisms.

<u>Hypothesis of How Smoking Accelerates Atherosclerosis</u>

Recent evidence supports the hypothesis that smoking-induced changes
in coronary vasomotor tone are a major component of the pathogenesis of
atherosclerosis (28). As noted above, smoking may initiate the
development of atherosclerosis by injuring the vascular intima as a
result of persistently increased coronary artery tone and frequent
episodes of hemodynamic stress. As has been observed in several
laboratories, smoking decreases pulse-wave transit time despite no change
in arterial dimension and with increased resistance, thereby increasing
arterial wall stiffness (40). The pattern of arterial flow appears to be
an important contributing factor in the initiation of the atherosclerotic
process by contributing to increased turbulence and shear forces, leading
to endothelial damage and denudation (28).

This induction of coronary intimal damage with loss of endothelial
integrity is the most likely initial event by which smoking accelerates
the development of atherosclerosis and predisposes to the occurrence of
subsequent coronary events (41). Damage may be initiated by inducing
local hypoxia or by some mechanical effect. For example, intraluminal
thrombosis with subsequent organization and thickening and intimal
reaction with resultant endothelial injury (42,43) have been
hypothesized. A desquamating effect on aortic vascular endothelium has
been shown in rabbits (27,44). Degenerative changes of the endothelium
have been found in umbilical arteries of infants whose mothers smoked
cigarettes. Increased lysosomal fragility and vacuolation (45) may
represent the initial steps in this process. Once vascular injury
occurs, chronic hyperlipidemia may then accelerate vascular lipid
accumulation (46).

In an elegant study, Oberai et al. showed coronary and renal
arteriolar thickening in chronic smokers (47). Increased collagen
content and hyperplasia of smooth muscle were responsible for the
increase in wall thickness in cardiac small vessels. The fact that only
arterioles in these two organs, which are particularly responsive to
vascular tone alterations, were affected and not those in other organ
systems is a very provocative finding.

The deposition of platelets at sites of endothelial damage in the
coronary arteries may be an initial stage in the development of
atherosclerotic plaques (41). Cigarette smoking and nicotine increase
platelet adhesiveness (48) and aggregability (49-53) and lead to a
shortened platelet survival time (52-53). These effects have been
ascribed to the stimulation of adrenal catecholamine secretion. Once
platelet deposition is initiated, the process tends to be
self-perpetuating, as more thromboxane A_2 is released locally. As
thrombosis (51,54) and coronary vasoconstriction represent the ultimate
mechanisms by which myocardial infarction occurs an entire pathway
linking smoking, atherosclerosis (55-58) and its clinical sequelae
becomes apparent. Indeed, the size of an infarction (59) may be
increased by daily exposure to nicotine before the acute event.

<u>Conclusion</u>

The acute systemic and coronary hemodynamic effects of smoking pose a
repetitive and constant stress on the vasculature. A person who smokes
one pack per day has been self-exposed to significant vasoconstrictor

stimuli for over 25% of any 24-hour time period. Smoking acutely alters
the steady state balance of local and systemic mediators of vascular
tone, resulting in measurable effects on coronary blood flow and
myocardial demand. It is highly likely that a number of interrelated
mechanisms act in combination to cause dynamic vascular alterations in
coronary vasomotor tone. Most importantly, the constant stimulation of
these mechanisms over the course of time can be linked by both theory and
anatomic study to early initiation of atherosclerosis, occurrence of
chronic vascular changes, progression of coronary artery disease and
acute coronary thrombosis.

REFERENCES

1. Klein LW. Effects of cigarette smoking on systemic and coronary
    hemodynamics, in Blum A (ed). The Cigarette Underworld.
    Secaucus, New Jersey, Lyle Steward, 24-26 (1985).
2. Rabinowitz BD, Thorp K, Huber GL. Acute hemodynamic effects of
    cigarette smoking in man assessed by sytolic time interval and
    echocardiography. Circulation, 60:752-760 (1979).
3. Jain AC, Bowyer AF, Marshall RJ. Left ventricular function after
    cigarette smoking by chronic smokers: Comparison of normal
    subjects and patients with coronary artery disease. Am J Cardiol,
    39:27-31 (1977).
4. Cellina GU, Honour AJ, Littler WA. Direct arterial pressure, heart
    rate, and electrocardiogram during cigarette smoking in
    unrestricted patients. Am J Cardiol, 89:18-25 (1975).
5. Klein LW. Cigarette smoking and coronary artery disease: Recent
    findings and implications for patient management. Card Rev Rep,
    8:74-78 (1987).
6. Cryer PE, Haymond MW, Santiago JV. Norepinephrine and epinephrine
    release and adrenergic mediation of smoking - associated
    hemodynamic and metabolic events. N Engl J Med, 295:573-577
    (1976).
7. Richards AM, Idram H, Nicholls MG. Ambulatory pulmonary arterial
    pressures in humans: relationship to arterial pressure and
    hormones. Am J Physiol, 251:H101-H108 (1986).
8. Fenton RA, Dobson JG. Nicotine increases heart adenosine release,
    oxygen consumption, and contractility. Am J Physiol, 249:H463-469
    (1985).
9. Turino GN. Effects of carbon monoxide on the cardio-respiratory
    system. Carbon monoxide toxicity. Physiology and biochemistry.
    Circulation, 63:253-259 (1981).
10. Klein LW, Ambrose J, Pichard A. Acute coronary hemodynamic response
    to cigarette smoking in patients with coronary artery disease. J
    Am Coll Cardiol, 3:879-886 (1984).
11. Nicod P, Reher R, Winniford MD. Acute systemic and coronary
    hemodynamic and serologic responses to cigarette smoking in long
    term smokers with atherosclerotic coronary artery disease. J Am
    Coll Cardiol, 4:964-971 (1984).
12. Martin JL, Wilson JR, Ferraro N. Acute coronary vasoconstrictive
    effects of cigarette smoking in coronary heart disease. Am J
    Cardiol, 54:56-60 (1984).
13. Klein LW, Pichard AD, Holt J. Effects of chronic tobacco smoking
    on the coronary circulation. J Am Coll Cardiol, 1:421-426 (1983).
14. Kaijser L, Berglund B. Does prostacyclin synthesis inhibition
    significantly contribute to nicotine effects on coronary blood
    flow? Adv Prost Thromb Leuk Res, 13:97-100 (1985).
15. Richardson DR. Effects of habitual tobacco smoking on reactive
    hyperemia in the human hand. Arch Env Health, 40:114-119 (1985).

16. Fox KM, Jonathan A, Williams H. Interaction between cigarettes and propranolol in the treatment of angina pectoris. Br Med J, 281:191-193 (1980).
17. Deanfield JE, Shea MJ, Wilson RA. Direct effects of smoking on the heart: silent ischemic disturbances of coronary flow. Am J Cardiol, 57:1005-1009 (1986).
18. Hussain MK, Frantz AG, Ciarochi F. Nicotine stimulated release of neurophysin and vasopressin in humans. J Clin Encocrin Metab, 41:1113-1117 (1975).
19. DeLorgeril M, Reinharz A, Busslinger B. Acute influence of cigarette smoke in platelets, catecholamines and neurophysins in the normal conditions of daily life. Eur Heart J, 6:1063-1068 (1985).
20. Nicod P, Winniford MD, Campbell WB. Alterations in coronary blood flow induced by cigarette smoking: lack of relation to plasma arginine vasopressin concentrations. Am J Cardiol, 54:667-668 (1984).
21. Jeremy JY, Mikhailidis DP, Dandona P. Cigarette smoke extracts, but not nicotine, inhibit prostacyclin (PgI2) synthesis in human, rabbit and rat vascular tissue. Prostaglandins, Leukotrienes, and Medicine, 19:261-270 (1985).
22. Wennmalm A and Alster P. Nicotine inhibits vascular prostacyclin but not platelet thromboxane formation. Gen Pharmac, 14:189-191 (1983).
23. Wennmalm A. Cigarette smoking prostaglandins, and reactive hyperemia. Prostaglandins and Medicine, 3:321-326 (1979).
24. Nadler JL, Velasco JS, Horton R. Cigarette smoking inhibits prostacyclin formation. Lancet, 1:1248-1250 (1983).
25. Nadler JL, Velasco JS, Duke R. Evidence that nicotine alters the balance of prostacyclin and thromboxane in man. Clinical Research, 32:192A (1984).
26. Mikhailidis DP, Barradas MA, Jeremy JY. Cigarette smoking inhibits prostacyclin formation. Lancet, 2:627-628 (1983).
27. Pittilo RM, Mackie JJ, Rowles PM. Effects of cigarette smoking on the ultrastructure of rat thoracic aorta and its ability to produce prostacyclin. Thromb Haemost, 48:173-176 (1982).
28. Klein LW. Cigarette smoking, atherosclerosis and the coronary hemodynamic response: A unifying hypothesis. J Am Coll Cardiol, 4:972-974 (1984).
29. van der Giessen WJ, Serruys PW, Stoel I. Acute effect of cigarette smoking on cardiac prostaglandin synthesis and platelet behavior in patients with coronary heart disease, in Samuelson B, Paolette R, Ramwell PW (eds): Advances in Prostaglandin, Thromboxane, and Leukotriene Research. New York, Raven Press, 359-364 (1983).
30. Nowak J, Murray JJ, Oates JA. Biochemical evidence of a chronic abnormality in platelet and vascular function in healthy individuals who smoke cigarettes. Circulation, 76:6-14 (1987).
31. Laustiola KE, Lassila R, Kaprio J. Decreased beta-adrenergic receptor density and catecholamine response in male cigarette smokers. Circulation, 78:1234-1240 (1988).
32. Winniford MD, Wheelan KR, Kremers MS. Smoking-induced coronary vasoconstriction in patients with atherosclerotic coronary artery disease: evidence for adrenergically mediated alterations in coronary artery tone. Circulation, 73:662-667 (1986).
33. Winniford MD, Jansen DE, Reynolds GA. Cigarette smoking-induced coronary vasoconstriction in atherosclerotic coronary artery disease and prevention by calcium antagonists and nitroglycerin. Am J Cardiol, 59:203-207 (1987).
34. Klein LW, Segal BL, Helfant RH. Dynamic coronary stenosis behavior in classic angina pectoris: Active process or passive response? J Am Coll Cardiol, 10:311-313 (1987).

35. Judgutt BI, Stevens GF, Zacks DJ. Myocardial infarction, oral contraception, cigarette smoking, and coronary artery spasm in young women. Am Heart J, 106:757-761 (1983).
36. School JM, Benacerrat A, Ducimetiere P. Comparison of risk factors in vasopastic angina without significant fixed coronary narrowing and no vasopastic angina. Am J Cardiol, 57:199-202 (1986).
37. Maouad J, Fernandez F, Barrillon A. Diffuse or segmental narrowing (spasm) of the coronary arteries during smoking demonstrated on angiography. Am J Cardiol, 53:354-355 (1984).
38. Cox RH, Santamore WP, Talenko TN. Alterations in canine coronary arteries produced by chronic cigarette smoking. Proc Soc Exp Biol Med, 181:151-162 (1986).
39. Folts JD and Bonebrake FC. The effects of cigarette smoke and nicotine on platelet thrombus formation in stenosed dog coronary arteries: inhibition with phentolamine. Circulation, 65:465-470 (1982).
40. Caro CG, Parker KH, Lever MJ. Effect of cigarette smoking on the pattern of arterial blood flow: possible insight into mechanisms underlying the development of arteriosclerosis. The Lancet, 11-13 (1987).
41. Fuster V. Pathogenesis of atherosclerosis, in Spittel JA (ed): Clinical Medicine, Vol 6, NY, Harper-Row, 1-22 (1981).
42. McGill HC, Jr. Potential mechanisms for augmentation of atherosclerosis and atherosclerotic disease by cigarette smoking. Prev Med, 8:390-403 (1979).
43. Sieffert GF, Keown K, Moore WS. Pathologic effect of tobacco smoke inhalation on arterial intima. Surgical Forum, 32:233-335 (1981).
44. Booyse FM, Osikowcz G. Quarfoot AJ. Effects of chronic oral consumption of nicotine on the rabbit aortic endothelium. Am J Pathol, 102:229-238 (1981).
45. Wenzel DG, Reed BL. Measurement of lysosomal fragility produced by hypoxia, nicotine, carbon monoxide and hyperoxia in cultured endothelioid cells. Res Commun Chem Pathol Pharmacol, 7:745-754 (1974).
46. Ross R. Harker L. Chronic hyperlipidemia initiates and maintains lesions by endothelial cell desquamation and lipid accumulation. Science, 193:1094-1100 (1976).
47. Oberai B, Adams CW, High OB. Myocardial and renal arteriolar thickening in cigarette smokers. Atherosclerosis, 52:185-190 (1984).
48. Murchison LE, Fyfe T. Effects of cigarette smoking on serum lipids, blood glucose and platelet adhesiveness. Lancet, 2:182-184 (1966).
49. Levine PH. An acute effect of cigarette smoking on platelet function. Circulation, 48:619-623 (1973).
50. Glynn MF, Mustard JF, Buchanan MR. Cigarette smoking and platelet aggregation. Can Med Assoc J, 95:549-553 (1966).
51. Hawkins RI. Smoking, platelets and thrombosis. Nature, 236:450-452 (1972).
52. Fuster V, Chesebro JH, Frye RL. Platelet survival and development of coronary artery disease in the young adult: Effects of cigarette smoking, strong family history, and medical therapy. Circulation, 63:546-551 (1981).
53. Mustard JF, Murphy EA. Effect of smoking on blood coagulation and platelet survival in man. Br Med J, 1:846-849 (1963).
54. Packham MS and Mustard JF. The role of platelets in the development and complications of atherosclerosis. Sem Hem, 23:8-26 (1986).
55. Rogers WR, Boss RL, Johnson DE. Atherosclerosis related responses to cigarette smoking in the baboon. Circulation, 61:1188-1193 (1980).
56. Strong JP, Richards ML. Cigarette smoking atherosclerosis in autopsied men. Atherosclerosis, 23:451-476 (1976).

57. Auerbach O, Carter HW, Garfinkle L. Cigarette smoking and coronary
     artery disease: A macroscopic and microscopic study. Chest,
     70:697–705 (1976).
58. Auerbach O, Hammond EC, Garfinkle L. Smoking in relation to
     atherosclerosis of the coronary arteries. N Eng J Med,
     273:775–779 (1965).
59. Sridharan MR, Flowers NC, Hand RC, Hand JW, Horan LG. Effect of
     various regimens of chronic and acute nicotine exposure on
     myocardial infarct size in the dog. Am J Cardiol, 55:1407–1411
     (1985).

CIGARETTE SMOKING AND CORONARY ARTERY DISEASE

Graham A. Colditz

Channing Laboratory
Department of Medicine
Harvard Medical School and
Brigham and Women's Hospital
Boston, MA

As many speakers have reviewed during this conference, the pathogenesis of coronary heart disease, which includes the clinical manifestations of myo-cardial infarction, angina pectoris, and sudden death, is extremely complex and mediated by multiple mechanisms and etiologic factors (1). At least five interrelated processes are likely to contribute to the clinical manifestations of myocardial infarction: atherosclerosis, thrombosis, coronary artery spasm, cardiac arrythmia, and reduced capacity of the blood to deliver oxygen.

The effects of smoking on coronary heart disease are probably mediated by multiple mechanisms, several of which are well established. Where some of these effects of smoking appear to be reversible within days or weeks, others may be irreversible or only slowly reversible, such as the development of atherosclerosis due to smooth muscle proliferation and lipid deposition in the arterial intima. However, the relative contribution of these mechanisms to the increase in incidence of cardiovascular disease caused by smoking is unknown.

After an exhaustive review of the data, the 1983 Report of the Surgeon General, which was devoted to cardiovascular disease, concluded that "cigarette smoking is a major cause of coronary heart disease in the United States for both men and women" and that cigarette smoking should be considered the most important of the known modifiable risk factors for CHD (2). Overall, smokers had about a 70% excess death rate from CHD; heavier smokers had an even greater excess risk. Additional evidence accumulated since the 1983 Report of the Surgeon General was presented in the 1989 Report of the Surgeon General (3). For the year 1985, cigarette smoking was estimated to be responsible for 21% of all CHD deaths in the U.S. among men 65 years of age or older, and 45% of CHD deaths among younger men. For women, 12% of the CHD deaths among those aged 65 or more and 41% of those in younger women were attributable to cigarette smoking (see Table 1). In that year, 115,000 deaths from coronary disease were attributable to cigarette smoking.

Overwhelming and conclusive data support the view that cigarette smoking substantially increases the risk of CHD. The data are also quite persuasive that former smokers have a lower risk of CHD than do current

Tobacco Smoking and Atherosclerosis
Edited by J. N. Diana
Plenum Press, New York, 1990

Table 1. Proportion of CHD deaths in the United States
attributable to cigarette smoking

Men 65 or more	21 percent
Men <65	45 percent
Women 65 or more	12 percent
Women <65	41 percent

Source: U.S. Department of Health and Human Services. Reducing the
Health Consequences of Smoking: 25 Years of Progress. A Report of the
Surgeon General. U.S. Department of Health and Human Services, Public
Health Service, Office of the Assistant Secretary for Health, Office on
Smoking and Health. DHHS Publication No. (CDC) 89-8411, 1989.

smokers. Despite methodological differences, the studies are remarkably
consistent in demonstrating a reduced risk of CHD among former smokers.
Much of this literature has been reviewed in the reports of the Surgeon
General of 1979 (4) and 1983 (2).

In this brief presentation I will review the data from the Nurses'
Health Study showing an interaction between cigarette smoking and other
risk factors for CHD among women. I will then briefly discuss data
pertaining to risk of CHD following cessation from cigarette smoking.

Among men, it has been clear for many years that cigarette smoking is
associated with an elevated risk of coronary heart disease (5). However,
for some time it was believed that cigarette smoking was not associated
with coronary heart disease among women (6).

THE NURSES' HEALTH STUDY

The Nurses' Health Study cohort was identified when 121,700 women
returned a mailed questionnaire in 1976. The study population consisted
of female, registered nurses, 30 to 55 years of age at that time, who
identified themselves as married in 1972 and resided in one of eleven
states of the United States (California, Connecticut, Florida, Maryland,
Massachusetts, Michigan, New Jersey, New York, Ohio, Pennsylvania and
Texas). They were identified from the 1972 files provided by the state
boards of nursing and the American Nurses Association.

In June 1976, Dr. Speizer and colleagues sent each nurse an
introductory letter, a two-page questionnaire, and a prepaid return
envelope. Identical materials were mailed in September and December 1976
in an attempt to enlist the participation of previous non-respondents.
Overall, 70% of those invited to participate in the study returned
questionnaires (7). A slightly elevated response rate was found among
younger nurses and nurses with academic degrees, and the response rate
was generally similar across the eleven states.

The proportion of women using oral contraceptives and smoking
cigarettes in 1976 was almost identical to that observed for white women
of the same age in other national samples. Overall, 48.5% of the
participants reported using oral contraceptives at some time. Of the
cohort, 5.5% were currently using oral contraceptives in 1976 (8). In
1976, 34.4% of the women free from heart disease, cancer and diabetes,
reported that they currently smoked cigarettes. A further 23.4% were
former smokers. Although the prevalence of current smoking was similar

for each 5-year age group, patterns of initiation and cessation differed
by age (9). Within each age group, women who started to smoke at an
earlier age had a lower rate of stopping. The proportion who stopped
smoking in the interval 1948 to 1958 was 10% and in the interval 1963 to
1973, 25%. This result is consistent with an impact of the Surgeon
General's Report on Smoking and Health, and the increasing information on
the health hazards of smoking.

The prevalence of cigarette smoking in this cohort is similar to that
among women in general (10), and reflects that of registered nurses
estimated from Health Interview Survey data included in the 1985 Report
of the Surgeon General (11). Thus we have evidence that cigarette
smoking among registered nurses is not elevated above that of women in
general, as is popularly believed.

FOLLOW-UP OF PARTICIPANTS

Follow-up questionnaires were mailed to all cohort members every two
years. These questionnaires inquire about a number of exposures as well
as the development of cancer, cardiovascular disease, and other major
medical conditions. We use up to five mailings over a 12-month period to
obtain a response from the participants.

Most deaths are reported by the subjects' next of kin or by postal
authorities. After each follow-up questionnaire cycle, we search state
vital records for deaths among the non-respondents. The ascertainment of
deaths from independent sources compared to that obtained through the
National Death Index allowed us to estimate that more than 96% of deaths
are identified in this cohort (12).

In 1982, we added a telephone follow-up to reach those women who had
not responded to any of the five mailings. More than 14,000 women were
successfully contacted and completed a brief telephone interview focused
on new diagnoses during 1983. This same telephone follow-up method was
repeated after the 1986 follow-up cycle.

Overall, participation has been very high. In 1982, 92% of women
responded and by 1986 we had questionnaire information from 85% of the
original cohort. These numbers underestimate the follow-up data
available for analysis, because we can carry forward several exposure
classifications; for example, a women who responds in 1980, then misses
the 1982 questionnaire, and responds again in 1984 can be included in
each two-year follow-up interval for the purpose of analyses. When we
calculate person-years of follow-up for non-fatal endpoints, a woman
contributes time up to the return of her latest questionnaire. As a
percentage of total possible follow-up time (i.e., the whole cohort) this
follow-up time was 95.4% in 1982 and 92% in 1986.

CORONARY HEART DISEASE

For any report of coronary heart disease we seek written permission
from study participants to review their medical records. Myocardial
infarction is classified as confirmed if the records meet the criteria of
the World Health Organization symptoms and either typical
electrocardiogram changes or elevations of serum cardiac enzymes (13).
Incident myocardial infarction reported by participants is classified as
probable if we cannot obtain hospital records but the event required
hospitalization and was corroborated by additional information from a
letter or telephone interview of the participant.

CIGARETTE SMOKING

To address the relation between cigarette smoking and the incidence
of coronary heart disease in women, we analyzed the data provided by the
119,404 women who were free from diagnosed coronary disease in 1976
(14). During six years of follow-up, 65 of the women died of fatal
coronary heart disease and 242 had a non-fatal myocardial infarction.

In 1976, approximately 30% of the cohort members were current
cigarette smokers. During the six years of follow-up these women had
higher rates of non-fatal myocardial infarction and fatal coronary heart
disease than women who had never smoked, and the rates increased with the
number of cigarettes smoked daily. As compared with women who never
smoked, even the lightest smokers (1 to 4 cigarettes per day) had a
significantly elevated risk of coronary heart disease (relative risk =
2.4) and those who smoked most heavily (45 or more cigarettes per day)
had a relative risk of 10.8 (see Table 1). In this table, we present
relative risks and also attributable risks; the attributable risk (or
absolute excess risk) is simply the rates of disease among nonsmokers
subtracted from the corresponding rates among smokers, and provides a
direct estimate of the effect of cigarette smoking on public health, in
terms of the number of cases of coronary heart disease that would be
avoided by elimination of smoking. Overall, in this population of women,
approximately 46 percent of fatal coronary heart disease and 54% of
non-fatal myocardial infarction was attributable to cigarette use.

The relative risks for cigarette smokers were slightly lower among
women with hypertension, hypercholesterolemia or diabetes than among
those without these risk factors (see Table 2). However, the
attributable risks were two or more times higher among women with these

Table 2

Age-adjusted relative and attributable risks of fatal coronary
heart disease and nonfatal myocardial infarction (combined)
in relation to daily cigarette consumption

Cigarette consumption	Person-years	Cases	Relative risk (95% CI)†	Attributable risk* (95% CI)†
Nonsmoker	302,375	63	1.0	0
Ex-smoker	174,237	55	1.5(1.0,2.1)	10(1,19)
Smoker				
1-4/day	15,765	7	2.4(1.1,5.0)	32(4,61)
5-14/day	45,635	19	2.1(1.2,3.4)	22(7,38)
15-24/day	95,430	82	4.2(3.1,5.6)	67(53,81)
25-34/day	40,790	44	5.4(3.8,7.7)	90(72,109)
35-44/day	20,093	31	7.1(4.9,10.2)	120(97,142)
≥45/day	2,476	6	10.8(5.5,21.1)	205(147,262)

*Cases per 100,000 person-years.   †CI denotes confidence interval.

Reproduced with permission from Willett WC, Green A, Stampfer MJ, et al.
Relative and absolute excess risks of coronary heart disease among women
who smoke cigarettes. N Engl J Med 1987: 317: 1303-9.

predisposing factors than among women without them (see Figs. 1-6). This
contrast between decreasing relative risks and increasing attributable
risks occurred because the baseline rates among non-smokers (the
denominators for calculating the relative risks) increased substantially
in the presence of the above predisposing factors. Thus, the use of
relative risks alone obscures an important increase in the absolute
effect of cigarette smoking when combined with other risk factors (Table
3). Without exception, we found that the attributable risks among women
smokers were elevated in the presence of other risk factors for coronary
heart disease. This indicates that the absolute effect of cigarette
smoking is substantially greater among women who are already at higher
risk because of these factors. Thus, although all women would benefit,
it is these women who will benefit the most from stopping smoking.

SMOKING CESSATION AND RISK OF MI

1. <u>Cross Sectional Studies</u>. These studies include a detailed study of
coronary atherosclerosis, by Auerbach, Carter, and Garfinkel et al.
(15). Among a total of 1,056 autopsy specimens from patients of the East
Orange Veterans Administration Hospital and the extent of coronary
atherosclerosis determined by arteriography Ramsdale, Faragher, and Bray
et al. (16)

Weintraub, Klein, and Seelaus et al. (17) evaluated smoking history
in a series of 1,349 patients in whom coronary arteriography was
performed. Of these, 984 had significant coronary disease defined as 75%
obstruction or more. Current amount of smoking was not a significant
predictor of serious obstruction after total pack-years were considered.
On average, the risk of such obstruction increased by about 1% per pack
year.

Studies of this design can be difficult to interpret because patients
with angiography who do not have serious obstruction are clearly not
representative of the general healthy population, and pack-years is a
complex variable including both amount and duration. Nonetheless, this
study supports the view that smoking increases atherosclerosis, and that
very recent quitting would therefore have little impact on coronary
stenosis.

2. <u>Case-control Studies</u>. Although prospective studies are generally
considered less prone to bias than case-control studies, case-control
studies are probably less susceptible to the misclassification due to
resumption of smoking among former smokers. Thus, for example, a person
with a recent myocardial infarction can probably recall his or her
smoking status just prior to the infarction with considerable accuracy.
Hence, case-control studies may be quite valuable in assessing the time
course for the decline in risk.

In a nested case-control study of women participating in the Nurses'
Health Study, Willett, et al. (18) identified 263 women who reported a
nonfatal MI on the baseline Nurses' Health Study questionnaire in 1976
when they were 30-55 years of age. Their smoking histories were compared
with randomly selected controls matched on age in a ratio of 20:1. Women
who were former smokers had no increase in risk of MI, with a relative
risk compared with never smokers of 1.0 (95% confidence interval 0.7-1.6).
In contrast, current smokers had a significantly elevated threefold
higher risk. When duration of abstinence was assessed, it appeared that
those who quit either one to four, or five to nine years before had a
nonsignificantly elevated risk of 1.5, and those who quit ten or more
years before had a relative risk of 0.6. Because there were only 29

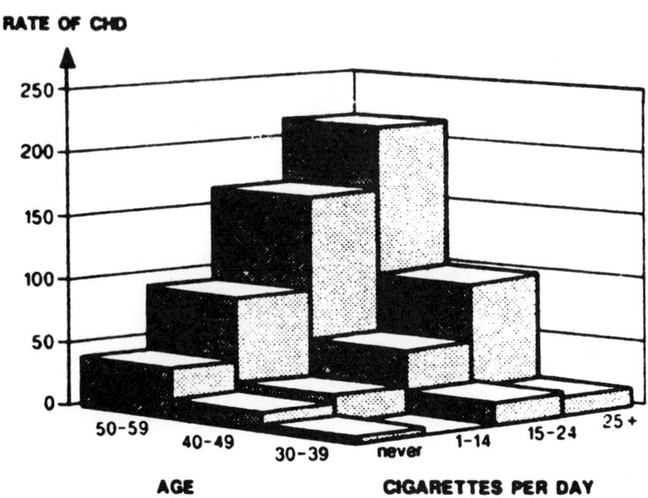

Figure 1. Rates of Total Coronary Heart Disease (CHD) per 100,000 Person-Years (Vertical Scale) among Women, According to Cigarette Use and Age.

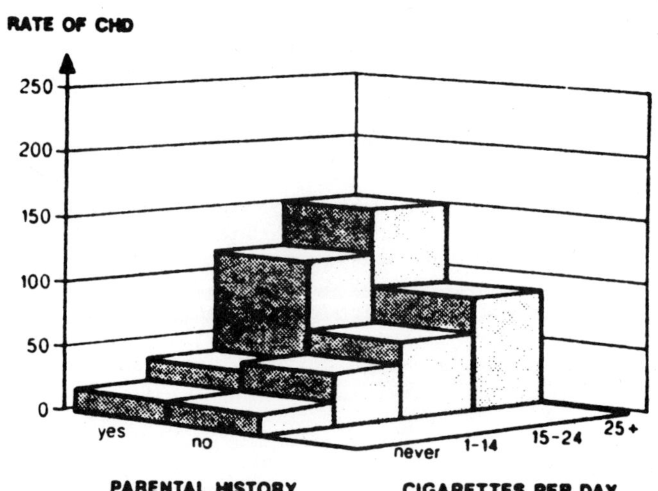

Figure 2. Age-Standardized Rates of Coronary Heart Disease per 100,000 Person-Years (Vertical Scale) among Women, According to Cigarette Use and Parental History of Myocardial Infarction.

316

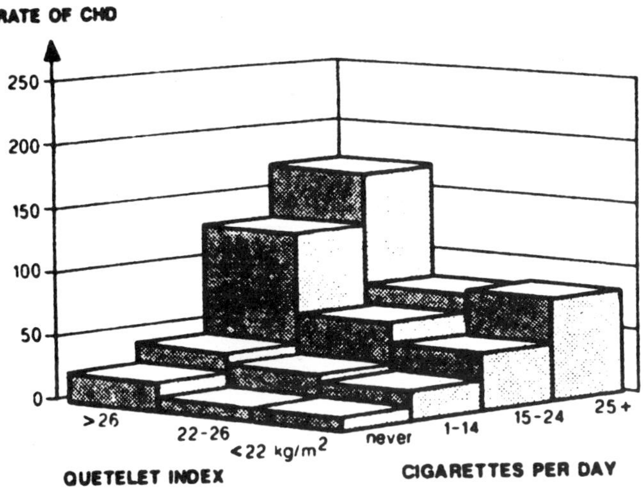

Figure 3. Age-Standardized Rates of Coronary Heart Disease per 100,000 Person-Years (Vertical Scale) among Women, According to Cigarette Use and Quetelet Index.

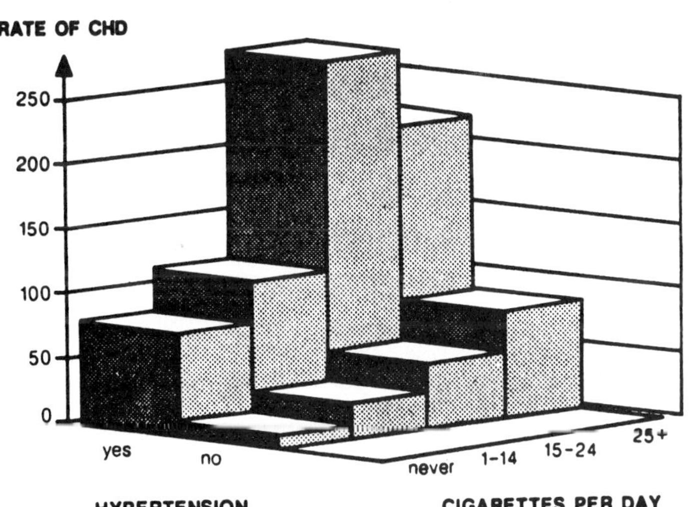

Figure 4. Age-Standardized Rates of Coronary Heart Disease (CHD) per 100,000 Person-Years (Vertical Scale) among Women, According to Cigarette Use and History of Hypertension.

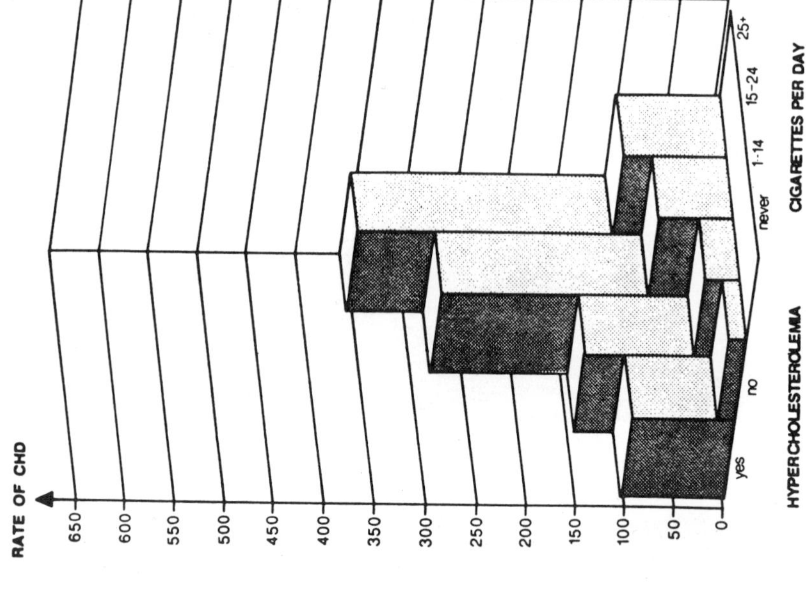

Figure 5. Age-Standardized Rates of Coronary Heart Disease (CHD) per 100,000 Person-Years (Vertical Scale) among Women, According to Cigarette Use and History of Hypercholesterolemia.

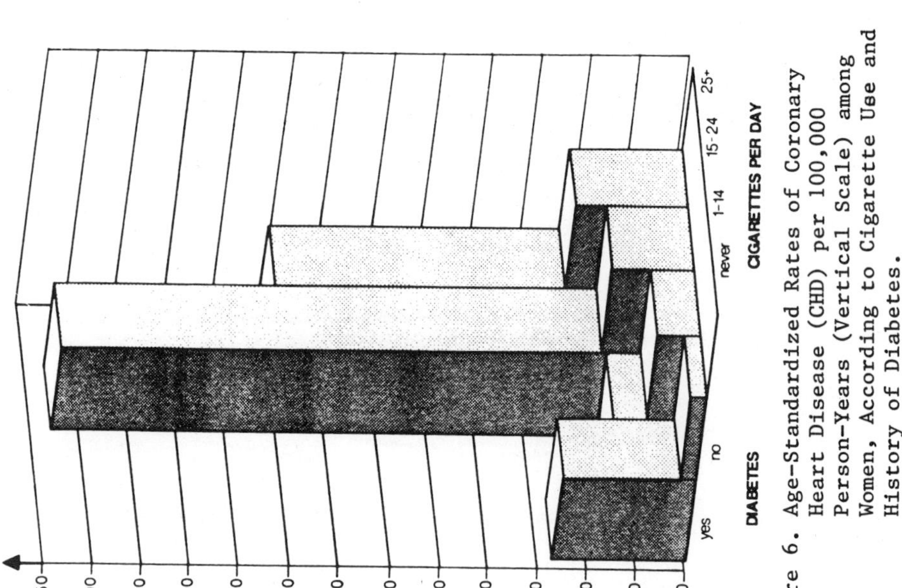

Figure 6. Age-Standardized Rates of Coronary Heart Disease (CHD) per 100,000 Person-Years (Vertical Scale) among Women, According to Cigarette Use and History of Diabetes.

318

Table 3

Age-adjusted relative risks (RR) and attributable risks (AR) of fatal coronary heart
disease and nonfatal myocardial infarction (combined) in relation to cigarette
smoking, as modified by hypertension, hypercholesterolemia, or diabetes*

| | Women who never smoked | | | Current smokers | | | | | | | | |
| | | | | 1-14/day | | | 15-24/day | | | $\geq$ 25/day | | |
Variable	Rate	RR	AR	Rate	RR	AR	Rate	RR	AR	Rate	RR	AR
Hypertension												
No	10	1.0	0	28	2.8	18	50	5.0	40	86	8.6	76
Yes	79	1.0	0	109	1.4	30	274	3.5	195	222	2.8	143
Hypercholesterolemia												
No	19	1.0	0	38	2.0	19	76	4.0	57	102	5.4	83
Yes	106	1.0	0	148	1.4	42	283	2.7	177	359	3.4	253
Diabetes												
No	18	1.0	0	42	2.3	24	81	4.5	63	117	6.5	99
Yes	139	1.0	0	79	0.6	-60	637	4.6	498	401	2.9	262

*Rates (cases per 100,000 person-years) were directly standardized to the age distribution
on nonsmoking women without the specified risk factor.

cases among former smokers, the estimates for risk by duration of
abstinence are not precise (Table 4).

Rosenberg, Kaufman, Helmrich and Shapiro (19) analyzed the impact of
smoking cessation on risk of a first MI among men under age 55, using a
hospital-based case-control design. Men with known preexisting heart
disease were excluded. The controls were mostly cases of fracture or
sprain, disk disorders, and gastrointestinal disorders that were thought
not to be related to cigarette smoking. There were 1,873 cases and 2,775
controls. For current smokers defined as men who smoked within the past
year, the relative risk was 2.9 (95% confidence interval 2.4–3.4), and
for past smokers overall, it was 1.1 (95% confidence interval 0.9–1.4).
The relative risk for those who had not smoked for 12–23 months was 2.0
(95% confidence interval 1.1–3.8). For those with longer durations of
cessation, the relative risk was 1.1 (95% confidence interval 0.9–1.4;
see Fig. 7).

The same research group also conducted a case-control study among
young women (20). This was a hospital-based case-control study of first
nonfatal MI among women under age 50. Women who smoked in the year
before admission were classified as current smokers. Participants
consisted of 555 cases and 1,864 controls, who were hospitalized for
trauma, orthopedic disorders, and other conditions thought to be
unrelated to smoking. Current smokers had relative risks increasing from
1.4 to 7.0 depending upon number of cigarettes per day. In contrast,
former smokers (of at least one year of abstinence) had the same risk as
never smokers, with a relative risk of 1.0 (95% confidence interval
0.7–1.6).

3. <u>Cohort Studies</u>. Many cohort studies have addressed the relation
between cessation from smoking and subsequent risk of coronary heart
disease. Among the larger is the cohort followed by the American Cancer
Society. The first large-scale American Cancer Society cohort was
assembled in 1952 when 187,783 men living in nine states, aged 50 to 69
years, completed a questionnaire related primarily to smoking (21,22).
The men were enrolled by over 22,000 American Cancer Society volunteers,
each of whom was asked to enroll ten individuals, excluding those who
were seriously ill. There was no further update of cigarette use. These
men were followed for fatal outcomes for an average of 44 months, for a
total of 667,753 person-years. Cause of death was determined by death
certificate for 11,870 men. Compared with never smokers, the relative
risk for current smokers of less than one pack per day was 1.75. For
former smokers of less than one pack per day, those who quit within the
previous year had a relative risk of 2.09, for quitters from 1–10 years
before, the risk was 1.54, and for quitters of more than ten years'
duration, the relative risk was 1.09. A similar pattern was observed for
smokers of one or more packs per day. For current smokers, the relative
risk was 2.2; for quitters within the past year, 3.00; for quitters of
1–10 years' duration , 2.06; and for quitters of more than 10 years'
duration, 1.60. The elevated risk among the recent quitters may reflect
the inclusion of men who stopped smoking because of early symptoms of
heart disease.

A second cohort study, the Cancer Prevention Study I (formerly called
the 25-State study), was undertaken by the American Cancer Society
between 1959 and 1972. Recruitment was by family. Eligible families had
at least one person over age 45, and all family members over age 35 were
asked to participate; in all, over one million persons were enrolled. In
a six-year follow-up of 358,534 men free of diagnosed serious illness,
clear reductions in risk of CHD mortality were observed in former smokers

Table 4. Case-Control Studies of Risk of CHD in Former Smokers.

First Authors	Population	No. Cases	No. Controls	No. Cases in Former Smokers	Relative Risk vs. Never Smokers	
					Former Smokers	Current Smokers
Willett, et al. (1981)	Partici- pants in Nurses' Health Study, women aged 30-55	263	5260	29	overall 1.0 (0.7-1.6) quit 1-4 yrs 1.5 (0.7-3.1) quit 5-9 yrs 1.5 (0.8-3.0) quit 10+ yrs 0.6 (0.3-1.3)	3.0 (2.3-4.0)
Rosenberg, et al. (1985)	Eastern US men under age 55	1873	2775	348	1.2 (1.0-1.5)	2.9 (2.4-3.4)
Rosenberg, et al. (1985)	Eastern US wonen under age 50	555	1864	35	1.0 (0.7-1.6)	1.4-7.0 depending on cigarettes per day

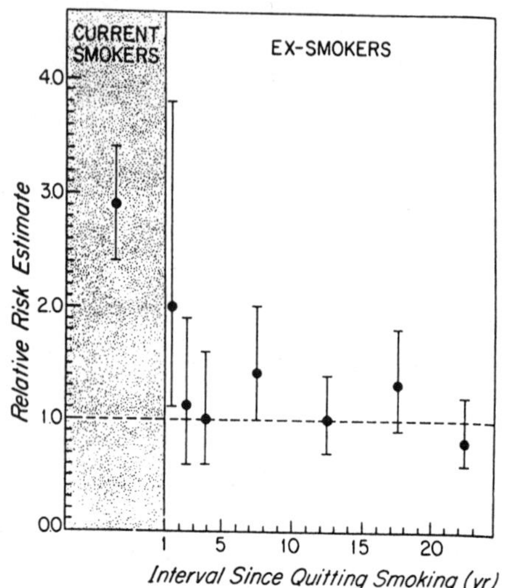

Figure 7. Estimated Relative Risk of Myocardial Infarction after Quitting
Smoking. Relative-risk estimates are adjusted for age; 95%
confidence intervals are indicated by vertical lines. The relative risk for men who never smoked is 1.0.

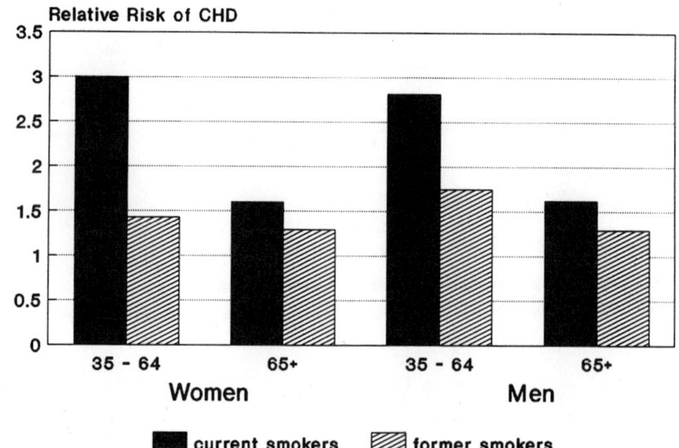

Figure 8. Relative risks for current and former smokers, 4 year follow-up
(1982-1986) of American Cancer Society 50-state study. Source:
Surgeon General's Report, 1989.

as compared with current smokers (23). Among those smoking less than one
pack per day the relative risk for current smokers was 1.9. For those
who stopped in the previous year, the relative risk was 1.62, and for
those with ten or more year's duration of cessation, the risk was nearly
the same as for never smokers. A similar pattern was observed for those
smoking a pack or more per day. In this period there was a substantial
decline in cigarette smoking, especially in the predominantly white,
middle- to upper-class groups, such as are represented by this study
population. Hence, there is likely to have been some misclassification
of the current smoking group, but the relative risks among the past
smokers, apart from the most recent quitters (some of whom inevitably
resumed smoking), are likely to be accurate.

In 1982, a third American Cancer Society cohort was begun in all 50
states, the Cancer Prevention Study II. The methods for recruitment and
the population enrolled were similar to CPS I, but the cohort was larger,
with over 1.2 million participants. Preliminary data based on four years
of follow-up have been published in the 1989 Surgeon General's Report
(3). In men, former smokers had relative risks of 1.41 for those 35 or
under, 1.75 for 36-64, and 1.29 for 65+. The corresponding relative
risks for current smokers were 1.94, 2.81 and 1.62. A generally similar
pattern was seen among women (see Fig. 8).

The large size and careful methodology of the three American Cancer
Society cohorts provides considerable evidence for the apparent benefit
of quitting to reduce CHD risk. It also provides strong evidence that
there is some residual risk attributable to the past smoking which
persists for a considerable duration after cessation.

4. _Randomized Trials_. An attempt has been made to evaluate the effect of
altering risk factors for CHD, including smoking, in several clinical
trials (24-31). Another potential difficulty is in assessing self-report
of smoking cessation or decrease in smoking level. There may be a
tendency of subjects to seek approval and avoid negative feedback by
reporting less cigarette use than is really the case. Such a tendency
would have the effect of misclassification and yield an underestimate of
the benefits of cessation.

Only one trial has attempted to assess the effect of advice for
smoking cessation without intervening on other risk factors
simultaneously. In theory, trials of this design can provide the
clearest indication of the effect of such advice in the absence of other
effects. However, such studies are difficult to conduct. Participants
were drawn from the Whitehall cohort of 16,016 British Civil Servants
(32). From this group, 1,445 high risk (based on a variety of coronary
risk factors) smoking men aged 40-59 were randomized to a normal care
group and to the intervention group which received anti-smoking advice.
At year one, 51% of the intervention group reported not to be smoking,
and at year three, 36%. A third of the quitters reported smoking cigars
or a pipe. In the normal care group, the corresponding percentages were
10% and 14%. It is important to note that the questionnaire response
rate at three years was 64% in the intervention group and 70% in the
normal care group (33). The nine year response rate was 83% overall. At
that point, 55% of responders in the intervention group reported quitting
as did 41% in the normal care group. Despite the similarity of smoking
prevalence of the two groups, at ten years, there was an 18% decrease in
CHD mortality in the intervention group. However, this difference did
not attain statistical significance (34).

CONCLUSION

From these data, we conclude that cigarette smoking is a major cause of coronary artery disease among men and women. The absolute effect of smoking is particularly strong among women already at higher risk because of obesity, hypercholesterolemia, diabetes, or their parents' history. Furthermore, major benefits can be achieved through cessation of smoking which reduces the risk of coronary heart disease in both men and women.

REFERENCES

1.   Munro, J.M., Cotran, R.S. Biology of Disease, the pathogenesis of atherosclerosis: atherogenesis and inflammation. Laboratory Investigation, 58(3): 249-261 (1988).
2.   U.S. Department of Health and Human Services. The Health Consequences of Smoking: Cardiovascular Disease. A Report of the Surgeon General. U.S. Department of Health and Human Services, Public Health Service, Office of the Assistant Secretary for Health, Office on Smoking and Health. DHHS Publication No. (PHS) 84-50204 (1983).
3.   U.S. Department of Health and Human Services. Reducing the Health Consequences of Smoking: 25 Years of Progress. A Report of the Surgeon General. U.S. Department of Health and Human Services, Public Health Service, Office of the Assistant Secretary for Health, Office on Smoking and Health. DHHS Publication No. (CDCO 89-8411 (1989).
4.   U.S. Department of Health, Education and Welfare. Smoking and Health: A Report of the Surgeon General. U.S. Department of Health, Education, and Welfare, Public Health Service, Office of the Assistant Secretary for Health, Office on Smoking and Health, DHEW Publication No. (PHS) 79-500667 (1979).
5.   A Report of the Surgeon General. The Health Consequences of Smoking: Cardiovascular Disease. Rockville, MD: Office of Smoking and Health (1983).
6.   Dawber, T.R. The Framingham Study: The epidemiology of atherosclerotic disease. Cambridge MA: Harvard University Press (1980).
7.   Barton J., Bain C., Hennekens, C.H., Rosner, B., Belanger, C., Roth A., Speizer, F.E. Characteristics of respondents and non-respondents to a mailed questionnaire. Am. J. Public Health, 70: 823-25 (1980).
8.   Belanger, C.F., Hennekens, C.H., Rosner, B., Speizer, F.E. The Nurses' Health Study. Am. J. Nurs., 78: 1039-40 (1978).
9.   Myers, A.H., Rosner, B., Abbey, H., Willett, W.C., Stampfer, M.J., Bain, C., Lipnick, R., Hennekens, C., Speizer, F. Smoking behavior among participants in the Nurses' Health Study. Am. J. Pub. Health, 77: 628-30 (1987).
10. US DHHS. The Health Consequences of Smoking for Women. Report of the Surgeon General. DHHS Pub. No. 0-326-003. Washington, D.C., Govt. Printing Office (1980).
11. United States Department of Health and Human Services. The Health Consequences of Smoking. Cancer and Chronic Lung Disease in the Workplace. A Report of the Surgeon General. U.S. Department of Health and Human Services, Public Health Service, Office of the Assistant Secretary for Health, Office on Smoking and Health. DHHS Publication No. PHS 85-50207 (1985).
12. Stampfer, M.M., Willett, W.C., Speizer, F.E., Dysert, D.C., Lipnick R., Rosner, B., Hennekens, C.H. Test of the National Death Index. Am. J. Epidemiol., 119: 837-39 (1984).

13. Rose, G.A., Blackburn, H. Cardiovascular Survey Methods. Geneva, WHO Monograph Series No. 58 (1982).

14. Willett, W.C., Green, A., Stampfer, M.J., Speizer, F.E., Colditz, G.A., Rosner, B., Monson, R., Stason, W., Hennekens, C.H. Relative and absolute excess risks of coronary heart disease among women who smoke cigarettes. N. Engl. J. Med., 317: 1303-9 (1987).

15. Auerback, O., Garfinkel, L. Atherosclerosis and aneurysm of the aorta in relation to smoking habits and age. Chest, 78(6): 805-809 (December 1980).

16. Ramsdale, D.R., Faragher, E.B., Bray, C.L., Bennett, D.H., Ward, C., Beton, D.C. Smoking and coronary artery disease assessed by routine coronary arteriography. British Medical Journal, 290(6463): 197-200 (January 19, 1985).

17. Weintraub, W.S., Klein, L.W., Seelaus, P.A., Agarwal, J.B., Helfant, R.H. Importance of total life consumption of cigarettes as a risk factor for coronary artery disease. American Journal of Cardiology, 55(6): 669-672 (March 1, 1985).

18. Willett, W.C., Hennekens, C.H., Bain, C., Rosner, B., Speizer, F.E. Cigarette smoking and non-fatal myocardial infarction in women. American Journal of Epidemiology, 113(5): 575-582 (May 1981).

19. Rosenberg, L., Kaufman, D.W., Helmrich, S.P., Shapiro, S. The risk of myocardial infarction after quitting cigarette smoking in men under 55 years of age. New England Journal of Medicine, 313(24): 1511-1514 (December 12, 1985).

20. Rosenberg, L., Kaufman, D.W., Helmrich, S.P., Miller, D.R., Stolley, P.D., Shapiro, S. Myocardial infarction and cigarette smoking in women younger than 50 years of age. Journal of the American Medical Association, 253(20): 2965-2969 (May 24/31, 1985).

21. Hammond, E.C., Horn, D. Smoking and death rates- report on forty-four months of follow-up of 187,783 men. I. Total mortality. Journal of the American Medical Association, (a) 166(10): 1159-1172 (March 8, 1958a).

22. Hammond, E.C., Horn, D. Smoking and death rates- report on forty-four months of follow-up of 187,783 men. II. Death rates by cause. Journal of the American Medical Association, (b) 166(11): 1294-1308 (March 15, 1958b).

23. Hammond, E.C., Garfinkel, L. Coronary heart disease, stroke, and aortic aneurysm: factors in etiology. Archives of Environmental Health, 19(2): 167-182 (August 1969).

24. Hughes, G.H., Hymowitz, N., Ockene, J.K., Simon, N., Vogt, T.M. The Multiple Risk Factor Intervention Trial (MRFIT). Intervention on smoking. Preventive medicine, 10(4): 476-500 (July 1981).

25. Multiple Risk Factor Intervention Trial Research Group. Coronary heart disease death, nonfatal acute myocardial infarction and other clinical outcomes in the Multiple Risk Factor Intervention Trial. American Journal of Cardiology, 58(1): 1-13 (July 1, 1986).

26. Grimm, R.H., Jr. The drug treatment of mild hypertension in the Multiple Risk Factor Intervention Trial. A Review. Drugs, (Supplement 1): 13-21 (1986).

27. Ockene, J.K., Kuller, L.H., Svensden, K.H., Meilahn, E. The relationship of smoking cessation to coronary heart disease and lung cancer in the Multiple Risk Factor Intervention Trial (MRFIT). Paper presented at the American Heart Association Annual Scientific sessions, Washington, D.C. (November 1988).

28. Hjermann, I., Velve, Byrne, K., Holme, I., Leren, P. Effect of diet and smoking intervention on the incidence of coronary heart disease. Report from the Oslo Study Group of a randomized trial in healthy women. Lancet, 2(8259): 1303-1310 (December 12, 1981).

29. World Health Organization European Collaborative Group. Multi-
     factorial trial in the prevention of coronary heart disease. 3
     Incidence and mortality results. European Heart Journal, 4(3):
     141-147 (March 1983).
30. Rose, G., Tunstall-Pedoe, H.D., Heller, R.F. UK Heart Disease
     Prevention Project: incidence and mortality results. Lancet,
     1(8333): 1062-1066 (May 14, 1983).
31. Kornitzer, M., De Backer, G., Dramaix, M., Kittel, F., Thilly, C.,
     Graffar, M., Vuylsteek, K. Belgian Heart Disease Prevention
     Project: incidence and mortality results. Lancet, 1(8333):
     1066-1070 (May 14, 1983).
32. Fuller, J.H., Shipley, M.J., Rose, G., Jarrett, R.J., Keen, H.
     Mortality from coronary heart disease and stroke in relation to
     degree of glycaemia: the Whitehall Study. British Medical
     Journal, 287(6396): 867-870 (September 24, 1983).
33. Rose, G., Hamilton, P.J.S. A randomized controlled trial of the
     effect on middle-aged men of advice to stop smoking. Journal of
     Epidemiology and Community Health, 32(4): 275-281 (December 1978).
34. Rose, G., Hamilton, P.J.S., Colwell, L., Shipley, M.J. A randomized
     controlled trial of anti-smoking advice: 10-year results.
     Journal of Epidemiology and Community Health, 36(2): 102-108 (June
     1982).

ADRENAL RELEASE OF CATECHOLAMINES IN THE CORONARY AND MYOCARDIAL RESPONSE TO NICOTINE

H. Fred Downey

Department of Physiology, Texas College of Osteopathic Medicine
Fort Worth, Texas 76107

Nicotine increases myocardial contractility in the in situ heart (12,15,17,22). Several mechanisms might account for this positive inotropic effect: 1) Stimulation of the myocardium by sympathetic nerves (1,3); 2) Release of catecholamines from cardiac tissue (6,23); 3) Stimulation of the myocardium by nicotine independent of catecholamines (4,12,19); 4) Release of catecholamines from the adrenal glands (2,26,27,28). Each of these mechanisms can be activated under appropriate experimental conditions, but their relative potency and contribution to the integrated cardiac response to intravenous nicotine was uncertain. Thus, we investigated the extent to which the increase in myocardial contractile function could be accounted for independently by cardiac nerve activation or by direct action of intracoronary nicotine. When we found these mechanisms incapable of accounting for the observed cardiac responses to intravenous nicotine, we examined the role of the adrenal glands (11).

METHODS

Experiments were conducted in three groups of adult mongrel dogs anesthetized with sodium pentobarbital, 30 mg/kg i.v. initially and supplemented as needed to maintain a constant state of anesthesia. After tracheotomy and insertion of a cuffed endotracheal tube, the dogs were ventilated with room air plus oxygen. The tidal volume and respiratory rate were adjusted to maintain blood gas parameters within normal physiological ranges (PO_2, 100-135 mm Hg; PCO_2, 28-42 mm Hg; pH, 7.36-7.44). Catheters were inserted into 1) the left femoral artery and advanced into the aorta to measure aortic blood pressure, 2) the right femoral artery to sample arterial blood and to supply a coronary perfusion system, 3) the right femoral vein to return coronary venous blood and to inject drugs.

A left thoracotomy was performed, and the heart was suspended in a pericardial cradle. A catheter was positioned in the left atrium to measure left atrial blood pressure. A 5F Millar catheter-tip pressure transducer was inserted into the left atrial appendage and advanced across the mitral valve to measure left ventricular blood pressure and its first derivative, dP/dt, an index of global left ventricular function.

The left anterior descending coronary artery (LAD) was isolated distal to its first major diagonal branch for cannulation or placement of an electromagnetic flow transducer. The interventricular vein was cannulated at this level to provide samples of venous blood from the LAD region. Coronary venous blood was allowed to drain freely into a beaker. Shed venous blood was intermittently returned to the animal through the femoral venous catheter. The oxygen content of arterial and coronary venous blood was measured with a Lexington Instrument LEX-O_2-CON. Regional myocardial oxygen consumption was calculated by multiplying the arteriovenous oxygen content difference by LAD blood flow.

Regional myocardial contractile function was evaluated with ultrasonic length gauges (25) implanted in the perfusion fields of the LAD and left circumflex (LC) coronary arteries. The integrity of cardiac innervation was verified by an increase in myocardial segment shortening in the LAD and LC regions during electrical stimulation of the left stellate ganglion (10 Hz, 8-10 V, 5 ms). When all surgical procedures were completed, heparin (500 U/kg) was injected intravenously to prevent blood coagulation. Aortic, left atrial, left ventricular, and LAD blood pressures, LAD blood flow, left ventricular dP/dt, and regional myocardial segment lengths were recorded on a Beckman R611 polygraph.

In Groups 1 and 2, the LAD was cannulated and perfused by an extracorporeal circuit with blood shunted from the left subclavian artery or from a pressurized reservoir of arterial blood (Fig. 1). LAD perfusion pressure was measured through a small polyethylene tube (PE 50) inserted into the perfusion line and advanced to the orifice of the coronary cannula. Coronary flow was measured with an electromagnetic flow transducer in the perfusion line and a Carolina Medical Electronics flowmeter, model FM501. In Group 3, the LAD was not cannulated. LAD blood flow was measured with

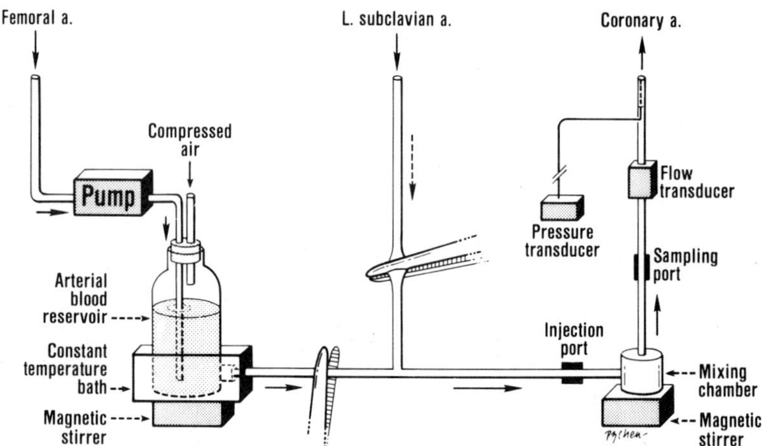

FIG. 1. System used to perfuse the left anterior descending coronary artery (LAD) with blood from a reservoir during intravenous infusion of nicotine (Group 1) and to perfuse the LAD with blood containing a known concentration of nicotine (Group 2). The shunt from the left subclavian artery perfused the LAD when the reservoir was not in use (from ref 11, by permission of Raven Press).

an electromagnetic flow transducer applied directly around the artery. In all groups, the zero flow reference point was checked frequently by mechanically occluding the perfusion line or the LAD. The flow transducers were calibrated at the end of the experiments by timed collections of blood.

<u>Experimental Protocols</u>

<u>Group 1: Role of cardiac nerves (n = 10)</u>. Hemodynamic measurements were recorded, and arterial and coronary venous blood samples were collected. Nicotine was then infused intravenously at 36 μg/kg/min, a dose we had previously found to increase cardiac function, coronary blood flow, and peripheral vasoconstriction (8,9,10). To avoid regional cardiac effects of nicotine and other circulating agents, the LAD was perfused from a previously filled, pressurized reservoir of arterial blood. LAD perfusion pressure was adjusted to reflect mean aortic pressure during the nicotine infusion. Hemodynamic variables, regional myocardial segment shortening, and LAD arteriovenous oxygen content differences were measured at the peak hypertensive response to nicotine, which occurred 2-3 min after starting the infusion. Systemic arterial blood was also sampled at this time for nicotine assay.

Following these measurements in five dogs, the LAD was perfused with systemic arterial blood while continuing the nicotine infusion. A third set of measurements were taken when contractile function in the LAD region reached a new steady state.

<u>Group 2: Intracardiac effects of nicotine (n = 6)</u>. While the LAD was perfused with systemic arterial blood from a left subclavian shunt (Fig. 1), nicotine was added to arterial blood in the LAD perfusate reservoir to achieve concentrations of 0.5, 1.0, 2.5, and 5.0 μg/ml blood. Nicotine equilibrates quickly between plasma and red cells with a plasma to cell ratio of approximately 1.0 : 0.9 (16,18), so the plasma concentration of nicotine was approximately 10% greater than that of whole blood. After control hemodynamic and myocardial function measurements were obtained, the LAD was perfused from the reservoir. Values of hemodynamic and cardiac function variables were recorded 20 s after initiating the perfusion with nicotine-containing blood, since preliminary experiments had demonstrated a stable increase in myocardial segment length shortening at 20 s after perfusion with 5.0 μg nicotine/ml coronary blood. The LAD was perfused with nicotine-containing blood for 90 s at each dose rate to allow sufficient time for collection of the coronary venous blood sample. The LAD was perfused with normal arterial blood from the left subclavian shunt for 30 min between each intracoronary administration of nicotine to allow variables to return to control values.

<u>Group 3: Role of adrenal glands (n = 7)</u>. In this group the myocardial responses to intravenous nicotine were examined before and after adrenalectomy. After control determinations, nicotine was infused intravenously at 36 μg/kg/min and hemodynamic variables, regional myocardial segment shortening, and regional coronary blood gases were measured at the peak hypertensive response to nicotine. Bilateral adrenalectomy was then performed through an abdominal subcostal incision, and a 30-min recovery period was allowed for the preparation to stabilize. Control measurements were repeated and then a second dose of nicotine (36 μg/kg/min) was administered intravenously. Hemodynamic variables, regional myocardial segment shortening, and regional coronary blood gases were again measured at the peak hypertensive response to nicotine.

<u>Nicotine Assay</u>

The concentration of nicotine in arterial plasma of Group 1 dogs was measured with a high pressure liquid chromatographic procedure developed in our laboratory (16).

<u>Catecholamine Assay</u>

Plasma concentrations of epinephrine and norepinephrine in systemic arterial blood were measured in three dogs. These dogs were prepared in a manner similar to those of Group 1. Nicotine was infused intravenously at 36 μg/kg/min, and arterial blood samples were collected at the peak hypertensive response. Six milliliter blood samples were immediately transferred to iced centrifuged tubes containing 100 μl of stabilizing solution (90 mg/ml ethylene-bis(-aminoethyl ester)-N,N-tetraacetic acid and 60 mg/ml glutathione). The blood was centrifuged at 0°C, and the plasma removed for catecholamine analysis by HPLC (5).

<u>Statistical Analyses</u>

Data are reported as mean ± standard error of the mean. In Groups 1 and 2, the effects of nicotine were compared to respective control values with the Student's t-test for paired observations. In Group 3, the statistical significance of the effects of nicotine before and after adrenalectomy was evaluated with a randomized block analysis of variance and the Student-Newman-Keuls test. Differences with $P < 0.05$ were considered statistically significant. Percentage changes described in the text are significant at $P < 0.05$ unless otherwise noted.

RESULTS

<u>Group 1: Role of Cardiac Nerves</u>

Under control, pre-nicotine conditions, mean aortic pressure was 97 ± 6 mm Hg, mean left atrial pressure was 4.3 ± 0.4 mm Hg, LAD perfusion pressure was 99 ± 8 mm Hg, and heart rate was 140 ± 6 beats/min. LAD blood flow was 31.5 ± 4.1 ml/min. Intravenous nicotine increased mean aortic blood pressure 102%, left atrial blood pressure 216%, and LAD blood flow 160%. Heart rate did not change significantly. Values for left ventricular (LV) +dP/dt$_{max}$ and oxygen consumption (MVO$_2$) in LAD-perfused myocardium are presented in Fig. 2. LV dP/dt$_{max}$ increased 274% and MVO$_2$ increased 129% during intravenous nicotine. At the maximum hypertensive response to intravenous nicotine, plasma nicotine concentration was 0.69 ± 0.08 μg/ml.

Regional myocardial segment length shortening is illustrated in Fig. 3. In the normally perfused LC region, myocardial segment length shortening increased from 10.8% to 19.2% during intravenous nicotine. In the vascularly isolated LAD region, myocardial contractile function did not increase, but instead was reduced to essentially zero. Although myocardial oxygen consumption increased significantly in the LAD region, this region was unable to contract against the increased afterload without an increase in inotropy.

In five of the Group 1 animals, the LAD was perfused with systemic arterial blood at the peak hypertensive response to intravenous nicotine. This resulted in a dramatic

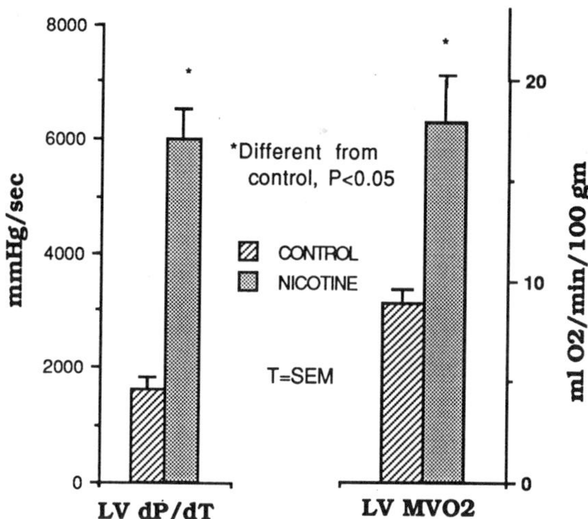

FIG. 2.   Effect of intravenous nicotine (36 μg/kg/min) on left ventricular dP/dt$_{max}$ and oxygen consumption (MVO$_2$) of the LAD region in Group 1 (n = 10).

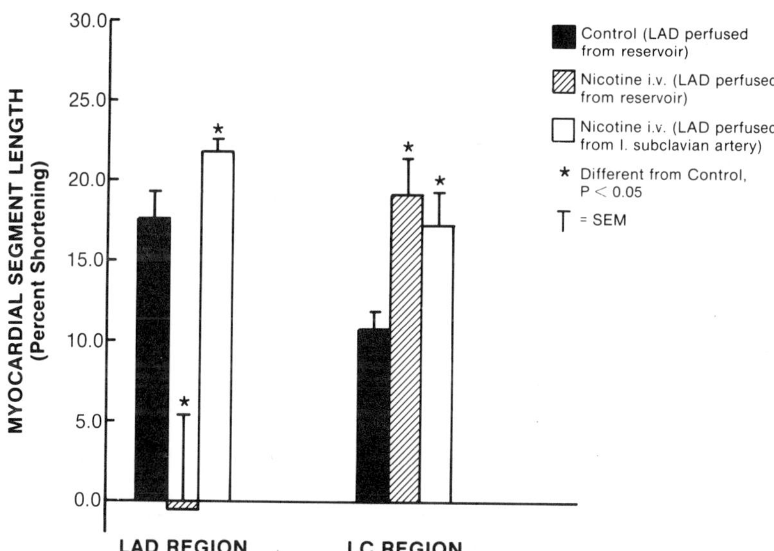

FIG. 3.   Effect of intravenous nicotine (36 μg/kg/min) on myocardial contractile function in the reservoir-perfused left anterior descending (LAD) region and in the systemically perfused left circumflex (LC) region (Group 1, n = 10). Switching the LAD perfusate from reservoir blood to systemic arterial blood caused a marked increased in contractile function of the LAD region (n = 5; from ref 11, by permission of Raven Press).

increase in contractile function of the LAD region, which had previously been perfused from the reservoir. As shown in Fig. 3, LAD myocardial segment shortening increased from -0.5 ± 5.9% to 21.8 ± 0.8%, a value similar to that observed in the systemically perfused LC region during intravenous nicotine. Oxygen consumption in the LAD region further increased by 30%. The increase in regional function improved total left ventricular performance as indicated by a 15% increase in dP/dt_{max}, and a 45% decrease in left atrial pressure. LAD blood flow increased by 16%, but aortic blood pressure and heart rate were not altered by changing the source of LAD perfusate.

Under control conditions, arterial plasma concentrations of epinephrine and norepinephrine in three dogs averaged 0.11 ± 0.03 and 0.17 ± 0.08 ng/ml, respectively. At the peak hypertensive response to intravenous nicotine, these concentrations increased to 7.95 ± 4.57 and 1.40 ± 0.42 ng/ml, respectively.

Group 2: Intracardiac Effects of Nicotine

Increasing concentrations of nicotine were infused directly into the LAD to determine the intracoronary concentration necessary to increase myocardial contractile function to approximately the same degree caused by systemic nicotine in the dogs of Group 1. Control values of hemodynamic and myocardial function variables were similar to those of Group 1. Brief intracoronary infusion of nicotine had no detectable systemic hemodynamic effects. Only at the highest nicotine concentration, 5 μg/ml, was a significant increase in LAD blood flow observed. The effect of intracoronary nicotine on shortening of myocardial segments is shown in Fig. 4. Only at a nicotine concentration of 5 μg/ml was a significant increase in contractile function observed. This 80% increase in segment length shortening was similar to that observed in the normally perfused regions of Group 1 dogs during intravenous infusion of nicotine at 36 μg/kg/min.

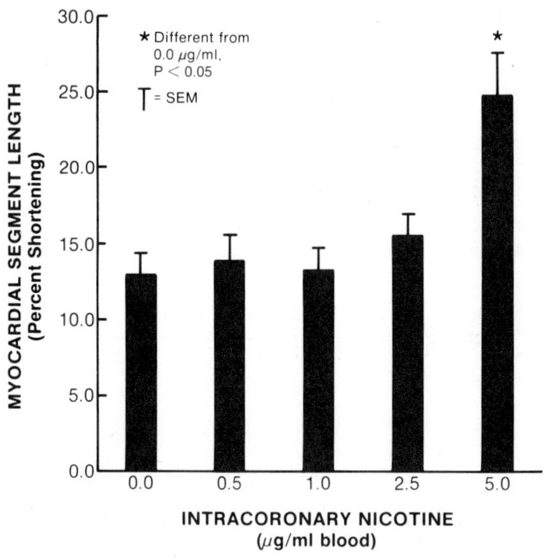

FIG. 4. Dose response relationship between intracoronary nicotine and myocardial contractile function (Group 2, n = 6; from ref 11, by permission of Raven Press).

<u>Group 3: Role of the Adrenal Glands</u>

Hemodynamic and myocardial contractile responses to intravenous nicotine before and after adrenalectomy are illustrated in Fig. 5. Hemodynamic values under control conditions and during infusion of nicotine before adrenalectomy were similar to those of Group 1. Before adrenalectomy, intravenous nicotine caused a significant increase in myocardial contractile function, as reflected by a 44% increase in segment length shortening (Fig. 6) and by a 182% increase in left ventricular dP/dt_{max}.

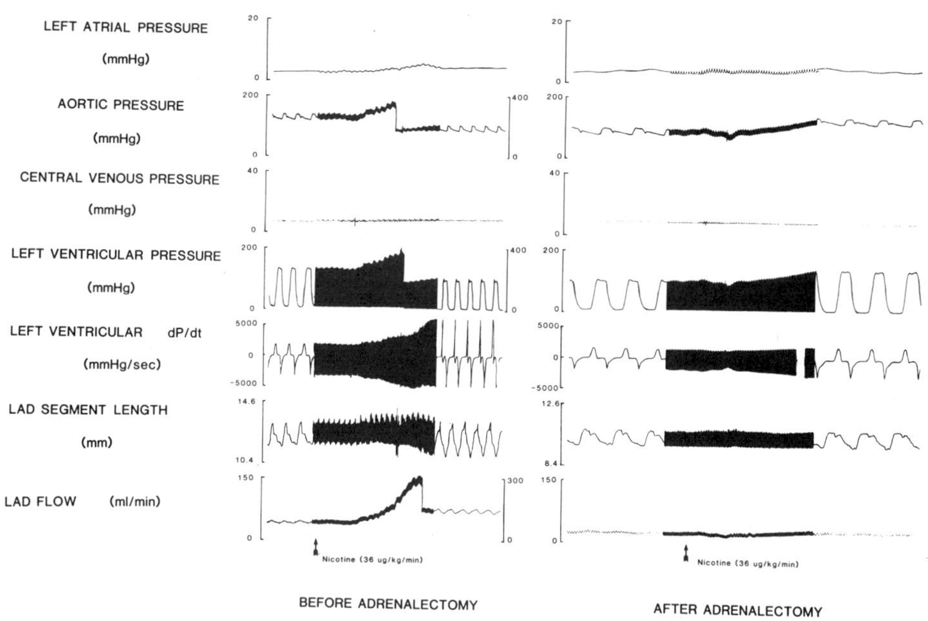

FIG. 5. Record of hemodynamic and contractile responses to intravenous nicotine (36 μg/kg/min) before and after adrenalectomy.

After adrenalectomy, control mean aortic pressure was reduced 20%, and myocardial oxygen consumption was reduced 25% (Fig. 7), but other hemodynamic variables were not significantly altered. Adrenalectomy caused no significant changes in control regional cardiac function as assessed by segment length shortening, or in global cardiac function as assessed by left ventricular dP/dt_{max} (Fig. 6).

Intravenous nicotine after adrenalectomy caused a 22% increase in aortic pressure ($P < 0.1$) and a 30% increase in LAD blood flow ($P < 0.1$). Regional contractile function as indicated by myocardial segment length shortening (Fig. 6) and global cardiac contractile function as indicated by dP/dt_{max} was not altered by intravenous nicotine after adrenalectomy. Myocardial oxygen consumption did not increase during intravenous nicotine after adrenalectomy (Fig. 7).

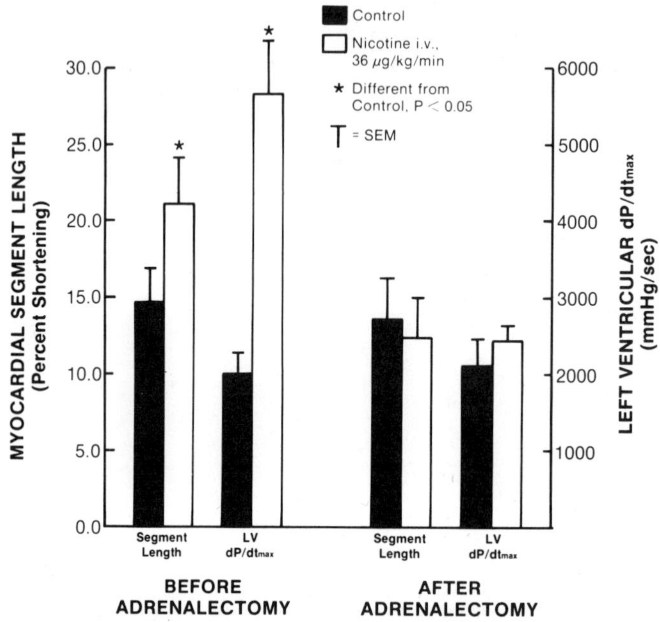

FIG. 6. Myocardial contractile function during intravenous nicotine (36 µg/kg/min) before and after adrenalectomy (Group 3, n = 7; from ref 11, by permission of Raven Press).

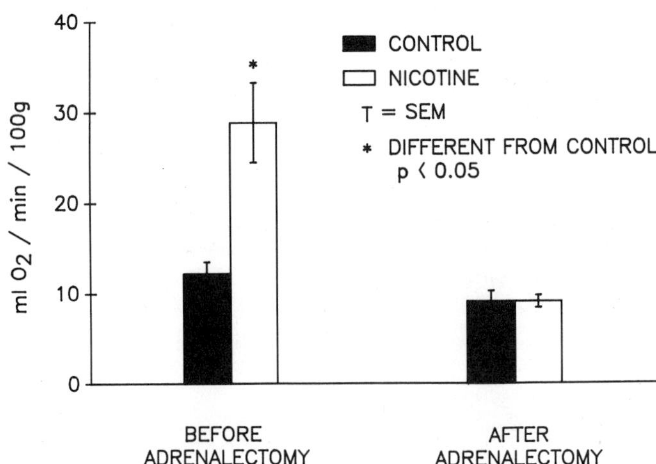

FIG. 7. Myocardial oxygen consumption during intravenous nicotine (36 µg/kg/min) before and after adrenalectomy (Group 3, n = 7).

DISCUSSION

The most important findings of this investigation were: 1) Only in myocardium perfused with systemic arterial blood did intravenous nicotine increase myocardial contractile function. 2) A direct intracardiac effect of nicotine could not account for the increase in myocardial contractile function caused by intravenous nicotine. 3) Adrenalectomy abolished the increase in myocardial contractile function caused by intravenous nicotine.

Since nicotine activates chemoreceptors (7,20), the vasomotor center (24), and sympathetic ganglia (7,13,21), we had anticipated an increase in contractile function of an innervated region of myocardium (14) during intravenous nicotine even if nicotine and other circulating factors were excluded from that region. The marked hypertensive response and the increase in contractile function of the normally perfused myocardium are consistent with activation of sympathetic vasomotor and cardiac nerves. However, contractile function in the reservoir-perfused region decreased. This indicated that a circulating factor was responsible for the increased cardiac function of the normally perfused region. Introduction of systemic arterial blood into the previously isolated region produced a marked increase in contractility and confirmed the action of a circulating factor.

We then investigated the possibility that a direct intracardiac action of nicotine was responsible for the increased contractility caused by intravenous nicotine, i.e., that the required circulating factor was nicotine. Previous investigators had demonstrated positive inotropic responses to intracoronary injections of nicotine (23,29), but the dose response relationship had not been defined. We confirmed that sufficiently high concentrations of intracoronary nicotine could produce a marked increase in myocardial contractile function, but the concentration required to increase myocardial contractility to the extent we observed in Group 1 was approximately seven-fold the arterial concentration of nicotine in that group. Thus, a direct intracardiac effect of nicotine was not responsible for the increase in contractile function caused by intravenous nicotine.

Nicotine causes release of catecholamines from the adrenal medullae (2,7,26,27,28), and we measured large increases in circulating catecholamines in intact dogs during intravenous nicotine. To determine the role of circulating catecholamines, we determined the response to nicotine before and after bilateral adrenalectomy. Adrenalectomy markedly attenuated but did not abolish the pressor response to intravenous nicotine. The residual increase in pressure may have been due to activation of the sympathetic vasoconstrictor nerves (8), or to a non-adrenergic mechanism (9). Adrenalectomy did abolish the effect of intravenous nicotine on myocardial contractile function and on myocardial oxygen consumption. These findings are consistent with the absence of any increase in myocardial contractile function in this group. Thus, we conclude circulating catecholamines released from the adrenal medullae rather than neural stimulation or intracardiac action of nicotine is responsible for increased myocardial contractile function during intravenous nicotine.

The dose of nicotine used in this study produced pronounced hemodynamic responses. The relative importance of circulating catecholamines and sympathetic stimulation of myocardium might differ at lower doses of nicotine. This question merits further study.

SUMMARY

The role of cardiac neural stimulation in the absence of direct effects of nicotine and other circulating factors was evaluated by perfusing a region of left ventricular myocardium with blood from a reservoir during systemic nicotine infusion. In normally perfused myocardium, contractile function increased, but in reservoir-perfused myocardium, function deteriorated. To evaluate direct effects of nicotine on myocardial contractile function, nicotine was infused directly into a perfused coronary artery. An intracoronary concentration of nicotine equivalent to that caused by intravenous administration (0.69 ± 0.08 μg per ml plasma) had no effect on myocardial contractile function. An intracoronary nicotine concentration of 5 μg/ml was required to directly increase contractile function to approximately the same extent observed during intravenous nicotine. To evaluate the role of circulating catecholamines in the myocardial contractile response to intravenous nicotine, observations were made before and after bilateral adrenalectomy. Adrenalectomy markedly attenuated the pressor response to intravenous nicotine and abolished the positive cardiac inotropic response. We conclude that the adrenal release of catecholamines is primarily responsible for increased myocardial contractile function during intravenous nicotine.

ACKNOWLEDGMENTS

The author thanks Arthur Williams, Jr., and Wendi Daniels for technical assistance in performing the experiments, Janet Mattern for nicotine assays, Hong Gu for catecholamine assays, and Rana Burt for word processing. This study was supported by Grant 0090 from the Smokeless Tobacco Research Council.

REFERENCES

1. Antonaccio JM. Central transmitters: physiology, pharmacology, and effects on the circulation. In: Antonaccio JM. ed. <u>Cardiovascular Pharmacology</u> New York: Raven Press. 1984:155-96.
2. Armitage AK. Effects of nicotine and tobacco smoke on blood pressure and release of catecholamines from the adrenal glands. <u>Brit J Pharmacol</u> 1965;25:515-26.
3. Ball K, Turner R. Smoking and the heart: the basis for action. <u>Lancet</u> 1974;2:822-6.
4. Bassett AL, Wiggins JR, Danilo P, Nilsson K, Gelband H. Direct and indirect inotropic effects of nicotine on cat ventricular muscle. <u>J Pharmacol Exp Ther</u> 1974;188:148-56.
5. Bioanalytical System, Inc. Plasma Catecholamines LCEC Application Note No. 14, 1984.
6. Burn JH, Rand MJ. Action of nicotine on the heart. <u>Brit Med J</u> 1958;1:137-9.
7. Comroe Jr, JH. The pharmacological actions of nicotine. <u>Ann NY Acad Sci</u> 1960;90:48-51.
8. Downey HF, Bashour CA, Boutros IS, Bashour FA, Parker PE. Regional myocardial blood flow during nicotine infusion: Effects of <u>beta</u> adrenergic blockade and acute coronary artery occlusion. <u>J Pharmacol Exp Ther</u> 1977;202:55-68.
9. Downey HF, Crystal GJ, Bashour FA. Regional renal and splanchnic blood flows during nicotine infusion: Effects of <u>beta</u> adrenergic blockade. <u>J Pharmacol Exp Ther</u> 1981;216:363-367.

10. Downey HF, Crystal GJ, Bashour FA. Regional renal and splanchnic blood flows during nicotine infusion: Effects of <u>alpha</u> and of combined <u>alpha</u> and <u>beta</u> adrenergic blockade. <u>J Pharmacol Exp Ther</u> 1982;220:375-81.

11. Downey HF, Williams Jr, AG, Yonekura S, Watanabe N, Lawrence ME. Dominant role of circulating catecholamines in the myocardial contractile response to intravenous nicotine. <u>J Cardiovasc Pharmacol</u> 1988;12:58-64.

12. Fenton RA, Dobson Jr, JG. Nicotine increases heart adenosine release, oxygen consumption, and contractility. <u>Am J Physiol (Heart Circ Physiol</u> 18) 1985;249:H463-70.

13. Gebber GL. Neurogenic basis for the rise in blood pressure evoked by nicotine in the cat. <u>J Pharmacol Exp Ther</u> 1969;166:255-63.

14. Granata L, Olsson RA, Huvos A, Gregg DE. Coronary inflow and oxygen usage following sympathetic nerve simulation in unanesthetized dogs. <u>Circ Res</u> 1965;16:114-20.

15. Greenspan K, Edmands RE, Knoebel SB, Fisch C. Some effects of nicotine on cardiac automaticity, conduction, and inotropy. <u>Arch Intern Med</u> 1969;123:707-12.

16. Hefner JE, Williams, Jr, AG, Robinson EJ, Downey HF. Measurement of nicotine in plasma by high-performance liquid chromatography. <u>J Liquid Chrom</u> 1988;2375-2389.

17. Ilebekk A, Lekven J. Cardiac effects of nicotine in dogs. <u>Scand J Clin Lab Invest</u> 1974;33:153-9.

18. Isaac PF, Rand MJ. Blood levels of nicotine and physiological effects after inhalation of tobacco smoke. <u>Eur J Pharmacol</u> 1969;8:269-83.

19. Koley J, Saha JK, Koley BN. Positive inotropic effect of nicotine on electrically evoked contraction of isolated toad ventricle. <u>Arch Int Pharmacodyn Ther</u> 1984;267:269-78.

20. Murray PA, Lavallee M, Vatner SF. Alpha-adrenergic-mediated reduction in coronary blood flow secondary to carotid chemoreceptor reflex activation in conscious dogs. <u>Circ Res</u> 1984;54:96-106.

21. Paton WDM, Perry WLM. The relationship between depolarization and block in the cats' superior cervical ganglion. <u>J Physiol (London)</u> 1953;119:43-57.

22. Puri PS, Alamy D, Bing RJ. Effect of nicotine on contractility of the intact heart. <u>J Clin Pharmacol</u> 1968;8:295-301.

23. Ross G, Blesa MI. The effect of nicotine on the coronary circulation of dogs. <u>Amer Heart J</u> 1970;79:96-101.

24. Schaeppi U. Nicotine treatment of selected areas of the cat brain: effects upon EEG and autonomic system. <u>Int J Neuropharmacol</u> 1968;7:207-20.

25. Theroux P, Franklin D, Ross Jr. J, Kemper WS. Regional myocardial function during acute coronary artery occlusion and its modification by pharmacologic agents in the dog. <u>Circ Res</u> 1974;35:896-908.

26. Tsujimoto A, Nishikawa T. Comparison of the effects of nicotine on catecholamine release from isolated adrenal glands of dogs and monkeys. <u>Eur J Pharmacol</u> 1974;29:316-9.

27. Van Loon GR, Kiritsy-Roy JA, Brown LV, Bobbitt FA. Nicotinic regulation of sympathoadrenal catecholamine secretion: cross-tolerance to stress. <u>Adv Behav Biol</u> 1985;31:263-276.

28. Watts DT. The effect of nicotine and smoking on the secretion of epinephrine. <u>Ann NY Acad Sci</u> 1961;90:74-80.

29. West JW, Guzmann SV, Bellet S. Cardiac effects of intracoronary arterial injections of nicotine. <u>Circ Res</u> 1958;6:389-95.

EFFECTS OF CIGARETTE SMOKE AND NICOTINE ON PLATELETS AND EXPERIMENTAL

CORONARY ARTERY THROMBOSIS

J.D. Folts, S.A. Gering, S.W. Laibly, B.G. Bertha,
F.C. Bonebrake, and J.W. Keller
Department of Medicine
University of Wisconsin
Madison, WI

INTRODUCTION

The first study linking smoking with predicted risk of cardiovascular
disease was published in 1958 (1), and this has been confirmed by
numerous epidemiologic studies from around the world. The major forms of
coronary artery disease and peripheral vascular disease associated with
smoking are the result of progressive development of atherosclerosis.
Atherosclerotic plaques and fatty streaks can occur early in life as
shown in the Korean war autopsy studies in 1953 (2), and confirmed by
several other studies (3,4). By some unknown mechanism smoking appears
to accelerate the development of the atherosclerotic process.

The contributing role of platelets to the development of atherosclerosis is well accepted (5,6). Experimentally it has been shown that
the presence of Von Willebrands disease with impaired platelet-vessel
wall interactions (7) or administering platelet inhibitors (8) to animals
predisposed to developing atherosclerosis will retard the rate of
atherosclerotic development. Conversely, it is thought that mechanisms
which increase platelet activity or hemostasis, such as elevated plasma
fibrinogen (6) or Homocysteinemia (9), would increase the rate of
development of atherosclerosis.

Platelet Aggregation Studies

The many constituents of cigarette smoke that may influence
hemostasis remain to be clearly identified. Nicotine has been reported
to influence both thrombin time and the platelet response to a variety of
aggregating agonists in vitro (10). However, ex vivo platelet
aggregation studies using platelet rich plasma (PRP) done before and
after smoking (11), or done in anesthetized animals before or after
ventilation with cigarette smoke (12), have not shown clear results. It
is thought that platelets may indeed be activated in vivo by a direct
effect of some constituent of cigarette smoke, such as nicotine, as well
as the increased catecholamine levels associated with smoking. However,
these activated platelets may then become attached to red cells or form
platelet aggregates during the collection and centrifugation needed to
make platelet rich plasma. There is some evidence that the activated
platelets then come down with the red cells or as platelet aggregates and
are lost from the supernatant PRP. Older or less active platelets may

remain in the PRP and thus the increased in vivo platelet activation due
to cigarette smoke would be missed (12).

<u>Platelet Turnover Studies</u>

Mustard (13) first reported a shorter platelet life-span in chronic
smokers, which has been confirmed by others (14). Although platelet
turnover studies demonstrate that an abnormality exists between platelets
and vascular endothelium they do not discriminate between a direct effect
of a constituent of cigarette smoke on the platelet or altered platelet
function due to endothelial damage caused by smoking. In addition,
moderately elevated levels of platelet derived Beta-Thromboglobulin in
the plasma, or increased numbers of circulating platelet aggregates have
been demonstrated in the blood of smokers (15). However, this is
controversial and these may have been generated during collection in the
blood sampling catheter or syringe and thus may be artifactual. The
acute effects of cigarette smoking on the development of atherosclerosis
and its sequelae, coronary arterial occlusion, myocardial infarction, or
sudden death may be transient. Methods are needed to rapidly assess
hemodynamic and hemostatic effects of smoking in animals and man, and an
in vivo method may be most useful.

We have developed an animal model of coronary artery stenosis with
moderate intimal damage and with periodic acute platelet aggregation and
intracoronary thrombus formation (12,16,17). We and others have shown
that, in our model, the primary factor determining the rate and frequency
of intravascular platelet accumulation is the intrinsic moment to moment
level of in vivo platelet activity (17-19). We have used this model to
demonstrate transient increases in platelet activity due to epinephrine
infusions (20), ventilating the animals with cigarette smoke (12) or
infusing nicotine intravenously (12). Conversely, we have shown that in
vivo platelet activity can be transiently diminished with continuous
infusions of platelet inhibitors such as RGD peptides (21) PGI$_2$ (20) or
nitroglycerin, 10 µg/kg/min for 30 minutes (22).

<u>Animal Model for Measuring In Vivo Platelet Aggregation and Periodic
Acute Platelet Thrombus Formation</u>

We felt that since our animal model provides a form of 'on line'
continuous bioassay of in vivo platelet activity, it might be a better
method for detecting transient increases in platelet activity, such as
caused by cigarette smoke or elevated plasma catecholamines than ex vivo
platelet aggregation or coagulation studies.

Anesthetized open chest dogs are prepared as previously described
(16-23). The heart is exposed and the left circumflex coronary artery
dissected out for a distance of 2-3 cm. An electromagnetic flow probe is
placed on it proximally and in some dogs a Silastic nonobstructing
catheter (21 gauge) is placed distally, for measuring pressure distal to
the stenosis, using the technique of Khouri and Gregg (Fig. 1). A second
larger electromagnetic flow probe can be placed on the ascending aorta
for measurement of cardiac output. A catheter is passed down the carotid
artery for systemic blood pressure measurements. Epicardial leads are
attached for surface electrograms and pairs of ultrasonic crystals are
placed in the myocardium for measuring myocardial segment length (MSL) as
a measure of regional 'contractility' (23). Encircling plastic cylinders
are constructed with a range of internal diameters as shown in the upper
right of Fig. 1. One of these is placed around the outside of the
coronary artery, constricting it a known controlled amount. One is
selected which produces a 60-80 percent fixed mechanical stenosis of the

coronary artery. The 36" long tapered nylon monofilament fishline, 0.20"
at one end and tapered to .006" at the other end, can be placed between
the outside of the stenosed coronary artery wall and the inside wall of
the rigid plastic cylinder (23). Then if slight increases in stenosis
are needed, it can be pulled in one direction. If less stenosis is
needed it can be pulled in the other direction. This technique was
demonstrated to be a reliable means of producing fixed controlled
stenosis (23). In this and another study it was determined that a 70-80
percent stenosis just abolished the reactive hyperemic response to a
20-second complete occlusion (12,23). This has been called 'critical
stenosis' and represents the amount of stenosis whereby the distal
arterioles should be near maximally dilated. Thus any increase in the
amount of stenosis caused by increased in vivo platelet aggregation and
by a developing intracoronary platelet thrombus should produce a decrease
in coronary blood flow.

With this model, platelets periodically collect in the stenosed
lumen, adding to the original fixed 70% stenosis and causing coronary
blood flow to decline, as measured with the EMF flow probe. The artery
gradually becomes occluded with a platelet thrombus and flow drops to
zero (Fig. 2a). Often the thrombus embolizes distally, due to the
pressure gradient which builds up across the stenosis (Fig. 2a). If the
thrombus does not embolize, one gently shakes the plastic constricting

ANIMAL MODEL FOR PRODUCING CONDITIONS SIMILAR
TO A PATIENT WITH CORONARY ARTERY DISEASE

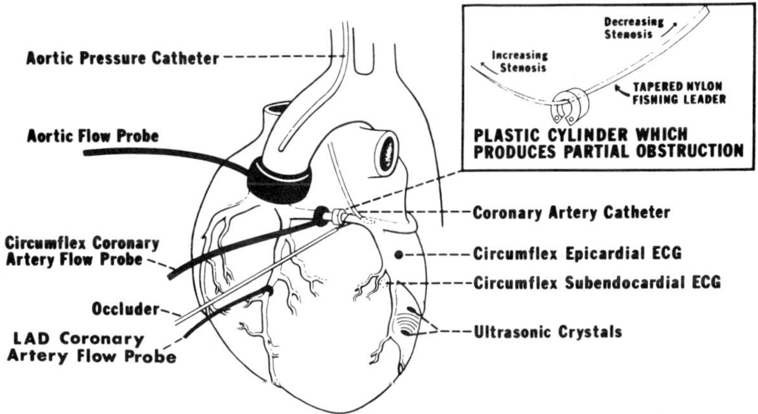

Fig. 1. Technique for producing fixed partial obstruction in either the
left circumflex or LAD coronary artery. The vessel diameter is
measured with a caliper and a plastic cylinder of appropriate inside diameter, shown in the upper right hand corner, is selected
which will produce a 60-80 percent reduction in coronary artery
diameter when placed around the outside of the vessel. A tapered
monofilament fishing leader is placed between the inside lumen of
the plastic cylinder and the outside of the coronary artery. Then
this leader is pulled in either direction to make small controlled increments in the amounts of stenosis.

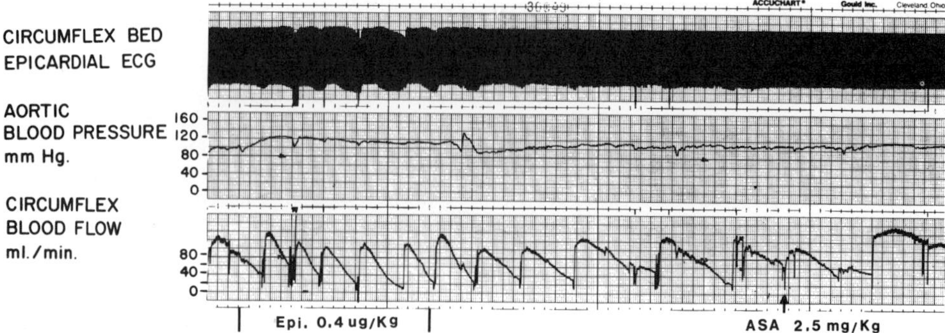

Fig. 2a. Control cyclical flow reductions in coronary blood flow shown
on the far left show an increase in frequency and rate of flow
decline when epinephrine is infused i.v. for 15 minutes. In
the center the rate of cyclical flow reductions return to pre-
epinephrine flow rates. On the right 2.5 mg/kg of aspirin (ASA)
is given i.v. and cyclical flow reductions due to periodic acute
platelet thrombus formation are abolished. Reproduced by permission Thrombosis Research.

cylinder, and the thrombus will be dislodged and embolize into the distal
microcirculation. This permits rapid restoration of blood flow, back to
control levels. As this is an ongoing process, the periodic formation
and embolization of thrombi produces cyclical flow reductions (CFRs) in
the measured coronary blood flow as shown in Fig. 2a. If one infuses
epinephrine to raise the plasma level, there will be an increase in the
size and frequency of the CRF's (Fig. 2a). These CFR's can be prevented
with aspirin 2.5 mg/kg or other platelet inhibitors (Fig. 2a), but can be
made to reoccur with the infusion of epinephrine 0.4 µg/kg/min (Fig. 2b)
(24). Any mechanism or process such as smoking cigarettes which causes
elevated plasma epinephrine levels is likely to reactivate platelets and
cause renewal of CRF's in spite of pretreatment with aspirin (24).

With this model the slope of the coronary blood flow decline is
proportional to the rate at which platelets accumulate in the stenosed
lumen. One can calculate the rate of flow decline as shown in Fig. 3.
Thus an intervention which causes an increase in slope indicates an
increase in in vivo platelet activity. Conversely any intervention which
causes a decrease in slope, indicates a decrease in in vivo platelet
activity.

<u>Studies with Nicotine and Cigarette Smoke on In Vivo Platelet Activity
and Coronary Thrombus Formation</u>

Healthy adult mongrel dogs were anesthetized with sodium
pentobarbital, the chest opened and surgically prepared as in Fig. 1.
Then several experimental protocols were followed.

<u>Group 1 -- Administration of Smoke from Cigarettes (n=15)</u>. After an
observation period of 30 minutes with CFRs occurring, due to periodic
acute platelet thrombus formation (similar to those in Fig. 2a), arterial
blood samples were drawn from 15 dogs for PO_2, PCO_2, pH, hemoglobin,
hematocrit and control plasma catecholamine level determinations. A
'smoking machine,' was constructed consisting of a clear polyurethane box
that holds a lighted cigarette within the box, and this was connected to

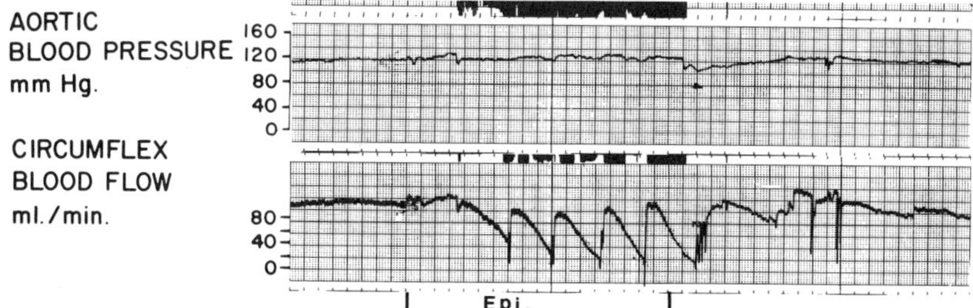

Fig. 2b. On the left the coronary blood flow is constant with no
cyclical flow reductions or periodic acute thrombus formation
due to the aspirin administration. However in the center,
cyclical flow reductions return during the infusion of
epinephrine for 15 min in spite of pretreatment with aspirin.
Reproduced by permission Thrombosis Research.

a controlled volume-flow ventilator that delivers a standard volume of
cigarette smoke during normal respiration. The smoke then goes from the
burning tip inside the box, through the tobacco and through tubing to the
trachea and lungs of the dog (Fig. 4).

At this time, a standard cigarette, in approximately the middle range
as determined by the Federal Trade Commission, April 1976, for tar and
nicotine levels (tar 17 mg, nicotine 1.0 mg) was lit and placed in

Calculation of the Rate of Cyclic Blood Flow Reductions (CFR's)

1. Draw best fit line through CFR.

2. Calculate the slope of the line: $m = \dfrac{y_2 - y_1}{x_2 - x_1} = \dfrac{\text{change in flow}}{\text{time elapsed}}$

3. Example: $y_2 - y_1$ = decline in coronary
blood flow from A to B =
$-8\,\text{ml/min} - 60\,\text{ml/min} = -68\,\text{ml/min}$

$x_2 - x_1$ = time elapsed from A to B = $\dfrac{30\,\text{mm}}{6\,\text{mm/min}} = 5\,\text{min}$

m = rate of CFR = $\dfrac{-68\,\text{ml/min}}{5\,\text{min}} = -13.6\,\text{ml/min}^2$

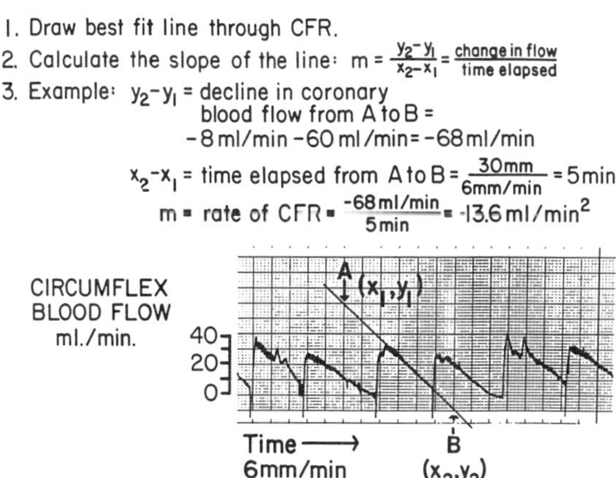

Fig. 3. Reproduced by permission Cardiovascular Research.

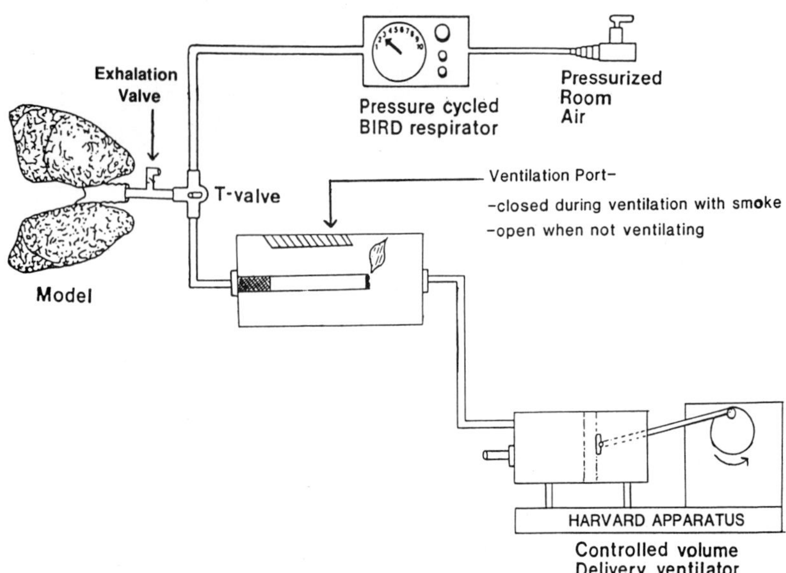

Fig. 4. Schematic describing how cigarette smoke can be delivered periodically, using the T-value, to the dogs lung.

the clear polyurethane box. The endotracheal tube was then switched by a T-valve to the control volume flow ventilator. Through manual cycling of the respirator, a 500-ml volume of room air and cigarette smoke was ventilated into the animal's lungs, and then the dog was switched back to just room air via the Bird respirator (Fig. 4). This procedure was repeated approximately twice a minute until the cigarette was finished (approximately 10 minutes). With this procedure we attempted to simulate human smoking patterns (12). At the conclusion of the cigarette, the CFRs were increased in size and frequency indicating increased in vivo platelet activity (Fig. 5) and a second set of blood samples was drawn to detect any changes in blood gases and plasma epinephrine measurements.

During the next 45 minute observation the CFR's gradually became smaller and returned to pre smoke levels after 10-15 minutes (Fig. 5). This suggests that the increase in platelet activity and acute intracoronary thrombus formation was transient in nature. Then 3 mg/kg of phentolamine (an alpha-adrenergic antagonist) dissolved in 40 ml of saline was administered over a 5-minute period. Fifteen minutes after the phentolamine was administered CFR's were abolished in all dogs (Fig. 6), and a third set of blood samples were drawn. Using the same brand of cigarette, smoke was again delivered to the dog as previously described. At the conclusion of the smoke administration, a fourth set of blood samples were drawn.

Group 2 -- Administration of Intravenous Nicotine (n=6). After the 30-minute observation period, when CFR's were occurring due to periodic intracoronary platelet thrombus formation, a control blood sample was drawn for blood gases and catecholamine level determinations. The dogs were then given 80 µg/kg of nicotine intravenously through an infusion pump, over 8-10 minutes (corresponding roughly to the period of time it took to ventilate the dog with smoke from one cigarette by our method).

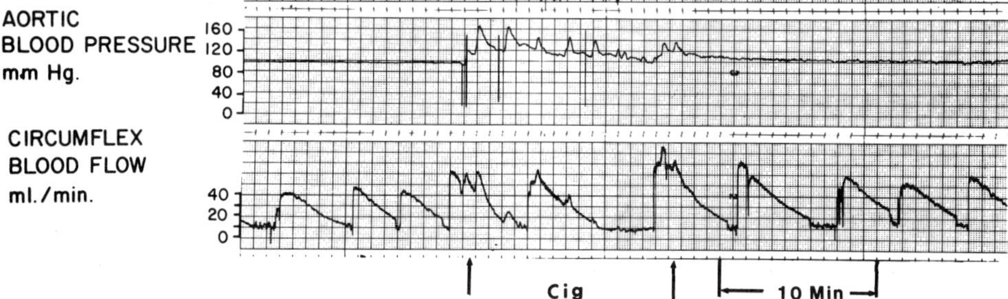

Fig. 5. Increased size, and slope of CFR's due to ventilating the dog
with cigarette smoke. Note the transient increase in mean
arterial blood pressure with some of the 'puffs' of smoke.

When the nicotine infusion was complete, CFR's were significantly
increased in size and frequency similar to Fig. 5, indicating that
nicotine also increased in vivo platelet activity and acute platelet
thrombus formation. Another blood sample was drawn to detect changes in
blood gasses and the plasma catecholamine levels. Again, the dogs were
observed for 45 minutes, and at this time 3 mg/kg of phentolamine was
administered as previously described. Fifteen minutes after the
phentolamine was administered, a third blood sample was drawn and again
80 µg/kg of nicotine was infused in the same manner. When the nicotine
administration was complete, a fourth blood sample was drawn.

Group 1. In all 15 dogs, plasma catecholamine levels and aortic
blood pressure increased in response to the administration of smoke from
one cigarette (Table 1). Arterial PO_2, pH and PCO_2 did not change
significantly, but the hematocrit and hemoglobin levels increased (Table
1). In all cases, the spontaneous CFRs were more severe after cigarette
smoke, as reflected by the greater flow decrease, the number of flow
reductions, and the size of the reductions in flow (Table 2). During the
45-minute observation period after cigarette smoke administration, the

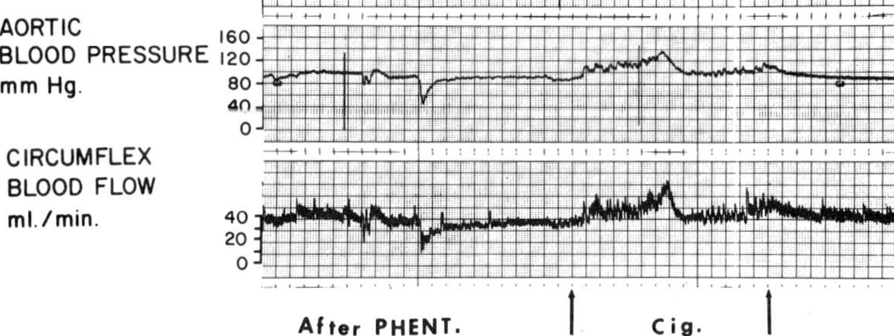

Fig. 6. CFR's in the coronary flow are abolished by the Phentolamine
(PHENT) and the coronary flow varies only with arterial pressure
changes. Note that there are no CFR's during ventilation with
cigarette smoke (marked Cig.) or afterwards.

Table 1. Hemodynamic Measurements Before and After the Administration
          of Cigarette Smoke (n=15)

	ABP (mm Hg)	Hct	HgB	Plasma epinephrine (pg/ml)
Before cigarette smoke	110±12	38±4	13±5	206±210
	p<0.001	p<0.01	p<0.02	p<0.001
After cigarette smoke	166±32	47±6	16±2.0	1817±1188

Values are mean±SEM. Abbreviations: ABP = arterial blood pressure;
Hct = hematocrit, HgB = Hemoglobin.

severity of the exacerbated flow reductions returned to presmoke levels
within 5-15 minutes. After phentolamine, the spontaneous flow reductions
were abolished in all dogs (Fig. 6).

   Group 2. In all six dogs, i.v. nicotine produced a blood pressure
increase and elevation of plasma catecholamines similar to that produced
by a cigarette as in Group 1 (Table 3).

   Similar exacerbations of coronary flow reductions were also recorded
in response to the intravenous nicotine (Table 4). Again, as observed in
the cigarette smoke trials, in the 45 minutes after the nicotine
infusion, the magnitude and frequency of the flow reductions gradually
returned to control levels in 5-10 minutes. Thus, any ex vivo test of
increased platelet activity such as platelet aggregometry may have missed
this transient increase in platelet activity due to smoke or nicotine.

   After phentolamine, spontaneous flow reductions and those produced by
nicotine were abolished in all dogs after 15 minutes. These studies

Table 2. Quantification of Spontaneous Coronary Blood Flow Reductions
          in the Circumflex Coronary Artery Before and After the
          Administration of Cigarette Smoke (n=15)

	Slope of CBF reduction (ml/min^2)	Number of CBF reductions/10 min	Size of CBF reductions (ml)
Before cigarette smoke	−1.75±2.23	1.6±0.9	15.3±13.9
	p<0.001	p<0.009	p<0.0006
After cigarette smoke (measured for 10 min after administration of cigarette smoke)	−10.57±4.8	3.0±1.13	36.3±15.8

Values are mean±SEM.   Abbreviation: CBF = coronary blood flow.

Reprinted by permission Circulation.

Table 3. Hemodynamic Measurements Before and After Intravenous Infusion of Nicotine (n=6)

	ABP ABP (mm Hg)	Hct	Plasma epinephrine (pg/ml)
Before i.v. nicotine	118±10	39±5	315±198
	p<0.001	p<0.05	p<0.005
After i.v. nicotine	156±28	44±4	1529±1310

Values are mean±SEM.

Abbreviations: ABP = arterial blood pressure; Hct = hematocrit.

Reprinted by permission Circulation.

showed that ventilating the anesthetized animals unaccustomed to smoking, with cigarette smoke or infusing nicotine elevated the plasma epinephrine, arterial blood pressure and hematocrit. In these animals phentolamine blocked the increase in vivo platelet activity and acute thrombus formation. Platelets are known to be activated by alpha adrenergic receptors and it is believed that the elevated epinephrine caused by cigarette smoke or elevated plasma nicotine is responsible. This is supported by the fact that phentolamine, an alpha adrenergic antagonist, blocked the response to cigarette smoke or nicotine.

These CFR's have been reported to occur in peripheral arteries of conscious patients with Peripheral Vascular Disease, with blood flow measured continuously for several hours by placing a doppler flow probe on the skin over the area of the stenosis (25). In addition, using a catheter tipped doppler flowmeter, CFR's have recently been observed in patients with unstable angina undergoing percutaneous coronary artery angioplasty (26). Thus if increased in vivo platelet activity, and size of CFR's occur in dogs with stenosed coronary arteries each time they are exposed to cigarette smoke or nicotine, this may also happen in

Table 4. Quantification of Spontaneous Coronary Blood Flow Reductions in the Circumflex Coronary Artery Before and After Intravenous Infusion of Nicotine (n=6).

	Slope of CBF reduction (ml/min^2)	Number of CBF reductions/10 min	Size of CBF reductions (ml)
Before cigarette smoke	-2.11±1.88	1.8±0.8	17.1±11.8
	p<0.005	p<0.01	p<0.001
After cigarette smoke (measured for 10 min after administration of nicotine)	-9.93±2.98	2.8±2.10	38.13±17.6

Values are mean±SEM.   Abbreviation: CBF = coronary blood flow.

Reprinted by permission Circulation.

patients. If there is indeed transiently increased platelet activity in
man with each cigarette smoked then this may represent a repetitive
mechanism for enhanced atherosclerotic development (5).

<u>Relative Effects of Cigarette Smoke and Ethanol on Acute Platelet
Thrombus Formation in Stenosed Coronary Arteries</u>

The risks of developing coronary artery disease, myocardial
infarction, and sudden cardiac death have been shown to be increased for
cigarette smokers (1,27,28), whereas the development of coronary artery
atherosclerosis and the occurrence of both fatal and non-fatal myocardial infarction have been negatively correlated with the consumption of
alcohol (28-32). Studies have shown that cigarette smoke increases
platelet activity (12,33,34), whereas alcohol decreases platelet
activity (35-38). Based on the above observations, it seemed logical
that the effects of cigarette smoking and alcohol consumption on the
incidence of coronary artery disease, myocardial infarction, and sudden
death may be due, in part, to their effects on platelet activity.
Cigarette smoking has been positively correlated with the consumption of
alcoholic beverages (39-41), and there is a general trend for cigarette
smokers to smoke more heavily while drinking (42,43). Therefore, this
study was conducted to determine the relative effects of cigarette smoke
and alcohol on in vivo platelet activity and acute thrombus formation in
our model. Changes in platelet activity were determined by changes in
the rates of cyclic reductions in blood flow, CFR's, through stenosed
canine coronary arteries. These changes in coronary flow were calculated as shown in Fig. 3.

After the initial observation period of the CFR's, each dog was
ventilated with a tidal volume of 300 ml of cigarette smoke from one
cigarette over an average period of 10 min in such a way as to simulate
human smoking patterns (44). The cyclic reductions in coronary flow
occurred at a calculated mean rate of -10 ± 3.3 ml·min^2. Ventilation with
cigarette smoke resulted in acute increases in arterial blood pressure
of 20-50 mm of Hg and the increase in slope of the cyclic flow reductions increased to -14 ± 5 ml/min^2 (P<.005). Thirty minutes after cigarette smoke the slope of the CFR's had returned to -10.1 ± 3.1 ml·min^2
indicating once again that the effect on in vivo platelet activity while
significant, was transient. The cigarettes used contained 16.5 mg tar
and 1.05 g nicotine as assessed by the Federal Trade Commission.

Approximately 30 min after the administration of cigarette smoke,
each dog was given 1.0 ml·kg^{-1} of 100% ethanol, diluted in 200 ml of
0.9% sodium chloride solution and given via a slow intravenous infusion
over a period of 10-15 min into a catheter placed in the right atrium
via the right femoral vein, thus preventing the ethanol from lysing red
blood cells. Blood samples were drawn immediately before and 20 min
after the administration of ethanol to be used for ex vivo platelet
aggregation studies. A blood sample was obtained 10 minutes after
ethanol administration to determine the steady state concentration
present in the blood.

Intravenous ethanol reached a steady state level of 1.9 ± 0.4
g·litre^{-1} and had no effect on systemic blood pressure. The ethanol
completely abolished the CFR's in all dogs (Fig. 7a) and repeat ventilation with cigarette smoke did not renew CFR's in any of the 10 dogs
(Fig. 7b).

SIGNIFICANCE

It has been demonstrated that cigarette smoke increases, whereas
alcohol decreases, platelet activity. The findings of this study are
consistent with these two general findings. The smoke from a single
cigarette significantly increased the rate of thrombus formation in
stenosed canine coronary arteries, whereas ethanol completely abolished
thrombus formation. Ex vivo platelet aggregation results were in
agreement with the observed effects of ethanol on in vivo platelet
activity (thrombus formation); aggregation stimulated by ADP and ADP
plus adrenaline was significantly inhibited after the administration of
ethanol (35). Although the peak blood alcohol concentrations achieved
in this study were high, we have observed in recent studies that one
half as much ethanol producing a blood alcohol concentration of 0.9
$g \cdot litre^{-1}$ will abolish the cyclical flow reductions (38).

Our evidence suggests that the exacerbating effect of cigarette
smoke on platelet thrombus formation is caused by stimulating the
release of endogenous adrenaline from the adrenal medulla of the dog
(12). Epinephrine is known to increase levels of platelet cytoplasmic
calcium by several proposed mechanisms including $alpha_2$ adrenergic
receptor stimulation which in turn mediates functional shape change and
primary aggregation (24). It has been shown that human platelets,
inhibited with aspirin could be reactivated with physiologic amounts of
epinephrine in an in vitro study (45). It has also been shown that
epinephrine reverses the inhibitory influence of aspirin on platelet-
vessel wall interactions (45). Thus, it is unfortunate that the only
agent, aspirin, which has been approved by the Food and Drug Administration may not protect patients from a coronary thrombosis if they raise
their plasma epinephrine by smoking or other means (24).

The ex vivo platelet aggregation response to ADP 50 $\mu mol \cdot litre^{-1}$,
and ADP 5 plus adrenaline 11 $\mu mol \cdot litre^{-1}$ was significantly inhibited by
the infusion of ethanol, but the platelet rich plasma platelet count was
not significantly changed. This is consistent with previous reports
(35-37.

The mechanisms through which ethanol affects platelet activity are
not well understood, but several theories have been suggested. Similar-
ities between ethanol and local anaesthesia have been noted, and local
anesthetics such as cocaine and tetracaine have been shown to inhibit
ADP induced platelet aggregation (46). It has been suggested that local
anesthetics inhibit platelet activity by stabilizing platelet membranes
(46) and by preventing calcium influx or the mobilization of intracellular calcium stores in the platelet (47).

A New, Possibly Safer, Cigarette?

Recently, the tobacco industry has introduced a new type of
cigarette with a burning charcoal tip which heats the tobacco as hot air
is drawn through it. This new type of cigarette is thought to be safer
because it heats the tobacco rather than burning it, as in conventional
cigarettes. It is thought that many of the toxic materials found in
cigarette smoke are created by combustion of the tobacco (48). We
elected to test the effects of smoke from this new type of cigarette,
and compared the rate of thrombus accumulation in stenosed canine
coronary arteries, before and after ventilation with smoke from both
regular cigarettes and this new non-tobacco-burning cigarette.

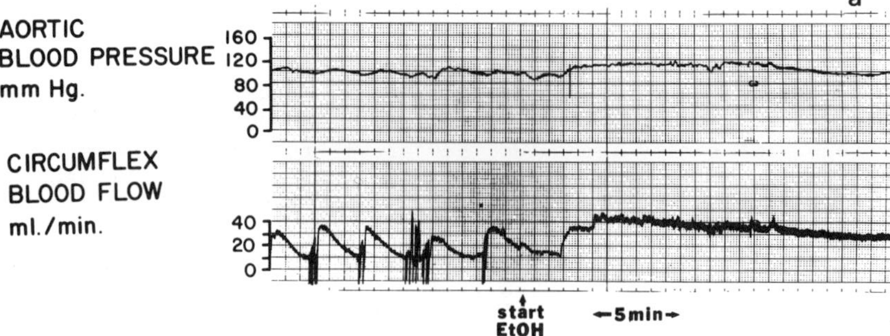

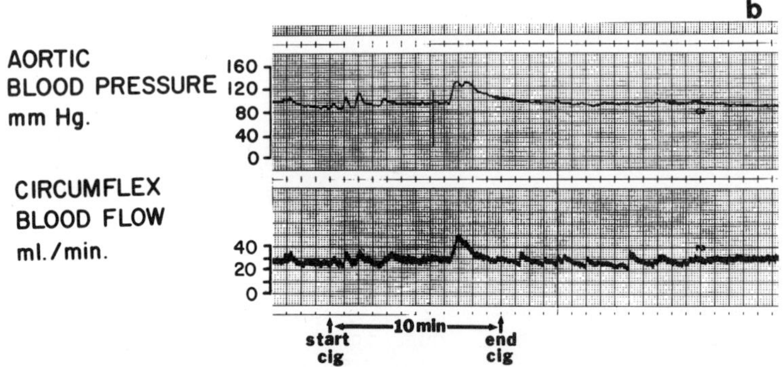

Fig. 7a,b. The CFR's on the left of Fig. 7a were abolished by
ethanol (ETOH) in all dogs. In Fig. 7b there are no
CFR's produced during or after the ventilation with
cigarette smoke.

Five dogs were prepared as described above and subject to the
following protocol.

Method of Smoking

Each of the 5 dogs showed CFR's in the circumflex coronary artery
during a 30 min observation period. In each dog the effects of ventilation with smoke from both the new non-tobacco-burning cigarette (NC) and
a regular cigarette (RC) were compared for their effects on in-vivo
platelet activity and the CFR's observed. This was accomplished by
first ventilating each dog with room air for approximately 30 minutes
which allowed for stabilization of the animal, while CFR's were occurring. A blood sample was drawn during this period to determine a pre-
smoke plasma epinephrine level, blood carbon monoxide (CO) saturation
(measured as carboxy-hemoglobin saturation), and hematocrit. Each dog
was then ventilated with either NC or RC cigarette smoke (both 0.4 mg
nicotine/cigarette) using the apparatus shown in Fig. 3. A 'puff' of
cigarette smoke was delivered by switching the 3-way valve from the Bird
pressure cycle respirator to the smoke machine. The control volume
ventilator was then manually triggered to push air through the cigarette
mounted in the smoke machine. Following the delivery of the smoke, the
3-way valve was switched back to the Bird respirator. This procedure
was repeated every 30 seconds for the duration of two cigarettes. The
volume of smoke delivered in each puff was maintained at approximately a

tidal volume of 500 ml. Immediately following the last 'puff' of the
second cigarette, a second blood sample was drawn to determine the post-
smoke plasma epinephrine level, blood CO saturation and hematocrit.
Following ventilation with smoke, the dog was allowed a one hour
recovery period.

Following the recovery period, a blood sample was drawn to again
determine the pre-smoke plasma epinephrine level, CO saturation, and
hematocrit. The same procedure was again used to test the effects of
the second type of cigarette smoke, either RC or NC depending upon which
was used in the first half of the protocol. Again, the ventilation was
altered between the Bird respirator and the smoke machine every 30
seconds to deliver a puff of smoke for the duration of two cigarettes.
A fourth blood sample was drawn following the last puff of the second
cigarette to determine the post-smoke physiological parameters mentioned
above. Each dog was monitored for an additional 45 minutes following
completion of the second type of cigarette. In two of the dogs tested
NC smoke was administered first followed by RC smoke in the second half
of the protocol, whereas the reverse order was used in the other three
dogs.

RESULTS

All five dogs instrumented with our model of canine coronary
artery stenosis and periodic acute platelet thrombus formation spontaneously produced cyclic flow reductions in their circumflex coronary
artery. The cyclic flow reductions (CFR's) were characterized by slowly
declining blood flow with periodic abrupt returns to control flow during
the pre-smoke stabilization period as shown in Figs. 8a and 8b. Prior
to ventilation with the non-tobacco-burning cigarette smoke (NC), the
average rate of flow decline was -4.35 ± 1.4 ml/min^2 (mean $\pm$ standard
deviation,) (n=5). In contrast, the rate of coronary blood flow decline
following ventilation with the NC smoke was significantly increased to
an average of -8.23 ± 1.5 ml/min^2. Similarly, prior to ventilation with
smoke from a regular cigarette (RC) the average rate of flow decline was
-5.19 ± 1.3 ml/min^2. Following ventilation with RC smoke, the average
rate of flow decline again significantly increased to -9.44 ± 3.1
ml/min^2. In both cases the increase in the rate of flow decline was
statistically significant (Anova matched pair analysis) with a P-value
<.01. In addition, a high percentage of the thrombus embolizations
prior to ventilation with either type of smoke were spontaneous,
indicating that the thrombi were not very adherent to the stenosed
artery. In contrast, following ventilation with either type of smoke
nearly all thrombi had to be gently shaken loose in order to restore
flow in the coronary artery. In one case, 5 minutes following ventilation with NC smoke, a thrombus could not be dislodged and the dog went
into ventricular fibrillation and died. This indicates that after smoke
with either type of cigarette, the thrombi formed were more adherent,
were more difficult to dislodge, and more likely to produce total
coronary occlusion.

Other physiological variables, including carbon monoxide blood
saturation, plasma epinephrine levels, hematocrit, arterial blood
pressure (ABP), and heart rate, were monitored both before and after
each type of smoke. These physiological variables, along with the flow
decline data, are recorded in Table 5. Both types of cigarette smoke
resulted in significant rises in CO blood saturation. In the pre-NC
smoke dogs, the CO saturation was $1.8 \pm 1.5\%$. Following ventilation

with NC smoke CO saturation increased to 14.2 ± 3.8%. Similarly, in the pre-RC smoke dogs initial CO saturation was 1.4 ± 0.5%. Following ventilation with RC smoke, CO saturation increased to 7.4 ± 1.8% which was less than with the NC smoke but still significant. The hematocrit, the arterial blood pressure, and the heart rate measurements all showed similar increases following ventilation with either type of smoke as shown in Table 5.

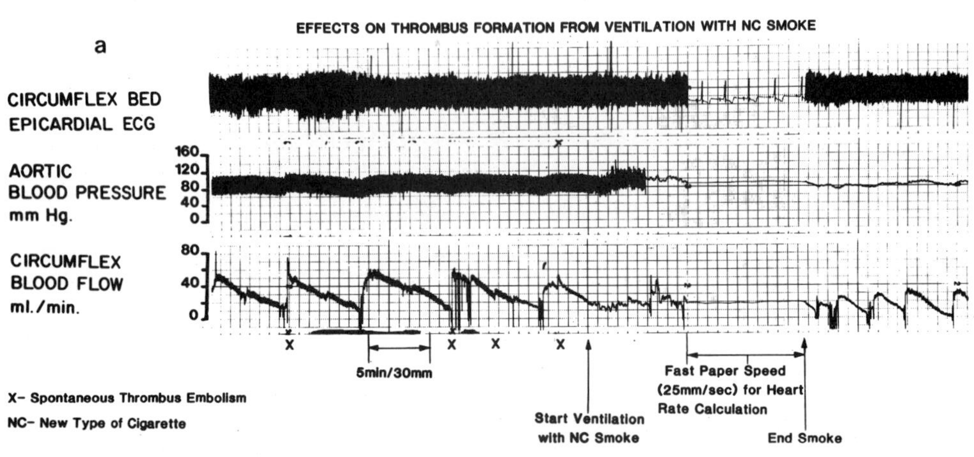

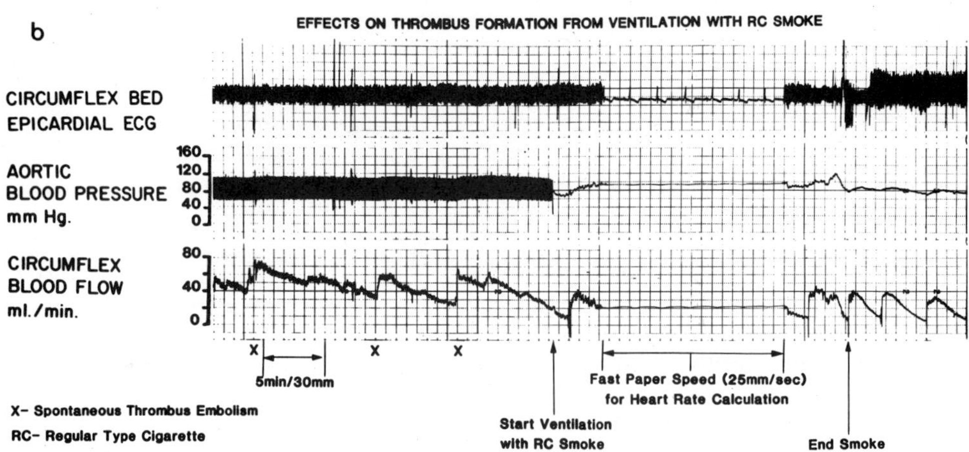

Fig. 8a,b.   The effect on acute platelet thrombus formation and the resulting cyclical reductions in coronary blood flow following ventilation of dogs with smoke from a nontobacco burning cigarette (NC) or regular cigarette (RC). The slope of the flow declines are significantly increased (P<.01) following ventilation with either type of smoke (0.4 mg nicotine/cigarette, 2 cigarettes over approximately 10 min.) which indicates an increase in the degree of in-vivo platelet wall interaction.

Table 5. The effect on specific physiological variables, including
rate of coronary blood flow decline (ROFD), carbon monoxide
(CO) blood saturation, arterial blood pressure (ABP), hematocrit (HCT), heart rate (HR), and plasma epinephrine (EPI)
levels, before and after ventilation with the smoke from both
non-tobacco-burning cigarettes (NC) and from regular cigarettes
(RC) in five dogs. All values are recorded as the mean ±
standard deviation, with the exception of post-smoke ABP data
which is a mean measure of the maximum arterial blood pressure
during ventilation with smoke. Each dog (n=5) was ventilated
with the smoke of 2 cigarettes (0.4 mg nicotine/cigarette).

	Before Smoke	After Smoke
ROFD ml/min^2 for NC	4.35 ± 1.4	8.23 ± 1.5*
ROFD ml/min^2 for RC	5.19 ± 1.3	9.44 ± 3.1*
CO saturation % for NC	1.8 ± 1.5	14.2 ± 3.8*
CO saturation % for RC	1.4 ± 0.6	7.4 ± 1.8*
Mean ABP mmHg for NC	106.8 ± 15.8	130.0 ± 13.3*
Mean ABP mmHg for NC	112.8 ± 13.0	155.0 ± 20.2*
HCT % for NC	34.0 ± 4.0	36.4 ± 4.1*
HCT % for RC	35.6 ± 5.0	40.1 ± 5.4*
HR bpm for NC	130.3 ± 33.1	144.0 ± 27.3*
HR bpm for RC	144.9 ± 33.3	135.3 ± 28.6*
Plasma epinephrine ng/L for NC	42.5 ± 38.8	80.0 ± 73.1 NS
Plasma epinephrine ng/L for RC	46.3 ± 49.3	823.7 ± 901.4*

* = P-value <.05.

The data on plasma epinephrine levels revealed a significant rise in
plasma epinephrine levels following ventilation with RC smoke, but not
following ventilation with NC smoke. The plasma epinephrine level in RC
smoke ventilated dogs rose from an average of 46.3 ± 49.3 ng/L to 823.7 ±
901.4 ng/L. In contrast, the data for pre- and post-NC smoke ventilated
dogs did not show a significant rise in plasma epinephrine levels: pre-
smoke epinephrine = 42.5 ± 38.3 ng/L and post-smoke epinephrine = 80.0 ±
73.1 ng/L. In fact, the plasma epinephrine levels failed to rise at all
following ventilation with smoke in 3 of 5 animals.

SIGNIFICANCE

Our results in this study, clearly indicate that the smoke from the
'New' proposed non-tobacco-burning cigarette is just as potent at
increasing in vivo platelet activity and exacerbating acute platelet
thrombus formation (APTF) in our model as in the smoke from conventional
tobacco burning cigarettes. The size of the CFR's and their greater
adherence suggests that the degree of platelet-vessel wall interaction,
was significantly increased (P<.01) following ventilation with either
type of smoke (NC or RC). Despite clear evidence that ventilation with
either RC or NC cigarette smoke causes an exacerbation of platelet-vessel
wall interaction and in-vivo thrombus formation, the exact mechanism by
which cigarette smoke leads to the exacerbation remains unclear.

In our previous studies, nicotine mediated release of epinephrine
from the adrenal medulla of the dog was suggested as the cause for the
exacerbation of acute platelet thrombus formation observed following
ventilation with regular cigarette smoke (12,49). Others have also
shown that cigarette smoke induces increased plasma epinephrine levels
in both dogs and man (26,50-53). We have previously shown that physiologic infusions of epinephrine transiently exacerbate acute platelet
thrombus formation in our model (20,23,24). Blood nicotine levels after
smoking have also been positively correlated with rises in plasma
epinephrine levels (54). It seems plausible that a nicotine mediated
epinephrine release may be responsible for the exacerbation of acute
platelet thrombus formation observed in our model following ventilation
with regular cigarette smoke. Unfortunately, this mechanism fails to
account for the exacerbation of acute platelet thrombus formation
following ventilation with NC smoke where no significant rise in plasma
epinephrine is observed (Table 5). In all 5 dogs, following RC smoke
ventilation, the plasma epinephrine levels significantly increased with
concomitant significant rises in ABP, hematocrit, and heart rate
(Table 5) which have all been associated with increased plasma epinephrine levels (12,52). In contrast, the change in plasma epinephrine
levels following ventilation with NC smoke is not significant. In fact,
3 of 5 dogs showed no increase in epinephrine levels and, in addition,
the rise in ABP, hematocrit, and heart rate were significantly lower
than the rise observed with RC smoke (Table 5). These data suggest an
alternative mechanism may be responsible for the exacerbation of acute
platelet thrombus formation observed following ventilation with NC
smoke.

It has been alternatively proposed that nicotine has a direct
effect on platelet-vessel wall interaction. In-vitro addition of
nicotine, in doses similar to those achieved through smoking, have been
shown to enhance platelet aggregation (10,55). It has also been suggested by several groups that nicotine may have a direct effect on the
platelet-vessel wall interaction by decreasing the capacity of vascular
endothelial cells to produce prostacyclin (PGI_2) (56-61) possibly by
inhibiting the cyclooxygenase enzyme (58). This type of inhibition
would allow the vessel endothelial lining to become 'more sticky', thus
enhance platelet adhesion and thrombi development. A direct effect of
nicotine may help to explain the exacerbation of acute platelet thrombus
formation we observed in this study and the resulting increase in
coronary flow decline, and the more 'adherent' or harder to dislodge
thrombi we observed after smoke. However, this possible mechanism has
been questioned and conflicting data have been presented (62,63).
Unfortunately, we were unable to measure blood nicotine levels, and
therefore could not speculate as to the role of nicotine in the exacerbation of platelet-vessel wall interaction observed, by the two types of
cigarettes.

Another mechanism that has been given consideration as a possible
link between acute coronary events and smoking is the increase in blood
carbon monoxide saturation. It has been demonstrated that carbon
monoxide transiently increases platelet activity, measured as an
increase in bleeding time, possibly by a transient inhibition of the
vessel endothelium PGI_2 synthesis (58). Increasing blood carboxyhemoglobin levels has also been correlated with enhanced platelet
aggregation, while increasing nicotine levels in the same study failed
to result in enhanced aggregation (64). The possible involvement of CO
in increased platelet-vessel wall interaction may help explain the NC-
smoke induced exacerbation of acute platelet thrombus formation in this
study. The elevation of blood CO levels following ventilation with NC
smoke, from 1.8 ± 1.5 to 14.2 ± 3.8, were significantly higher than the

rise observed following RC smoke, from 1.4 $\pm$ 0.6 to 7.4 $\pm$ 1.8 (P<.01).
Unfortunately, data on acute effects of carbon monoxide blood saturation
are limited, and it has been reported that enhanced aggregation after
smoking is not correlated with increasing CO blood saturation levels
(64). Thus, the contribution of CO to increased platelet activation and
exacerbation of acute platelet thrombus formation needs to be determined. It seems that nicotine, epinephrine, and carbon monoxide may all
play important roles in smoke induced exacerbation of platelet-vessel
wall interaction. Despite this lack of clarity, it is clear that smoke
from the new proposed non-tobacco-burning cigarette is just as potent at
exacerbating in-vivo acute platelet thrombus formation and platelet-
vessel wall interaction as smoke from conventional cigarettes.

SUMMARY

The causal link between smoking, atherosclerosis and an increased
risk for acute platelet mediated coronary events such as acute platelet
thrombus formation, myocardial infarction, and sudden coronary death is
not clear. Our studies suggest that there may be a transient increase
in in vivo platelet activity with each exposure to cigarette smoke or
elevated plasma nicotine. It is thought that platelets may contribute
to the acceleration of the atheroslcerotic process by several mechanisms
(5). Thus it may be that each time a person increases their platelet
activity, by smoking or some other means, and given other predisposing
conditions such as elevated lipids and/or acute intimal damage, such as
rupture of an atherosclerotic placque, the atherosclerotic process may
be enhanced. In addition it appears that cigarette smoke makes a
developing thrombus more adherent and less likely to embolize distally,
although the effect is transient. Finally, cigarette smoke may provide
the final stimulus for an occlusive coronary thrombus in a stenosed
coronary artery already predisposed to thrombosis by other risk factors.
This may account for the fact that the risk of coronary occlusion and
myocardial infarction decreases markedly in the first year after
quitting (65). It may be that smoking has a greater likelihood of
precipitating a fatal thrombus than it does of accelerating the althero-
sclerotic process. This would be reflected in the inability of aspirin
to prevent the smoking induced enhancement of CFR's in our model (23) or
the enhancement of platelet function in men with coronary artery disease
(66).

REFERENCES

1. E. C. Hammond, and D. Horn, Smoking and death rates--report on
    forty-four months of follow-up of 187,783 men. II. Death rates
    by cause, JAMA. 166:1294-308 (1958).
2. W. F. Enos, R. H. Holmes, and J. Beyer, Coronary disease among
    United States soldiers killed in action in Korea, JAMA.
    152:1090-3 1953.
3. J. P. Strong, and H. C. McGill Jr., The natural history of coro-
    nary atherosclerosis, Am J Pathol. 40:37-49 (1962).
4. C. Tejada, J. P. Strong, M. R. Montenegro, C. Restrepo, and L. A.
    Solberg, Distribution of coronary and aortic atherosclerosis by
    geographic location, race and sex, Lab Invest. 18:509-26 (1968).
5. J. L. Gordon and A. J. Milner, Blood platelets as multifunctional
    cells, in: 'Platelets in biology and pathology,' Gordon, ed.,
    Elsevier/North-Holland Biomedical Press, Cambridge, England
    (1976).
6. P. F. Davies, Biology of disease. Vascular cell interactions with
    special reference to the pathogenesis of atherosclerosis, Lab
    Invest. 55:5-24 (1986).

7. V. Fuster, J. T. Lie, L. Badimon, J. A. Rosemark, and W. Bowie, Spontaneous and diet induced coronary atherosclerosis in normal swine and swine with Von Willibrands disease, <u>Arteriosclerosis</u>., 5(1):67-73 (1985).

8. R. Ross, The pathogenesis of atherosclerosis - an update, <u>N Engl J Med</u>. 314(8):488-500 (1986).

9. K. S. McCully, Homocysteine theory of arteriosclerosis: development and current status, <u>in</u>: 'Atherosclerosis review', A. M. Gotto, R. Paoletti, eds., Raven Press, New York (1983).

10. S. Renaud, D. Blache, E. Dumont, T. Thevenan, and T. Wissendanger, Platelet function after cigarette smoking in relation to nicotine and carbon monoxide, <u>Clin Pharmacol Ther</u>. 36:389-95 (1984).

11. G. A. FizGerald, J. A. Oates, and J. Nowak, Cigarette smoking and hemostatic function, <u>Am Heart J</u>. 115:267 (1988).

12. J. D. Folts, and F. C. Bonebrake, The effects of cigarette smoke and nicotine on platelet plugging in stenosed dog coronary arteries: inhibition with phentolamine, <u>Circulation</u>. 65(3):465-9 (1982).

13. J. F. Mustard, Effect of smoking on blood coagulation and platelet survival in man, <u>Br Med J</u>. 1:846-9 (1963).

14. V. Fuster, J. H. Chesboro, R. Frye, and L. R. Elveback, Platelet survival and the development of coronary artery disease in the young adult: effects of cigarette smoking, strong family history and medical therapy, <u>Circulation</u>. 63:546-51 (1983).

15. G. A. FitzGerald, A. K. Pedesen, and C. Patrono, Analysis of prostacyclin and thromboxane A_2 biosynthesis in cardiovascular disease [Editorial], <u>Circulation</u>. 67:1174-7 (1983).

16. J. D. Folts, E. B. Crowell, and G. G. Rowe, Platelet aggregation in partially obstructed vessels and their elimination with aspirin, <u>Circulation</u>. 54:365-70 (1976).

17. R. Bolli, Ware J. A. Brandon T. A., D. G. Willacher, and M. L. Mace, Platelet mediated thrombosis in stenosed canine coronary arteries: Inhibition by a platelet-active alpha-adrenergic antagonist, <u>JACC</u>. 3(6):1417-26 (1984).

18. J. H. Ashton, J. M. Schmitz, W. B. Campbell, C. Fitzgerald, L. M. Buja, and J. T. Willerson, Inhibition of cyclic flow variations in stenosed canine coronary arteries by thromboxane A_2/prostaglandin H_2 receptor antagonists, <u>Circ Res</u>. 59(5):568-78 (1986).

19. P. Golino, M. Buja, J. H. Ashton, P. Kulkarni, A. Taylor, and J. T. Willerson, Effect of thromboxane and serotonin receptor antagonists on intracoronary platelet deposition in dogs with experimentally stenosed coronary arteries, <u>Circulation</u>. 78:701-11 (1988).

20. B. G. Bertha, and J.D. Folts, Inhibition of epinephrine exacerbated coronary thrombus formation by PGI_2 in the dog, <u>J Lab Clin Med</u>. 103(2):204-14 (1984).

21. J. T. Strony, J. D. Folts, and B. Adelman, In vivo antiplatelet effect of gly-arg-gly-asp-ser in the stenosed canine coronary artery occlusion model, <u>Clin Res</u>. 37(2):388A (1989).

22. J. D. Folts, J. Stamler, and J. Loscalzo, N-acetylcysteine potentiates i.v. nitroglycerin in inhibiting periodic platelet thrombus formation in stenosed dog coronary arteries, <u>JACC</u>. 13(2):145A (1989).

23. J. D. Folts, Experimental arterial platelet thrombosis, platelet inhibitors, and their possible clinical relevance, <u>Cardiovasc Rev Rep</u>. 3(3):370-82 (1982).

24. J. D. Folts, and G. G. Rowe, Epinephrine reverses aspirin inhibition of in vivo platelet thrombus formation in stenosed dog coronary arteries, <u>Thromb Res</u> 50:507-16 (1988).

25. J. D. Folts. D. Detmer, and R. Nadler, Possible acute platelet thrombus formation in dog and human femoral arteries, _Texas Heart Institute J._ 9:12-26 (1982).

26. E. Eichhorn, P. A. Grayburn, H. V. Anderson, J. B. Bedotto, M. M. Carry, and J. T. Willerson, Cyclic coronary artery flow variations before and after percutaneous transluminal coronary angioplasty in man, _Circulation._ (Suppl II) 80(4):371 (1989).

27. V. Fuster, J. H. Chesebro, R. L. Frye, and L. R. Elveback, Platelet survival and the development of coronary artery disease in the young adult: effects of cigarette smoking, strong family history and medical therapy, _Circulation_ 63:546-51 (1981).

28. J. Barboriak, A. J. Anderson, R. G. Hoffmann, Smoking, alcohol and coronary artery occlusion, _Atherosclerosis._ 43:277-82 (1982).

29. D. J. Kozarorevic, N. Vojvodic, and T. Dawber, Frequency of alcohol consumption and morbidity and mortality, _Lancet._ I:613-6 (1980).

30. M. G. Marmot, M. J. Shipley, G. Rose, and B. J. Thomas, Alcohol and mortality: a U-shaped curve, _Lancet._ I:580-3 (1981).

31. K. Yano, G. G. Rhoads, and A. Kagan, Coffee, alcohol and risk of coronary heart disease among Japanese men living in Hawaii, _N Engl J Med._ 297:405-9 (1977).

32. J. J. Barboriak, A. J. Anderson, and R. G. Hoffmann, Interrelationship between coronary artery occlusion, high-density lipoprotein cholesterol, and alcohol intake, _J Lab Clin Med._ 94:348-53 (1979).

33. P. H. Levine, An acute effect of cigarette smoking on platelet function-A possible link between smoking and arterial thrombosis, _Circulation._ 48:619-23 (1973).

34. R. I. Hawkens, Smoking, platelets, and thrombosis, _Nature._ 236:450-2 (1972).

35. J. W. Davis, and P. E. Phillips, The effect of ethanol on human platelet aggregation in vitro, _Atherosclerosis._ 11:473-7 (1970).

36. D. H. Cowman, The platelet defect in alcoholism. Ann NY Acad Sci 1975; 252:328-41.

37. C. G. Fenn, and J. M. Littleton, Inhibition of platelet aggregation by ethanol in vitro shows specificity for aggregating agent used and is influenced by platelet lipid composition, _Thromb Haemostas._ 48:49-53 (1982).

38. J. W. Keller, M. Saltzberg, and J. D. Folts, Synergistic inhibitor effects of aspirin and ethanol on epinephrine exacerbation of acute platelet thrombus formation in stenosed canine coronary arteries, _Cardiovasc Res._ (in press).

39. J. Ayers, C. F. Ruff, and D. I. Templer, Alcoholism, cigarette smoking, coffee drinking and extraversion, _J Stud Alcohol._ 37:983-5 (1976).

40. P. M. Moody, Drinking and smoking behavior of hospitalized medical patients, _J Stud Alcohol._ 37:1316-9 (1976).

41. R. O. Cummins, A. G. Shaper, M. Walker, and C. J. Wale, Smoking and drinking by middle aged British men: effects of social class and town of residence, _Br Med J._ 283:1497-502 (1981).

42. R. R. Griffiths, G. E. Bigelow, and I. Liebson, Facilitation of human tobacco self-administration by ethanol: a behavioral analysis, _J Exp Anal Behav_ 25:279-92 (1976).

43. N. K. Mello, J. H. Mendelson, M. L.Sellers, and J. L. Kuehnle, Effect of alcohol and marijuana on tobacco smoking, _Clin Pharmacol Ther._ 27:202-9 (1980).

44. E. Wynder and D. Hoffman, Some characteristics of tobacco smoke, in: 'Tobacco and tobacco smoke', Academic Press, New York (1976).

45. G. H. R. Rao, G. Escolar, and J. G. White, Epinephrine reverses the inhibitory influence of aspirin on platelet-vessel wall interactions, Thromb Res. 44:65-74 (1986).

46. D. H. Cowan, Effect of alcoholism in hemostasis, Semin Hematol. 17:137-47 (1980).

47. M. B. Feinstein, J. Fiekers, and C. Fraser, An analysis of the mechanism of local anesthetic inhibition of platelet aggregation and secretion, J Pharmacol Exp Ther. 197:215-28 (1976).

48. H. McGill, Potential mechanisms for the augmentation of atherosclerosis and atherosclerotic disease by cigarette smoking, Prev Med. 8:390-403 (1979).

49. J. W. Keller, and J. D. Folts, Relative effects of cigarette smoke and ethanol on acute platelet canine coronary arteries, Cardiovasc Res. 22(1):73-8 (1988).

50. A. Kershbaum, and S. Bellet, Cigarette smoking and blood lipids, JAMA. 187(1):32-6 (1964).

51. D. T. Watts, The effects of nicotine and smoking on the secretion of epinephrine, Ann NY Acad Sci. 90:74-80 (1960).

52. J. Trap-Jensen, Effects of smoking on the heart and peripheral circulation, Am Heart J. 115:263-7 (1988).

53. G. Grignani, G. Gamba, and E. Ascari, Cigarette-smoking effect on platelet function, Thrombos Haemostas. 37:223-8 (1977).

54. P. Hill, Effects of nicotine on the serum epinephrine and corticoids, Am Heart J. 87(4):491-6 (1974).

55. S. R. Saba, and R. G. Mason. Some effects of nicotine on platelets, Thromb Res. 7:819-24 (1975).

56. A. Wennmalm, Interaction of nicotine and prostaglandins in the cardiovascular system, Prostaglandins. 23:139-45 (1982).

57. A. Wennmalm, and P. Alster, Nicotine inhibits vascular prostacyclin but not platelet thromboxane formation, Gen Pharmacol. 14:189-91 (1983).

58. H. Madsen, and J. Dyerberg, Cigarette smoking and its effects on the platelet-vessel wall interaction, Scand J Clin Lab Invest. 44:203-6 (1984).

59. A. Boura, S. Hui, and W. Walters, Cigarette smoke inhalation specifically inhibits depressor responses to prostacyclin in the rat, Br J Pharmacol. 73:3 (1981).

60. R. M. Pittilo, J. M. Clark, D. Harris, I. J. Mackie, P. M. Rowles, S. J. Machin, and N. Woolf, Cigarette smoking and platelet adhesion, Br J Haematol. 58:627-32 (1984).

61. H. Sinzinger, and A. Kefalides, Passive smoking severely decreases platelet sensitivity to antiaggregating prostaglandins, Lancet. ii:392-3 (1982).

62. J. Hartiala, N Simberg, and P. Uotila, Exposure to CO or nicotine does not inhibit PGI_2 production by rat arterial rings incubated with human platelet rich plasma, Artery. 10:412-9 (1982).

63. P. Nicod, R. Rehr, M. D. Winniford, W. B. Campbell, B. G. Firth, and L. P. Hillis, Acute systemic and coronary hemodynamic and serologic response to cigarette smoking in chronic smokers with atherosclerotic coronary artery disease, JACC. 4:964-71 (1984).

64. M. L. Bierenbaum, A. I. Fleischman, and A. Stier, Effect of cigarette smoking upon in vivo platelet function in man, Thromb Res. 12:1051-7 (1978).

65. H. C. McGill Jr., The cardiovascular pathology of smoking, Am Heart J. 115:250-7 (1988).

66. J. W. Davis, C. R. Hartman, H. D. Lewis Jr., L. Shelton, D. A. Eigenberg, K. M. Hassanein, C. E. Hignite, and H. A. Ruttinger, Cigarette smoking-induced enhancement of platelet function: lack of prevention by aspirin in men with coronary artery disease, J Lab Clin Med. 105:479-83 (1985).

THE CENTRAL NERVOUS SYSTEM AND ATHEROGENESIS: INTERRELATIONSHIPS

William H. Gutstein[1] and Joseph M. Wu[2]

Departments of Pathology[1] and Biochemistry[2]
New York Medical College
Valhalla, N.Y. 10595

Despite almost a century of scientific study since Anitschkov and
Chalatov[1] observed that cholesterol feeding produced arterial lesions in
rabbits, the etiology and pathogenesis of atherosclerosis (AS) remain
unknown. In humans, clinical complications and sequelae occur when the
lesions have evolved to produce the fibrous plaque. The main histologic
features of this stage are lipid accumulation and fibroelastic and
fibromuscular thickening, which initially are present in a patchy distribution but later become more diffuse as individual plaques coalesce. A fatty
streak stage, characterized by lipid accumulation within intimal cells,
either of monocyte/ macrophage or arterial smooth muscle cell (ASMC)
origin, to give them a "foam cell" appearance in histologic sections, is
considered by some[2] to represent an intermediate step in the development of
the final lesion.

Because of the characteristic features of lipid accumulation and
intimal thickening, in humans it is generally believed that some form of
endothelial injury contributes strongly to the pathogenesis.[3] Much of the
work in endothelial biology today is an attempt to probe the more subtle
forms of endothelial dysfunction, since obvious evidence of damage, such as
morphologic stigmata at the microscopic level, is not easily found in human
observations or animal experiments. In humans, however, clinical and
epidemiological studies have uncovered statistically significant risk
factors, in particular, hypercholesterolemia, hypertension and cigarette
smoking, which are associated with advanced disease.[4-6] Despite these
associations however, relatively little light has been shed on the dark
shadows of pathogenesis. Moreover, even if the major risk factors are
taken into account, collectively, they are unable to predict the majority[7]
of new cases of the disease.

In the last 20-30 years, numerous clinical studies have shown that
neuropsychological mechanisms may be implicated in the development of
disease.[8] But it is not understood in what manner such events can translate into pathology at the level of the arterial wall which we recognize[10] as
that of atherosclerosis.[9] Except for certain behavioral experiments,
there has, in fact, been little work with animals attempting to explain
these relationships. The CNS itself, in this respect, has received little
if any attention. Yet, if there is a contribution to be made in the
development of atherosclerosis from the psychosocial influences which have

Tobacco Smoking and Atherosclerosis
Edited by J. N. Diana
Plenum Press, New York, 1990

become so pressing in our times, it must involve the CNS at some level
directly. This is an area which we believe has been overlooked in the
etiology and pathogenesis of atherosclerosis. Accordingly, in this paper
we will present some of our experimental results from such studies in
animals. These results, except for some departures, are consistent with
those of other investigators in the field of atherogenesis but are arrived
at through the technique of hypothalamic stimulation.

The hypothalamus has been chosen, not because it is specific in this
regard, which it cannot claim to be, but because of the strategic and
integral involvement of this area of the CNS in the emotional life of man,
a fact which is widely recognized in all of the psychological,
neurophysiological and behavioral disciplines which have studied it.[11]

After presenting the results, we will discuss the mechanisms by which
we believe they arise as a result of hypothalamic stimulation (HS), after
which an interpretation, based on these findings will be offered for the
etiology and pathogenesis of atherosclerosis induced in this manner. Since
some of the findings are new in terms of their relationship to HS, it may
be expected that the interpretation for the pathogenesis will also not be
totally orthodox.

Accordingly, in relation to HS, we will present data concerning
arterial and biliary tract spasm, hyperlipidemia involving special abnormal
lipoproteins, endothelial injury, arterial smooth muscle cell (ASMC)
proliferation and lipid accumulation by SMC of the vessel wall. These will
then be integrated, so far as possible, into a sequenced pattern of causal
events to lead into the final stages of the mature lesion which will also
be illustrated. It is hoped that such a denouement will provide a model,
which despite its limitations can serve a heuristic purpose for conducting
research in this field, which incorporates the role of the CNS, an area
strongly supported by clinical and epidemiological studies in the last two
to three decades.[12]

METHODS

Except for the initial experiments with the peripheral nervous system,
CNS activation was induced by electrical stimulation of the hypothalamus.
The lateral hypothalamic area (LHA) was used because of its close
behavioral relationship to aggression/anger in the awake subject. In some
cases stimulation was unilateral and in others bilateral. A range of
electrical parameters was used compatible with those employed in behavioral
studies in experimental psychology. At all times, however, attempts were
made to keep the stimulus as discreet as possible. Anesthetized animals
were generally stimulated by an external electronic stimulator while for
conscious, unrestrained animals a self-powered electronic stimulator
mounted on the skull or implanted under the skin was designed and
constructed. This unit was a miniatured computer-programmed impulse
deliverer and offered a wide range of parameters. All encapsulating
materials were biocompatible. Histologic examination of brain sections for
experiments up to 27 months duration showed no neurotoxic effects around
electrode tips which were 0.25-0.5 mm diameter, fashioned of stainless
steel and insulated except for the tips. In addition, the following
technical procedures were employed:

Serum/Plasma Lipid Assays
Serum Alkaline Phosphatase
Light Microscopy
Electron Microscopy
Morphometry by Light Microscopy

Vascular Corrosion Casting
ECG Recording
Anesthesia (Sodium Pentobarbital, Intraperitoneal)
Aseptic Surgical Techniques
Preparation of Plasma Derived Serum (PDS)
Cell Culture Techniques for Arterial Smooth Muscle Cells
Radiofrequency Stimulation Techniques
Gross Examination of Blood Vessels with Sudanophilic Staining
Dietary Manipulations
Blood Pressure Measurements (Extra and Intraarterial)
Gastric Intubation Techniques
Bile Duct Ligation Techniques
Ultracentrifugation for LDL, HDL, Etc.

The details of these procedures are to be found in the original
publications and are all state-of-the-art.

RESULTS

Vasospasm

One of the events which occurs consistently in this model and which
may have an important influence on the integrity of the endothelium is
vasospasm. The term "spasm" has unfortunately been difficult to define
with precision, yet is used extensively, both clinically and in experimental work. In this presentation we use it to denote an arterial constriction of sudden onset in the form of a sustained contraction, focal in
location, of variable duration and narrowing the lumen of the vessel by at
least 50% of its original diameter. Although physiologic narrowing of
vessel diameter occurs under appropriate metabolic conditions,[13] the terms
constriction and spasm as employed here will denote abnormal states of
contraction of the vessel musculature and will often be interchanged. In
general, the relationship of HS to constriction of various arterial beds is
well known.[14] In our work with the rat there are at least 3 lines of
evidence which indicate that arterial spasm occurs in the coronary circulation in association with HS: morphometric analysis with light
microscopy,[15] vascular corrosion casting[16] and electrocardiographic
documentation of acute myocardial ischemia.[16] In addition, arterial
constriction induced by HS has been found in various species, some in other
vascular beds.[17-25] There is no reason to believe that such a relationship
would not be true in humans as well. In the rat, although our data have
been obtained chiefly from the coronary circulation, earlier experiments
showed that neural impulses transmitted directly to the abdominal aorta via
the peripheral nervous system are also spasmogenic.[26]

Endothelial Dysfunction

Endothelial dysfunction or injury is a broad term and may signify
impairment of a variety of normal functions which have come to be
associated with the integrity of the endothelium, from antithrombogenicity
to regulation of permeability across the arterial wall.[27] In the last few
years, subtle changes of normal endothelial function have been probed,
especially in studies concerned with atherogenesis, since it is widely
believed, that the initial event in the pathogenesis of arterial lesions
begins with an interruption of this function, yet gross endothelial defects
are absent.[28] However, we have shown in both acute and chronic experiments
that severe endothelial damage at the microscopic level is to be found in
coronary arteries[29] and the aorta[30,31] following HS and is accompanied by
vasospasm.[15] This injury not infrequently takes the form of denudation of

the endothelium.[15,31] Its relevance for atherogenesis is easily
understood.

<u>Cellular Proliferation</u>

        Rats and squirrel monkeys which receive repeated episodes of HS over
a period of time disclose in vivo evidence of intimal migration of ASMC
(Fig. 1) and proliferation of these cells in developing lesions (Fig. 2)
even when maintained on basal fat-free or low fat diets. Endothelial
damage is usually present as well (Figs. 3 and 4). In addition, in vitro
studies, revealed that the serum of rats receiving HS was mitogenically
active against homologously cultured target ASMC in comparison with the
effects of the sera of nonstimulated controls,[32] effects that were enhanced
in aged rats.[33] Observations over 4 mos. made on the aortic cells of the
same animals, which were transferred as explants to a growth-supporting
medium after HS had been delivered, so that these cells were also exposed
to the sera while in vivo, revealed an interesting finding; ASMC from aged
control rats were unable to grow under these conditions, a situation which
was completely reversed, however, in subjects which had received HS (Table
1). Our interpretation of these findings in terms of regulation of ASMC
growth by the CNS will be discussed below as well as the putative
importance of this observation in the pathogenesis of AS related to a
contribution from the CNS.

<u>Table 1</u> Frequency of ASMC Outgrowth in FBS-Supplemented DMEM From
          Explants Of Rat Aortas In Different Experimental Groups

	Young Sham (Y)	Young Stimulated (YS)	Aged Sham (A)	Aged Stimulated (AS)
1 month	2/5 (40%)*	4/5 (80%)	0/5 (0)	3/5 (60%)
2 months	3/5 (60%)	5/5 (100%)	0/5 (0)	4/5 (80%)
3 months	4/5 (80%)	5/5 (100%)	0/5 (0)	5/5 (100%)
4 months	4/5 (80%)	5/5 (100%)	0/5 (0)	5/5 (100%)

*Fractions indicate the number of flasks showing positive outgrowth
per five flasks seeded; figures in parenthesis indicate percentage
positive

<u>Lipid Accumulation</u>

        Lipid accumulation was not searched for in arteries of rats which
received HS for a sustained period of time, a species usually resistant to
this type of change. These were also animals which were maintained on fat
free diets throughout the experimental procedure. On the other hand,
squirrel monkeys, even when placed on essentially fat-free diets revealed
considerable lipid accumulation in the ASMC of aortas as illustrated in
Fig. 2. Some of these cells were typically of "foam cell" appearance
(Fig. 5).

<u>Plasma Lipid Changes</u>

        The accumulation of lipids in the developing plaque, the other key

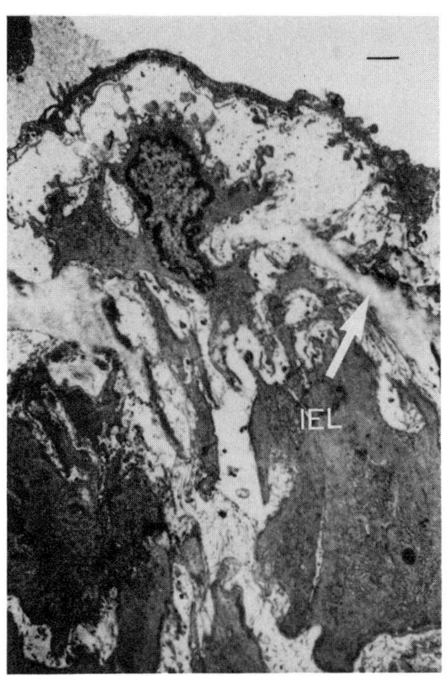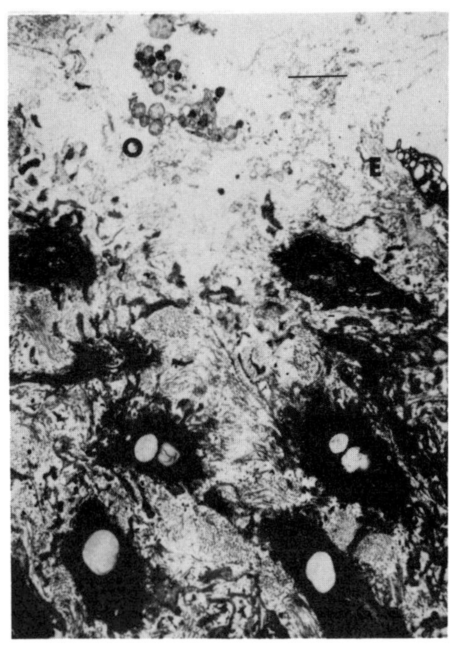

Fig. 1. Electron micrograph of coronary artery in a rat receiving acute hypothalamic stimulation, near a point of endothelial injury similar to that of Fig. <u>3</u>. Migration of a smooth muscle cell through the IEL from the media is clearly seen. Bar = 1 um

Fig. 2. Electronmicrograph of an aortic section in a squirrel monkey receiving hypothalamic stimulation for 12 months. At the upper surface the endothelium (E) has been denuded and extruded cytoplasmic organelles (O) are present. The remainder of the illustration reveals a vigorous proliferation of arterial smooth muscle cells in the intima, several of which have begun to accumulate lipid, as well as synthesis of collagen fibers. Bar = 1 um

element in AS is presumed to arise from events which occur in the circulating blood, i.e. elevation of plasma LDL levels. We therefore, investigated the status of blood lipids in this model. We found that fasted rats which were anesthetized and bled and then given HS did not show any elevation of plasma lipids. In fact losses were seen, presumably due to using these lipids as energy sources in order to compensate for such decreases arising from the stress of the surgical procedures. An entirely different picture emerged, however, if the animals were fed first, particularly if the subjects were given a lipid meal in the form of intubation. In this case plasma unesterified cholesterol[34], phospholipids and triglycerides were all significantly elevated. A hyperlipedemia of this type is characteristic of the presence of the abnormal lipoprotein LP-X as referred to in the work of Patsch below (ref 41). In a separate experiment plasma

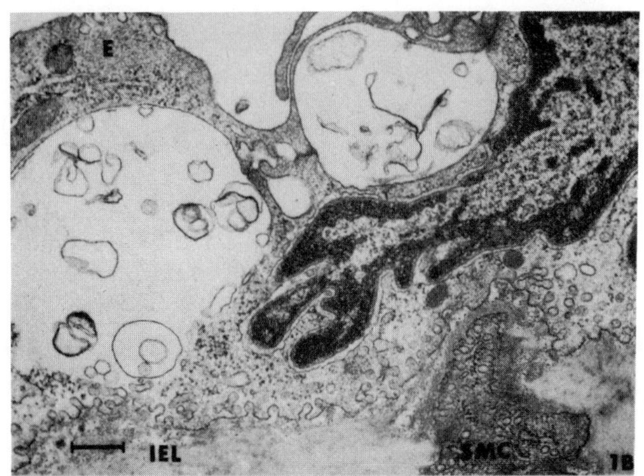

Fig. 3. Electron micrograph of coronary artery
of rat receiving hypothalamic stimulation
for 30 min showing severe injury of
endothelium, E, with vacuole formation and
plasma membrane degeneration. A vascular
smooth muscle cell, SMC can be seen
migrating into the area of the injury from
the internal elastic lamina, IEL, below.
Bar = 1 um

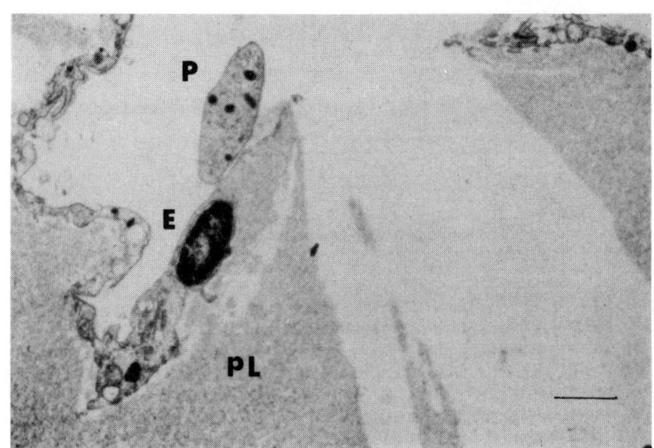

Fig. 4. Electron micrograph of aorta of squirrel
monkey which received intermittent
hypothalamic stimulation for an 18 month
period. Endothelium is broken at area of
attached platelet (p) and marked insudation
of plasma (PL) into the intima is present.
Bar = 1 um (Reproduced from Fig. 2, ref 31
with the permission of the publisher)

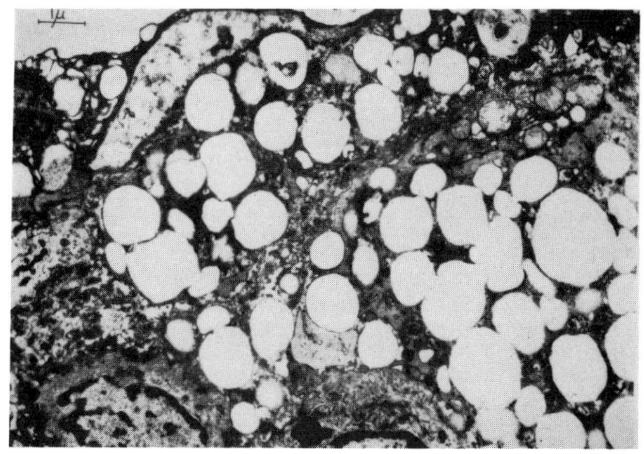

Fig. 5. Electron micrograph of aorta of conscious
unrestrained squirrel monkey receiving
hypothalamic stimulation for 18 months.
Beneath a damaged endothelium, the intima
shows accumulation of lipid within smooth
muscle cells converting them into "foam
cells".

concentrations of total cholesterol were increased by HS most by feeding
saturated fats, less so with[35] monounsaturated fats and remained unchanged
with polyunsaturated ones.[35] The mechanism for this hyperlipidemia was
uncovered[36] and its possible role in atherogenesis will be discussed below.

MECHANISMS

Vasospasm

It is well known from animal experiments that HS can cause arterial
constriction in a number of vascular beds, and in various species.[17-25]
The particular degree of constriction, whether considered as "spasm" or
not, will vary according to the region of the hypothalamus stimulated, the
vascular bed represented in the hypothalamus, the species investigated and
the actual mechanism involved. In general, once HS is administered, the
means of inducing the constriction reduces to those which are mediated
humorally or those which are transmitted by neural impulses directly to the
neuroeffector apparatus in the vascular wall. Of course, it is possible
that both are involved. Among the former are such well known mediators as
catecholamines[37] (norepinephrine and epinephrine) and various neuropep-
tides[38] (arginine vasopressin, angiotensin, etc). In the latter group,
impulses which arrive at the neuroeffector apparatus via nerve terminals
release vasoactive molecules, such as norepinephrine, serotonin, histamine,
etc.

An additional contribution to vasospasm can arise in the following
way. Recently it has been shown that endothelial cells if severely injured
can induce a constrictor response in vessels because of lack of endothelium
derived relaxing factor[39] and the release of endothelin[40]. In the present
model, circulating abnormal lipoproteins (LP-X), because of their high
phospholipid content[41] may remove lipids from plasma membranes of
endothelial cells when impinging upon them and induce such injury locally.
In support of this possibility we observed some years ago that erythrocytes
of rats receiving HS lost significant quantities of their membrane
lipids[42].

Whatever pathway for arterial constriction the neurobiology of HS dictates, two general conditions of arterial vessels predispose to the focal or patchy nature of the constrictor response, an event which is of seminal significance in the pathogenesis of AS -i.e. an event which correlates with the patchy distribution of the vascular lesions. In the case of neurohumoral mediation, the vessel wall contains a variety of receptors for circulating pressor substances which may be distributed along the vessel axis in a nonuniform fashion;[13] in the case of activation at neuroeffector sites, similarly, these are known to vary considerably in their density, even in the same vascular segment.[43] Thus, in either case, the constrictor response and its sequelae may be expected to exhibit a focal nature, an event which correlates well with the patchy distribution of subsequent AS plaques.

<u>Endothelial Injury</u> (EI)

Since EI is considered by many to be the initiating event in atherogenesis, it is important to understand how such a situation can arise in the present model. Our starting point here is the condition previously described, i.e. arterial spasm. The presence of a constricted area in a tube through which a fluid flows gives rise to local hemodynamic forces, the exact nature of which depend on the properties of the tube and fluid. These hemodynamic forces are in turn transmitted to the interface between wall and fluid. In the case of arterial flow where the vessel is viscoelastic and oscillating and the blood is viscous and pulsatile, complicated fluid patterns are generated[44] which also depend on the Reynolds number of the flow, the latter in turn influenced by the tube diameter. In blood flow experiments it is probable that the aorta and muscular arteries such as coronary, cerebral, mesenteric and femoral are subject to these local hemodynamic forces in the region of the constriction, with resulting damage to the lining composed of endothelial cells. Figs. 6 and 7, of fluid flowing through a glass tube to model this situation illustrates the damage that may occur under these conditions.

A second type of injury that may result from a severe arterial constriction at a focal point is due to strong elastodynamic rather than hemodynamic forces in the form of wall stresses developed by the spasm. The muscle fibers in the media of these vessels are arranged either in a circular or helical pattern and localized regions of spasm may generate strong intramural stresses, injuring both medial and endothelial cells in the vicinity (see below). That spasmogenic agents are capable of producing such endothelial injury was demonstrated earlier by Constantinides and Robinson[45] and more recently by Joris and Majno.[46] In addition, there are many other lines of indirect evidence which reveal that spasm or vigorous localized constriction severely damages endothelial cells, e.g., by inducing localized ischemia in the vascular wall.[47] Not only are endothelial cells capable of being injured in this manner, but so are ASMC of the intima which are in the same area as shown by Joris and Majno[48]. This is an important consideration for the process of atherogenesis which will be discussed below. Finally the endothelium may be injured in the manner described under mechanisms of vasospasm, i.e., through severe loss of membrane lipids brought on by contact with abnormal lipoproteins of the LP-X family because of the liposomal character of these abnormal lipoproteins. Inability of the membranes to reconstitute themselves appropriately with repeated insults via HS could even result in denudation, in this manner as well as from the hemodynamic and elastodynamic mechanical forces referred to above.

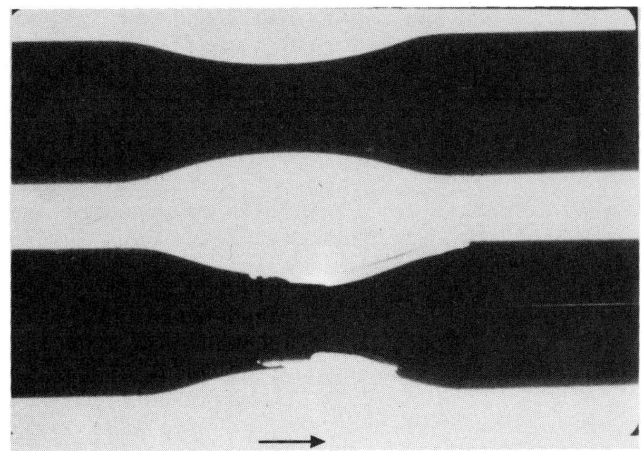

Fig. 6. Glass tube with constriction to model
"spasm" of artery. Fluid (water, 37°C)
flows through at relatively high speed.
Two hours after starting (top) a densely
adherent resin applied to the inner wall
is removed (bottom) in the area of the
constriction. Direction of flow is from
left to right.

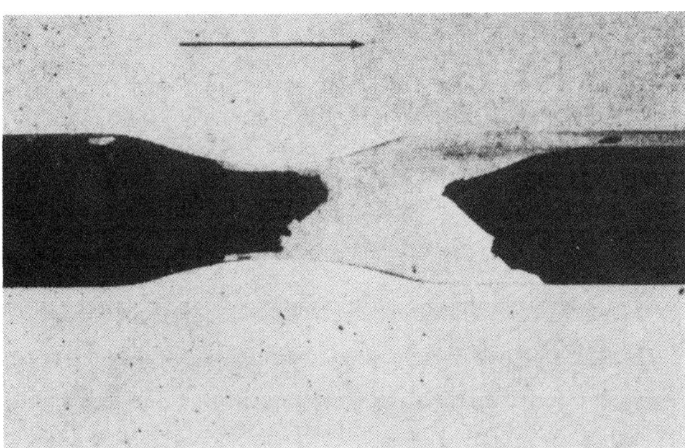

Fig. 7. Same tube, three hours after flow has
started. The lining has been almost
completely removed downstream of the
constriction.

<u>Plasma Lipid Changes</u>

It has been known for some time that emotional responses in humans may
be associated with elevation of blood lipids, including cholesterol and
LDL.[49] Similarly, in animals, emotionally charged situations have been
accompanied by increases in plasma lipids.[50] In both of these situations
it may be presumed that the hypothalamus was somehow involved. How does
the hypothalamus bring about such changes? In the model under discussion
we found a partial answer to this question, but this answer must be
qualified by the special nature of the hyperlipidemia which develops.

In order to understand the nature of this hyperlipidemia which is
related to HS, it is important to realize that a phenomenon quite similar
to the one resulting in arterial spasm also involves biliary smooth
muscle. During HS, the bile duct, – simplified in the rat since there is
no gallbladder – contracts to the point where a transient obstruction to
the flow of bile occurs Figs. 8 and 9 with regurgitation back into the
circulation from the liver[36]. This flow of bile is enhanced, of course,
by choleretic agents such as fats. Animals not fed fat prior to HS
generally experience little more than a trickle of bile from hepatic
sources, especially after an overnight fast, and so have little to
regurgitate. Even so, an occasional subject with some residue of
intestinal lipid left over from a previous meal may elicit a sufficient
choleretic flow to raise its plasma lipids in this manner, an event which
is expected to be further increased by lipid feeding.

The evidence that the hyperlipidemia in this model derives from
transient biliary obstruction (tbo) is straightforward. Accompanying the
elevation of plasma lipids there are: Elevation of direct bilirubin and
serum alkaline phosphatase, dilation of biliary canaliculi by examination
of hepatic slices with electron microscopy and most important of all, a
complete absence of these events including the hyperlipidemia, by
positioning a rigid cannula in the bile duct[36] or pretreatment of the
animal with an agent that dilates biliary smooth muscle, prior to
administering HS.[51]

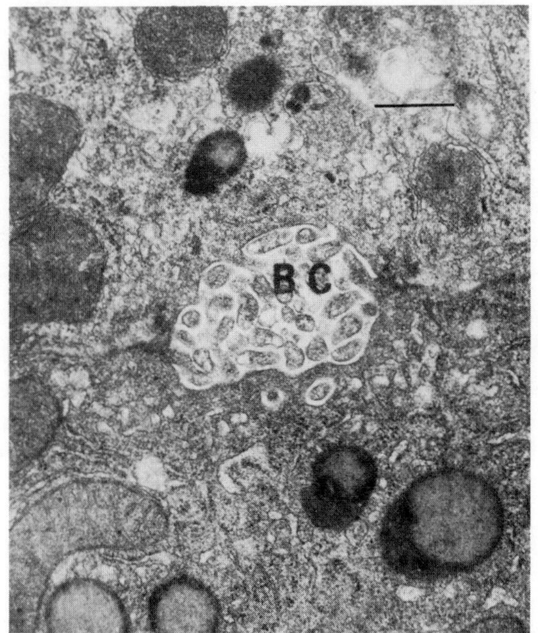

Fig. 8 Electron micrograph of
        section of liver from
        control rat showing
        normal bile canaliculus
        (BC). Bar = 0.5 um.
        (Reproduced from fig. 2,
        ref. 36, with the
        permission of the
        publisher)

One of the most important considerations for atherogenesis in relation to this "special hyperlipidemia" lies in the characteristics of the lipoproteins that are associated with it, especially the one isolated in the low density range, known as lipoprotein X.[52-54] Actually there is a family of these which differs in biochemical details.[41] This lipoprotein, synthesized by the liver under "obstructive" conditions is rich in free cholesterol and even more so in phospholipids and in its protein moiety contains aproprotein C as well as albumin.[41]

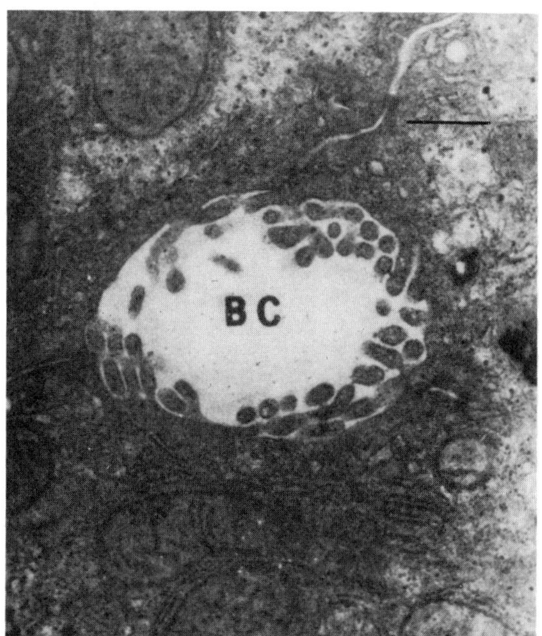

Fig. 9. Electron micrograph of section of liver from rat which received intermittent electrical stimulation in the lateral hypothalamic area for 30 sec. Note dilatation of bile canaliculus (BC) and flattening of villi. Bar = 0.5 um (Reproduced from fig. 5, ref. 36, with the permission of the publisher).

Cellular Proliferation

The proliferation of ASMC in the intima of susceptible vessels as a requirement for atherogenesis is now widely accepted.[55] These cells constitute the only source for the production of the fibrous tissue which is an integral component of the lesion.[56] Whereas animal experiments have brought to our attention numerous conditions, both in vivo and in vitro, which can influence proliferation of these cells, presumably after they have migrated to the intima from the media under chemotactic influences of some type, the stimulus for this myointimal hyperplasia in human AS remains unknown.

Since we had already noticed in long term HS experiments that AS plaques occurred which were rich in ASMC, together with their connective tissue elements,[30] we tested the sera of these animals after brief periods of stimulation as described in the preceding section. Analysis of the data suggested that there were two classes of mitogens present in the sera of HS rats; those only detectable in culture when the serum concentration was sufficiently high, and those associated with HS, observable even at low concentrations of the serum. We have termed these growth-supporting substances, "serum mitogens" and "stimulation mitogens" respectively. Both were enhanced with age. Referring to table 1, which reflects the subsequent growth history of the aortic cells in these experiments when transferred to an external growth supporting medium after having been exposed to the serum conditions just described, we see that the growth of cells in aged rats without HS is one of considerable indolence, there not being a sufficient level of mitogenic molecules to prompt them to divide, a condition which is significantly altered when the cells are exposed to "stimulation mitogens" as is the case in those aged animals which did receive HS. Whether this change reflects a qualitative difference or simply a quantitative one we do not know at this time. By comparison, cells derived from young animals grow well even if they have not been exposed to "stimulation mitogens" but proliferate also when the latter are available.

Thus far we have presented this interpretation of a CNS influence on cell proliferation from the standpoint of an effect on cells produced by mitogenic molecules in the serum elicited by HS. There are other possibilities, however, which could also explain these observations. For example, it is conceivable that HS induces substances in the serum which counteract normally-present growth inhibitors such as heparin,[57] rather than growth factors in the traditional sense. Still another possibility is that a direct neural influence is transmitted to the vessel wall from the hypothalamus with release of growth-inducing substances from nerve terminals at that location. Nor is it inconceivable that combinations of these influences are present. Until the putative mitogen is purified and identified these possibilities can not easily be sorted out. It is significant, however, that for our presentation a potential role for the CNS in the growth regulation of ASMC is revealed, an area that has not been dealt with to any great extent in the control of such finely tuned processes as cell division.

THE PATHOGENESIS OF ATHEROSCLEROSIS IN THE HS MODEL

It is often said that any theory concerning the etiology and pathogenesis of AS in humans must take into account its main features, i.e. lipid accumulation and cellular proliferation, the latter associated with fibroelastic thickening. To be sure, the amount of lipid in mature plaques is variable, some lesions having none at all, but in any case it is presumed that lipids have been present at some time. In the most prevalent theory currently held, "The response to injury",[3] it is believed that these events follow from some form of endothelial injury, or if not outright recognizable damage of the cells that line the vessel, at least some form of endothelial dysfunction. Two other aspects must enter the boundaries of any theory and these are the patchy nature of the lesions, at least in the early stages, and a resolution of the status of fatty streaks if possible, i.e., defining their role in atherogenesis.[2] To settle the questions raised and shore up the status of such a theory, enormous effort has been devoted with regard to both animal experimentation and human observations.

In view of the histologic findings in this model as presented in the preceding sections and the mechanisms which bring them about, we shall

attempt an interpretation for the pathogenesis of AS. For the most part,
the major features, such as endothelial injury, lipid accumulation and ASMC
proliferation in the intima are consistent with observations made by most
workers in the field, differing only in the way they are elicited, i.e.,
via the CNS, in the form of hypothalamic stimulation. Other observations,
such as vasospasm, transient biliary obstruction and the specially
characterized hyperlipidemia are additions in this model which interact
with the former and we shall attempt to tie them together to offer an
explanation for the evolution of the mature lesion. Of course this
interpretation is preliminary and much work needs to be done to refine it
and to obtain answers to additional questions that must arise, but we
believe it may stimulate an interest in a new approach to an old problem,
i.e., involvement of the CNS in atherogenesis, which is strongly supported
by developments in the field of the last 20-30 years by cardiologists and
epidemiologists.

Given the response to injury hypothesis for atherogenesis, the
greatest difficulty appears to lie in a clear understanding of conditions
which perturb the endothelium. Do they relate to excess circulating
lipids? Are they caused by naturally occurring hemodynamic forces in
geometrically strategic parts of the arterial tree? Are toxic or immune
factors in the blood involved? These and similar questions are often
asked. Because it has been difficult to find evidence of significant
damage to the endothelium on a microscopic level, current interest in
endothelial biology has focused on subtle forms of cellular and molecular
damage. This has contributed to the discovery of such important substances
as endothelium derived relaxing factor (EDRF) and endothelin, both of which
exert an influence on the vasoactive properties of arteries, a significant
component of the present theory.

In terms of the HS model and therefore, the contribution of the CNS to
atherogenesis, we are in total agreement that endothelial injury is a sine
qua non for lesion development. Because of the consistent coexistence of
vasospasm and endothelial injury we believe them to be related events, not
just coincidental ones and that they reinforce each others presence,
vasospasm, because it can generate the mechanical forces (hemodynamic and
elastodynamic) necessary to damage the endothelium and endothelial injury
because once initiated, EDRF and endothelin may be involved.

In this model, contraction of smooth muscle cells occurs almost
simultaneously in the biliary as well as in the vascular system, probably
due to release of free calcium from intracellular stores in both systems.
This leads to transient biliary obstruction in the biliary system and
arterial spasm in the vascular. As a result of tbo, as noted earlier, LP-X
joins the native lipoproteins in circulation and contributes further to the
damage of the endothelium as well as intimal lipid accumulation both of
which are discussed in detail below.

Because of the damage to the endothelium, its permeability is altered.
Our HS model shows this rather well (Figs. 4 and 10). The insudating
plasma carries in with it, a variety of constituents which inundate the
intimal tissue. Chemotactic influences for ASMC may arise to induce
migration of cells into the intima from the media. These influences may be
due to the presence of the "stimulation mitogens" described earlier,
perhaps preparing the cells for coming proliferation and providing them
with an expanded space in which to carry out this function since the media
is a dense and compact territory. Of course this is conjectural, but is in
keeping with the histologic findings.

Upon contact with plasma containing "stimulation mitogens", ASMC of
the intima are induced to proliferate just as they did in culture. The

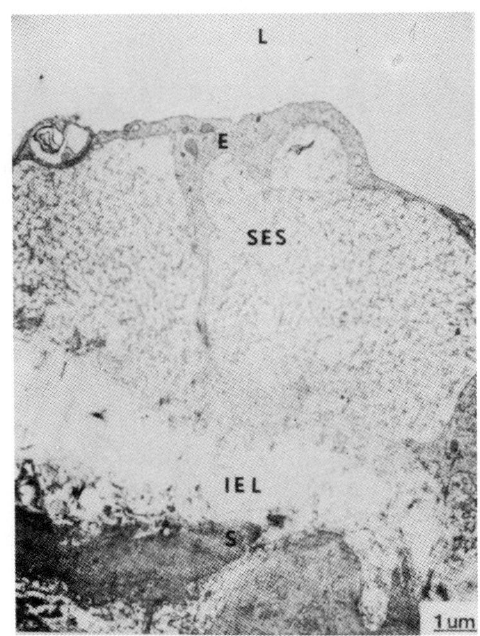

Fig. 10. Electron micrograph
of aorta of conscious,
unrestrained rat re-
ceiving hypothalamic
stimulation for 62
days. The subendo-
lial space (SES)
shows marked insu-
dation of plasma.
The endothelium (E)
at the top shows damage
at the upper left.
L=lumen; LEL=internal
elastic lamina.
(Reproduced from fig.
1, ref 30 with the
permission of the publisher).

degree of the proliferation is variable, but likely to be cumulative with
repeated episodes. Proliferation may also occur in migrated ASMC because
the release of free calcium ions from intracellular stores acts as a signal
for this event. A further contribution may arise from the release of PDGF-
like substances from injured endothelial and arterial smooth muscle cells[28]
which are present in this model. In addition, the traditionally found
"serum mitogens" can also influence this hyperplasia of ASMC. We suspect
that the greatest effect in this model, however, comes from exposure of the
cells to the "stimulation mitogens".

Inflowing plasma also contains lipids as stated. Some of these are in
the form of native LDL particles. Others, however, as we have seen from
the specific characteristics of the hyperlipidemia described above, are
undoubtedly present as abnormal lipoproteins, i.e., lipoprotein X, related
to the transient biliary obstruction which is associated with HS. Given,

that ASMC demonstrate limited uptake of native LDL,[58,59] unless these are
chemically modified in some way,[60] or derived from special dietary regimes,
and that there is a clearly evident lipid uptake in the ASMC of squirrel
monkeys on normal diets, which have received HS, (Figs. 2 and 5), it
appears likely that these LP-X particles make a significant contribution to
this uptake. In confirmation of this possibility we exposed cultured ASMC
of rats to such particles in cholestatic serum obtained from animals whose
bile ducts had been ligated and found after 24 hours that the cholesterol
content was three times as high as in cells exposed to normal rat serum
(unpublished observations). Whether a similar mechanism for injury also
occurs can only be speculated upon at this time.

The question arises at this point as to whether this mechanism for
lipid accumulation in the HS animal is realistic for man and can be
involved in the pathogenesis of AS. We believe that it can if viewed in
the proper perspective. To do this it is necessary, e.g., to conceive of
repeated subclinical bouts of transient biliary obstruction related to
emotional stimuli and stressful conditions over the life of the individual.
As in the rat, total sustained obstruction of the bile duct need not occur
to result in regurgitation of abnormal lipoproteins into the circulation,
especially if preceeded by lipid rich meals so as to insure adequate bile
flow. In our animals which exhibited this phenomenon[36], direct bilirubin
values were elevated only a few tenths of a milligram per deciliter of
blood (unpublished data), scarcely enough to impart an icteric tint to
mucus surfaces and sclerae yet biliary obstructive phenomena were clearly
evident as the data already presented show.

Moreover, psychogenic forms of <u>obvious</u> obstructive jaundice[61] and
neurogenic causes for spasm of the sphincter of Oddi[62] have been known from
at least the 1940s on and it is of interest to read that as far back as the
17th century, if not earlier, regurgitation of bile from the liver was
associated with powerful emotional events[63], all of which indicate that
biliary smooth muscle, like arterial smooth muscle responds most
sensitively to neuropsychological stimuli with release of LP-X into the
circulation, even in the absence of obvious "jaundice".

From a more exaggerated point of view, biliary disease when
unequivocally present may be closely associated with atherosclerosis[64] as a
multicenter European study based on 36,000 autopsies shows (Fig.11), and
may have major risk factors in common with the latter.[65] Furthermore,
xanthomas which occur in biliary disease and fatty streaks which may
represent a stage of AS, bear certain histologic resemblances to each
other. We could visualize in these details, an irregular stream of
lipoprotein X particles leaving the liver and mixing with the normal flux
of native LDL in the circulation.

When these LP-X particles as well as native LDL arrive at the site of
the damaged vessel they enter the wall readily. We know from the work of
Goldstein and Brown,[58,59] that LDL is restricted from entering the resident
cells of the arterial wall in excessive amounts by the specific high-
affinity receptor in ASMC, but LP-X on the other hand, may not be bound by
these restrictions since they lack apoB, a necessary ligand for binding to
the receptor. The rich contents of free cholesterol and phospholipids of
these particles also endow them with the characteristics of liposomal
vesicles[66,67] This in turn enables them to fuse with cell membranes of the
ASMC, possibly perturbing the function of the specific LDL receptor
which results in a loss of preventing excessive entrance of these
particles, or entering the cells on their own by passive diffusion.

In addition to the intracellular accumulation of lipids in this
manner, which may result in "foam cell" generation, the extracellular

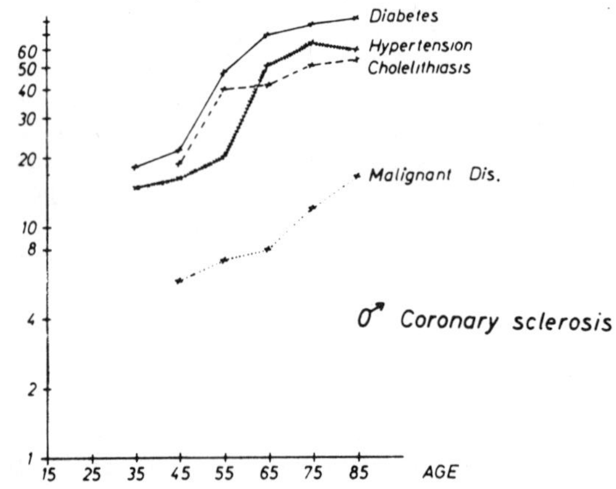

Fig. 11. Epidemiologic data based
on postmortem studies showing
prevalence of severe coronary
atherosclerosis in males related
to age. Associated conditions
include biliary tract disease as
evidenced by cholelithiasis.
(Reproduced from fig 4, p. 91,
ref 64 with the permission of
the publisher).

spaces of the intima also retain such a mixture of LDL and LP-X particles.
The presence of the latter, we believe, can explain the observations of
Simionescu et al.[68] and Mora et al.[69] concerning "prelesional events in
atherogenesis", that the earliest morphologically detectable changes in the
intima of dietary-induced lesions in experimental animals consists of
extracellular liposomes high in phospholipids and free cholesterol and very
low or devoid of translamellar proteins, characteristics remarkably well
suited for the role of LP-X.[41] These investigators also found an
associations of the liposomes with apoprotein B, the only protein of native
LDL, but of course, did not search for apoprotein C, the constituent
protein of LP-X. In addition, the delivery of LP-X to the arterial wall in
the manner described can explain the finding of significant quantities of
free cholesterol in the AS lesions of both humans and experimental
animals.[70,71] The concepts described above, may need some elaboration as
more information develops in this area, but the account presented, at least
provides some basis for connecting the CNS with one of the cardinal
features of the AS plaque, its intra- and extracellular lipid accumulation.

We have now arrived at the stage where the interaction of lipid
accumulation and ASMC proliferation result in the histologically processed
plaque. Overloaded "foam cells" which undergo necrosis and liberate their
contents and extracellular lipid which has been arrested in its transit
through the arterial wall may serve as signals to the proliferating cells
to synthesize connective tissue elements, such as collagen and newly formed
proteoglycans, which serve to isolate these lipid substances and "protect"
the remaining intimal tissue from their irritating effects. The sclerosis
that results, terminating in the fibrous plaque, Fig. 12 and ultimately
calcification, stiffen the arterial wall, enabling it to resist further

374

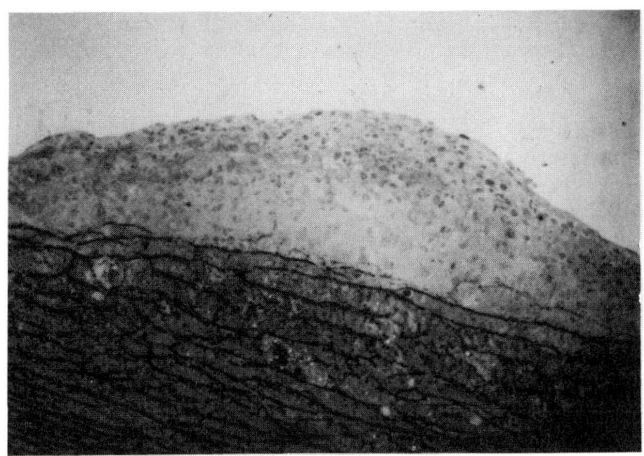

Fig. 12. Light micrograph section of abdominal
aorta of squirrel monkey with hypothalamic stimulation for 22 months showing
fibrous plaque. Many of the cells in the
plaque were ASMC by electron microscopic
examination but some resembled macrophages.
Toluidine Blue X 150.

vasopasm, the condition which was instrumental in initiating these events
at the outset. Newly formed elastic fibers may be regarded as a teleo-
logical affair to restore the normal function of arterial tissue, i.e.
perfusion. While conjectural, this description is consistent with evidence
that spasm, especially when repeated, can play a significant role in the
pathogenesis of AS, whereas arresting the spasm by calcium antagonists can
also arrest its progression, even in human populations, thus holding out
promise for a new era in treatment[72].

To complete the interpretation of the pathogenesis in this report we
have only to briefly consider the neurobiology of HS as it may elicit the
phenomenon of regional vasospasm and to implicate its role in the emotional
life of animals and man alike. While electrical stimulation of the brain
may seem to be spurious and artificial, it is known to reproduce the
neurobiology of natural or spontaneous stimulation remarkably well[73]. The
role of the hypothalamus and areas of the brain which connect intimately
with it, in emotional responses, is well known[11]. This has been amply
confirmed in animals since the pioneer work of Hess and Brugger[74] and
demonstrated in humans as well[75-78]. Vasoactive influences on different
vascular beds both dilator and constrictor are well represented in the
hypothalamus[14]. Arterial constriction, of coronary artery, aorta,
mesenteric artery, femoral artery, renal artery, etc. may have its origin
in the CNS as a response to conditions of a severely stressful nature,
giving rise to damage at the level of the arterial wall, especially if
repeated, as demonstrated in this paper. Much of the damage may be
repaired by the rapidity of endothelial regeneration[79,80] but continuing
episodes may take their toll in the pathology we have come to know as AS.
As indicated above, the pathologic changes which take place in the artery,
especially those of sclerosis and calcification tend to offset these insults by altering the compliance of the vessel, i.e., making it more rigid
so that it may resist further spastic events. In this sense, AS, may

be considered as an evolutionary development or adjustment equipping the
organism with a strategy for survival, rather than a disease threatening
to extinguish the species.

In this paper it has been seen that the biliary system also
occupies an important position in experimental atherogenesis by giving
rise through hypothalamic stimulation to circulating LP-X particles,
which in turn comes about through the induction of spasm in the biliary
system (bile duct) similar to that occurring in arteries. As previously
indicated a relationship between disease of the biliary system and
atherosclerosis is not a new concept, having been given serums
consideration, e.g., nearly forty years ago by Edward H. Ahrens, Jr.,
writing from his eminent position at the Rockefeller Institute for
Medical Research, in the March 1950 issue of the Bulletin of the New York
Academy of Medicine. [81] If this relationship returns as a viable
concept, its link to the CNS may also be considered based on the material
in this paper.

REFERENCES

1. N. N. Anitschkov, and S. Chalatov, Uber Experimentalle
    Cholesterinsteatose und Ihre Bedeutung fur die Entstehung Einiger
    Pathologischer Prozesse. _Zbl_. _allg_. _Path_. _path_. _Anat_. 24:1 (1913).
2. H. C. McGill, Jr., Fatty Streaks in the Coronary Arteries and
    Aorta. Lab. Invest. 18:560 (1968).
3. R. Ross, and J. Glomset, Medical Progress. The Pathogenesis of
    Atherosclerosis, _New_ _Eng_. _J_. _Med_. 295:369 (1976).295:420 (1976).
4. A. Keys, Coronary heart disease in seven countries. _Circulation_
    41:I-1 (1970).
5. R. Doll, and R. Peto, Mortality in Relation to Smoking - 20 Years
    Observation in Male British Doctors, _Br_. _Med_. _J_. 2:1525 (1976).
6. D. S. Bloom, B. H. Eidelman, and T. A. McCalden, Modification of the
    Cerebrovascular Response to Noradrenaline by Bile Duct Ligation,
    Gut 16:1723 (1975).
7. R. S. Eliot, Coronary Artery Heart Disease: Biobehavioral Factors,
    Overview, in: _Circulation_, Suppl. (Part 2, Vol. 76,) J. T.
    Shepherd, and S. M. Weiss, eds. (1987).
8. R. B. Williams, Jr., Psychological Factors in Coronary Artery
    Disease: Epidemiologic Evidence, IBID., ref. 7, I-117.
9. C. D. Jenkins, J. Zyzanski, T. J. Ryan, A. Flessas, and S. I.
    Tannenbaum, Social Insecurity and Coronary-Prone Type A Responses
    as Identifiers of Severe Atherosclerosis, _J_. _Consult_. _Clin_.
    Phyiol. 45:1060 (1977).
10. T. B. Clarkson, J. R. Kaplan, M. R. Adams, and S. B. Manuck,
    Phychosocial Influences on the Pathogenesis of Atherosclerosis
    Among Nonhuman Primates, IBID., ref. 7, I-29.
11. P. D. MacLean, The Hypothalamus and Emotional Behavior, in: "The
    Hypothalamus," W. Haymaker, E. Anderson, and W. J. Nauta, eds.,
    Charles C. Thomas, Springfield, Illinois (1969).
12. P. T. Costa, (Chairman), Task Force 2: Psychological Risk Factors
    in Coronary Artery Disease, IBID, ref. 7, I-145.
13. A. Maseri, and S. Chierchia, Coronary Artery Spasm: Demonstration,
    Definition, Diagnosis, and Consequences, _Prog_. _Cardiovasc_. _Dis_.
    25:169 (1982).
14. R. F. Rushmore, "Structure and Function of the Cardiovascular
    System," Second Edit. W. B. Saunders Company, Phila. (1976).
15. W. H. Gutstein, P. Anversa, C. Beghi, G. Kiu, and D. Pacanovsky,
    Coronary Artery Spasm in the Rat Induced by Hypothalamic
    Stimulation, _Atherosclerosis_ 51:135 (1984).

16. W. H. Gutstein, P. Anversa, and G. Guideri, Spasm of Small Coronary
    Arteries and Ischemic Myocardial Injury Induced by Hypothalamic
    Stimulation, Am. J. Pathol. 129:287 (1987).
17. N. Weckman, Local Constriction and Spasm of Large Arteries Elicited
    by Hypothalamic Stimulation, Experientia 16:34 (1960).
18. K. I. Melville, B. Blum, H. E. Shister, and M. D. Silver, Cardiac
    Ischemic Changes and Arrhythmias Induced by Hypothalamic
    Stimulation, Amer. J. Cardiol. 12:781 (1963).
19. C. C. Chi, and J. P. Flynn, The Effects of Hypothalamic and
    Reticular Stimulation on Evoked Responses in the Visual System of
    the Cat, Electroenceph. Clin. Neurophysiol. 24:343 (1968).
20. H. L. Garvey, and I. I. Melville, Cardiovascular Effects of Lateral
    Hypothalamic Stimulation in Normal and Coronary–Ligated Dogs, J.
    Cardiovasc. Surg. 10:377 (1969).
21. H. L. Garvey, and E. I. Melville, Effects of Verapamil on
    Cardiovascular Responses to Lateral Hypothalamic Stimulation in
    Normal and Coronary–Ligated Dogs, Canad. J. Physiol. Pharmacol.
    47:675 (1969).
22. M. H. Evans, An Effect of Hypothalamic Stimulation On Cardiac Output
    in the Rabbit, Quart. J. Exp. Physiol. 65:217 (1980).
23. M. H. Evans, Vasoactive Sites in the Diencephalon of the Rabbit,
    Brain Res. 183:329 (1980).
24. B. Blum, J. Israeli, M. Dujovny, A. Davidovich, and M. Farchi,
    Angina–Like Cardiac Disturbances of Hypothalamic Etiology in Cat,
    Monkey, and Man, Israel J. Med. Sci. 18:127 (1982).
25. A. C. Bonham, D. D. Gutterman, J. M. Arthur, M. L. Marcus, G. F.
    Gebhart, and M. J. Brody, Electrical Stimulation in Perifornical
    Lateral Hypothalmus Decreases Coronary Blood Flow in Cats. Am. J.
    Physiol. 252:H474 (1987).
26. W. H. Gutstein, J. N. LaTaillade, and L. Lewis, Role of
    Vasoconstriction In Experimental Arteriosclerosis, Circ. Res.
    10:925 (1962).
27. C. J. Schwartz, J. L. Kelley, R. M. Nerem, E. A. Sprague, M. M.
    Rozek, A. J. Valente, E. H. Edwards, A. R. S. Prasad, J. J.
    Kerbacher, and S. A. Logan, Pathophysiology of the Atherogenic
    Process, Amer. J. Cardiol. 64:23G (1989).
28. S. M. Schwartz, G. R. Campbell, and J. H. Campbell, Replication of
    Smooth Muscle Cells in Vascular Disease, Circ. Res. 58:427 (1986).
29. W. H. Gutstein, A. R. D'Aguillo, F. Parl, and G. Kiu, Influence of
    Hypothalamic Stimulation on Large Endothelial Space Formation in
    the Coronary Artery of the Rat, Artery 1:385 (1975).
30. W. H. Gutstein, J. Harrison, F. Parl, G. Kiu, and M. Avitable,
    Nerual Factors Contribute to Atherogenesis, Science 199:449 (1978).
31. W. H. Gutstein, P. Anversa, G. K. Turi, L. Korcek, J. E. Harrison,
    and G. Kiu, Effect of Hypothalamic Stimulation on the Endothelial
    Morphology of the Aorta in the Conscious Squirrel Monkey,
    Atherosclerosis, 39:329 (1981).
32. W. H. Gutstein, C-H. Wang, L. Korcek, J. E. Harrison, and D.
    Pacanovsky, Proliferation of Arterial Smooth Muscle Cells
    Incubated in Serum From Brain–Stimulated Rats, Life Sci. 34:2627
    (1984).
33. W. H. Gutstein, J. M. Wu, Unpublished Observations.
34. W. H. Gutstein, D. J. Schneck, and H. Appleton, Association of
    Increased Plasma Lipid Levels with Brain Stimulation, Metabolism
    17:535 (1968).
35. W. H. Gutstein, and G. A. Farrell, Serum Cholesterol Responses to
    Hypothalamic Stimulation and Fatty Acid Administration in the Rat,
    Proc. Soc. Exp. Biol. Med. 141:137 (1972).
36. W. H. Gutstein, D. J. Schneck, and H. Appleton, Mechanism of Plasma
    Lipid Increases Following Brain Stimulation, Metabolism 17:535
    (1968).

37. N. Blaes, and J. P. Boissel, Growth Stimulating Effects of
    Catecholamines on rat Aortic Smooth Muscle Cells in Culture. J.
    Cell Physiol. 116:167 (1983).
38. M. Campbell-Boswell, and A. Lazzarini-Robertson, Jr. Effects of
    Angiotensin II and Vasopressin on Human Smooth Muscle Cells in
    Vitro, Exper. Molec. Pathol. 35:265 (1981).
39. R. F. Furchgott, and J. V. Zawadzki, The Obligatory Role of
    Endothelial Cells in the Relaxation of Arterial Smooth Muscle by
    Acetylcholine, Nature 288:373 (1980).
40. M. Yanagisawa, H. Kurihara, S. Kimura, Y. Tomobe, M. Kobayashi, Y.
    Mitsui, Y. Yazaki, K. Gote and T. Masaki, A Novel Potent
    Vasoconstrictor Peptide Produced by Vascular Endothelial Cells,
    Nature 332:411 (1988).
41. J. R. Patsch, K. C. Aune, A. M. Gotto, Jr. and J. D. Morrisett,
    Isolation, Chemical Characterization, and Biophysical Properties
    of Three Different Abnormal Lipoproteins: J. Bio. Chem. 252:2113
    (1977).
42. F. Parl, L. D. Bjornson, G. Kiu, and W. H. Gutstein, Effects of
    Electrical Brain Stimulation on Erythrocyte Membrane Lipids, Life
    Sci. 20:1983 (1977).
43. F. M. Abboud, and M. D. Thames, Interaction of Cardiovascular
    Reflexes in Circulatory Control, in: "Handbook of Physiology,
    Section 2: Circulation III, Peripheral Circulation and Organ Blood
    Flow, Part 2," Amer. Physiol. Soc., Bethesda (1983).
44. B. L. Languille, M. Reidy, and R. L. Kline, Injury and Repair of
    Endothelium at Sites of Flow Disturbances Near Abdominal Aortic
    Coarctations in Rabbits, Atherosclerosis. 6:146 (1986).
45. P. Constantinides, and M. Robinson, Ultrastructural Injury of
    Arterial Endothelium, II. Effect of Vasoactive Amines, Arch
    Pathol. 88:106 (1969).
46. I. Joris, and G. Majno, Endothelial Changes Induced by Arterial
    Spasm, Am. J. Pathol. 102:346 (1981).
47. S. D. Gertz, M. S. Forbes, S. Tunaga, J. Kawamura, M. L. Rennels, T.
    Shimamoto, and E. Nelson, Ischemic Carotid Endothelium,
    Transmission Electron Microscopic Studies, Arch. Pathol. Lab. Med.
    100:346 (1976).
48. I. Joris, and G. Majno, Medial Changes in Arterial Spasm Induced by
    L-Norepinephrine, Am. J. Pathol. 105:212 (1981).
49. J. E. Dimsdale, and J. A. Herd, Variability of Plasma Lipids in
    Response to Emotional Arousal, Psychosom. Med. 44:413 (1982).
50. N. Schneiderman, Psychophysiologic Factors in Atherogenesis and
    Coronary Artery Disease, IBID., ref 7, I-41.
51. W. H. Gutstein, D. J. Schneck, F. A. Farrell, and W. Long, Jr.
    Hypothalamically Induced Hyperlipidemia, Protection by
    Pentaerythritol Tetranitrate, Metabolism 19:230 (1970).
52. D. Seidel, P. Alaupovic, and R. H. Furman, A Lipoprotein
    Characterizing Obstructive Jaundice, I. Method for Quantitative
    Separation and Identification of Lipoproteins in Jaundiced
    Subjects, J. Clin. Invest. 48:1211 (1969).
53. D. Seidel, P. Alaupovic, R. H. Furman, and W. J. McConathy, A
    Lipoprotein Characterizing Obstructive Jaundice, II. Isolation
    and Partial Characterization of the Protein Moieties of Low
    Density Lipoproteins, J. Clin. Invest. 49:2396 (1970).
54. D. Seidel, B. Agostini, and P. Muller, Structure of an Abnormal
    Plasma Lipoprotein (LP-X) Characterizing Obstructive Jaundice,
    Biochem. Biophys. Acta 260:146 (1972).
55. G. R. Campbell, and J. H. Chamley-Campbell, Invited Review, The
    Cellular Pathobiology of Atherosclerosis, Pathol. 13:423 (1981).
56. J. C. Geer, and M. D. Haust, "Monographs on Atherosclerosis, Vol. 2,
    Smooth Muscle Cells in Athersoclerosis, S. Karger, Basel (1972).

57. A. W. Clowes, and M. J. Karnovsky, Suppression by Heparin of Smooth Muscle Cell Proliferation in Injured Arteries, _Nature_ (London) 265:625 (1977).

58. J. L. Goldstein, and M. S. Brown, The Low Density Lipoprotein Pathoway and its Relation to Atherosclerosis, _Annu. Rev. Biochem._ 46:897 (1977).

59. M. S. Brown, J. R. Faust, and J. L. Goldstein, Role of the Low Density Lipoprotein Receptor in Regulating the Content of Free and Esterified Cholesterol in Human Fibroblasts, _J. Clin. Invest._ 55:783 (1975).

60. J. L. Goldstein, R. G. W. Anderson, L. M. Buja, S. K. Basu, and M. S. Brown, Overloading Human Aortic Smooth Muscle Cells with Low Density Lipoprotein-Cholesteryl Esters Reproduces Features of Atherosclerosis in Vitro, _J. Clin. Invest._ 59:1196 (1977).

61. E. Gaither, Gallbladder Dis. Chap. XIV in: "Diseases of the Digestive System," ed. S. A. Portis, Lea and Febiger, Phila. (1944).

62. S. A. Portisi, Jaundice, Chap. XVIII, IBID., ref. 61.

63. J. Webster, The Dutchess of Malfi, in: "Critical Temper", Great Britain (1623).

64. G. Schettler, Factors that Modify, Aggravate and Prevent the Atheromatous Process, in: "Advances in Experimental Medicine and Biology, Vol. 16B, The Artery and the Process of Arteriosclerosis," S. Wolf, ed., Plenum Press, New York (1972).

65. A. K. Diehl, S. M. Haffner, H. P. Hazuda, and M. P. Stern, Coronary Risk Factors and Clinical Gallbladder Disease: An Approach to the Prevention of Gallstones? _Amer. J. Publ. Health_ 77:841 (1987).

66. G. Weissmann, D. Bloomgarden, R. Kaplan, C. Cohen, S. Hoffstein, T. Collins, A. Gotlieb, and D. Nagle, A General Method for the Introduction of Enzymes, by Means of Immunoglobulin-Coated Liposomes, into Lysosomes of Deficient Cells, _Proc. Nat. Acad. Sci. USA_, 72:88 (1975).

67. D. Papahadjopoulous, E. Mayhew, G. Poste, and S. Smith, Incorporation of Lipid Vesicles by Mammalian Cells Provide a Potential Method for Modifying Cell Behavior, _Nature_ 252:163 (1974).

68. N. Simionescu, E. Vasile, F. Lupu, G. Popescu, and M. Simionescu, Prelesional Events in Atherogenesis, Accumulation of Extracellular Cholesterol-Rich Liposomes in the Arterial Intima and Cardiac Valves of the Hyperlipidemic Rabbit, _Am. J. Pathol._ 123:109 (1986).

69. R. Mora, F. Lupu, and N. Simionescu, Prelesional Events in Atherogenesis, Colocalization of Apolipoprotein B, Unesterified Cholesterol and Extracellular Phospholipid Liposomes in the Aorta of Hyperlipidemic rabbit, _Atherosclerosis_ 67:143 (1987).

70. H. S. Kruth, Filipin-Positive, Oil Red O-Negative Particles in the Atherosclerotic Lesions Induced by Cholesterol Feeding, _Lab. Inves._ 50:87 (1983).

71. H. S. Kruth, Localization of Unesterified Cholesterol in Human Atherosclerotic Lesions, Demonstration of Filipin Positive, Oil-Red-O- Negative Particles, _Amer. J. Pathol._ 114:201 (1983).

72. G. Kober, W. Schneider, and M. Kaltenbach, Can the Progression of Coronary Sclerosis be Influenced by Calcium Antagonists? _J. Cardio. Pharm._ 13:S2.

73. S. A. Hilton, Hypothalamic Regulation of the Cardiovascular System, _Br. Med. Bull._ 22:243 (1966).

74. W. R. Hess, and M. Brugger, Das Subkortikale Zentrum der Affektiven Abwehrreaktion, _Helv. Physiol. Acta_ 1:33 (1943).

75. C. W. Sem-Jacobsen, Effects of Electrical Stimulation on the Human Brain, _Electroenceph. Clin. Neurophysiol._ 11:379 (1959).

76. R. G. Heath, and W. A. Mickle, Evaluation of Seven Years Experience
     with Depth Electrode Studies inHuman Patients, in: "Electrical
     Studies on the Unanesthetized Brain," E. R. Ramey, and D. S. O.
     Doherty, eds., Paul B. Hoeber, New York (1960).
77. C. W. Sem-Jacobsen, and A. Torkildsen, Depth Recording and
     Electrical Stimulation in the Human Brain, Ibid, ref. 76, p. 275.
78. R. G. Heath, Pleasure Response of Human Subjects to Direct
     Stimulation of the Brain, Physiologic and Psychodynamic
     Consideration, in: "The Role of Pleasure in Behavior," R. G.
     Heath, ed., Harper and Row, New York (1964).
79. S. M. Schwartz, C. C. Haudenschid, E. M. Eddy, Endothelial
     Regeneration I. Quantitative Analysis of Initial Stages of
     Endothelial Regeneration in Rat Aortic intima, Lab. Invest. 38:568
     (1978).
80. C. C. Haudenschild, and S. M. Schwartz, Endothelial Regeneration,
     II. Restitution of Endothelial Continuity, Lab. Invest. 41:407
     (1979).
81. E. H. Ahrens, Jr., The Lipid Disturbance in Biliary Obstruction and
     its Relationship to the Genesis of Arteriosclerosis. Bull. N.Y.
     Acad. Med. :151 (1950).

CONTRIBUTORS

Allen, David R., M.D., F.R.C.S.
Department of Surgery
Southhampton General Hospital
Graham Road
Southhampton, S09 4PE England

Anderson, James W., M.D.
Metabolic Research Group
V.A. Medical Center
University of Kentucky
Lexington, KY 40511

Armstrong, Mark L., M.D.
Department of Internal Medicine
University of Iowa
College of Medicine
Iowa City, Iowa 52242

Bertha, Brian G., M.D.
Department of Medicine
University of Wisconsin
  Medical School
Clinical Science Center H6
Room 379
600 Highland Ave.
Madison, Wisconsin 53792

Bondjers, Goran, M.D.
Department of Medicine
University of Goteborg
Goteborg, Sweden

Bonebrake, Frank C., M.D.
Department of Medicine
University of Wisconsin
  Medical School
Clinical Science Center H6
Room 379
600 Highland Ave.
Madison, Wisconsin 53792

Bonnevie-Nielsen, Vagn, M.D.
Departments of Nuclear Medicine
  and Clinical Chemistry
Odense University Hospital
Odense, Denmark

Browse, N.L., M.D.
Department of Surgery
St. Thomas Hospital
London, England

Campbell, Gordon R., Ph.D.
Cardiovascular Research Unit
University of Melbourne
Parkville, 3052, Victoria
Australia

Campbell, Julie H., Ph.D.
Cardiovascular Research Unit
University of Melbourne
Parkville, 3052, Victoria
Australia

Caro, Colin G., M.D.
Physiological Flow Studies Unit
Imperial College of Science and
Technology, Prince Consort Road
London, SW7-2AZ England

Colditz, Graham A., M.D.
Channing Laboratory
Department of Medicine
Harvard Medical School
180 Longwood Ave.
Boston, MA 02115

Davies, Peter F., Ph.D.
Department of Pathology
University of Chicago
School of Medicine
5841 S. Maryland Avenue
Chicago, IL 60637

Davis, James W., M.D.
Department of Medicine
University of Kansas
School of Medicine
V.A. Medical Center
4801 Linwood Boulevard
Kansas City, MO 64128

Diana, John N., Ph.D.
Director, Tobacco and Health
Research Institute, University of
Kentucky, Cooper and University
Drives, Lexington, KY 40546-0236

Downey, H. Fred, Ph.D.
Department of Physiology
Texas College of Osteopathic
Medicine, Camp Bowie Boulevard
Fort Worth, TX 76107

Dull, Randal O., Ph.D.
Department of Pathology
Pritzker School of Medicine
University of Chicago
Chicago, Illinois

Effeney, David J., M.D., F.R.A.C.S.
Department of Surgery
University of Queensland
Princess Alexandria Hospital
Ipswich Road Woolloongabba 4102
Queensland, Australia

Ernst, Edzard, M.D., Ph.D.
MHH (Medizische Hochschule
Hannover) Department of Physical
Medicine, 3000 Hannover 61
West Germany

Evans, Gregory, M.S.
Department of Public
   Health Sciences
Bowman Gray School of Medicine
300 S. Hawthorne Road
Winston-Salem, NC 27103

FitzGerald, Garret A., M.D.
Division of Clinical Pharmacology
Departments of Medicine
 and Pharmacology
Vanderbilt Medical Center
Nashville, TN 37232

Floore, Tammy L., R.N., B.S.N.
Metabolic Research Group
V.A. Medical Center
University of Kentucky
Lexington, KY 40511

Folts, John D., Ph.D.
Department of Medicine
Cardiology Section
University of Wisconsin
Medical School
Clinical Science Center H6
Room 379
600 Highland Avenue
Madison, WI 53792

Gering, Scott A., B.S.
Department of Medicine
University of Wisconsin
   Medical School
Clinical Science Center H6
Room 379
600 Highland Ave.
Madison, Wisconsin 53792

Gutstein, William H., M.D.
Basic Science Building
Department of Pathology
New York Medical College
Valhalla, NY 10595

Hansson, Goran, M.D.
Wallenberg Laboratory for
   Cardiovascular Research
University of Goteborg
Goteborg, Sweden

Hayashi, Takuji, M.D.
National Heart, Lung
   and Blood Institute
Honolulu Heart Program
Kuakini Medical Center
Honolulu, Hawaii

Heistad, Donald, M.D.
Department of Internal Medicine
University of Iowa, College
of Medicine, E-318-1 GH
Iowa City, IA 52242

Howard, George, Ph.D.
Department of Public
   Health Sciences
Bowman Gray School of Medicine
300 S. Hawthorne Road
Winston-Salem, NC 27103

Jeremy, Jamie Y., M.D.
Department of Clinical
Pathology and Human Metabolism
Royal Free Hospital and School
of Medicine
Pond Street, Hampstead
London NW3-2QG, England

Jerome, Gray, W., Ph.D.
Department of Pathology
Bowman Gray School of Medicine
Wake Forest University
Winston Salem, NC

Keller, Jeff W., M.D.
Department of Medicine
University of Wisconsin
  Medical School
Clinical Science Center H6
Room 379
600 Highland Ave.
Madison, Wisconsin 53792

Klein, Lloyd W., M.D.
Director of Interventional
Cardiology, Division of
Cardiology, Nortwestern
School of Medicine and
Northwestern Memorial Hospital
Wesley Pavillion - Room 510
710 North Fairbanks Street
Chicago, IL 60611

Koenig, Wolfgang, M.D.
Department of Internal Medicine
Medical School
300 Hannover 61
P.O. Box 610180 FRG
University of Ulm
Germany

Kutti, Jack, M.D.
Department of Medicine
Ostra Hospital
University of Goteborg
Goteborg, Sweden

Laibly, Shawn W., B.S.
Department of Medicine
University of Wisconsin
  Medical School
Clinical Science Center H6
Room 379, 600 Highland Ave.
Madison, Wisconsin 53792

Lewis, Jon C., Ph.D.
Department of Pathology
Bowman Gray School of Medicine
Wake Forest University
Winston Salem, NC

Lopez, J. Antonio M.D.
Department of Internal Medicine
University of Iowa, College
of Medicine, E-318-1 GH
Iowa City, IA 52242

Marcus, Ellen, M.D.
National Heart, Lung
  and Blood Institute
Honolulu Heart Program
Kuakini Medical Center
Honolulu, Hawaii

McGeachie, John K., Ph.D.
Department of Anatomy and
Human Biology
University of Western Australia
Nedlands (Perth)
WA 6009, Australia

McGill, Henry C., Jr., M.D.
Department of Pathology
The University of Texas
Health Science Center
7703 Floyd Curl Drive
San Antonio, TX 78284-7750

McKinney, William, M.D.
Department of Neurology
Bowman Gray School of Medicine
300 S. Hawthorne Road
Winston-Salem, NC 27103

Mikhailidis, D.P., Ph.D.
Department of Chemical Pathology
  and Human Metabolism
Royal Free Hospital
  and School of Medicine
University of London
Pond Street
London NW3 2QG, England

Moffatt, Robert J., Ph.D.
Department of Movement Science
and Physical Education
Florida State University
Tallahassee, FL 32306

Molgaard, Jorgen, M.D.
Department of Internal Medicine
Faculty of Health Sciences
Linkoping Medical School
S-581 85 Linkoping, Sweden

Murray, John, M.D., Ph.D.
Division of Clinical Pharmacology
Vanderbilt University
School of Medicine
Nashville, TN 37232

Nowak, Jacek, M.D.
Division of Clinical Pharmacology
Departments of Medicine
  and Pharmacology
Vanderbilt Medical Center
Nashville, TN 37232

Oates, John A., M.D.
Division of Clinical Pharmacology
Departments of Medicine
  and Pharmacology
Vanderbilt Medical Center
Nashville, TN 37232

Olsson, Anders G., M.D.
Department of Internal Medicine
Linkoping University Medical
School, S-581 85 Linkoping, Sweden

Olsson, Gun, M,D.
Wallenberg Laboratory for
  Cardiovascular Research
University of Goteborg
Goteborg, Sweden

Pettersson, Knut, M.D.
Wallenberg Laboratory for
  Cardiovascular Research
University of Goteborg
Goteborg, Sweden

Pittilo, Robert M., Ph.D.
School of Life Sciences
Kingston Polytechnic Institute
Penrhyn Road
Kingston-upon-Thames
Surrey KT1-2EE
England

Rasmussen, Jens W., M.D.
Department of Radiophysics
Odense University Hospital
DK-5000 Odense C
Denmark

Reed, Dwayne M., M.D.
Honolulu Heart Program
347 North Kuakini Street
Honolulu, HI   96817

Renaud, Serge, Ph.D.
Director, INSERM Unit 63
22 Ave du Doyen Lepine
69675 Bron Cedex, France

Riddle, Jeannie M., M.D.
Heart and Vascular Institute
Henry Ford Hospital
Detroit, Michigan 48202

Rival, Jan, M.D.
Heart and Vascular Institute
Henry Ford Hospital
Detroit, Michigan 48202

Schmidt, Kai Gerloff, M.D.
Department of Radiophysics
Odense University Hospital
DK-5000 Odense C
Denmark

Siess, Wolfgang, M.D.
Hiltensbergerstr. 36
D8 Muenchen 40
West Germany

Smith, Michael L.
Department of Public
  Health Sciences
Bowman Gray School of Medicine
300 S. Hawthorne Road
Winston-Salem, NC   27103

Stein, Paul D., M.D.
Director, Cardiovascular Research
Department of Medicine, Henry
Ford Heart and Vascular Institute
Room 4016 E & R Bldg.
2799 West Grand Blvd.
Detroit, MI   48202

Tai, Hsin-Hsiung, Ph.D.
Department of Medicinal Chemistry
University of Kentucky, College
of Pharmacy, 527 907-909 Rose Street
Lexington, Kentucky   40536-0081

Taylor, Richard G., Ph.D.
Department of Pathology
Bowman Gray School of Medicine
300 S. Hawthorne Road
Winston-Salem, NC   27103

Tell, Grethe S., Ph.D.
Department of Public Health Sciences
The Bowman-Gray School of Medicine
300 S. Hawthorne Road
Winston-Salem, NC   27103

Toole, James F., M.D.
Department of Neurology
Bowman Gray School of Medicine
300 S. Hawthorne Road
Winston-Salem, NC   27103

Volgman, Annabelle S., M.D.
Northwestern Memorial Hospital
Northwestern Medical School
250 E. Superior, Suite 150
Chicago, Illinois

Weintraub, William S., M.D.
Department of Medicine
Emory University
School of Medicine
1365 Clifton Road
N.E. Suite F410
Atlanta, GA   30322

Wu, Joseph M., Ph.D.
Department of Biochemistry
New York Medical College
Basic Science Building - Rm 127
Valhalla, New York   10595

Zimmerman, Matthew, M.B.B.S.
Queen Elizabeth II Medical Centre
Perth, Western Australia

Phytosteroid, 4
Pig
  and cholesterol feeding, 91
  and Von Willebrand disease,
      120
Pigeon *var*. White Carneau, 89–93
  and atherosclerosis, 89–93
  and endothelium, 89–93
  as model, 89–93
Pinocytosis, 284, 285
Plaque
  atheromatous, 96, 221
  atherosclerotic, 121
  fibrous, 9, 10, 61, 91
    formation, 13
Plasma, 296–300
  and hemoglobin, 297
  and protein, 296, 297
  variables, 298, 300
  viscosity, 296, 297
Plasminogen, 14
  activator, 285, 286
Platelet, 3, 13, 62, 65, 96,
      339–358
  activated
    and atherosclerosis, 173–180
    pathway, intracellular, 120
    role of, 174–176
  activation, 199–209
    in chronic smokers, 190–194
    mechanism of, 119–120
  adhesion, 5, 51, 68
  aggregate ratio, 108–115,
      199–202, 204, 206
  aggregation, 4, 5, 176, 354
    and ADP, 116
    circulating, 183
    due to collagen, 116
    ratio, 130–131
    and smoking, 107, 130–131
    studies, 130–132, 339–342
  aspirin, 124
  atherosclerosis, 119–127
  bleeding time studies, 130
  and cigarette smoking, 107–118,
      199–209
    and aggregation, 107
  cytosol, 212
  and electron microscopy, 181–183
  and ethanol, 348
  factor *4*, 108, 111, 115, 130,
      184, 199–202, 207
  on formvar film, 182–185
  function, 119
  and α-granule, 123
  growth factor release, 51, 96
  and hypercholesterolemia, 122
  indium-labelled, 52
  inhibition strategy, 123–124
    and aspirin, 124
    and exercise, physical, 124

Platelet (continued)
  lipoprotein, 122
  macrophage, 122
  monocyte, 122
  pathogenesis, 5, 121
    of atherosclerosis, 121
  preparation, 212, 226
  protein, 129
  reactivity, 161–171
    and smoking, 129–132, 161–162
  receptor, 119
  release factor, 183, 184
  response to injury hypothesis,
      121
  in smoker, chronic, 181–187
  and stimuli, 119
  survival studies, 130
  turnover studies, 340
  -vessel wall interaction,
      189–199
    and cigarette smoking, 189–198
Platelet-activating factor, 119, 138
Platelet-derived growth factor
      3, 123, 147, 247
Polyinositol phosphate, 6
Potassium chloride, 148
Propanolol, 305
Properdin factor B, 132
Propionaldehyde, 218
Prostacyclin (PGI$_2$), 3, 5, 62,
      69–71, 121, 136, 137, 189
      –192, 199, 211–215, 220,
      221, 225–235, 285, 289,
      304, 340, 354
  and AMP, 121
  assay, 227
  and cigarette smoking, 69–71
  discovery in *1976*, 225
  in endothelium, 69, 70
  function, 225
  of leukocyte, 137
Prostaglandin, 13, 69–71, 231
  E$_2$, 174, 175
  endoperoxidase, 119, 120, 211
  I$_2$, *see* Prostacyclin
Prostanoids, 135–141
  and cigarette smoking, 138–141
  function, 135
  radioimmunoassay, 139
  smoking, 135, 138–141
Protease, 3
Protein glycosylation, 6
Protein kinase C, 6, 119–122, 136, 290
Proteoglycan, 3
Psyllium (fiber), 249
Pyrene, 4
Pyridylcarbinol, 238

Rabbit, 91, 92, 120, 225–235
  atherosclerosis discovered
      in *1913*, 359